W0018299

Analysis of Variance for Functional Data

MONOGRAPHS ON STATISTICS AND APPLIED PROBABILITY

General Editors

F. Bunea, V. Isham, N. Keiding, T. Louis, R. L. Smith, and H. Tong

1. Stochastic Population Models in Ecology and Epidemiology *M.S. Barlett* (1960)
2. Queues *D.R. Cox and W.L. Smith* (1961)
3. Monte Carlo Methods *J.M. Hammersley and D.C. Handscomb* (1964)
4. The Statistical Analysis of Series of Events *D.R. Cox and P.A.W. Lewis* (1966)
5. Population Genetics *W.J. Ewens* (1969)
6. Probability, Statistics and Time *M.S. Barlett* (1975)
7. Statistical Inference *S.D. Silvey* (1975)
8. The Analysis of Contingency Tables *B.S. Everitt* (1977)
9. Multivariate Analysis in Behavioural Research *A.E. Maxwell* (1977)
10. Stochastic Abundance Models *S. Engen* (1978)
11. Some Basic Theory for Statistical Inference *E.J.G. Pitman* (1979)
12. Point Processes *D.R. Cox and V. Isham* (1980)
13. Identification of Outliers *D.M. Hawkins* (1980)
14. Optimal Design *S.D. Silvey* (1980)
15. Finite Mixture Distributions *B.S. Everitt and D.J. Hand* (1981)
16. Classification *A.D. Gordon* (1981)
17. Distribution-Free Statistical Methods, 2nd edition *J.S. Maritz* (1995)
18. Residuals and Influence in Regression *R.D. Cook and S. Weisberg* (1982)
19. Applications of Queueing Theory, 2nd edition *G.F. Newell* (1982)
20. Risk Theory, 3rd edition *R.E. Beard, T. Pentikäinen and E. Pesonen* (1984)
21. Analysis of Survival Data *D.R. Cox and D. Oakes* (1984)
22. An Introduction to Latent Variable Models *B.S. Everitt* (1984)
23. Bandit Problems *D.A. Berry and B. Fristedt* (1985)
24. Stochastic Modelling and Control *M.H.A. Davis and R. Vinter* (1985)
25. The Statistical Analysis of Composition Data *J. Aitchison* (1986)
26. Density Estimation for Statistics and Data Analysis *B.W. Silverman* (1986)
27. Regression Analysis with Applications *G.B. Wetherill* (1986)
28. Sequential Methods in Statistics, 3rd edition *G.B. Wetherill and K.D. Glazebrook* (1986)
29. Tensor Methods in Statistics *P. McCullagh* (1987)
30. Transformation and Weighting in Regression *R.J. Carroll and D. Ruppert* (1988)
31. Asymptotic Techniques for Use in Statistics *O.E. Bandorff-Nielsen and D.R. Cox* (1989)
32. Analysis of Binary Data, 2nd edition *D.R. Cox and E.J. Snell* (1989)
33. Analysis of Infectious Disease Data *N.G. Becker* (1989)
34. Design and Analysis of Cross-Over Trials *B. Jones and M.G. Kenward* (1989)
35. Empirical Bayes Methods, 2nd edition *J.S. Maritz and T. Lwin* (1989)
36. Symmetric Multivariate and Related Distributions *K.T. Fang, S. Kotz and K.W. Ng* (1990)
37. Generalized Linear Models, 2nd edition *P. McCullagh and J.A. Nelder* (1989)
38. Cyclic and Computer Generated Designs, 2nd edition *J.A. John and E.R. Williams* (1995)
39. Analog Estimation Methods in Econometrics *C.F. Manski* (1988)
40. Subset Selection in Regression *A.J. Miller* (1990)
41. Analysis of Repeated Measures *M.J. Crowder and D.J. Hand* (1990)
42. Statistical Reasoning with Imprecise Probabilities *P. Walley* (1991)
43. Generalized Additive Models *T.J. Hastie and R.J. Tibshirani* (1990)
44. Inspection Errors for Attributes in Quality Control *N.L. Johnson, S. Kotz and X. Wu* (1991)
45. The Analysis of Contingency Tables, 2nd edition *B.S. Everitt* (1992)

46. The Analysis of Quantal Response Data *B.J.T. Morgan* (1992)
47. Longitudinal Data with Serial Correlation—A State-Space Approach *R.H. Jones* (1993)
48. Differential Geometry and Statistics *M.K. Murray and J.W. Rice* (1993)
49. Markov Models and Optimization *M.H.A. Davis* (1993)
50. Networks and Chaos—Statistical and Probabilistic Aspects
O.E. Barndorff-Nielsen, J.L. Jensen and W.S. Kendall (1993)
51. Number-Theoretic Methods in Statistics *K.-T. Fang and Y. Wang* (1994)
52. Inference and Asymptotics *O.E. Barndorff-Nielsen and D.R. Cox* (1994)
53. Practical Risk Theory for Actuaries *C.D. Daykin, T. Pentikäinen and M. Pesonen* (1994)
54. Biplots *J.C. Gower and D.J. Hand* (1996)
55. Predictive Inference—An Introduction *S. Geisser* (1993)
56. Model-Free Curve Estimation *M.E. Tarter and M.D. Lock* (1993)
57. An Introduction to the Bootstrap *B. Efron and R.J. Tibshirani* (1993)
58. Nonparametric Regression and Generalized Linear Models *P.J. Green and B.W. Silverman* (1994)
59. Multidimensional Scaling *T.F. Cox and M.A.A. Cox* (1994)
60. Kernel Smoothing *M.P. Wand and M.C. Jones* (1995)
61. Statistics for Long Memory Processes *J. Beran* (1995)
62. Nonlinear Models for Repeated Measurement Data *M. Davidian and D.M. Giltinan* (1995)
63. Measurement Error in Nonlinear Models *R.J. Carroll, D. Rupert and L.A. Stefanski* (1995)
64. Analyzing and Modeling Rank Data *J.J. Marden* (1995)
65. Time Series Models—In Econometrics, Finance and Other Fields
D.R. Cox, D.V. Hinkley and O.E. Barndorff-Nielsen (1996)
66. Local Polynomial Modeling and its Applications *J. Fan and I. Gijbels* (1996)
67. Multivariate Dependencies—Models, Analysis and Interpretation *D.R. Cox and N. Wermuth* (1996)
68. Statistical Inference—Based on the Likelihood *A. Azzalini* (1996)
69. Bayes and Empirical Bayes Methods for Data Analysis *B.P. Carlin and T.A Louis* (1996)
70. Hidden Markov and Other Models for Discrete-Valued Time Series *I.L. MacDonald and W. Zucchini* (1997)
71. Statistical Evidence—A Likelihood Paradigm *R. Royall* (1997)
72. Analysis of Incomplete Multivariate Data *J.L. Schafer* (1997)
73. Multivariate Models and Dependence Concepts *H. Joe* (1997)
74. Theory of Sample Surveys *M.E. Thompson* (1997)
75. Retrial Queues *G. Falin and J.G.C. Templeton* (1997)
76. Theory of Dispersion Models *B. Jørgensen* (1997)
77. Mixed Poisson Processes *J. Grandell* (1997)
78. Variance Components Estimation—Mixed Models, Methodologies and Applications *P.S.R.S. Rao* (1997)
79. Bayesian Methods for Finite Population Sampling *G. Meeden and M. Ghosh* (1997)
80. Stochastic Geometry—Likelihood and computation
O.E. Barndorff-Nielsen, W.S. Kendall and M.N.M. van Lieshout (1998)
81. Computer-Assisted Analysis of Mixtures and Applications—Meta-Analysis, Disease Mapping and Others
D. Böhning (1999)
82. Classification, 2nd edition *A.D. Gordon* (1999)
83. Semimartingales and their Statistical Inference *B.L.S. Prakasa Rao* (1999)
84. Statistical Aspects of BSE and vCJD—Models for Epidemics *C.A. Donnelly and N.M. Ferguson* (1999)
85. Set-Indexed Martingales *G. Ivanoff and E. Merzbach* (2000)
86. The Theory of the Design of Experiments *D.R. Cox and N. Reid* (2000)
87. Complex Stochastic Systems *O.E. Barndorff-Nielsen, D.R. Cox and C. Klüppelberg* (2001)
88. Multidimensional Scaling, 2nd edition *T.F. Cox and M.A.A. Cox* (2001)
89. Algebraic Statistics—Computational Commutative Algebra in Statistics
G. Pistone, E. Riccomagno and H.P. Wynn (2001)
90. Analysis of Time Series Structure—SSA and Related Techniques
N. Golyandina, V. Nekrutkin and A.A. Zhigljavsky (2001)
91. Subjective Probability Models for Lifetimes *Fabio Spizzichino* (2001)
92. Empirical Likelihood *Art B. Owen* (2001)

93. Statistics in the 21st Century *Adrian E. Raftery, Martin A. Tanner, and Martin T. Wells* (2001)
94. Accelerated Life Models: Modeling and Statistical Analysis
Vilijandas Bagdonavicius and Mikhail Nikulin (2001)
95. Subset Selection in Regression, Second Edition *Alan Miller* (2002)
96. Topics in Modelling of Clustered Data *Marc Aerts, Helena Geys, Geert Molenberghs, and Louise M. Ryan* (2002)
97. Components of Variance *D.R. Cox and P.J. Solomon* (2002)
98. Design and Analysis of Cross-Over Trials, 2nd Edition *Byron Jones and Michael G. Kenward* (2003)
99. Extreme Values in Finance, Telecommunications, and the Environment
Bärbel Finkenstädt and Holger Rootzén (2003)
100. Statistical Inference and Simulation for Spatial Point Processes
Jesper Møller and Rasmus Plenge Waagepetersen (2004)
101. Hierarchical Modeling and Analysis for Spatial Data
Sudipto Banerjee, Bradley P. Carlin, and Alan E. Gelfand (2004)
102. Diagnostic Checks in Time Series *Wai Keung Li* (2004)
103. Stereology for Statisticians *Adrian Baddeley and Eva B. Vedel Jensen* (2004)
104. Gaussian Markov Random Fields: Theory and Applications *Håvard Rue and Leonhard Held* (2005)
105. Measurement Error in Nonlinear Models: A Modern Perspective, Second Edition
Raymond J. Carroll, David Ruppert, Leonard A. Stefanski, and Ciprian M. Crainiceanu (2006)
106. Generalized Linear Models with Random Effects: Unified Analysis via H-likelihood
Youngjo Lee, John A. Nelder, and Yudi Pawitan (2006)
107. Statistical Methods for Spatio-Temporal Systems
Bärbel Finkenstädt, Leonhard Held, and Valerie Isham (2007)
108. Nonlinear Time Series: Semiparametric and Nonparametric Methods *Jiti Gao* (2007)
109. Missing Data in Longitudinal Studies: Strategies for Bayesian Modeling and Sensitivity Analysis
Michael J. Daniels and Joseph W. Hogan (2008)
110. Hidden Markov Models for Time Series: An Introduction Using R
Walter Zucchini and Iain L. MacDonald (2009)
111. ROC Curves for Continuous Data *Wojtek J. Krzanowski and David J. Hand* (2009)
112. Antedependence Models for Longitudinal Data *Dale L. Zimmerman and Vicente A. Núñez-Antón* (2009)
113. Mixed Effects Models for Complex Data *Lang Wu* (2010)
114. Intoduction to Time Series Modeling *Genshiro Kitagawa* (2010)
115. Expansions and Asymptotics for Statistics *Christopher G. Small* (2010)
116. Statistical Inference: An Integrated Bayesian/Likelihood Approach *Murray Aitkin* (2010)
117. Circular and Linear Regression: Fitting Circles and Lines by Least Squares *Nikolai Chernov* (2010)
118. Simultaneous Inference in Regression *Wei Liu* (2010)
119. Robust Nonparametric Statistical Methods, Second Edition
Thomas P. Hettmansperger and Joseph W. McKean (2011)
120. Statistical Inference: The Minimum Distance Approach
Ayanendranath Basu, Hiroyuki Shioya, and Chanseok Park (2011)
121. Smoothing Splines: Methods and Applications *Yuedong Wang* (2011)
122. Extreme Value Methods with Applications to Finance *Serguei Y. Novak* (2012)
123. Dynamic Prediction in Clinical Survival Analysis *Hans C. van Houwelingen and Hein Putter* (2012)
124. Statistical Methods for Stochastic Differential Equations
Mathieu Kessler, Alexander Lindner, and Michael Sørensen (2012)
125. Maximum Likelihood Estimation for Sample Surveys
R. L. Chambers, D. G. Steel, Suojin Wang, and A. H. Welsh (2012)
126. Mean Field Simulation for Monte Carlo Integration *Pierre Del Moral* (2013)
127. Analysis of Variance for Functional Data *Jin-Ting Zhang* (2013)

Monographs on Statistics and Applied Probability 127

Analysis of Variance for Functional Data

Jin-Ting Zhang

CRC Press
Taylor & Francis Group
Boca Raton London New York

CRC Press is an imprint of the
Taylor & Francis Group, an **Informa** business
A CHAPMAN & HALL BOOK

MATLAB® is a trademark of The MathWorks, Inc. and is used with permission. The MathWorks does not warrant the accuracy of the text or exercises in this book. This book's use or discussion of MATLAB® software or related products does not constitute endorsement or sponsorship by The MathWorks of a particular pedagogical approach or particular use of the MATLAB® software.

First published in paperback 2024

First published 2014 by Chapman & Hall/CRC Press
2385 NW Executive Center Drive, Suite 320, Boca Raton FL 33431
4 Park Square, Milton Park, Abingdon, Oxon, OX14 4RN

CRC Press is an imprint of Taylor & Francis Group, LLC

© 2014, 2024 Taylor & Francis Group, LLC

Reasonable efforts have been made to publish reliable data and information, but the author and publisher cannot assume responsibility for the validity of all materials or the consequences of their use. The authors and publishers have attempted to trace the copyright holders of all material reproduced in this publication and apologize to copyright holders if permission to publish in this form has not been obtained. If any copyright material has not been acknowledged please write and let us know so we may rectify in any future reprint.

Except as permitted under U.S. Copyright Law, no part of this book may be reprinted, reproduced, transmitted, or utilized in any form by any electronic, mechanical, or other means, now known or hereafter invented, including photocopying, microfilming, and recording, or in any information storage or retrieval system, without written permission from the publishers.

For permission to photocopy or use material electronically from this work, access www.copyright.com or contact the Copyright Clearance Center, Inc. (CCC), 222 Rosewood Drive, Danvers, MA 01923, 978-750-8400. For works that are not available on CCC please contact mpkbookspermissions@tandf.co.uk

Trademark notice: Product or corporate names may be trademarks or registered trademarks and are used only for identification and explanation without intent to infringe.

Publisher's Note
The publisher has gone to great lengths to ensure the quality of this reprint but points out that some imperfections in the original copies may be apparent.

ISBN: 978-1-4398-6273-5 (hbk)
ISBN: 978-1-03-292039-9 (pbk)
ISBN: 978-0-429-07008-2 (ebk)

DOI: 10.1201/b15005

Visit the Taylor & Francis Web site at
http://www.taylorandfrancis.com

and the CRC Press Web site at
http://www.crcpress.com

To my Parents and Teachers
To Tian-Hui, Tian-Yu, and Yan

Contents

List of Figures

List of Tables

Preface

Functional data analysis has been a popular statistical research topic for the past three decades. Functional data are often obtained by observing a number of subjects over time, space, or other continua densely. They are frequently collected from various research areas, including audiology, biology, children's growth studies, ergonomics, environmentology, meteorology, and women's health studies among others in the form of curves, surfaces, or other complex objects. Statistical inference for functional data generally refers to estimation and hypothesis testing about functional data. There are at least two monographs available in the literature that are devoted to estimation and classification of functional data. This book mainly focuses on hypothesis testing problems about functional data, with topics including reconstruction of functional observations, functional ANOVA, functional linear models with functional responses, ill-conditioned functional linear models, diagnostics of functional observations, heteroscedastic ANOVA for functional data, and testing equality of covariance functions, among others. Although the methodologies proposed and studied in this book are designed for curve data only, it is straightforward to extend them to the analysis of surface data.

The main purpose of this book is to provide the reader with a number of simple methodologies for functional hypothesis testing. Pointwise, L^2-norm based, F-type, and bootstrap tests are discussed. The key ideas of these methodologies are stated at a relatively low technical level. The book is self-contained and assumes only a basic knowledge of statistics, calculus, and matrix algebra. Real data examples from the aforementioned research areas are provided to motivate and illustrate the methodologies. Some bibliographical notes and exercises are provided at the end of each chapter. A supporting website is provided where most of the real data sets analyzed in this book may be downloaded. Some related MATLAB® codes for analyzing these real data examples are also available and will be updated whenever necessary. Statistical researchers or practitioners analyzing functional data may find this book useful.

Chapter 1 provides a brief overview of the book, and in particular, it presents real functional data examples from various research areas that have motivated various models and methodologies for statistical inferences about functional data. Chapters 2 and 3 review four popular nonparametric smoothing techniques for reconstructing or interpolating a single curve or a group of curves in a functional data set. Chapters 4–8 present the core contents of this book. Chapters 9 and 10 are devoted to heteroscedastic ANOVA and tests of equality of covariance functions.

Most of the contents of this book should be comprehensible to readers with some basic statistical training. Advanced mathematics and technical skills are not necessary for understanding the key ideas of the methodologies and for applying them to real data analysis. The materials in Chapters 1–8 may be used in a lower- or medium-level graduate course in statistics or biostatistics. Chapters 9 and 10 may be used in a higher-level graduate course or as reference materials for those who intend to do research in this field.

I have tried my best to acknowledge the work of the many investigators who have contributed to the development of the models and methodologies for hypothesis testing in the context of analysis of variance for functional data. However, it is beyond the scope of this book to prepare an exhaustive review of the vast literature in this active research field, and I regret any oversight or omissions of particular authors or publications.

I am grateful to Rob Calver, Rachel Holt, Jennifer Ahringer, Karen Simon, and Mimi Williams at Chapman and Hall who have made great efforts in coordinating the editing, review, and finally the publishing of this book. I would like to thank my colleagues, collaborators, and friends, Mingyen Cheng, Jengmin Chiou, Jianqing Fan, James S. Marron, Heng Peng, Naisyin Wang, and Chongqi Zhang, for their help and encouragement. Special thanks go to Professor Wenyang Zhang, Department of Mathematics, the University of York, United Kingdom and Professor Jeff Goldsmith, Department of Biostatistics, Columbia University, New York and an anonymous reviewer who went through the first draft of the book and made some insightful comments and suggestions which helped improve the book substantially. Thanks also go to some of my PhD students, including Xuehua Liang, Xuefeng Liu, and Shengning Xiao, for their reading some chapters of the book. I thank my family who provided strong support and encouragement during the writing process of this book.

The book was partially supported by the National University of Singapore Academic Research grants R-155-000-108-112 and R-155-000-128-112. Most of the chapters were written when I visited the Department of Operational Research and Financial Engineering, Princeton University. I thank Professor Jianqing Fan for his hospitality and partial financial support. Some chapters of the book were written when I visited Professor Jengmin Chiou at the Institute of Statistical Science, Academia Sinica, Taipei, Taiwan; and Professor Ming-Yen Cheng at the Department of Mathematics, National Taiwan University, Taipei, Taiwan.

Jin-Ting Zhang
Department of Statistics & Applied Probability
National University of Singapore
Singapore

MATLAB® is a registered trademark of The MathWorks, Inc. For product information, please contact:

The MathWorks, Inc.
3 Apple Hill Drive
Natick, MA 01760-2098 USA
Tel: 508 647 7000
Fax: 508-647-7001
E-mail: info @ mathworks.com
Web: www.mathworks.com

Chapter 1

Introduction

This chapter aims to give an introduction to the book. We first introduce the concept of functional data in Section 1.1. We then present several motivating real functional data sets in Section 1.2. The difference between classical multivariate data analysis and functional data analysis is briefly discussed in Section 1.3. Section 1.4 gives an overview of the book. The implementation of the methodologies, the options for reading this book, and some bibliographical notes are given in Sections 1.5, 1.6, and 1.7, respectively.

1.1 Functional Data

Functional data are a natural generalization of multivariate data from finite dimensional to infinite dimensional. To get some feeling about functional data, Figure 1.1(a) displays a functional data set, consisting of twenty curves, and Figure 1.1(b) displays one of the curves. These curves were simulated without adding measurement errors and they look smooth and continuous.

In practice, functional data are obtained by observing a number of subjects over time, space, or other continua. The resulting functional data can be curves, surfaces, or other complex objects; see next section for several real curve data sets and see Zhang (1999) for some surface data sets. One may note that real functional data can be accurately observed so that the measurement errors can be ignored, or can be observed subject to substantial measurement errors. For example, measurements of the heights of children over a wide range of ages in the Berkeley growth data presented in Figure 1.4 in Section 1.2.2 have an error level so small as to be ignorable. However, the progesterone data presented in Figures 1.2 and 1.3 in Section 1.2.1 and the daily records of the Canadian temperature data at a weather station presented in Figures 1.6 and 1.7 in Section 1.2.4 are very noisy.

1.2 Motivating Functional Data

Functional data arise naturally in many research areas. Classical examples are provided by growth curves in longitudinal studies. Other examples can be found in Rice and Silverman (1991), Kneip (1994), Faraway (1997), Zhang (1999), Shen and Faraway (2004), Ramsay and Silverman (2002, 2005), Fer-

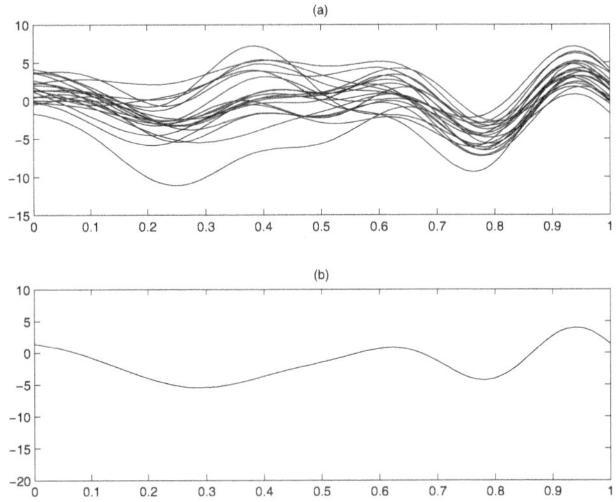

Figure 1.1 *A simulated functional data set with (a) twenty simulated curves and (b) one of the simulated curves.*

raty and Vieu (2006), and references therein. In this section, we present a number of curve data examples. For surface data examples, see Zhang (1999). These curve data sets are from various research areas, including audiology, biology, children's growth studies, ergonomics, environmentology, meteorology, and women's health studies, among others. They will be used throughout this book for motivating and illustrating the functional hypothesis testing techniques described in this book. Although these functional data examples are from only a few research areas, the proposed methodologies in this book are also applicable to functional data collected from other scientific fields. All these curve data sets and the computer codes for the corresponding analysis in this book are accessible at the website: *http://www.stat.nus.edu.sg/~zhangjt/books/Chapman/FANOVA.htm.*

1.2.1 Progesterone Data

The progesterone data were collected in a study of early pregnancy loss conducted by the Institute for Toxicology and Environmental Health at the Reproductive Epidemiology Section of the California Department of Health Services, Berkeley. Each observation shows the levels of urinary metabolite progesterone over the course of the women's menstrual cycles (in days). The observations came from patients with healthy reproductive function enrolled in an artificial insemination clinic where insemination attempts were well-timed for each menstrual cycle. The data had been aligned by the day of ovulation (Day 0), determined by serum luteinizing hormone, and truncated at each end to present curves of equal length. Measurements were recorded once per day per cycle from eight days before the day of ovulation and until fifteen days after the ovulation. A woman may have one or several cycles. The length of the observation period was twenty-four days. Some measurements from some subjects are missing due to various reasons. For more details about the progesterone data, see Munro et al. (1991), Yen and Jaffe (1991), Brumback and Rice (1998), and Fan and Zhang (2000), among others.

The data set consisted of two groups: nonconceptive progesterone curves (sixty-nine menstrual cycles) and conceptive progesterone curves (twenty-two menstrual cycles), as displayed in Figures 1.2 (a) and (b), respectively. The nonconceptive and conceptive progesterone curves came from those women without or with viable zygotes formed. From Figures 1.2 (a) and (b), it is seen that the raw progesterone data are rather noisy, showing that some nonparametric smoothing techniques may be needed to remove some amount of measurement errors and to reconstruct the progesterone curves. Figure 1.3 displays four individual nonconceptive progesterone curves. It is seen that some outlying measurements appear in the progesterone curves 10 and 20. To reduce the effects of these outliers, some robust smoothing techniques may be desirable. To this end, four major nonparametric smoothing techniques for a single curve and for all the curves in a functional data set will be briefly reviewed in Chapters 2 and 3, respectively.

From Figure 1.2 (a), it is also seen that the nonconceptive progesterone curves are quite flat before the ovulation day. A question arises naturally. Is the mean nonconceptive curve a constant before the ovulation day? This is a one-sample problem for functional data. It will be handled in Chapter 4. The statement "the mean conceptive curve is a constant before the ovulation day" can be tested similarly.

Other questions also arise. For example, is there a conception effect? That is, is there a significant difference between the nonconceptive and conceptive mean progesterone curves? Over what period? These are two-sample problems for functional data and will be handled in Chapter 5.

The progesterone data have been used for illustrations of nonparametric regression methods by several authors. Brumback and Rice (1998) used them to illustrate a smoothing spline modeling technique for estimating both mean and individual functions; Fan and Zhang (2000) used them to illustrate their two-step method for estimating the underlying mean function for functional data; and Wu and Zhang (2002) used them to illustrate a local polynomial mixed-effects modeling approach.

1.2.2 Berkeley Growth Curve Data

The Berkeley growth curve data were collected in the Berkeley growth study (Tuddenham and Snyder 1954). The heights of thirty-nine boys and fifty-four girls were recorded at thirty-one not equally spaced ages from Year 1 to Year 18. The growth curves of girls and boys are displayed in Figures 1.4 (a) and

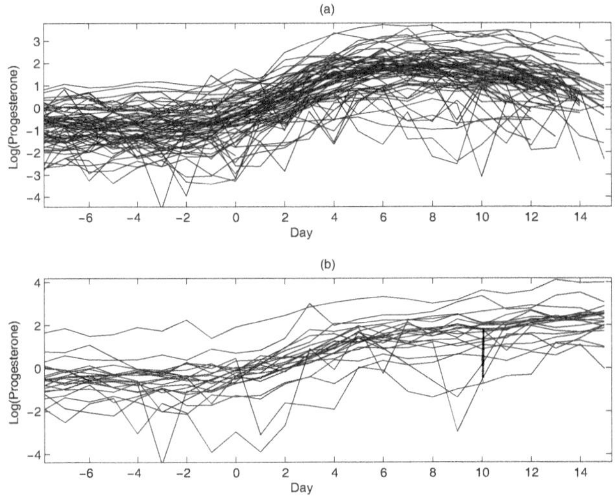

Figure 1.2 *The raw progesterone data with (a) nonconceptive progesterone curves and (b) conceptive progesterone curves. The progesterone curves are the levels of urinary metabolite progesterone over the course of the women's menstrual cycles (in days). They are rather noisy so that some smoothing may be needed.*

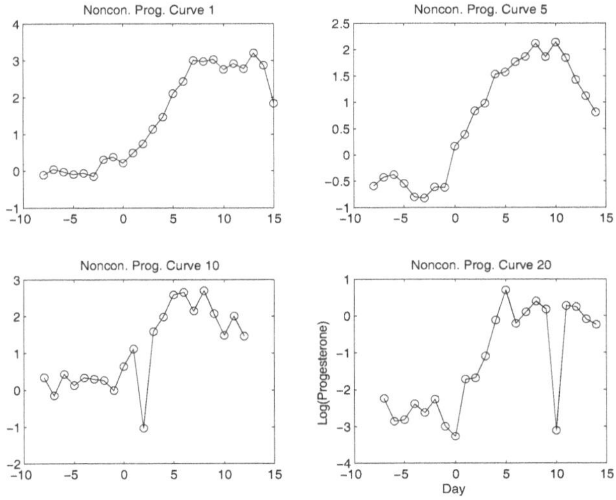

Figure 1.3 *Four individual nonconceptive progesterone curves. Outlying measurements appear in the progesterone curves 10 and 20.*

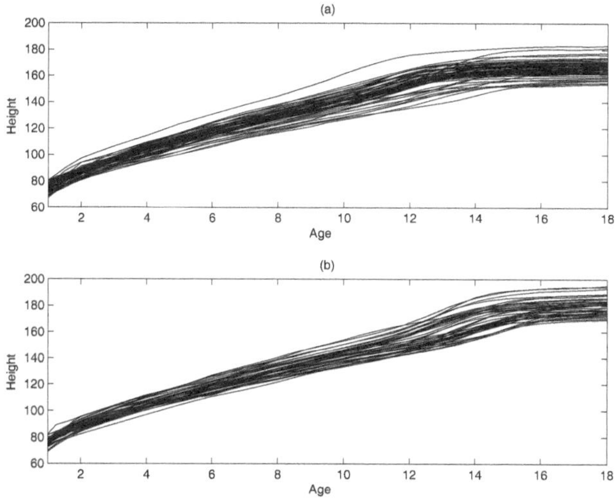

Figure 1.4 *The Berkeley growth curve data with (a) growth curves for fifty-four girls and (b) growth curves for thirty-nine boys. The growth curves are the heights of the children recorded at thirty-one not equally spaced ages from Year 1 to Year 18. The measurement errors of the heights of the children are rather small and may be ignored for many purposes.*

(b), respectively. It is seen that these growth curves are rather smooth, indicating that the measurement errors of the heights are rather small and may be ignored for many purposes. However, to evaluate the individual growth curves at a desired resolution, these growth curves may be interpolated or smoothed using some nonparametric smoothing techniques reviewed in Chapter 2.

It is of interest to check if gender difference has some impact on the growth process of a child. That is, one may want to test if there is any significant difference between the mean growth curves of girls and boys. This is a two-sample problem for functional data. It can be handled using the methods developed in Chapter 5. Another question also arises naturally. Is there any significant difference between the covariance functions of girls and boys? This problem will be handled in an exercise problem in Chapter 10. Then comes a related question. How to test if there is any significant difference between the mean growth curves of girls and boys when there is no knowledge about if the covariance functions of girls and boys are equal. This latter problem may be referred to as a two-sample Behrens-Fisher problem that can be handled using the methodologies developed in Chapter 9.

The Berkeley growth curve data have been used by several authors. Ramsay and Li (1998) and Ramsay and Silverman (2005) used it to illustrate their curve registration techniques. Chiou and Li (2007) used it to illustrate their k-centers functional clustering approach. Zhang, Liang, and Xiao (2010) used it to motivate and illustrate their testing approaches for a two-sample Behrens-Fisher problem for functional data.

1.2.3 Nitrogen Oxide Emission Level Data

Nitrogen oxides (NOx) are known to be among the most important pollutants, precursors of ozone formation, and contributors to global warming (Febrero, Galeano, and Gonzalez-Manteiga 2008). NOx is primarily caused by combustion processes in sources that burn fuels such as motor vehicles, electric utilities, and industries among others. Figures 1.5 (a) and (b) show NOx emission levels for seventy-six working days and thirty-nine non-working days, respectively, measured by an environmental control station close to an industrial area in Poblenou, Barcelona, Spain. The control station measured NOx emission levels in $\mu g/m^3$ every hour per day from February 23 to June 26 in 2005. The hourly measurements in a day (twenty-four hours) form a natural NOx emission level curve of the day. It is seen that within a day, the NOx levels increase in morning, attain their extreme values around 8 a.m., then decrease until 2 p.m. and increase again in evening. The influence of traffic on the NOx emission levels is not ignorable as the control station is located at the city center. It is not difficult to notice that the NOx emission levels of working days are generally higher than those of non-working days. This is why these NOx emission level curves were divided into two groups as pointed out by Febrero, Galeano, and Gonzalez-Manteiga (2008). In both the upper and lower panels, we highlighted one NOx emission level curve, respectively, to emphasize that they have very large NOx emission levels compared with the remaining curves in each group. It is important to identify these and other curves whose NOx emission levels are significantly large and try to figure out the sources that produced these abnormally large NOx emissions. This task will be handled in Chapter 8.

The NOx emission level data were kindly provided by the authors of Febrero, Galeano, and Gonzalez-Manteiga (2008) who used these data to illustrate their functional outlier detection methods.

1.2.4 Canadian Temperature Data

The Canadian temperature data (Canadian Climate Program 1982) were downloaded from "ftp://ego.psych.mcgill.ca/pub/ramsay/FDAfuns/Matlab/" at the book website of Ramsay and Silverman (2002, 2005). The data are the daily temperature records of thirty-five Canadian weather stations over a year (365 days), among which fifteen are in Eastern, another fifteen in Western, and the remaining five in Northern Canada. Panels (a), (b), and (c) of Figure 1.6 present the raw Canadian temperature curves for these thirty-five weather stations.

From these figures, we see that at least at the middle of the year, the temperatures at the Eastern weather stations are comparable with those at the

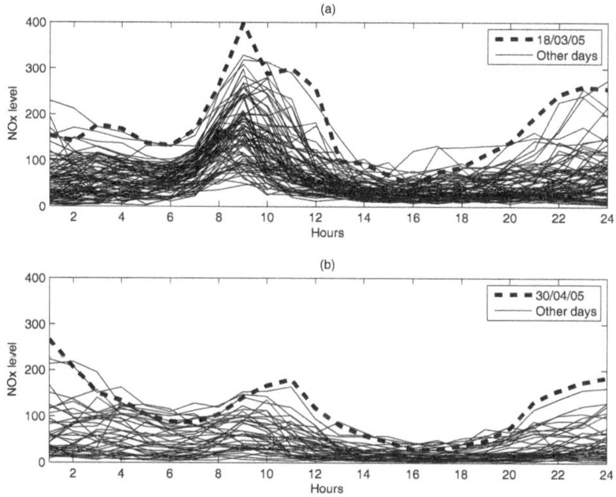

Figure 1.5 *The NOx emission level data for (a) seventy-six working days and (b) thirty-nine non-working days, measured by an environmental control station close to an industrial area in Poblenou, Barcelona, Spain. Each of the NOx emission level curve consists of the hourly measurements in a day (24 hours). In each panel, an unusual NOx emission level curve is highlighted.*

Western weather stations, but they are generally higher than those temperatures at the Northern weather stations. This observation seems reasonable as the Eastern and Western weather stations are located at about the same latitudes while the Northern weather stations are located at higher latitudes. Statistically, we can ask the following question. Is there a location effect among the mean temperature curves of the Eastern, Western, and Northern weather stations? This is a three-sample or one-way ANOVA problem for functional data. This problem will be treated in Chapters 5 and 9, respectively, when the three samples of the temperature curves are assumed to have the same and different covariance functions.

The figures in Figure 1.6 may indicate that the covariance functions of the temperature curves of the Eastern, Western, and Northern weather stations are different. In fact, we can see that at the beginning or at the end of the year, the temperatures at the Eastern weather stations are less variable than those at the Western weather stations, but they are generally more variable than those temperatures at the Northern weather stations. To statistically verify if the covariance functions of the temperature curves of the Eastern, Western, and Northern weather stations are the same, we will develop some tests about the equality of the covariance functions in Chapter 10.

Four individual temperature curves are presented in Figure 1.7. It is seen that these temperature curves are quite similar in shape and the temperature

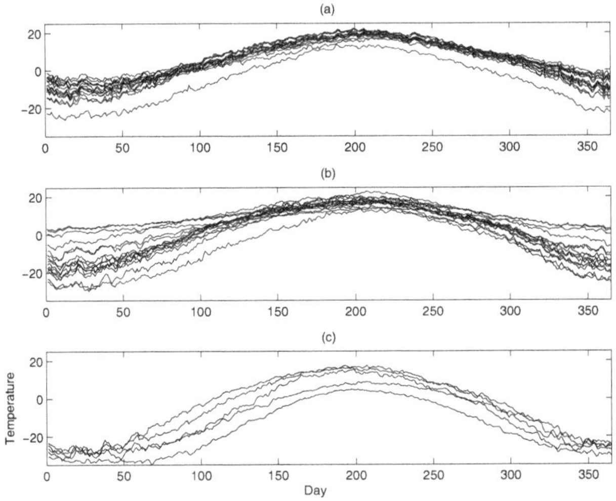

Figure 1.6 *The Canadian temperature data for (a) fifteen Eastern weather stations, (b) fifteen Western weather stations, and (c) five Northern weather stations. It is seen that at least at the middle of the year, the temperatures recorded at the Eastern weather stations are comparable with those recorded at the Western weather stations, but they are generally higher than those temperatures recorded at the Northern weather stations.*

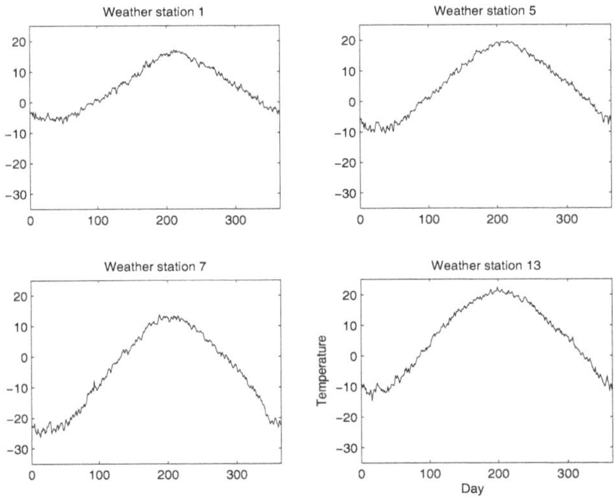

Figure 1.7 *Four individual Canadian temperature curves. They are similar in shape. It seems that the temperature records have large measurement errors.*

records are with large measurement errors. To remove some of these measurement errors, some smoothing techniques may be needed. As mentioned earlier, we will introduce four major smoothing techniques for a single curve and for all the curves in a functional data set in Chapters 2 and 3, respectively.

The Canadian temperature data have been analyzed in a number of papers and books in the literature, including Ramsay and Silverman (2002, 2005), Zhang and Chen (2007), and Zhang and Liang (2013) among others.

1.2.5 Audible Noise Data

The audible noise data were collected in a study to reduce audible noise levels of alternators, as reported in Nair et al. (2002). When an alternator rotates, it generates some amount of audible noise. With advances in technology, the engine noise can be reduced greatly so that the alternator noise levels become more remarkable, leading to an increasing quality concern. A robust design study conducted by an engineering team aiming to investigate the effects of seven process assembly factors and their potential for reducing noise levels. The seven factors under consideration are "Through Bolt Torque," "Rotor Balance," "Stator Varnish," "Air Gap Variation," "Stator Orientation," "Housing Stator Slip Fit," and "Shaft Radial Alignment." For simplicity, these seven factors are represented by A, B, C, D, E, F, and G respectively. In these seven factors, D is the noise factor while the others are control factors. Each factor has only two levels: low and high, denoted as "-1" and "$+1$," respectively.

The study adopted a 2_{IV}^{7-2} design, supplemented by four additional replications at the high levels of all factors, resulting in an experiment with thirty-six runs. Nair et al. (2002) pointed out that the fractional factorial design was a combined array, allowing estimation of the main effects of the noise and control factors and all two-factor control-by-noise interactions by assuming that all three factors and higher-order interactions are negligible. For each run, audible noise levels were measured over a range of rotating speeds. The audible sound was recorded by the microphones located at several positions near the alternator. The response was a transformed pressure measurement, known as sound pressure level. For each response curve, forty-three measurements of sound pressure levels (in decibels) were recorded with rotating speeds ranging from $1,000$ to $2,500$ revolutions per minute.

Figure 1.8 displays the thirty-six response curves of the audible noise data. It appears that the sound pressure levels were accurately measured with the measurement errors ignorable as shown by the four individual curves of the audible noise data presented in Figure 1.9. Nevertheless, Figure 1.8 indicates that these response curves are rather noisy although the measurement errors are rather small.

It is of interest to test if the main effects and the control-by-noise interaction effects of the seven factors on the response curves are significant. This problem will be handled in Chapter 7.

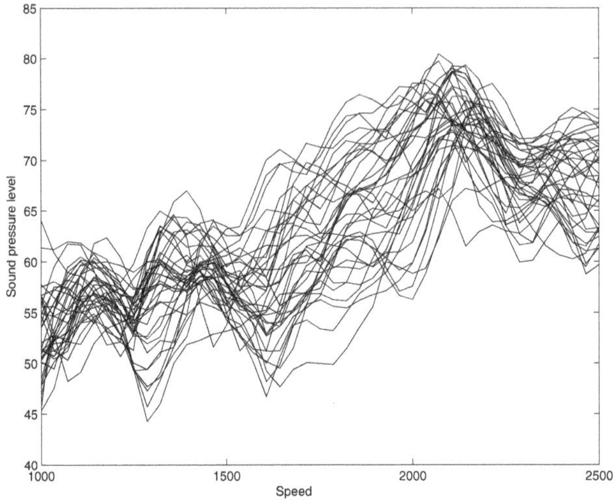

Figure 1.8 *The audible noise data with thirty-six sound pressure level curves. Each curve has forty-three measurements of sound pressure levels (in decibels), recorded with rotating speeds ranging from* 1,000 *to* 2,500 *revolutions per minute. These sound pressure level curves are rather noisy although the measurement errors of each curve are rather small.*

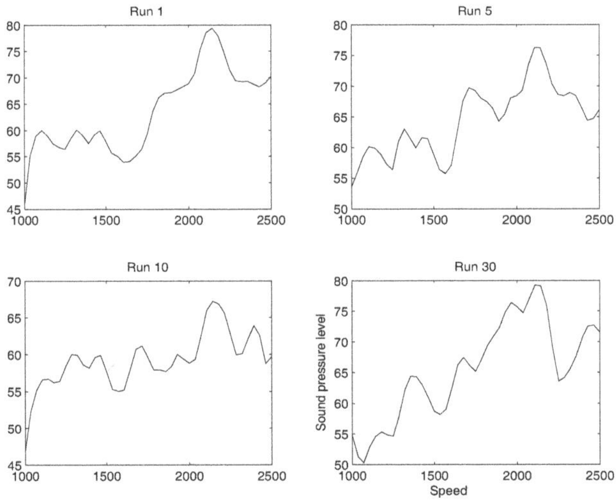

Figure 1.9 *Four individual curves of the audible noise data. It appears that the sound pressure levels were accurately measured with the measurement errors ignorable.*

The audible noise data have been analyzed by Nair et al. (2002) who computed the pointwise estimators of the main effects and control-by-noise interaction-effects, and constructed the pointwise confident intervals with standard errors estimated by the four additional replications at the high levels of all factors. They were also analyzed by Shen and Xu (2007) using their diagnostic methodologies.

1.2.6 *Left-Cingulum Data*

The cingulum is a collection of white-matter fibers projecting from the cingulate gyrus to the entorhinal cortex in the brain, allowing for communication between components of the limbic system. It is a prominent white-matter fiber track in the brain that is involved in emotion, attention, and memory, among many other functions. To study if the Radial Diffusibility (RD) in the left-cingulumn is affected by age and family of a child, the left-cingulum data were collected for thirty-nine children from 9 to 19 years old over arc length from -60 to 60. The response variable is "RD" while the covariates include "GHR" and "AGE" where GHR stands for Genetic High Risk and AGE is the age of a child in the study. The GHR variable is a categorical variable, taking two values. When GHR $= 1$, it means that the child is from a family with at least 1 direct relative with schizophrenia disease and when GHR $= 0$, the child is from a normal family. The AGE variable is a continuous variable. Figure 1.10 displays the thirty-nine left-cingulum curves over arc length. For each curve, there are 119 measurements. It is seen that the left-cingulum curves are very noisy and an outlying left-cingulum curve is also seen. Figure 1.11 displays two individual left-cingulum curves, showing that the left-cingulum curve was measured with very small measurement errors. The main variations of the left-cingulum curves come from the between-curve variations.

In Chapter 5, the left-cingulum data will be used to motivate and illustrate unbalanced two-way ANOVA models for functional data by transforming the AGE variable into a categorical variable with two levels: children with AGE < 15 and children with AGE ≥ 15.

The left-cingulum data were kindly provided by Dr. Hongbin Gu, Department of Psychiatry, School of Medicine, University of North Carolina at Chapel Hill during email discussions.

1.2.7 *Orthosis Data*

An orthotic is an orthopedic device applied externally to limb or body to provide support, stability, prevention of deformity from getting worse, or replacement of lost function. Depending on the diagnosis and physical needs of the individual, a large variety of orthosis is available. According to Abramovich, Antoniadis, Sapatinas, and Vidakovic (2004), the orthosis data were acquired and computed in an experiment by Dr. Amarantini David and Dr. Martin Luc (Laboratoire Sport et Performance Motrice, EA 597, UFRAPS, Grenoble

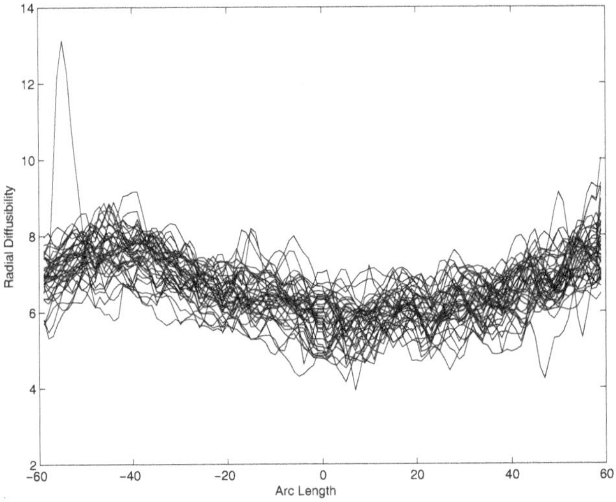

Figure 1.10 *The left-cingulumn data for thirty-nine children from 9 to 19 years old. Each curve has 119 measurements of the Radial Diffusibility (RD) in the left-cingulum of a child, recorded over arc length from −60 to 60.*

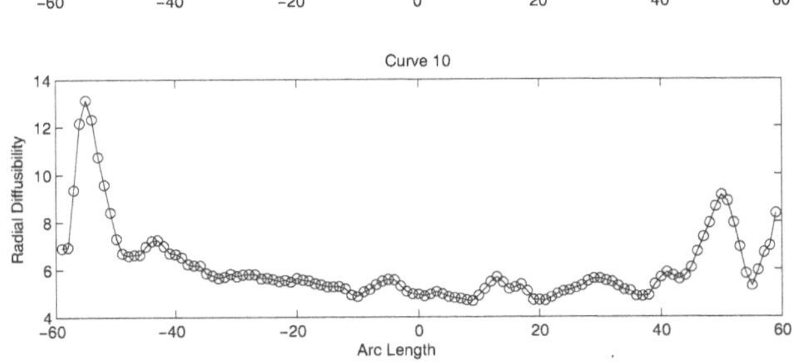

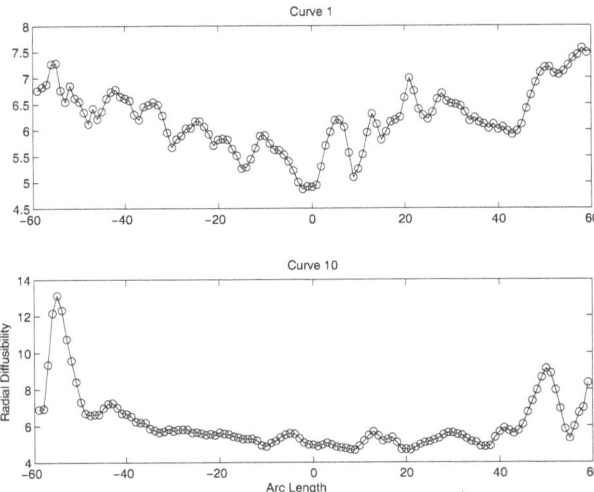

Figure 1.11 *Two left-cingulum curves. It appears that the measurements of the left-cingulum curve were measured with very small measurement errors.*

University, France). The aim of the experiment was to analyze how muscle copes with an external perturbation.

The experiment recruited seven young male volunteers. They wore a spring-loaded orthosis of adjustable stiffness under four experimental conditions: *a control condition* (without orthosis), *an orthosis condition* (with the orthosis only), and *two spring conditions* (spring 1, spring 2) in which stepping-in-place was perturbed by fitting a spring-loaded orthosis onto the right knee joint. All the seven subjects tried all the four conditions ten times for twenty seconds each while only the central ten seconds were used in the study in order to avoid possible perturbations in the initial and final parts of the experiment. The resultant moment of force at the knee is derived by means of body segment kinematics recorded with a sampling frequency of 200 Hz. For each stepping-in-place replication, the resultant moment was computed at 256 time points equally spaced and scaled so that a time interval corresponds to an individual gait cycle. A typical observation is a curve over time $t \in [0, 1]$. Therefore, the orthosis data set consists of

$$7 \text{ (Subjects)} \times 4 \text{ (Treatments)} \times 10 \text{ (Replications)} = 280 \text{ (Curves)}.$$

Figure 1.12 presents the 280 raw orthosis curves with each panel displaying ten of them.

The orthosis data have been previously studied by a number of authors, including Abramovich et al. (2004), Abramovich and Angelini (2006), Antoniadis and Sapatinas (2007), and Cuesta-Albertos and Febrero-Bande (2010), among others. In this book, this data set will be used to motivate and illustrate a heteroscedastic two-way ANOVA model for functional data in Chapter 9. It will also be used to motivate and illustrate multi-sample equal-covariance function testing problems in Chapter 10. The orthosis data were kindly provided by Dr. Brani Vidakovic by email communication.

1.2.8 Ergonomics Data

To study the motion of drivers of automobiles, the researchers at the Center for Ergonomics at the University of Michigan collected data on the motion of an automobile driver to twenty locations within a test car. Among other measures, the researchers measured three times the angle formed at the right elbow between the upper and lower arms of the driver. The data recorded for each motion were observed on an equally spaced grid of points over a period of time, which were rescaled to $[0, 1]$ for convenience, but the number of such time points varies from observation to observation. See Faraway (1997) and Shen and Faraway (2004) for detailed descriptions of the data. Figure 1.13 displays the right elbow angle curves of the ergonomics data downloaded from the website of the second author of Shen and Faraway (2004). These right elbow angle curves have been smoothed so that no further smoothing is needed but some interpolation or smoothing may be needed if one wants to evaluate these curves at a desired level of resolution.

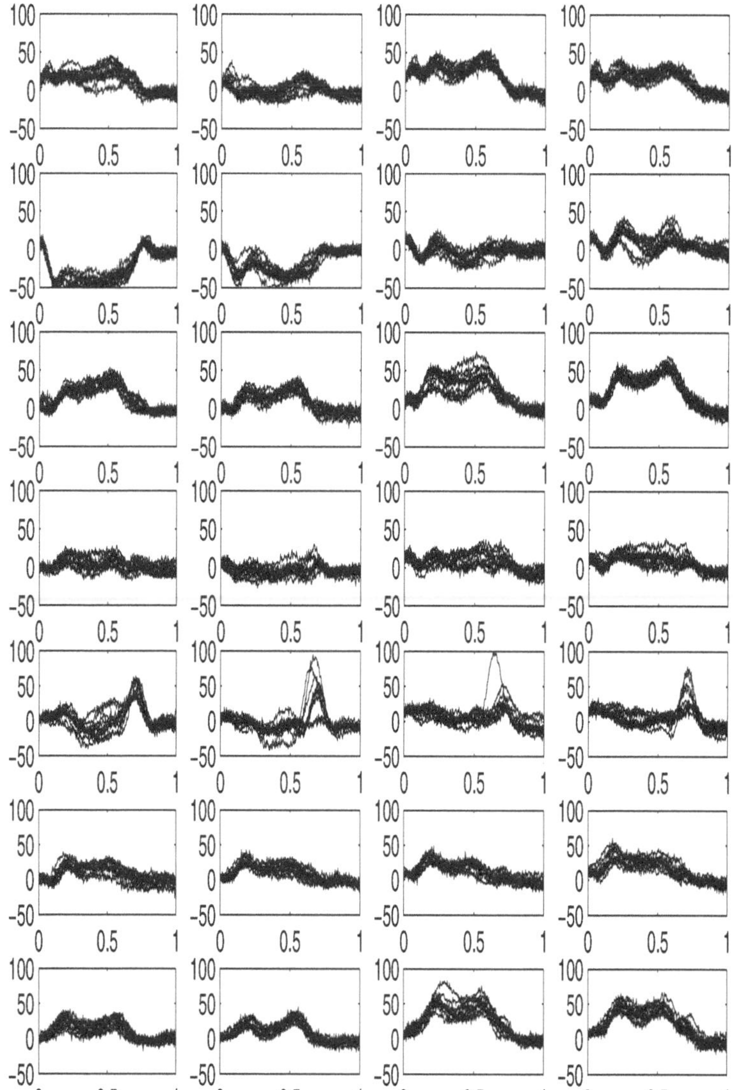

Figure 1.12 *The orthosis data. The row panels are associated with the seven subjects and the column panels are associated with the four treatment conditions. There are ten orthosis curves displayed in each panel. An orthosis curve has* 256 *measurements of the resultant moment of force at the knee, recorded over a period of time (ten seconds), scaled to* $[0, 1]$.

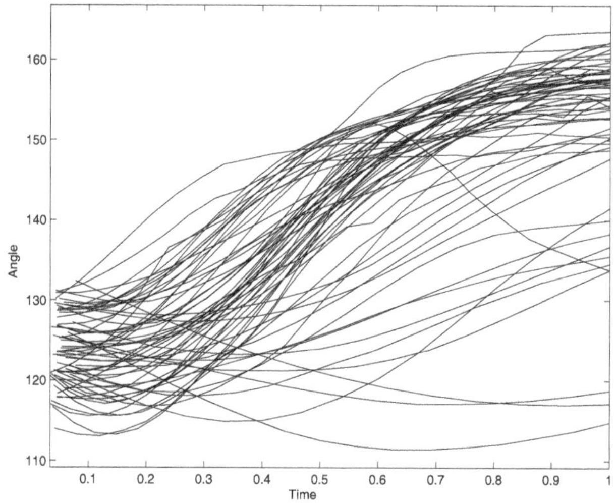

Figure 1.13 *The ergonomics data collected on the motion of an automobile driver to twenty locations within a test car, measured three times. The measurements of a curve are the angles formed at the right elbow between the upper and lower arms of the driver. The measurements for each motion were recorded on an equally spaced grid of points over a period of time, rescaled to* $[0, 1]$ *for convenience.*

There are totally sixty right elbow angle curves. They can be classified into a number of groups according to "location" and "area," where "location" stores the locations of the targets, taking twenty values while "area" stores the areas of the targets, taking four values. The categorical variables "location" and "area" are highly correlated as they are different only in the ways for grouping the locations of the targets.

One can also find a functional regression model to predict the right elbow angle curve $y(t), t \in [0, 1]$ using the coordinates (a, b, c) of a target, where a represents the coordinate in the left-to-right direction, b represents the coordinate in the close-to-far direction, and c represents the coordinate in the down-to-up direction. Shen and Faraway (2004) found that the following quadratic model is adequate to fit the data:

$$\begin{aligned} y_i(t) &= \beta_0(t) + a_i\beta_1(t) + b_i\beta_2(t) + c_i\beta_3(t) + a_i^2\beta_4(t) \\ &\quad +b_i^2\beta_5(t) + c_i^2\beta_6(t) + a_ib_i\beta_7(t) + a_ic_i\beta_8(t) \\ &\quad +b_ic_i\beta_9(t) + v_i(t), i = 1, \cdots, 60, \end{aligned} \tag{1.1}$$

where $y_i(t)$ and $v_i(t)$ denote the ith response and location-effect curves over time, respectively, (a_i, b_i, c_i) denotes the coordinates of the target associated with the ith angle curve, and $\beta_r(t), r = 0, 1, \cdots, 9$ are unknown coefficient functions.

Some questions then arise naturally. Is each of the coefficient functions significant? Can the quadratic model (1.1) be further reduced? These two questions and other similar questions can be answered easily after we study Chapter 6 where this ergonomics data set will be used to motivate and illustrate linear models for functional data with functional responses.

1.3 Why Is Functional Data Analysis Needed?

In many situations curve or surface data can be treated as multivariate data so that classical multivariate data analysis (MDA) tools (for example, Anderson 2003) can be applied directly. This treatment, however, ignores the fact that the underlying object of the measurements of a subject is a curve or a surface or any continuum as indicated by the real functional data sets presented in Section 1.2. In addition, direct MDA treatment may encounter difficulties in many other situations, including

- The sampling time points of the observed functional data are not the same across various subjects. For example, the nonconceptive and conceptive progesterone curves have different numbers of measurements for different curves.
- The sampling time points are not equally spaced. For example, the sampling time points of the Berkeley growth curve data are not equally spaced.
- The number of sampling time points is larger than the number of subjects in a sample of functional data. For example, the number of sampling time points for a Canadian temperature curve is 365 while the total number of the Canadian temperature curves is only fifteen.

In the first situation, direct MDA treatment may not be possible or reasonable; in the second situation, MDA inferences may be applied directly to the data but wether the observed data really represent the underlying curves or surfaces may be questionable; and in the third situation, standard MDA treatment fails as the associated sample covariance matrix is degenerated so that most of the inference tools in MDA, for example, the Hotelling T^2-test or the Lawley-Hotelling trace test (Anderson 2003), will not be well-defined. A remedy is to reduce the dimension of the functional data using some dimension reduction techniques. These dimension reduction techniques often work well. However, in many situations, dimension reduction techniques may also fail to reduce the dimension of the data sufficiently without loss of too much information.

Therefore, in these situations, functional data analysis (FDA) is more natural. In fact, Ramsay and Silverman (2002, 2005) provide many nice FDA tools to solve the aforementioned problems. In Chapters 2 and 3 of this book, we also provide some tools to overcome difficulties encountered in the first two situations while other chapters of the book provide methodologies to overcome difficulties encountered in the third situation.

1.4 Overview of the Book

In this book, we aim to conduct a thorough survey on the topics of hypothesis testing in the context of analysis of variance for functional data and give a systematic treatment of the methodologies. For this purpose, the remaining chapters are arranged as follows.

With science and technology development, functional data can be observed densely. However, they are still discrete observations. Fortunately, continuous versions of functional data can be reconstructed from discrete functional data by some smoothing techniques. For this purpose, we review four major nonparametric smoothing techniques for a single curve in Chapter 2 and present the methods for reconstructing a whole functional data set in Chapter 3.

In classical MDA, inferences are often conducted based on normal distribution and Mahalanobis distance. In FDA, the Mahalanobis distance often cannot be well-defined as the sample covariance matrices of discretized functional data are often degenerated. Instead we have to use the L^2-norm distance (and its modifications). In fact, the Gaussian process and the L^2-norm distance in FDA play the roles of normal distribution and Mahalanobis distance in classical MDA. Thus, we shall discuss the properties of the L^2-norm of a Gaussian process in Chapter 4. We also discuss the properties of Wishart processes (a natural extension of Wishart matrices), chi-squared mixtures, and F-type mixtures there. These properties are very important for this book and will be used in successive chapters.

ANOVA models for functional data are handled in Chapter 5, including two-sample problems, one-way ANOVA, and two-way ANOVA for functional data. Chapters 6 and 7 study functional linear models with functional responses when the design matrices are full rank and ill-conditioned, respectively. Chapter 8 studies how to detect unusual functions. In all these chapters, different groups are assumed to have the same covariance function. When this assumption is not satisfied, we deal with the heteroscedastic ANOVA for functional data in Chapter 9. The last chapter (Chapter 10) is devoted to testing the equality of the covariance functions. Examples are provided in each chapter. Technical proofs and exercises are given in almost every chapter.

1.5 Implementation of Methodologies

Most methodologies introduced in this book can be implemented using existing software such as S-PLUS, SAS, and MATLAB®, among others. We shall publish our MATLAB codes for most of the methodologies proposed in this book and the data analysis examples on our website: *http://www.stat.nus.edu.sg/~zhangjt/books/Chapman/FANOVA.htm.* We will keep updating the codes when necessary. We shall also make the data sets used in this book available through our website. The reader may try to apply the methodologies to the related data sets. We thank all the people who kindly provided us with these functional data sets.

1.6 Options for Reading This Book

Readers who are particularly interested in one or two of nonparametric smoothing techniques for functional data analysis may read Chapters 2 and 3. For a lower-level graduate course, Chapters 4–8 are recommended. If students already have some background in nonparametric smoothing techniques, Chapters 2 and 3 may be briefly reviewed or even skipped. Chapters 9 and 10 may be included in a higher-level graduate course or can be used as individual research materials for those who want to do research in a related field.

1.7 Bibliographical Notes

There are two major monographs about FDA methodologies currently available. One is by Ramsay and Silverman (2005). This book mainly focuses on description statistics in FDA, with topics such as curve registration, principal components analysis, canonical correlation analysis, pointwise t-test and F-test, fitting functional linear models, among others. Another book is by Ferraty and Vieu (2006). This book mainly focuses on nonparametric kernel estimation, functional prediction, and classification of functional observations. There are another three books, accompanying the above two books, by Ramsay and Silverman (2002), Ramsay et al. (2009), and Ferraty and Romain (2011), respectively.

In recent years, there has been a prosperous period in the development of functional hypothesis testing procedures. In terms of test statistics constructed, these testing procedures include the L^2-norm-based tests (Faraway 1997, Zhang and Chen 2007, Zhang, Peng, and Zhang 2010, Zhang, Liang, and Xiao 2010, etc.), F-type tests (Shen and Faraway 2004, Zhang 2011a, Zhang and Liang 2013, etc.), basis-based tests (Zhang 1999), thresholding tests (Fan and Lin 1998, Yang and Nie 2008), bootstrap tests (Faraway 1997, Cuevas, Febrero, and Fraiman 2004, Zhang, Peng, and Zhang 2010, Zhang and Sun 2010, etc.), penalized tests (Mas 2007), among others. In terms of the regression models considered, these testing procedures study the models including one-sample problems (Mas 2007), two-sample problems (Zhang, Peng, and Zhang 2010), one-way ANOVA (Cuevas, Febrero, and Fraiman 2004), functional linear models with functional responses (Faraway 1997, Shen and Faraway 2004, Zhang and Chen 2007, Zhang 2011a), two-sample Behrens-Fisher problems for functional data (Zhang, Peng, and Zhang 2010, Zhang, Liang, and Xiao 2010), k-sample Behrens-Fisher problems (Cuevas, Febrero, and Fraiman 2004), analysis of surface data (Zhang 1999), among others. However, all these important results and recent developments in this area are spread out in various statistical journals. This book aims to conduct a systematic survey of these results, to develop a common framework for the newly developed models and methods, and to provide a comprehensive guideline for applied statisticians to use the methods. However, the book is not intended to and is actually impossible to cover all the statistical inference methodologies available in the literature.

Chapter 2

Nonparametric Smoothers for a Single Curve

2.1 Introduction

In the main body of this book, we aim to survey various hypothesis testing methodologies for functional data analysis. In the development of these methodologies, we essentially assume that continuous functional data are available or can be evaluated at any desired resolution. In practice, however, the observed functional data are discrete, probably with large measurement errors as indicated by the curve data sets presented in the previous chapter. To overcome this difficulty, in this chapter, we briefly review four well-known nonparametric smoothing techniques for a single curve. These smoothing techniques may allow us to achieve the following goals:

- To reconstruct the individual functions in a real functional data set so that any reconstructed individual function can be evaluated at any desired resolution.
- To remove measurement errors as much as possible so that the variations of the reconstructed functional data mainly come from the between-subject variations.
- To improve the power of a proposed hypothesis testing procedure by the removal of measurement errors.

In Chapter 3, we focus on how to apply these smoothing techniques to reconstruct the functions in a functional data set so that the reconstructed functional data can be used directly by the methodologies surveyed in this book and study the effects of the substitution of the underlying individual functions with the reconstructed individual functions in a functional data set.

In this chapter, we focus on smoothing an individual function in a functional data set. For convenience, we generally denote the observed data of an individual function in a functional data set as

$$(t_i, y_i), \quad i = 1, 2, \cdots, n, \tag{2.1}$$

where $t_i, i = 1, 2, \cdots, n$ denote the design time points, and $y_i, i = 1, 2, \cdots, n$ are the responses at the design time points. The design time points may be equally spaced in an interval of interest, or be regarded as a random sample from a continuous design density, namely, $\pi(t)$. For simplicity, let us denote the interval of interest or the support of $\pi(t)$ as $\mathcal{T}$, which can be a finite interval $[a, b]$ or the whole real line $(-\infty, \infty)$. The responses $y_i, i = 1, 2, \cdots, n$ are often observed with measurement errors. For the observed data (2.1), a standard nonparametric regression model is usually written as

$$y_i = f(t_i) + \epsilon_i, \quad i = 1, \cdots, n, \tag{2.2}$$

where $f(t)$ models the underlying regression function and $\epsilon_i, i = 1, 2, \cdots, n$ denote the measurement errors that cannot be explained by the regression function $f(t)$. Mathematically, $f(t)$ is the conditional expectation of y_i, given $t_i = t$. That is,

$$f(t) = \mathrm{E}(y_i | t_i = t), \quad i = 1, 2, \cdots, n.$$

For an individual function in a functional data set, the regression function $f(t)$ is the associated underlying individual function.

There are a number of existing smoothing techniques that can be used to smooth the regression function $f(t)$ in (2.2). Different smoothing techniques have different strengths in one aspect or another. For example, smoothing splines may be good for handling sparse data, while local polynomial smoothers may be computationally advantageous for handling dense designs. In this chapter, as mentioned previously, we review four well-known smoothing techniques, including local polynomial kernel smoothing (Wand and Jones 1995, Fan and Gijbels 1996), regression splines (Eubank 1999), smoothing splines (Wahba 1990, Green and Silverman 1994), and P-splines (Ruppert, Wand and Carroll 2003), respectively, in Sections 2.2, 2.3, 2.4, and 2.5. We conclude this chapter with some bibliographical notes in Section 2.6.

2.2 Local Polynomial Kernel Smoothing

2.2.1 *Construction of an LPK Smoother*

The main idea of local polynomial kernel (LPK) smoothing is to locally approximate the underlying function $f(t)$ in (2.2) by a polynomial of some degree. Its foundation is Taylor expansion, which states that any smooth function can be locally approximated by a polynomial of some degree.

Specifically, let t_0 be an arbitrary fixed time point where the function $f(t)$ in (2.2) will be estimated. Assume $f(t)$ has a $(p+1)$st continuous derivative for some integer $p \geq 0$ at t_0. By Taylor expansion, $f(t)$ can be locally approximated by a polynomial of degree p. That is,

$$f(t) \approx f(t_0) + (t-t_0)f^{(1)}(t_0)/1! + (t-t_0)^2 f^{(2)}(t_0)/2! + \cdots + (t-t_0)^p f^{(p)}(t_0)/p!,$$

in a neighborhood of t_0 that allows the above expansion where $f^{(r)}(t_0)$ denotes the rth derivative of $f(t)$ at t_0.

Set $\beta_r = f^{(r)}(t_0)/r!$, $r = 0, \cdots, p$. Let $\hat{\beta}_r, r = 0, 1, 2, \cdots, p$ be the minimizers of the following weighted least squares (WLS) criterion:

$$\sum_{i=1}^{n} \{y_i - [\beta_0 + (t_i - t_0)\beta_1 + \cdots + (t_i - t_0)^p \beta_p]\}^2 K_h(t_i - t_0), \tag{2.3}$$

where $K_h(\cdot) = K(\cdot/h)/h$, obtained by rescaling a kernel function $K(\cdot)$ with a constant $h > 0$, called the bandwidth or smoothing parameter. The bandwidth h is mainly used to specify the size of the local neighborhood, namely,

$$I_h(t_0) = [t_0 - h, t_0 + h], \tag{2.4}$$

where the local fit is conducted. The kernel function, $K(\cdot)$, determines how observations within $[t_0 - h, t_0 + h]$ contribute to the fit at t_0.

Remark 2.1 *The kernel function $K(\cdot)$ is usually a probability density function. For example, the uniform density $K(t) = 1/2, t \in [-1, 1]$ and the standard normal density $K(t) = \exp(-t^2/2)/\sqrt{2\pi}, t \in (-\infty, \infty)$ are two well-known kernels, namely, the uniform kernel and the Gaussian kernel. Other useful kernels can be found in Gasser, Müller, and Mammitzsch (1985), Marron and Nolan (1988), and Zhang and Fan (2000), among others.*

Denote the estimate of the rth derivative $f^{(r)}(t_0)$ as $\hat{f}_h^{(r)}(t_0)$. Then

$$\hat{f}_h^{(r)}(t_0) = r!\hat{\beta}_r, r = 0, 1, \cdots, p.$$

In particular, the resulting pth degree LPK estimator of $f(t_0)$ is $\hat{f}_h(t_0) = \hat{\beta}_0$.

An explicit expression for $\hat{f}_h^{(r)}(t_0)$ is useful and can be made by matrix notation. Let

$$\mathbf{X} = \begin{bmatrix} 1 & (t_1 - t_0) & \cdots & (t_1 - t_0)^p \\ \vdots & \vdots & \ddots & \vdots \\ 1 & (t_n - t_0) & \cdots & (t_n - t_0)^p \end{bmatrix},$$

and

$$\mathbf{W} = \text{diag}(K_h(t_1 - t_0), \cdots, K_h(t_n - t_0)),$$

be the design matrix and the weight matrix for the LPK fit around t_0. Then the WLS criterion (2.3) can be re-expressed as

$$(\mathbf{y} - \mathbf{X}\boldsymbol{\beta})^T \mathbf{W} (\mathbf{y} - \mathbf{X}\boldsymbol{\beta}), \tag{2.5}$$

where $\mathbf{y} = (y_1, \cdots, y_n)^T$ and $\boldsymbol{\beta} = (\beta_0, \beta_1, \cdots, \beta_p)^T$. It follows that

$$\hat{f}_h^{(r)}(t_0) = r!\mathbf{e}_{r+1,p+1}^T \mathbf{S}_n^{-1} \mathbf{T}_n \mathbf{y},$$

where $\mathbf{e}_{r+1,p+1}$ denotes a $(p+1)$-dimensional unit vector whose $(r+1)$st entry is 1 and the other entries are 0, and

$$\mathbf{S}_n = \mathbf{X}^T \mathbf{W} \mathbf{X}, \quad \mathbf{T}_n = \mathbf{X}^T \mathbf{W}.$$

When t_0 runs over the whole support $\mathcal{T}$ of the design time points, a whole range estimation of $f^{(r)}(t)$ is obtained. The derivative estimator $\hat{f}_h^{(r)}(t), t \in \mathcal{T}$ is usually called the LPK smoother of the underlying derivative function $f^{(r)}(t)$. The derivative smoother $\hat{f}_h^{(r)}(t_0)$ is usually calculated on a grid of t's in $\mathcal{T}$.

In this section, we only focus on the LPK curve smoother

$$\hat{f}_h(t_0) = \mathbf{e}_{1,p+1}^T \mathbf{S}_n^{-1} \mathbf{T}_n \mathbf{y} = \sum_{i=1}^{n} K_h^n(t_i - t_0) y_i, \tag{2.6}$$

where $K_h^n(t_i - t_0) = \mathbf{e}_{1,p+1}^T \mathbf{S}_n^{-1} \mathbf{T}_n \mathbf{e}_{i,n}$ with $K^n(t)$ known as the empirical equivalent kernel for the p-order LPK; see Fan and Gijbels (1996). From (2.6), it is seen that for any $t \in \mathcal{T}$,

$$\hat{f}_h(t) = \sum_{i=1}^{n} K_h^n(t_i - t) y_i = \mathbf{a}(t)^T \mathbf{y}, \tag{2.7}$$

where $\mathbf{a}(t) = [K_h^n(t_1 - t), K_h^n(t_2 - t), \cdots, K_h^n(t_n - t)]^T$, which can be obtained from $\mathbf{T}_n^T \mathbf{S}_n^{-1} \mathbf{e}_{1,p+1}$ after replacing t_0 with t. This general formula allows us to evaluate $\hat{f}_h(t)$ at any resolution. Set $\hat{y}_i = \hat{f}_h(t_i)$ as the fitted value of $f(t_i)$. Let $\hat{\mathbf{y}}_h = [\hat{y}_1, \cdots, \hat{y}_n]^T$ denote the fitted values at all the design time points. Then $\hat{\mathbf{y}}_h$ can be expressed as

$$\hat{\mathbf{y}}_h = \mathbf{A}_h \mathbf{y}, \tag{2.8}$$

where

$$\mathbf{A}_h = (\mathbf{a}(t_1), \cdots, \mathbf{a}(t_n))^T \tag{2.9}$$

is known as the smoother matrix of the LPK smoother.

Remark 2.2 *As $\mathbf{A}_h$ does not depend on the response vector $\mathbf{y}$, the LPK smoother $\hat{f}_h(t)$ is known as a linear smoother, which is a linear combination of the responses. The regression spline, smoothing spline, and P-spline smoothers introduced in the next three sections are also linear smoothers.*

2.2.2 *Two Special LPK Smoothers*

Local constant and linear smoothers are the two simplest and most useful LPK smoothers. The local constant smoother is known as the Nadaraya-Watson estimator (Nadaraya 1964, Watson 1964). This smoother results from the LPK smoother $\hat{f}_h(t_0)$ (2.6) by simply taking $p = 0$:

$$\hat{f}_h(t_0) = \frac{\sum_{i=1}^{n} K_h(t_i - t_0) y_i}{\sum_{i=1}^{n} K_h(t_i - t_0)}. \tag{2.10}$$

Within a local neighborhood $I_h(t_0)$, it fits the data with a constant. That is, it is the minimizer $\hat{\beta}_0$ of the following WLS criterion:

$$\sum_{i=1}^{n}(y_i - \beta_0)^2 K_h(t_i - t_0).$$

The Nadaraya-Watson estimator is simple to understand and easy to compute. Let $I_A(t)$ denote the indicator function of some set A.

Remark 2.3 *When the kernel function K is the uniform kernel $K(t) = 1/2, t \in [-1, 1]$, the Nadaraya-Watson estimator (2.10) is exactly the local average of y_i's that are within the local neighborhood $I_h(t_0)$ (2.4):*

$$\hat{f}_h(t_0) = \frac{\sum_{i=1}^{n} I_{I_h(t_0)}(t_i) y_i}{\sum_{i=1}^{n} I_{I_h(t_0)}(t_i)} = \left\{ \sum_{t_i \in I_h(t_0)} y_i \right\} / m_h(t_0),$$

where $m_h(t_0)$ denotes the number of the observations falling into the local neighborhood $I_h(t_0)$.

Remark 2.4 *When t_0 is a boundary point of $\mathcal{T}$, fewer design points are within the neighborhood $I_h(t_0)$ so that $\hat{f}_h(t_0)$ has a slower convergence rate than the case when t_0 is an interior point of $\mathcal{T}$. For a detailed explanation of this boundary effect, the reader is referred to Müller (1991, 1993), Fan and Gijbels (1996), Cheng, Fan, and Marron (1997), and Müller and Stadtmuller (1999), among others.*

The local linear smoother (Stone 1984, Fan 1992, 1993) is obtained by fitting a data set locally with a linear function. Let $(\hat{\beta}_0, \hat{\beta}_1)$ minimize the following WLS criterion:

$$\sum_{i=1}^{n}[y_i - \beta_0 - (t_i - t_0)\beta_1]^2 K_h(t_i - t_0).$$

Then the local linear smoother is $\hat{f}_h(t_0) = \hat{\beta}_0$. It can be easily obtained from the LPK smoother $\hat{f}_h(t_0)$ (2.6) by simply taking $p = 1$.

Remark 2.5 *The local linear smoother is known as a smoother with a free boundary effect (Cheng, Fan, and Marron 1997). That is, it has the same convergence rate at any point in $\mathcal{T}$. It also exhibits many good properties that the other linear smoothers may lack. Good discussions on these properties can be found in Fan (1992, 1993), Hastie and Loader (1993), and Fan and Gijbels (1996, Chapter 2), among others.*

A local linear smoother can be simply expressed as

$$\hat{f}_h(t_0) = \frac{\sum_{i=1}^{n}[s_2(t_0) - s_1(t_0)(t_i - t_0)] K_h(t_i - t_0) y_i}{s_2(t_0)s_0(t_0) - s_1^2(t_0)}, \tag{2.11}$$

where

$$s_r(t_0) = \sum_{i=1}^{n} K_h(t_i - t_0)(t_i - t_0)^r \ , \ r = 0, 1, 2.$$

Remark 2.6 *The choice of the degree of the local polynomial used in LPK smoothing, p, is usually not as important as the choice of the bandwidth, h. A local constant ($p = 0$) or a local linear ($p = 1$) smoother is often good enough for most applications if the kernel function K and the bandwidth h are properly determined.*

Remark 2.7 *Fan and Gijbels (1996, Chapter 3) pointed out that for curve estimation (not valid for derivative estimation), an odd p is preferable. This is true as an LPK fit with $p = 2q + 1$ introduces an extra parameter compared to an LPK fit with $p = 2q$, but does not increase the variance of the associated LPK estimator. However, the associated bias may be significantly reduced especially in the boundary regions (Fan 1992, 1993; Hastie and Loader 1993; Fan and Gijbels 1996; Cheng, Fan, and Marron 1997).*

Thus, the local linear smoother is strongly recommended for most nonparametric smoothing problems in practice. For fast computation of the local linear smoother, the reader is referred to Fan and Marron (1994).

2.2.3 Selecting a Good Bandwidth

In most applications, the kernel function $K(\cdot)$ can be simply chosen as the uniform kernel or the Gaussian kernel. It is more crucial to choose the bandwidth h (Fan and Gijbels 1996).

Example 2.1 *We now employ the LPK smoother (2.6) to smooth the first nonconceptive progesterone curve by the LPK method with $p = 2$ and $K(t) = \exp(-t^2/2)/\sqrt{2\pi}$. That is, we employ the local quadratic kernel smoother with the Gaussian kernel. The data are shown as circles in Figure 2.1, which are the same as those circles presented in the left upper panel of Figure 1.3. Three different bandwidths $h^*/4, h^*, 4h^*$ are used to show the effect of different choices of bandwidth, where $h^* = 1.3$ is the optimal bandwidth selected by the GCV rule defined below in (2.12). It is seen that the three different local quadratic kernel fits are very different from each other, as shown in Figure 2.1. The dot-dashed curve is associated with $h = h^*/4 = 0.32$ and it is rather rough and almost interpolates the data. Therefore, when the bandwidth h is too small, the fitting bias is small; but as only a few data points are involved in the local fit, the associated fitting variance is large. The dashed curve is associated with $h = 4h^* = 5.22$ and is rather smooth but it does not fit the data adequately. Therefore, when the bandwidth h is too large, the fitting variance is small; but as the local neighborhood is large, the associated fitting bias is also large. A good fit will be obtained with a bandwidth that is not too small or too large.*

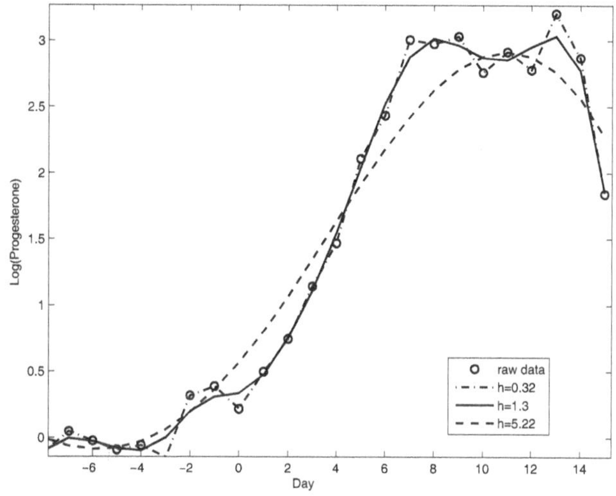

Figure 2.1 *Fitting the first nonconceptive progesterone curve by the LPK method with $p = 2$. The circles represent the data. The dot-dashed curve, obtained using a small bandwidth $h = 0.32$, nearly interpolates the data while the dashed curve, obtained using a large bandwidth $h = 5.22$, does not provide an adequate fit to the data. The solid curve, obtained using a bandwidth $h = 1.3$ selected by the GCV rule (2.12), provides a nice fit to the data.*

The solid curve is associated with $h = h^* = 1.3$ [$\log_{10}(1.3) = 0.1139$], which is not too small or too large according to the GCV rule described below and the solid curve is indeed a nice fit to the data.

Thus, for a given p and a fixed kernel function $K(\cdot)$, a good bandwidth h should be chosen to trade off between the goodness of fit and the model complexity of the LPK fit (2.8). Here, better goodness of fit often means less bias while smaller model complexity usually means less rough or smaller variance. To better choose the bandwidth h in LPK smoothing, we can use the following generalized cross-validation (GCV) rule (Wu and Zhang 2006, Chapter 3):

$$\text{GCV}(h) = \frac{\|\mathbf{y} - \hat{\mathbf{y}}_h\|^2}{(1 - \text{tr}(\mathbf{A}_h)/n)^2}, \tag{2.12}$$

where the fitted response vector $\hat{\mathbf{y}}_h$ and the LPK smoother matrix $\mathbf{A}_h$ are defined in (2.8) and (2.9), respectively. In the literature, many other bandwidth selectors are available. Famous bandwidth selectors can be found in Fan and Gijbels (1992), Herrmann, Gasser, and Kneip (1992), and Ruppert, Sheather, and Wand (1995), among others.

Remark 2.8 *With h increasing, the goodness of fit of the LPK fit measured*

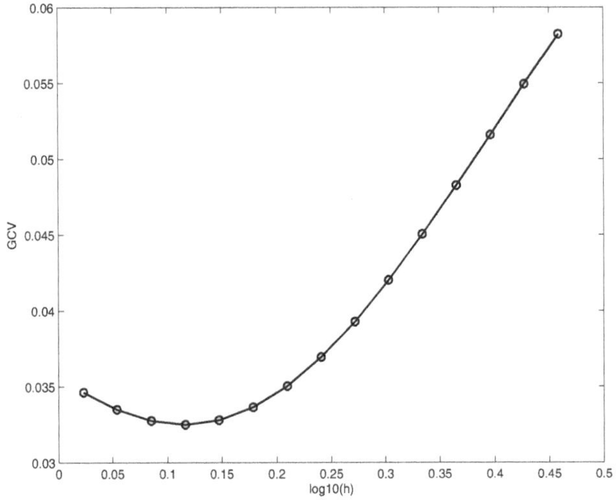

Figure 2.2 *Plot of GCV score against bandwidth h (in $\log_{10}$-scale) for fitting the first nonconceptive progesterone curve. The optimal bandwidth that minimizes the GCV rule (2.12) is obtained as $h^* = 1.3$.*

by the usual squared L^2-norm $\|\mathbf{y} - \hat{\mathbf{y}}_h\|^2$ is usually increasing while the model complexity of the LPK fit measured by the trace $tr(\mathbf{A}_h)$ is usually decreasing so that a good choice of h will give a nice trade off between the goodness of fit and the model complexity of the LPK fit. The optimal h is then obtained by minimizing the right-hand side of (2.12) with respect to a grid of possible values for h.

Example 2.2 *Figure 2.2 displays the GCV curve versus bandwidth h in $\log_{10}$-scale for selecting a good bandwidth for the local quadratic kernel fit of the first nonconceptive progesterone curve with the Gaussian kernel. The associated optimal bandwidth is $h^* = 1.3$, associated with $\log_{10}(h^*) = 0.1139$.*

2.2.4 Robust LPK Smoothing

It is well known that the LPK smoother is not robust in the presence of outliers. To show this, we provide the following example.

Example 2.3 *Figure 2.3 presents the usual local quadratic kernel fit of the twentieth nonconceptive progesterone curve as a dot-dashed curve. It was obtained with the Gaussian kernel and with the bandwidth chosen by the GCV rule (2.12). The data are presented as circles. It is seen that there is an outlier near Day 10. The usual local quadratic kernel fit is strongly dragged away from*

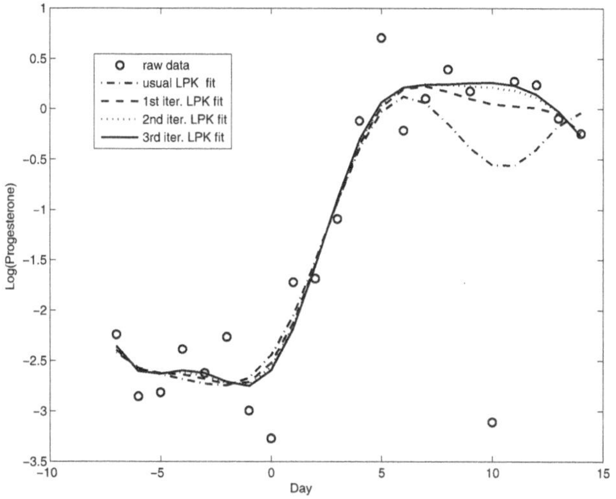

Figure 2.3 *Fitting the twentieth nonconceptive progesterone curve by the robust LPK method with* $p = 2$. *The usual LPK fit, first-iteration fit, second-iteration fit, and third-iteration fit are shown as dot-dashed, dashed, dotted, and solid curves, respectively. The data are shown as circles. It seems that the robust LPK smoothing algorithm converges rather quickly.*

the majority of the data around the neighborhood of the outlier, showing that the usual LPK smoother is indeed not robust against the presence of outliers.

To overcome the aforementioned problem, Cleveland (1979) proposed a robust locally weighted linear smoother that generally works well. For LPK smoothing, we can use the following robust LPK smoothing algorithm:

Step 0. Set $y_i^0 = y_i, i = 1, 2, \cdots, n$.

Step 1. Compute the LPK fit to the data $(t_i, y_i^0), i = 1, 2, \cdots, n$. Record the fitted responses $\hat{y}_i^0, i = 1, 2, \cdots, n$ and the associated residuals $\hat{\epsilon}_i = y_i - \hat{y}_i^0, i = 1, 2, \cdots, n$.

Step 2. Compute the standard deviation, $\hat{\sigma}$, of the residuals $\hat{\epsilon}_i, i = 1, 2, \cdots, n$.

Step 3. Set $y_i^0 = y_i$ when $|\hat{\epsilon}_i| < 2\hat{\sigma}$ and set $y_i^0 = \hat{y}_i^0$ when $|\hat{\epsilon}_i| \geq 2\hat{\sigma}$.

Step 4. Repeat Steps 1, 2, and 3 until convergence.

Remark 2.9 *Step 3 is used to identify the potential outliers using the* 2σ *rule (the* 3σ *rule may also be used) and to reduce the impact of the outliers on the LPK fit by replacing the values of the outlying responses* y_i^0 *with their fitted values* $\hat{y}_i^0$, *which are usually closer to the true response values than those* *outlying responses y_i. Usually, a few iterations are sufficient to make the robust LPK smoothing algorithm converge.*

Remark 2.10 *In the above robust LPK smoothing algorithm, the residuals computed in Step 1 assure that in Step 3, only the response values of the outliers identified in Step 3 will be changed.*

Remark 2.11 *The standard deviation $\hat{\sigma}$ of the residuals computed in Step 2 is rather crude. Other sophistical approaches may be employed to improve the algorithm.*

Example 2.4 *Figure 2.3 shows an application of the robust LPK smoothing algorithm to fit the twentieth nonconceptive progesterone curve. It is seen that the first-iteration LPK fit (dashed curve) is indeed closer to the majority of the data than the usual LPK fit (dot-dashed curve). One more iteration leads to the second-iteration LPK fit (dotted curve). The second-iteration LPK fit is indeed better than the first-iteration LPK fit. As the third-iteration LPK fit (solid curve) is very close to the second-iteration LPK fit (dotted curve), no further iterations are needed. The robust LPK smoothing algorithm is then terminated at the third iteration. It is seen that the impact of the outliers to the second- and third-iteration LPK fits are very small.*

2.3 Regression Splines

It is well-known that polynomials are not flexible in their ability to model data over a large range of the design time points. However, this is not the case when the range is small enough or when the big range is divided into some small subintervals or local neighborhoods. For LPK smoothing presented in the previous section, the local neighborhood is specified by a bandwidth and a fitting location. In regression spline smoothing, however, the local neighborhoods are specified by a group of locations, say,

$$\tau_0, \tau_1, \tau_2, \cdots, \tau_K, \tau_{K+1}, \tag{2.13}$$

in the range of interest, say, an interval $[a, b]$ where $a = \tau_0 < \tau_1 < \cdots < \tau_K < \tau_{K+1} = b$. These locations are known as knots, and $\tau_r, r = 1, 2, \cdots, K$ are called interior knots or simply knots. These knots divide the interval of interest, $[a, b]$, into K subintervals: $[\tau_r, \tau_{r+1}),\ \ r = 0, 1, \cdots, K$, so that within any two neighboring knots, a Taylor expansion up to some degree is valid. Mathematically, a regression spline is defined as a piecewise polynomial that is a polynomial of some degree within any two neighboring knots τ_r and τ_{r+1} for $r = 0, 1, \cdots, K$ and is joined together at knots properly but allows discontinuous derivatives at the knots.

2.3.1 Truncated Power Basis

A regression spline can be constructed using the following so-called kth degree truncated power basis with K interior knots $\tau_1, \tau_2, \cdots, \tau_K$:

$$1, t, \cdots, t^k, (t-\tau_1)_+^k, \cdots, (t-\tau_K)_+^k, \tag{2.14}$$

where $w_+^k = [w_+]^k$ denotes the power k of the positive part of w with $w_+ = \max(0, w)$.

Remark 2.12 *It is seen that the first $(k+1)$ basis functions of the truncated power basis (2.14) are polynomials of degree up to k, and the others are all the truncated power functions of degree k.*

Remark 2.13 *When $K = 0$, there are no interior knots involved. In this case, the truncated power basis reduces to a usual polynomial basis of degree k. Therefore, a kth degree truncated power basis is a generalization of a kth degree polynomial basis.*

Conventionally, the truncated power basis of degree "$k = 1, 2,$ or 3" is called the "linear, quadratic," or "cubic" truncated power basis, respectively.

Using the truncated power basis (2.14), a regression spline can be expressed as

$$f(t) = \sum_{s=0}^{k} \beta_s t^s + \sum_{r=1}^{K} \beta_{k+r}(t-\tau_r)_+^k, \tag{2.15}$$

where $\beta_0, \beta_1, \cdots, \beta_{k+K}$ are the associated coefficients. For convenience, it may be called a regression spline of degree k with knots $\tau_1, \tau_2, \cdots, \tau_K$. The regression splines (2.15) associated with $k = 1, 2,$ and 3 are usually called linear, quadratic, and cubic regression splines, respectively.

We can see that within any subinterval or local neighborhood $[\tau_r, \tau_{r+1})$, we have

$$f(t) = \sum_{s=0}^{k} \beta_s t^s + \sum_{l=1}^{r} \beta_{k+l}(t-\tau_l)^k,$$

which is a kth degree polynomial. However, for $r = 1, 2, \cdots, K$, we have

$$f^{(k)}(\tau_r-) = k!(\beta_k + \sum_{l=1}^{r-1} \beta_{k+l}), \text{ and } f^{(k)}(\tau_r+) = k!(\beta_k + \sum_{l=1}^{r} \beta_{k+l}).$$

It follows that $f^{(k)}(\tau_r+) - f^{(k)}(\tau_r-) = k!\beta_{k+r}$. That is, $f^{(k)}(t)$ jumps at the knot τ_r with amount $k!\beta_{k+r}$ for $r = 1, 2, \cdots, K$.

Remark 2.14 *In other words, a regression spline of degree k with knots $\tau_1, \cdots, \tau_K$ has continuous derivatives up to $(k-1)$ times everywhere, and has a discontinuous k-times derivative at the knots; the coefficient β_{k+r} of the rth truncated power basis function measures how large the jump is (up to a constant multiplicity of $k!$).*

2.3.2 *Regression Spline Smoother*

For convenience, it is often useful to denote the truncated power basis (2.14) as a basis vector:

$$\mathbf{\Phi}_p(t) = [1, t, \cdots, t^k, (t-\tau_1)_+^k, \cdots, (t-\tau_K)_+^k]^T, \tag{2.16}$$

where $p = K + k + 1$ denotes the number of the basis functions involved. Similarly, we can collect the associated coefficients into a coefficient vector: $\boldsymbol{\beta} = [\beta_0, \cdots, \beta_k, \beta_{k+1}, \cdots, \beta_{k+K}]^T$. Then the regression spline (2.15) can be re-expressed as $f(t) = \mathbf{\Phi}_p(t)^T\boldsymbol{\beta}$, so that the model (2.2) can be approximately expressed as

$$\mathbf{y} = \mathbf{X}\boldsymbol{\beta} + \boldsymbol{\epsilon}, \tag{2.17}$$

where $\mathbf{y} = (y_1, \cdots, y_n)^T$, $\mathbf{X} = (\mathbf{\Phi}_p(t_1), \cdots, \mathbf{\Phi}_p(t_n))^T$, and $\boldsymbol{\epsilon} = (\epsilon_1, \cdots, \epsilon_n)^T$. As $\mathbf{\Phi}_p(t)$ is a basis vector, $\mathbf{X}$ is of full rank, and hence $\mathbf{X}^T\mathbf{X}$ is invertible whenever $n > p$. A natural estimator of $\boldsymbol{\beta}$, which solves the approximation linear model (2.17) by the ordinary least squares method, is

$$\hat{\boldsymbol{\beta}} = (\mathbf{X}^T\mathbf{X})^{-1}\mathbf{X}^T\mathbf{y}. \tag{2.18}$$

It follows that the regression spline fit of the function $f(t)$ in (2.2) is

$$\hat{f}_p(t) = \mathbf{\Phi}_p(t)^T(\mathbf{X}^T\mathbf{X})^{-1}\mathbf{X}^T\mathbf{y} = \mathbf{a}(t)^T\mathbf{y}, \tag{2.19}$$

which is often called a regression spline smoother of $f(t)$ and obviously $\mathbf{a}(t) = \mathbf{X}(\mathbf{X}^T\mathbf{X})^{-1}\mathbf{\Phi}_p(t)$. In particular, the values of $\hat{f}_p(t)$ evaluated at the design time points $t_i, i = 1, 2, \cdots, n$ are collected in the following fitted response vector

$$\hat{\mathbf{y}}_p = \mathbf{X}(\mathbf{X}^T\mathbf{X})^{-1}\mathbf{X}^T\mathbf{y} = \mathbf{A}_p\mathbf{y}, \tag{2.20}$$

where $\hat{\mathbf{y}}_p = (\hat{y}_1, \cdots, \hat{y}_n)^T$ with $\hat{y}_i = \hat{f}_p(t_i), i = 1, 2, \cdots, n$, and $\mathbf{A}_p = \mathbf{X}(\mathbf{X}^T\mathbf{X})^{-1}\mathbf{X}^T$ is called the regression spline smoother matrix.

Remark 2.15 *The regression spline smoother matrix $\mathbf{A}_p$ is an idempotent matrix, satisfying $\mathbf{A}_p^T = \mathbf{A}_p, \mathbf{A}_p^2 = \mathbf{A}_p$ and $tr(\mathbf{A}_p) = p$. The trace of the smoother matrix $\mathbf{A}_p$ is often called the degrees of freedom of the regression spline smoother. It measures the complexity of the fitted regression spline model.*

2.3.3 *Knot Locating and Knot Number Selection*

Good performance of the regression spline smoother (2.19) strongly depends on good knot location, $\tau_1, \tau_2, \cdots, \tau_K$, and good choice of the number of knots, K. The degree of the regression spline basis (2.14), k, is usually less crucial, and it is often taken as 1, 2, or 3 for computational convenience. In this subsection, we first introduce two widely used methods for locating the knots and then mention briefly how to select the number of knots.

Uniform Knot Spacing Method. This method takes K equally spaced points in the range of interest, say, $[a, b]$, as knots. That is, the K knots are defined as

$$\tau_r = a + (b - a)r/(K + 1), r = 1, 2, \cdots, K.$$

Remark 2.16 *The uniform knot placing method is independent of the design time points. It is usually employed when the design time points are believed to be uniformly scattered in the range of interest.*

Quantiles as Knots Method. This method uses equally spaced sample quantiles of the design time points $t_i, i = 1, 2, \cdots, n$ as knots. Let $t_{(1)}, \cdots, t_{(n)}$ be the order statistics of the design time points. Then the K knots are defined as

$$\tau_r = t_{(1+[rn/(K+1)])}, r = 1, 2, \cdots, K,$$

where $[a]$ denotes the integer part of a.

Remark 2.17 *This quantiles as knots method is design adaptive. It locates more knots where more design time points are scattered. When the design time points are uniformly scattered, it is approximately equivalent to the uniform knot spacing method.*

After a knot placing method is specified, the number of knots, K, can be chosen by the following GCV rule:

$$\text{GCV}(K) = \frac{\|\mathbf{y} - \hat{\mathbf{y}}_p\|^2}{(1 - \text{tr}(\mathbf{A}_p)/n)^2}, \tag{2.21}$$

where $p = K + k + 1, \text{tr}(\mathbf{A}_p) = p$, and $\mathbf{y}$ and $\hat{\mathbf{y}}_p$ are the response vector (2.17) and the fitted response vector (2.20), respectively.

Remark 2.18 *The GCV rule (2.21) is essentially the same as the GCV rule (2.12) for the LPK smoother. It is used to trade off between the goodness of fit of the regression spline smoother, measured by the squared L^2-norm $\|\mathbf{y} - \hat{\mathbf{y}}_p\|^2$, and the model complexity, measured by p, the trace of $\mathbf{A}_p$. The optimal K is the one that minimizes the right-hand side of (2.21) with respect to a sequence of possible values of K.*

Example 2.5 *We now apply the regression spline method to fit the fifth nonconceptive progesterone curve. We use the regression spline basis (2.14) with $k = 2$. The individual measurements are shown as circles in Figure 2.4. The quadratic regression spline fit is shown as a solid curve. It is obtained with $K = 5$ interior knots. The number of knots $K = 5$ is chosen by the GCV rule (2.21) as shown in Figure 2.5. It seems that the resulting quadratic regression spline fit is indeed adequate for the data.*

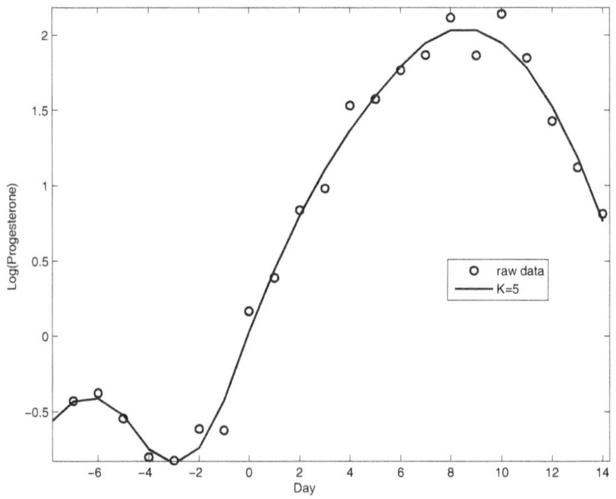

Figure 2.4 *Fitting the fifth nonconceptive progesterone curve by the regression spline method using the truncated power basis (2.14) with $k = 2$. The $K = 5$ interior knots are used. The resulting regression spline fit (solid curve) is indeed adequate for the data (circles).*

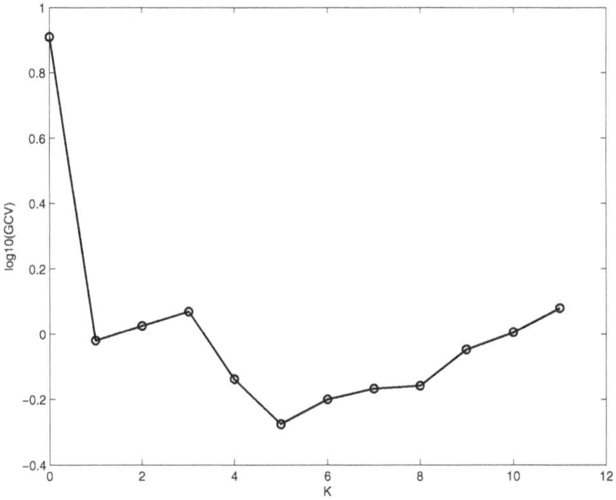

Figure 2.5 *Plot of GCV score (in $\log_{10}$-scale) against number of knots K for fitting the fifth nonconceptive progesterone curve by the regression spline method using the truncated power basis (2.14) with $k = 2$. The optimal number of knots that minimizes the GCV rule (2.21) is obtained as $K = 5$.*

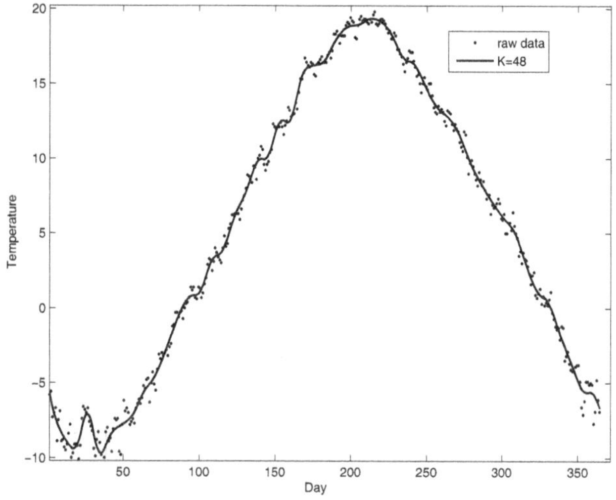

Figure 2.6 *Fitting the fifth Canadian temperature curve by the regression spline method using the truncated power basis (2.14) with* $k = 2$*. The* $K = 48$ *interior knots are used. The resulting regression spline fit (solid curve) is indeed adequate for the data (dots).*

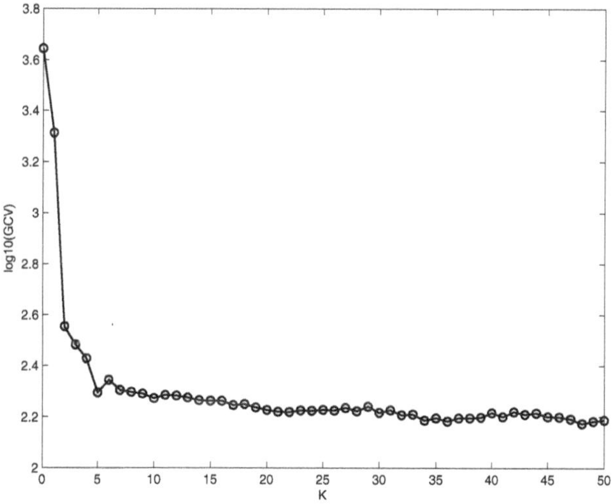

Figure 2.7 *Plot of GCV score (in* $\log_{10}$*-scale) against number of knots* K *for fitting the fifth Canadian temperature curve by the regression spline method using the truncated power basis (2.14) with* $k = 2$*. The optimal number of knots which minimizes the GCV rule (2.21) is obtained as* $K = 48$*. It seems that the GCV rule favors a large number of interior knots in this regression spline fit.*

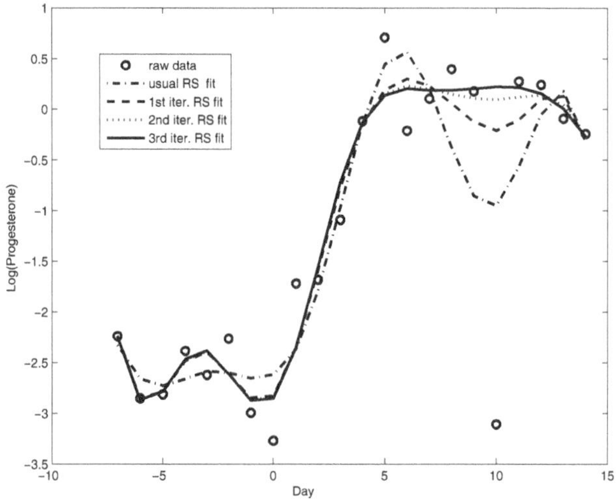

Figure 2.8 *Fitting the twentieth nonconceptive progesterone curve by the robust regression spline method using the truncated power basis (2.14) with $k = 2$. The usual regression spline fit, first-iteration fit, second -iteration fit, and third-iteration fit are shown as the dot-dashed, dashed, dotted, and solid curves, respectively. The data are shown as circles. The robust regression spline smoothing algorithm also converges rather quickly.*

Remark 2.19 *When the number of the observations for an underlying individual curve is large, a large number of interior knots are often needed to fit the data adequately.*

Example 2.6 *We now apply the regression spline method to fit the fifth Canadian temperature curve. We continue to use the regression spline basis (2.14) with $k = 2$. For each of the Canadian temperature curves, there are* 365 *measurements. These individual measurements are presented as dots in Figure 2.6. Figure 2.7 shows the GCV curve for such a quadratic regression spline fit. It is seen that the GCV rule favors a large number of interior knots in this regression spline fit and suggests that the optimal number of knots is $K = 48$. The MATLAB® code "rsfit.m" limits the large number of interior knots to* 50*; otherwise, a larger number of interior knots is also possible. Figure 2.6 shows the quadratic regression spline fit as a solid curve. It is seen that the quadratic regression spline fit with* 48 *interior knots is indeed adequate for the data.*

2.3.4 Robust Regression Splines

Like the LPK smoother, the regression spline smoother is also not robust. It can be made robust by modifying the robust LPK smoothing algorithm presented in Section 2.2.4 by replacing the LPK smoother with the regression spline smoother (2.20).

Example 2.7 *Figure 2.8 shows an application of a robust regression spline smoothing algorithm to fit the twentieth nonconceptive progesterone curve. The measurements are shown as circles, and the standard regression spline fit, the first-, second-, and third-iteration regression spline fits are shown as the dot-dashed, dashed, dotted, and solid curves, respectively. It is seen that the usual regression spline fit is indeed affected by the outlier, the first-iteration regression spline fit improves the usual regression spline fit and it is improved by the second-iteration regression spline fit. The second- and third-iteration regression spline fits are close to each other. The robust regression spline smoothing algorithm is then terminated at the third iteration.*

2.4 Smoothing Splines

2.4.1 Smoothing Spline Smoothers

In this section, we introduce the smoothing spline smoother. As mentioned in the previous section, for the regression spline smoother, when the knot locating method is specified, the remaining task is to choose the number of knots, K. In general, K is smaller than the number of the measurements. As K must be an integer, choices for K are limited. Alternatively, we can use all the distinct design time points as knots. This may result in undersmoothing when there are too many distinct design time points. The resulting fit is usually quite rough. To overcome this problem, we then introduce a penalty to control the roughness of the fitted curve.

To be specific, without loss of generality, let us again assume that the range of interest of $f(t)$ in (2.2) is a finite interval, say, $[a, b]$ for some finite numbers a and b. The roughness of $f(t)$ is usually defined as the integral of its squared k-times derivative

$$\int_a^b \left\{f^{(k)}(u)\right\}^2 du \tag{2.22}$$

for some $k \geq 1$. This quantity is large when the function $f(\cdot)$ is rough. The smoothing spline smoother of the function $f(t)$ in (2.2) is defined as the minimizer $\hat{f}_\lambda(t)$ of the following penalized least squares (PLS) criterion:

$$\sum_{i=1}^{n} [y_i - f(t_i)]^2 + \lambda \int_a^b \left\{f^{(k)}(t)\right\}^2 dt, \tag{2.23}$$

over the following kth order Sobolev space $\mathcal{W}_2^k[a, b]$:

$$\left\{f : f^{(r)} \text{ absolute continuous for } 0 \leq r \leq k-1, \int_a^b \{f^{(k)}(t)\}^2 dt < \infty\right\},$$

where $\lambda > 0$ is a smoothing parameter controlling the size of the roughness penalty, and it is usually used to trade off the goodness of fit, represented by the first term in (2.23), and the roughness of the resulting curve. The $\hat{f}_\lambda(t)$ is known as a natural smoothing spline of degree $(2k-1)$. In particular, when $k = 2$, the associated $\hat{f}_\lambda(t)$ is a natural cubic smoothing spline. For a detailed description of smoothing splines, see for example, Eubank (1999), Wahba (1990), Green and Silverman (1994), and Gu (2002), among others.

2.4.2 *Cubic Smoothing Splines*

To minimize the PLS criterion (2.23), we need to compute the integral that defines the roughness. This is a challenging issue for computing a smoothing spline. When $k = 2$, however, the associated cubic smoothing spline is less computationally challenging. Actually, there is a way to compute the roughness term quickly, as stated in Green and Silverman (1994). It is one of the reasons why cubic smoothing splines are popular in statistical applications. Let

$$\tau_1, \cdots, \tau_K, \tag{2.24}$$

be all the distinct design time points and be sorted in increasing order. They are all the knots of the cubic smoothing spline $\hat{f}_\lambda(t)$ that minimizes (2.23) when $k = 2$. Let $\mathbf{f} = (f_1, \cdots, f_K)^T$, where $f_r = f(\tau_r), r = 1, 2, \cdots, K$. Then by Green and Silverman (1994), it is known that the roughness (2.22) of $f(t)$ for $k = 2$ can be expressed as

$$\int_a^b [f''(t)]^2 dt = \mathbf{f}^T \mathbf{G} \mathbf{f}, \tag{2.25}$$

where the matrix $\mathbf{G} : K \times K$ is known as a roughness matrix of the cubic smoothing spline smoother. Given the K distinct knots (2.24), a detailed method for computing $\mathbf{G}$ can be found in Green and Silverman (1994, p. 12–13). It follows that the PLS criterion (2.23) can be written as

$$\|\mathbf{y} - \mathbf{X}\mathbf{f}\|^2 + \lambda \mathbf{f}^T \mathbf{G} \mathbf{f}, \tag{2.26}$$

where $\mathbf{X} = (x_{ir})$ is an $n \times K$ incidence matrix with $x_{ir} = 1$ if $t_i = \tau_r$ and 0 otherwise, and $\| \cdot \|$ denotes the usual L^2-norm. Therefore, an explicit expression for the cubic smoothing spline $\hat{\mathbf{f}}_\lambda$, evaluated at the knots (2.24), is as follows:

$$\hat{\mathbf{f}}_\lambda = (\mathbf{X}^T \mathbf{X} + \lambda \mathbf{G})^{-1} \mathbf{X}^T \mathbf{y}. \tag{2.27}$$

For any given $t \in \mathcal{T}$, a general formula for computing $f(t)$ is available (Green and Silverman 1994) and it can be written as

$$\hat{f}_\lambda(t) = \mathbf{a}(t)^T \mathbf{y}, \tag{2.28}$$

but the expression of $\mathbf{a}(t)$ is rather involved. In general, for any fixed t, the value of $\hat{f}_\lambda(t)$ can be interpolated based on the fitted response vector at the design time points which is

$$\hat{\mathbf{y}}_\lambda = \mathbf{A}_\lambda \mathbf{y}, \tag{2.29}$$

where

$$\mathbf{A}_\lambda = \mathbf{X}(\mathbf{X}^T\mathbf{X} + \lambda\mathbf{G})^{-1}\mathbf{X}^T \tag{2.30}$$

is known as the cubic smoothing spline smoother matrix. When all the design time points are distinct, $\hat{\mathbf{y}}_\lambda = \hat{\mathbf{f}}_\lambda = (\mathbf{I}_n + \lambda\mathbf{G})^{-1}\mathbf{y}$, as $\mathbf{X} = \mathbf{I}_n$, an identity matrix of size n. For more details about cubic smoothing splines, see Green and Silverman (1994), among others.

2.4.3 Smoothing Parameter Selection

For regression spline smoothing, we need to locate the knots and select the number of the knots to get a better regression spline fit to the data. Similarly, we need to select the smoothing parameter λ, which plays an important role in the smoothing spline smoother (2.27). It trades off between the goodness of fit (the first term in (2.23)) and the roughness of the smoothing spline smoother $\hat{\mathbf{f}}_\lambda$ (2.27). For this purpose, we can again use the GCV rule (Craven and Wahba 1979) defined as

$$\text{GCV}(\lambda) = \frac{\|\mathbf{y} - \hat{\mathbf{y}}_\lambda\|^2}{(1 - \text{tr}(\mathbf{A}_\lambda)/n)^2}. \tag{2.31}$$

The optimal λ is chosen to minimize the right-hand side of (2.31) with respect to a grid of possible values for λ.

Remark 2.20 *It is not difficult to notice that the GCV rule (2.31) for smoothing spline smoothing is essentially the same as the one defined in (2.21) for regression spline smoothing, and the one defined in (2.12) for LPK smoothing.*

Example 2.8 *We now apply the cubic smoothing spline smoother to fit the fifth audible noise curve. As mentioned in Section 1.2.5 of Chapter 1, the measurement errors of the audible noise data are very small so that an individual audible noise curve can be smoothly fitted with very small bias. The individual data are presented as circles in Figure 2.9. Figure 2.10 shows the GCV curve for such a cubic smoothing spline fit. It suggests that among the specified values of smoothing parameter, the smallest one is optimal, suggesting that the fitting bias is very small. Figure 2.9 displays the cubic smoothing spline fit (solid curve) which is indeed adequate for the data.*

Remark 2.21 *When the measurement errors are small, like other smoothers, the cubic smoothing spline smoother favors a small smoothing parameter. This property can be used to interpolate the individual curve. From Section 1.2, we can see that the measurement errors of the Berkeley growth curves are rather small; in this case, the cubic smoothing spline smoother will favor small smoothing parameter and the resulting cubic smoothing spline fit will interpolate the data. This is actually true, as shown in Figures 2.12 and 2.11.*

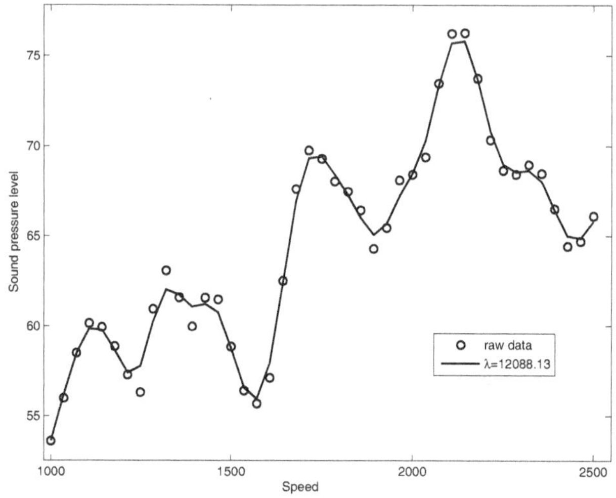

Figure 2.9 *Fitting the fifth audible noise curve by the cubic smoothing spline method. The smoothing parameter* $\lambda = 12,088.13$ *is used. The resulting cubic smoothing spline fit (solid curve) looks adequate for the data (circles).*

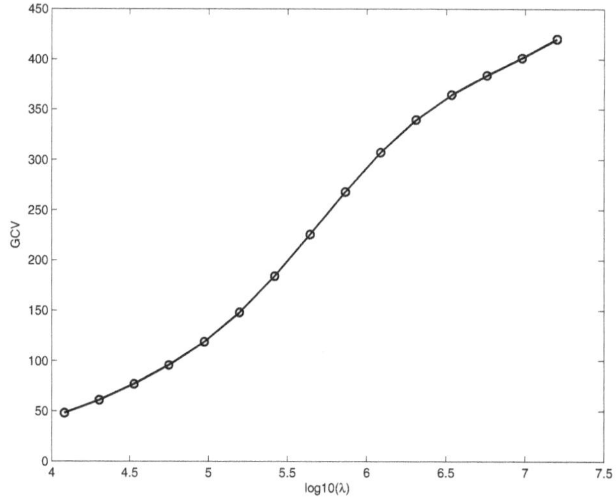

Figure 2.10 *Plot of GCV score against smoothing parameter* λ *(in* $\log_{10}$*-scale) for fitting the fifth audible noise curve by the cubic smoothing spline method. The optimal smoothing parameter that minimizes the GCV rule (2.31) is obtained as* $\lambda = 12,088.13$. *The GCV rule favors the smallest smoothing parameter in this cubic smoothing spline fit.*

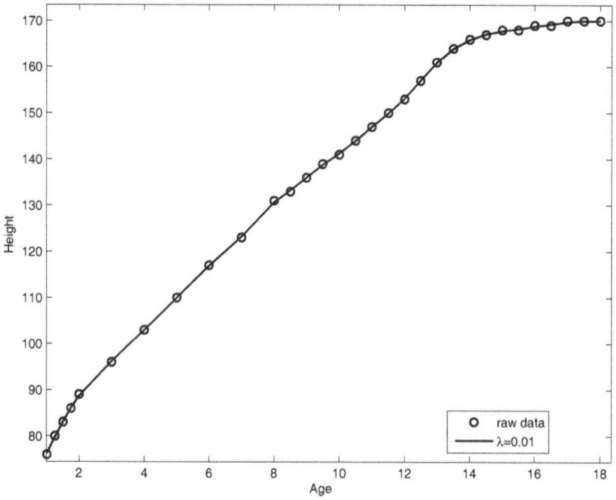

Figure 2.11 *Fitting the fifth Berkeley growth curve by the cubic smoothing spline method. The smoothing parameter* $\lambda = 0.01$ *is used. The resulting cubic smoothing spline fit (solid curve) nearly interpolates the data (circles).*

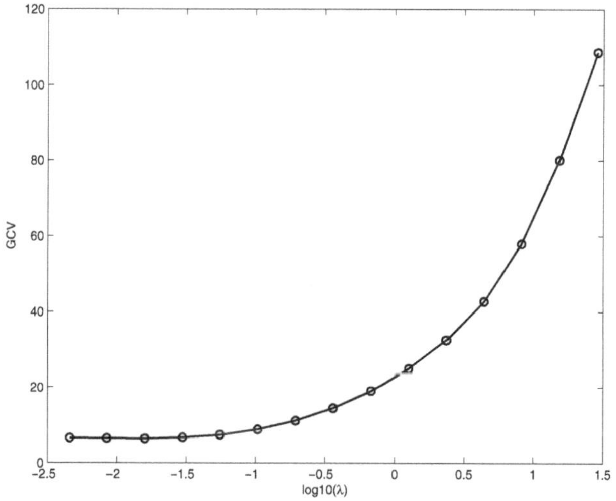

Figure 2.12 *Plot of GCV score against smoothing parameter* λ *(in* $\log_{10}$*-scale) for fitting the fifth Berkeley growth curve by the cubic smoothing spline method. The optimal smoothing parameter that minimizes the GCV rule (2.31) is obtained as* $\lambda = 0.01$. *The GCV rule favors the smallest smoothing parameter in this cubic smoothing spline fit due to the fact that the Berkeley growth curves were recorded with very small measurement errors.*

Remark 2.22 *Like the LPK and regression spline smoothers, the cubic smoothing spline smoother is also not robust. It can be made robust by modifying the robust LPK smoothing algorithm presented in Section 2.2.4 by replacing the LPK smoother with the cubic smoothing spline smoother (2.27).*

2.5 P-Splines

2.5.1 P-Spline Smoothers

Consider the penalized least squares criterion (2.23) for smoothing splines. When $k \neq 2$, it is difficult to compute the integral that defines the roughness of the underlying function $f(t)$. To overcome this difficulty, one can apply a basis vector $\mathbf{\Phi}_p(t)$ to compute the integral and the associated roughness matrix. Instead of using a basis vector $\mathbf{\Phi}_p(t)$ and computing the associated roughness matrix directly, a penalized spline (P-spline) is obtained using a special basis, for example, the truncated power basis (2.14) and a special matrix in place of the associated roughness matrix. Specifically, let $\mathbf{\Phi}_p(t)$ be the truncated power basis vector (2.16) of degree k with K interior knots $\tau_1, \tau_2, \cdots, \tau_K$. Then the underlying function $f(t)$ in (2.2) can be first written as a regression spline $\mathbf{\Phi}_p(t)^T\boldsymbol{\beta}$, where $\boldsymbol{\beta} = [\beta_0, \beta_1, \cdots, \beta_{k+K}]^T$ is the associated coefficient vector. Let $\mathbf{G}$ be a $p \times p$ diagonal matrix with its first $k+1$ diagonal entries being 0 and other diagonal entries 1. That is,

$$\mathbf{G} = \begin{pmatrix} \mathbf{0} & \mathbf{0} \\ \mathbf{0} & \mathbf{I}_K \end{pmatrix}. \tag{2.32}$$

Then a P-spline smoother of the function $f(t)$ in (2.2) is defined as $\hat{f}_\lambda(t) = \mathbf{\Phi}_p(t)^T\hat{\boldsymbol{\beta}}$, where $\hat{\boldsymbol{\beta}}$ is the minimizer of the following PLS criterion:

$$\sum_{i=1}^{n} \left(y_i - \mathbf{\Phi}_p(t_i)^T\boldsymbol{\beta}\right)^2 + \lambda\boldsymbol{\beta}^T\mathbf{G}\boldsymbol{\beta}. \tag{2.33}$$

Remark 2.23 *We have*

$$\boldsymbol{\beta}^T\mathbf{G}\boldsymbol{\beta} = \sum_{r=1}^{K} \beta_{k+r}^2,$$

and β_{k+r} is the coefficient of the rth truncated power basis function $(t-\tau_r)_+^k$ in the basis vector $\mathbf{\Phi}_p(t)$. The coefficient β_{k+r} measures the jump of the k-times derivative of $f(t)$ at the knot τ_r. It follows that the penalty in (2.33) is imposed just for the derivative jumps of the resulting P-spline at the knots. Therefore, the larger the coefficients $\beta_{k+r}, r = 1, \cdots, K$, the rougher the resulting P-spline. Thus, the term $\boldsymbol{\beta}^T\mathbf{G}\boldsymbol{\beta}$ is a measure of the roughness of the resulting P-spline.

For convenience, we call $\mathbf{G}$ the P-spline roughness matrix.

Minimizing (2.33) with respect to $\boldsymbol{\beta}$ leads to

$$\hat{\boldsymbol{\beta}} = (\mathbf{X}^T\mathbf{X} + \lambda\mathbf{G})^{-1}\mathbf{X}^T\mathbf{y},$$

where $\mathbf{X} = [\boldsymbol{\Phi}_p(t_1), \cdots, \boldsymbol{\Phi}_p(t_n)]^T$ and $\mathbf{y} = [y_1, \cdots, y_n]^T$. It follows that the P-spline smoother $\hat{f}_\lambda(t)$ can be expressed as

$$\hat{f}_\lambda(t) = \boldsymbol{\Phi}_p(t)^T(\mathbf{X}^T\mathbf{X} + \lambda\mathbf{G})^{-1}\mathbf{X}^T\mathbf{y} = \mathbf{a}(t)^T\mathbf{y}, \quad (2.34)$$

where $\mathbf{a}(t) = \mathbf{X}(\mathbf{X}^T\mathbf{X} + \lambda\mathbf{G})^{-1}\boldsymbol{\Phi}_p(t)$.

Based on (2.34), the fitted response vector at the design time points is

$$\hat{\mathbf{y}}_\lambda = \mathbf{A}_\lambda\mathbf{y}, \quad (2.35)$$

where the smoother matrix is

$$\mathbf{A}_\lambda = \mathbf{X}(\mathbf{X}^T\mathbf{X} + \lambda\mathbf{G})^{-1}\mathbf{X}^T. \quad (2.36)$$

Therefore, a P-spline is also a linear smoother as mentioned in Remark 2.2.

2.5.2 Smoothing Parameter Selection

Choice of the knots for P-splines is not so crucial as long as the number of knots K is large enough so that when $\lambda = 0$, the P-spline smoother (2.34) is undersmoothing. One can use the uniform knot placing method or the quantiles as knots method to locate the knots as described in Section 2.3. The number of knots K may be prespecified subjectively, for example, taken as a third or a quarter of the total number of distinct design time points. Like the smoothing spline smoother, however, it is crucial to select the smoothing parameter λ that plays a similar role in the P-spline smoother (2.34) as the one played by the smoothing parameter λ in the cubic smoothing spline smoother (2.27). The smoothing parameter λ trades off between the goodness of fit (the first term in (2.33)) and the roughness of the P-spline smoother $\hat{\mathbf{f}}_\lambda$ (2.34). Fortunately, we can again choose λ using the GCV rule (2.31) with $\hat{\mathbf{y}}_\lambda$ and $\mathbf{A}_\lambda$ given in (2.35) and (2.36), respectively. The optimal λ is chosen to minimize the resulting GCV rule with respect to a grid of possible values of λ.

Example 2.9 *We now apply the P-spline smoother to fit the fifth left-cingulum curve. The measurement errors of the left-cingulum data are very small so that an individual left-cingulum curve can be smoothly fitted with small bias. The individual measurements are presented as circles in Figure 2.13. Figure 2.14 shows the GCV curve for such a P-spline fit. We take* $K = 59$ *and use the quadratic truncated power basis. It is seen that the GCV rule favors the smallest smoothing parameter due to small measurement errors and a large number of observations. Figure 2.13 displays the P-spline fit (solid curve) that obviously fits the data well.*

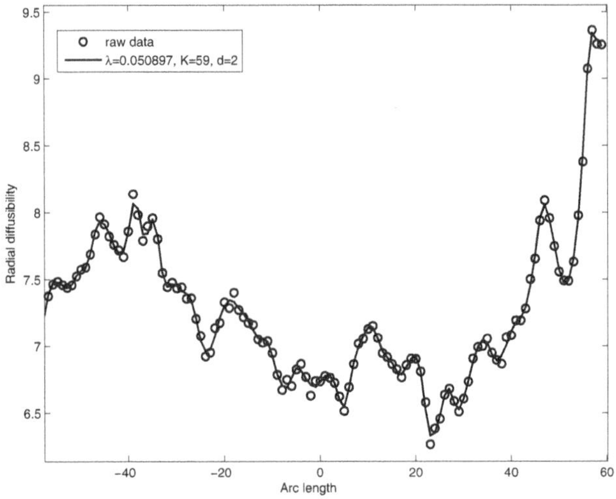

Figure 2.13 *Fitting the fifth left-cingulum curve by the P-spline method. The* $K = 59$ *interior knots and the truncated power basis (2.14) with* $k = 2$ *are used. The smoothing parameter* $\lambda = 0.050897$ *selected by the GCV rule is used. The resulting P-spline fit (solid curve) is adequate for the data (circles).*

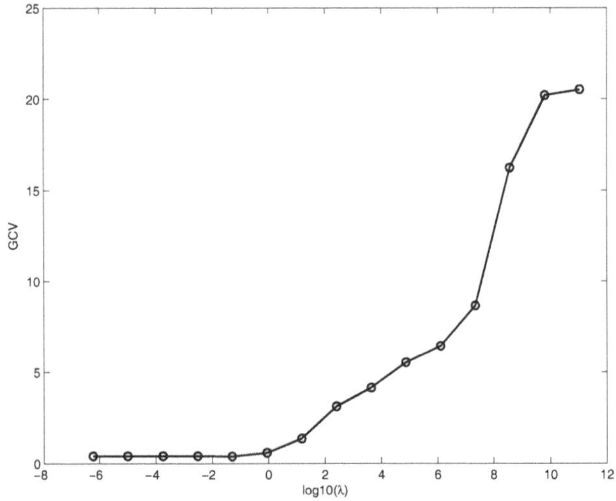

Figure 2.14 *Plot of GCV score against smoothing parameter* λ *(in* $\log_{10}$*-scale) for fitting the fifth left-cingulum curve by the P-spline method. The optimal smoothing parameter that minimizes the associated GCV rule is obtained as* $\lambda = 0.050897$*. The GCV rule favors the smallest smoothing parameter in this P-spline fit due to the fact that the left-cingulum curves were recorded with very small measurement errors.*

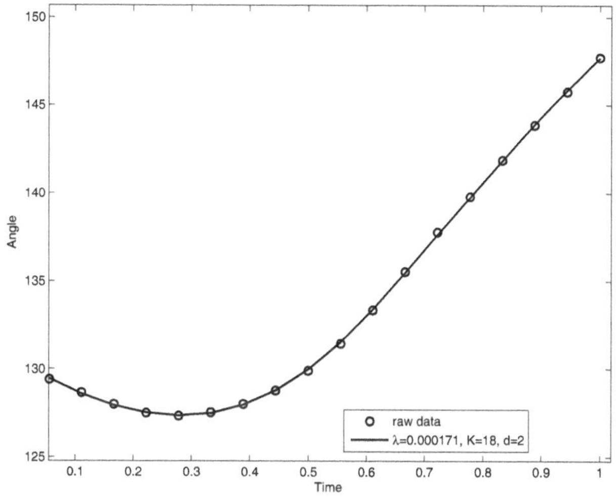

Figure 2.15 *Fitting the fifth ergonomics curve by the P-spline method. The* $K = 18$ *interior knots and the truncated power basis (2.14) with* $k = 2$ *are used. The smoothing parameter* $\lambda = 0.000171$ *selected by the GCV rule is used. The resulting P-spline fit (solid curve) actually interpolates the data (circles).*

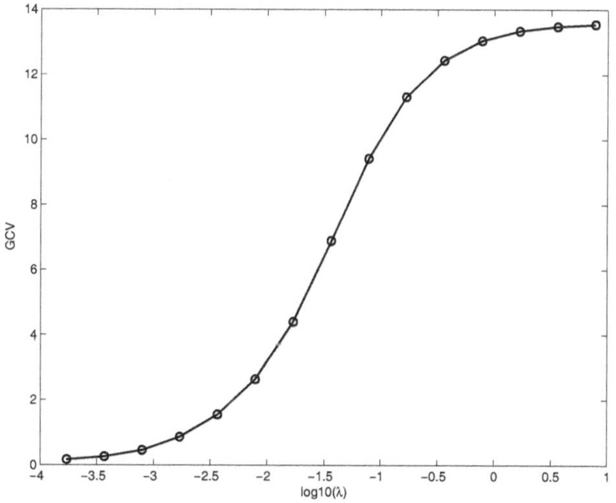

Figure 2.16 *Plot of GCV score against smoothing parameter* λ *(in* $\log_{10}$*-scale) for fitting the fifth ergonomics curve by the P-spline method. The optimal smoothing parameter that minimizes the associated GCV rule is obtained as* $\lambda = 0.000171$*. The GCV rule favors the smallest smoothing parameter in this P-spline fit due to the fact that the ergonomics curves are already very smooth.*

Example 2.10 *The ergonomics data presented in Figure 1.13 are already very smooth. No further smoothing is actually needed, but, as mentioned in Remark 2.21, we can use the P-spline smoother to interpolate the ergonomics curves if desired. Figure 2.16 displays the GCV curve for fitting the fifth ergonomics curve by the P-spline method with* $K = 18$ *interior knots and the quadratic truncated power basis, showing that the GCV rule favors the smallest value of the smoothing parameter. Figure 2.15 displays the associated P-spline fit which actually interpolates the data.*

Remark 2.24 *Like the LPK, regression spline, and smoothing spline smoothers, the P-spline smoother is also not robust. It can be made robust by modifying the robust LPK smoothing algorithm presented in Section 2.2.4 by replacing the LPK smoother with the P-spline smoother given in (2.34).*

2.6 Concluding Remarks and Bibliographical Notes

In this chapter, we briefly reviewed the four most popular linear smoothers: the local polynomial, regression spline, smoothing spline, and P-spline smoothers. We mainly focused on how they are constructed, and how the associated smoothing parameters are selected. These four linear smoothers are known to perform well for uncorrelated data, and we shall use them as tools to reconstruct the individual functions of a functional data set in the next chapter.

Good surveys on local polynomial smoothing include Eubank (1999), Härdle (1990), Wand and Jones (1995), Fan and Gijbels (1996), and Simonoff (1996), among others. Good surveys for smoothing splines include Eubank (1999), Wahba (1990), Green and Silverman (1994), and Gu (2002), among others. Eubank (1999) also provided a nice account of regression splines. A survey on P-splines was given by Ruppert, Wand, and Carroll (2003).

There are some other linear and nonlinear smoothers well developed in the statistical literature, and also widely used. For example, the Gasser-Müller smoother (Gasser and Müller 1979) was popular in the early statistical literature of nonparametric smoothing. For comparison of this smoother to the local constant and linear smoothers, see Fan (1992, 1993), and Fan and Gijbels (1996), among others. The B-spline bases (de Boor 1978, Eilers and Marx 1996) are very popular in spline smoothing as they have many good statistical properties that the truncated power bases discussed in this chapter do not admit.

Linear smoothers are simple to construct and easy to implement. They also have some disadvantages. For example, linear smoothers are usually not robust against unusual observations/outliers. This is a result of being weighted least-squares (WLS) based smoothers. Nonlinear smoothers were then introduced to overcome this drawback, paying a price for more computational efforts when unusual observations are present in the data set. An example of nonlinear smoothers is provided by Cleveland (1979). He used a locally WLS estimate to fit a nonparametric regression function in a re-weighting manner. The robust LPK smoothing, regression spline, smoothing spline, and P-spline algorithms provided in this chapter are simple to implement and generally work reasonably well. Wavelet-based thresholding smoothers (Donoho and Johnstone 1994, Fan 1996) are also examples of nonlinear smoothers.

Chapter 3

Reconstruction of Functional Data

3.1 Introduction

In the previous chapter, we reviewed four well-known smoothing techniques that can be used to smooth or interpolate the individual functions of a functional data set. In this chapter, we aim to reconstruct the whole functional data set using one of the four techniques.

Following Zhang and Chen (2007), we model a functional data set as independent realizations of an underlying stochastic process:

$$y_i(t) = \eta(t) + v_i(t) + \epsilon_i(t), \quad i = 1, 2, \cdots, n, \tag{3.1}$$

where $\eta(t)$ models the population mean function of the stochastic process; $v_i(t)$ the ith individual variation (subject-effect) from $\eta(t)$; $\epsilon_i(t)$ the ith measurement error process; and $y_i(t)$ the ith response process. Without loss of generality, throughout this chapter we assume that the stochastic process has a finite support $\mathcal{T} = [a, b], \quad -\infty < a < b < \infty$. Moreover, we assume that $v_i(t)$ and $\epsilon_i(t)$ are independent, and are independent copies of $v(t) \sim \text{SP}(0, \gamma)$ and $\epsilon(t) \sim \text{SP}(0, \gamma_\epsilon), \gamma_\epsilon(s,t) = \sigma^2(t)1_{\{s=t\}}$, respectively, where $\text{SP}(\eta, \gamma)$ denotes a stochastic process with mean function $\eta(t)$ and covariance function $\gamma(s,t)$. It follows that the underlying individual functions (trajectories),

$$f_i(t) = \text{E}\{y_i(t)|v_i(t)\} = \eta(t) + v_i(t), \ i = 1, 2, \cdots, n, \tag{3.2}$$

are i.i.d. copies of the underlying stochastic process $f(t) = \eta(t) + v(t) \sim \text{SP}(\eta, \gamma)$.

In practice, functional data are observed discretely. Let $t_{ij}, j = 1, 2, \cdots, n_i$ be the design time points of the ith subject. Then by (3.1), we have

$$y_i(t_{ij}) = \eta(t_{ij}) + v_i(t_{ij}) + \epsilon_i(t_{ij}), \ j = 1, 2, \cdots, n_i; i = 1, 2, \cdots, n. \tag{3.3}$$

Set $y_{ij} = y_i(t_{ij}), \epsilon_{ij} = \epsilon_i(t_{ij})$, and $f_i(t) = \eta(t) + v_i(t)$; then the above model can be written as

$$y_{ij} = f_i(t_{ij}) + \epsilon_{ij}, j = 1, 2, \cdots, n_i; i = 1, 2, \cdots, n. \tag{3.4}$$

For any fixed i, the model (3.4) reduces to the standard regression model defined in (2.2), which can be smoothed by one of the four smoothing techniques described in Chapter 2.

The reconstruction of a functional data set aims to recover the individual functions $f_i(t), i = 1, 2, \cdots, n$ with the following main goals in mind:

- A reconstructed individual function can be evaluated at any resolution.
- The reconstructed individual functions of a functional data set are nearly i.i.d. (identically and independently distributed).
- The reconstructed individual functions are asymptotically equivalent to the underlying individual functions of a functional data set so that the latter can be replaced with the former with very little loss.

The key is how to apply the smoothers presented in Chapter 2 and how to properly choose one or two smoothing parameters for reconstructing or smoothing all the individual functions in a functional data set simultaneously so that the above requirements can be approximately satisfied.

Reconstructing functional data properly is a first step for the method of "smoothing first, then estimation or inference" for functional data, adopted in this book. The key steps of this method include

- Reconstruct the individual functions in a functional data set by smoothing or interpolation using one of the four smoothing techniques discussed in Chapter 2. Denote the reconstructed individual functions as $\hat{f}_i(t), i = 1, 2, \cdots, n$.
- Construct the estimators $\hat{\eta}(t), \hat{\gamma}(s,t)$ and $\hat{\sigma}^2(t)$ of the functional parameters $\eta(t), \gamma(s,t)$ and $\sigma^2(t)$ in the model (3.1) based on $\hat{f}_i(t), i = 1, 2, \cdots, n$.
- Conduct further statistical inferences, principal component analysis or hypothesis testing, among others, based on the estimators $\hat{\eta}(t), \hat{\gamma}(s,t)$, and $\hat{\sigma}^2(t)$ or based on $\hat{f}_i(t), i = 1, 2, \cdots, n$ directly.

The "smoothing first, then estimation or inference" method is generally used in Ramsay and Silverman (2002, 2005) and the references therein. The main theoretical results of this chapter come from Zhang and Chen (2007), where the effect of the substitution of the underlying individual functions $f_i(t), i = 1, 2, \cdots, n$ by their local polynomial kernel (LPK) reconstructions for functional data analysis (FDA) is investigated. They show that under some mild conditions, the effect of the substitution is asymptotically ignorable. In particular, they show that under some mild conditions, $\hat{\eta}(t)$ and $\hat{\gamma}(s,t)$ are $\sqrt{n}$-consistent and asymptotically Gaussian; the $\sqrt{n}$-consistence of $\hat{\eta}(t)$ will not be affected by better choice of the bandwidth than the bandwidth selected by a GCV rule; the convergence rate of $\hat{\sigma}^2(t)$ is affected by the convergence rate of the LPK reconstructions; and the LPK reconstructions-based estimators are usually more efficient than those estimators involving no smoothing.

The chapter is organized as follows. Section 3.2 presents the nonparametric reconstruction methods based on the four well-known smoothers described in the previous chapter. In Section 3.3, the accuracy of the LPK reconstructions is investigated theoretically and by a simulation study. The accuracy of the LPK reconstructions in functional linear models is discussed in Section 3.4. Technical proofs of some asymptotical results are outlined in Section 3.5. The chapter concludes with some remarks and bibliographical notes in Section 3.6 and some exercises in Section 3.7.

3.2 Reconstruction Methods

3.2.1 Individual Function Estimators

As mentioned earlier, based on the model (3.4), for any given i, one of the four smoothing techniques described in Chapter 2 can be used. The associated estimators of the individual functions in a functional data set can be expressed as

$$\hat{f}_i(t) = \mathbf{a}_i(t)^T \mathbf{y}_i, i = 1, 2, \cdots, n, \tag{3.5}$$

where $\mathbf{y}_i = [y_{i1}, y_{i2}, \cdots, y_{in_i}]^T$ denotes the response vector for the ith individual function observed at the design time points

$$t_{i1}, t_{i2}, \cdots, t_{in_i}, \tag{3.6}$$

the vectors $\mathbf{a}_i(t)$ depends on the design time points (3.6) and the smoothing technique used. For example, for the LPK smoother, $\mathbf{a}_i(t)$ defined in (2.7) depends on the kernel function $K(t)$, the bandwidth h, the degree of the local polynomial used in the LPK smoother, and the design time points (3.6) for the ith individual function. For the regression spline, cubic smoothing spline, and P-spline smoothers, $\mathbf{a}_i(t)$ is defined in (2.19), (2.28), and (2.34), respectively, after replacing $t_1, t_2, \cdots, t_n$ there with the design time points (3.6) here for the ith individual function. More details about $\mathbf{a}_i(t)$, $i = 1, 2, \cdots, n$ will be given in Sections 3.2.3 through 3.2.6 of this chapter. Notice that (3.5) allows us to evaluate or interpolate $\hat{f}_i(t)$ at any desired resolution. For example, one can use (3.5) to evaluate or interpolate all the individual functions at a prespecified grid of locations in $\mathcal{T}$.

For estimating the ith individual function $f_i(t)$, the smoother matrix can be expressed as

$$\mathbf{A}_i = [\mathbf{a}_i(t_{i1}), \mathbf{a}_i(t_{i2}), \cdots, \mathbf{a}_i(t_{in_i})]^T, i = 1, 2, \cdots, n, \tag{3.7}$$

so that

$$\hat{\mathbf{y}}_i = \mathbf{A}_i \mathbf{y}_i, i = 1, 2, \cdots, n, \tag{3.8}$$

where $\hat{\mathbf{y}}_i$ is the fitted response vector of the ith individual function at the design time points (3.6). The respective expressions for $\mathbf{A}_i$ for the LPK, regression spline, cubic smoothing spline, and P-spline smoothers are given in (2.9), (2.20), (2.30), and (2.36) after replacing $t_1, t_2, \cdots, t_n$ there with the design time points (3.6) here for the ith individual function.

3.2.2 Smoothing Parameter Selection

Let δ denote the smoothing parameter used in the individual function reconstructions $\hat{f}_i(t), i = 1, 2, \cdots, n$. It is understood that for different smoothing techniques, δ has different meaning. For the LPK, regression spline, cubic smoothing spline, and P-spline smoothers, δ denotes h, K, λ, and λ, respectively. For a single individual function reconstruction, the smoothing parameter δ can be selected by the GCV rule defined in (2.12), (2.21), and (2.31) for the LPK, regression spline, cubic smoothing spline, and P-spline smoothers. Let $\text{GCV}_i(\delta)$ denote the associate GCV rule for the ith individual function estimation. Using (3.7) and (3.8), the $\text{GCV}_i(\delta)$ can be expressed in a unified manner as

$$\text{GCV}_i(\delta) = \frac{\|\mathbf{y}_i - \hat{\mathbf{y}}_i\|^2}{(1 - \text{tr}(\mathbf{A}_i)/n_i)^2}, \tag{3.9}$$

where again $\|\cdot\|$ denotes the usual L^2-norm and $\text{tr}(\mathbf{A})$ denotes the trace of the matrix $\mathbf{A}$. It is understood that for different i, the smoothing parameter δ selected by using $\text{GCV}_i(\delta)$ can be different. It is often desired that a common smoothing parameter δ should be used for all the individual functions generated by a common underlying stochastic process so that the reconstructed individual functions are nearly independently and identically distributed. For this purpose, Zhang and Chen (2007) proposed the use of the following GCV rule for fitting a functional data set:

$$\text{GCV}(\delta) = n^{-1} \sum_{i=1}^{n} \text{GCV}_i(\delta). \tag{3.10}$$

The above GCV rule allows us to select an appropriate smoothing parameter for all the individual functions in a functional data set. In practice, the optimal smoothing parameter δ^* can be obtained by minimizing $\text{GCV}(\delta)$ over a number of prespecified smoothing parameter candidates.

3.2.3 LPK Reconstruction

In the LPK reconstruction of the individual functions, the same kernel K and the same bandwidth h should be used for all the individual curves. This is because the individual functions in a functional data set often admit similar smoothness property and sometimes similar shape, it is then rather reasonable to use a common kernel and a common bandwidth for all of them. The advantages for using a common kernel and a common bandwidth at least include: it reduces the computational effort for bandwidth selection; the resulting LPK reconstructions are nearly i.i.d.; and it simplifies the asymptotic results of the LPK estimators. A brief description of the LPK reconstruction of a functional data set is as follows.

Based on the model (3.4), for any fixed time point t, assume $f_i(t)$ has a $(p+1)$st continuous derivative in a neighborhood of t for some positive integer

p. Then by Taylor expansion, $f_i(t_{ij})$ can be locally approximated by a p-order polynomial

$$f_i(t_{ij}) \approx f_i(t) + (t_{ij} - t)f_i^{(1)}(t) + \cdots + (t_{ij} - t)^p f_i^{(p)}(t)/p! = \mathbf{z}_{ij}^T \boldsymbol{\beta}_i,$$

in the neighborhood of t, where $\boldsymbol{\beta}_i = [\beta_{i0}, \beta_{i1}, \cdots, \beta_{ip}]^T$ with $\beta_{ir} = f_i^{(r)}(t)/r!$, and $\mathbf{z}_{ij} = [1, t_{ij} - t, \cdots, (t_{ij} - t)^p]^T$. Then the p-order LPK reconstructions of $f_i(t)$ are defined as $\hat{f}_i(t) = \hat{\beta}_{i0} = \mathbf{e}_{1,p+1}^T \hat{\boldsymbol{\beta}}_i$, where and throughout $\mathbf{e}_{r,s}$ denotes the s-dimensional unit vector whose rth component is 1 and others are 0, and $\hat{\boldsymbol{\beta}}_i$ are the minimizers of the following weighted least squares criterion:

$$\sum_{i=1}^{n} \sum_{j=1}^{n_i} \left[y_{ij} - \mathbf{z}_{ij}^T \boldsymbol{\beta}_i\right]^2 K_h(t_{ij} - t) = \sum_{i=1}^{n} (\mathbf{y}_i - \mathbf{Z}_i \boldsymbol{\beta}_i)^T \mathbf{K}_{ih} (\mathbf{y}_i - \mathbf{Z}_i \boldsymbol{\beta}_i), \quad (3.11)$$

where $\mathbf{y}_i = [\mathbf{y}_{i1}, \cdots, \mathbf{y}_{in_i}]^T, \mathbf{Z}_i = [\mathbf{z}_{i1}, \cdots, \mathbf{z}_{in_i}]^T$ and $\mathbf{K}_{ih} = \text{diag}(K_h(t_{i1} - t), \cdots, K_h(t_{in_i} - t))$, with $K_h(\cdot) = K(\cdot/h)/h$, obtained by rescaling a kernel function $K(\cdot)$ with bandwidth $h > 0$ that controls the size of the associated neighborhood. Minimizing (3.11) with respect to $\boldsymbol{\beta}_i, i = 1, 2, \cdots, n$ is equivalent to respectively minimizing the ith term in the summation of the right-hand side of (3.11) with respect to $\boldsymbol{\beta}_i$ for each $i = 1, 2, \cdots, n$. It follows that the p-order LPK reconstructions can be expressed as

$$\begin{aligned} \hat{f}_i(t) &= \mathbf{e}_{1,p+1}^T (\mathbf{Z}_i^T \mathbf{K}_{ih} \mathbf{Z}_i)^{-1} \mathbf{Z}_i^T \mathbf{K}_{ih} \mathbf{y}_i \\ &= \textstyle\sum_{j=1}^{n_i} K_h^{n_i}(t_{ij} - t) y_{ij}, \; i = 1, 2, \cdots, n, \end{aligned} \quad (3.12)$$

where $K^{n_i}(t)$ are known as the empirical equivalent kernels for the p-order LPK, as defined in (2.7). Thus, for the expression (3.5), we have

$$\begin{aligned} \mathbf{a}_i(t) &= [K_h^{n_i}(t_{i1} - t), K_h^{n_i}(t_{i2} - t), \cdots, K_h^{n_i}(t_{in_i} - t)]^T, \\ & \quad i = 1, 2, \cdots, n. \end{aligned} \quad (3.13)$$

Based on $\mathbf{a}_i(t)$, $i = 1, 2, \cdots, n$, we can define the LPK smoother matrix (3.7) and use the GCV rule (3.10) to select the optimal bandwidth h for the LPK reconstructions of a functional data set. Therefore, we have the following remark.

Remark 3.1 *For local polynomial smoothing, reconstructing a whole functional data set using the same kernel and the same bandwidth is equivalent to reconstructing each of the individual functions using the same kernel and the same bandwidth separately.*

Example 3.1 *We now apply the LPK reconstruction method to reconstruct the progesterone data introduced in Section 1.2.1 of Chapter 1. In this data set, there are about 8.3% missing data. In this example, for simplicity, we impute the missing values by the pointwise means at the distinct design time*

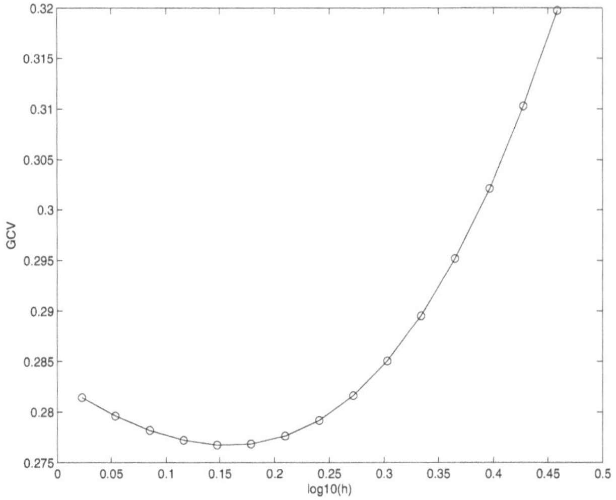

Figure 3.1 *Plot of GCV score against bandwidth h (in* $\log_{10}$*-scale) for reconstructing the progesterone data by the LPK reconstruction method with* $p = 2$. *The optimal bandwidth that minimizes the GCV rule (3.10) is obtained as* $h^* = 1.40$.

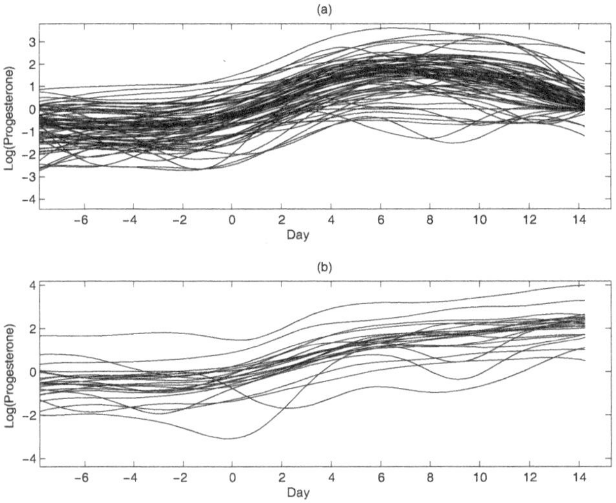

Figure 3.2 *Reconstructed progesterone curves with (a) nonconceptive and (b) conceptive by the LPK reconstruction method with* $p = 2$. *The optimal bandwidth* $h^* = 1.40$ *selected by the GCV rule (3.10) is used for reconstructing the whole functional data set.*

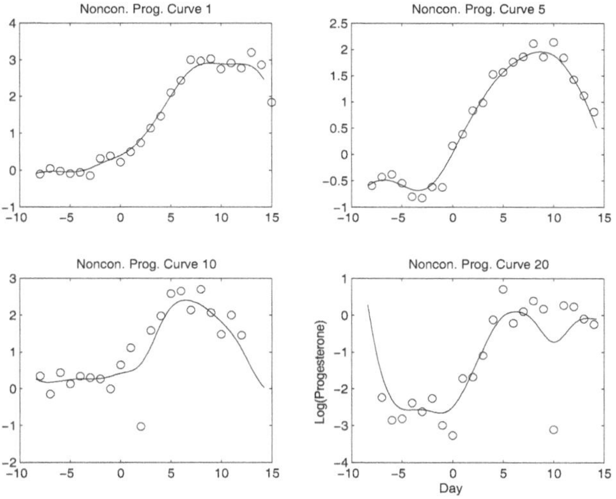

Figure 3.3 *Four reconstructed nonconceptive progesterone curves (solid) with the associated data (circles). The effects of the outliers in the two lower panels are spotted.*

points and hope that the effect of the imputation is ignorable. Figure 3.1 displays the associated GCV curve computed using (3.10). It is seen that when the bandwidth is too small or too large, the associated GCV scores are large. The optimal bandwidth is found to be $h^* = 1.40$ $[\log_{10}(1.40) = 0.147]$ *and the associated GCV score is* 0.277. *The reconstructed nonconceptive and conceptive progesterone curves with this optimal bandwidth are displayed in the upper and lower panels of Figure 3.2, respectively. Notice that one can also fit the nonconceptive progesterone curves and the conceptive progesterone curves separately using different optimal bandwidths, especially when one believes that the nonconceptive progesterone curves and the conceptive progesterone curves have very different covariance structures. To see how well the individual progesterone curves were fitted, Figure 3.3 displays the four reconstructed individual nonconceptive progesterone curves with the associated raw data presented as circles. These raw data were also displayed in Figure 1.3 in Chapter 1. It is seen that the reconstructed individual curves fit the associated data reasonably well although the same optimal bandwidth is used to fit all the nonconceptive and conceptive progesterone curves. However, the effects of the outliers in the two lower panels are also spotted.*

To reduce the effects of the outliers, we apply the robust LPK smoother proposed in Section 2.2.4 of Chapter 2 to fit the progesterone data again. The optimal bandwidth is now found to be $h^* = 1.87$, *which is larger than the one for usual LPK reconstructions. The associated GCV score is* 0.119 *which is smaller than the one for usual LPK reconstructions. The smaller GCV score is due to the fact that the effects of the outliers are excluded in*

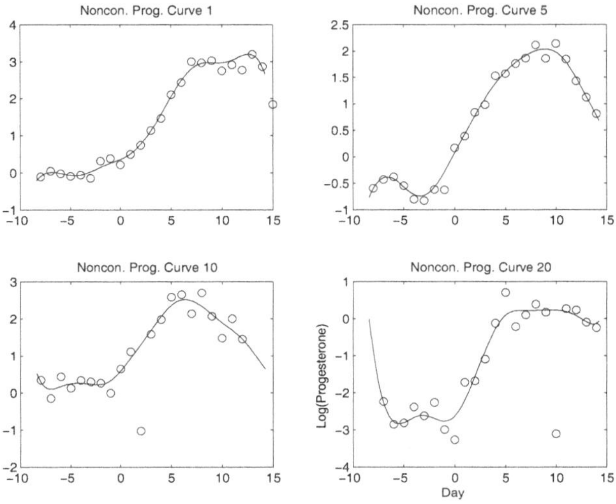

Figure 3.4 *Four robustly reconstructed nonconceptive progesterone curves (solid) with the associated data (circles). The effects of the outliers are largely reduced.*

the computation of the GCV scores. The four robustly reconstructed individual curves are presented in Figure 3.4. It is seen that the robustly reconstructed individual curves indeed fit the associated data much better than those usual LPK reconstructed individual curves presented in Figure 3.3.

3.2.4 Regression Spline Reconstruction

We here give some details for the regression spline reconstruction of a functional data set. The two knot locating methods presented in Section 2.3.3 of Chapter 2 can still be used to select a common sequence of knots

$$\tau_1, \tau_2, \cdots, \tau_K, \tag{3.14}$$

for the regression spline reconstructions of a functional data set, but we would like to emphasize that this common sequence of knots (3.14) should be based on the pooled design time points for all the individual functions:

$$t_{ij}, \; j = 1, 2, \cdots, n_i; i = 1, 2, \cdots, n. \tag{3.15}$$

Using the knots (3.14), we can construct the kth-order truncated power basis vector $\mathbf{\Phi}_p(t)$ using (2.14) where $p = K + k + 1$. Set $\mathbf{X}_i = [\mathbf{\Phi}_p(t_{i1}), \mathbf{\Phi}_p(t_{i2}), \cdots, \mathbf{\Phi}_p(t_{in_i})]^T$, $i = 1, 2, \cdots, n$. Then we have

$$\hat{f}_i(t) = \mathbf{\Phi}_p(t)^T \hat{\boldsymbol{\beta}}_i = \mathbf{\Phi}_p(t)^T (\mathbf{X}_i^T \mathbf{X}_i)^{-1} \mathbf{X}_i^T \mathbf{y}_i, \; i = 1, 2, \cdots, n, \tag{3.16}$$

where $\hat{\boldsymbol{\beta}}_i, i = 1, 2, \cdots, n$ are the fitted coefficient vectors of the individual functions. We then have the following remark.

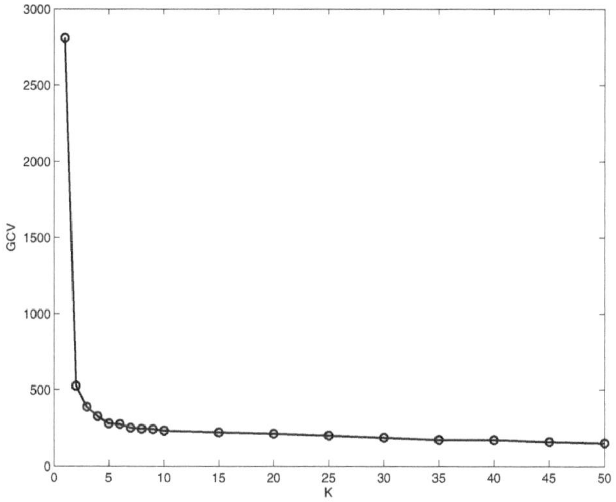

Figure 3.5 *Plot of GCV score against number of knots K for reconstructing the Canadian temperature data by the regression spline method. It seems that in this example the GCV rule (3.10) favors the largest number of knots.*

Remark 3.2 *For regression spline smoothing, reconstructing a whole functional data set using a common sequence of knots is equivalent to reconstructing each of the individual functions using the common sequence of knots separately.*

It is seen that in the expression (3.5) for the regression spline constructions of a functional data set, we have

$$\mathbf{a}_i(t) = \mathbf{X}_i(\mathbf{X}_i^T\mathbf{X}_i)^{-1}\mathbf{\Phi}_p(t),\ i = 1, 2, \cdots, n. \tag{3.17}$$

We generally choose the order of the truncated power basis, k, to be a small integer such as $1, 2, 3$, or 4 and we use one of the two knot locating methods described in Section 2.3.3 of Chapter 2 based on the pooled design time points (3.15). The number of knots, K, can then be chosen by minimizing the GCV rule (3.10).

Remark 3.3 *When p, the number of basis function in $\mathbf{\Phi}_p(t)$, is small, one can conduct some classical statistical inferences using some multivariate test statistics based on the estimated coefficient vectors, $\hat{\boldsymbol{\beta}}_i$, $i = 1, 2, \cdots, n$. However, when the number of knots, K, is large, the dimension p of the estimated coefficient vectors can also be large, and even much larger than the number of the individual functions in a functional data set so that the classical multivariate statistical inferences which use the inverse of the sample covariance matrix, for example, the well-known Hotelling T^2-test (Anderson 2003), will*

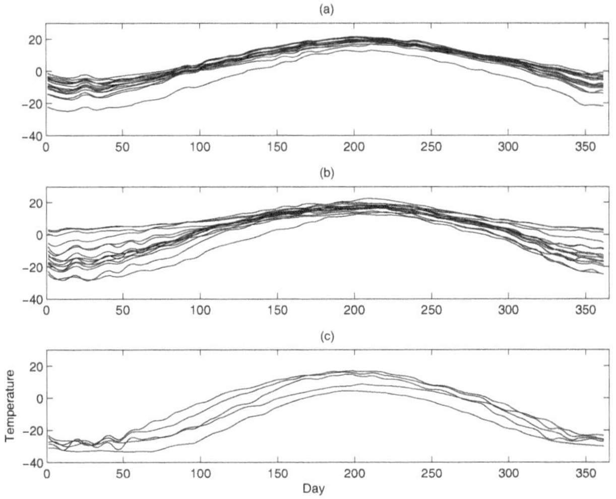

Figure 3.6 *Reconstructed Canadian temperature curves for (a) fifteen Eastern weather stations, (b) fifteen Western weather stations, and (c) five Northern weather stations by the regression spline method. The regression spline basis (2.14) with* $k = 2$ *is used. The knot sequence (3.14) is located by the "quantiles as knots" method based on all the design time points for the Canadian temperature data. The number of knots,* $K^* = 50$*, selected by the GCV rule (3.10), is used for all thirty-five temperature curves.*

no longer work. We now use the Canadian temperature data to illustrate such a situation.

Example 3.2 *Figure 1.6 displays the raw Canadian temperature curves for fifteen Eastern weather stations, fifteen Western weather stations, and five Northern weather stations. The temperature records involve large measurement errors. In order to remove some of these measurement errors, we now apply the regression spline method to fit these thirty-five temperature curves. We use the quadratic regression spline basis (2.14) with the knots located by the "quantiles as knots" method. For each Canadian temperature curve, there are 365 measurements. This implies that a lot of knots can be located to trade off between the goodness of fit and the model complexity of the regression spline reconstructions. Figure 3.5 displays the associated GCV curve. It is seen that in this example, the GCV rule (3.10) favors the largest number of knots,* $K^* = 50$*. Thus,* $p^* = K^* + k + 1 = 50 + 2 + 1 = 53$*, which is much larger than, 35, the number of the Canadian temperature curves. In this case, some classical multivariate statistical inferences such as the well-known Hotelling* T^2*-test cannot be conducted.*

The reconstructed Canadian temperature curves with $K^* = 50$ *are displayed in Figure 3.6. These thirty-five temperature curves are reconstructed*

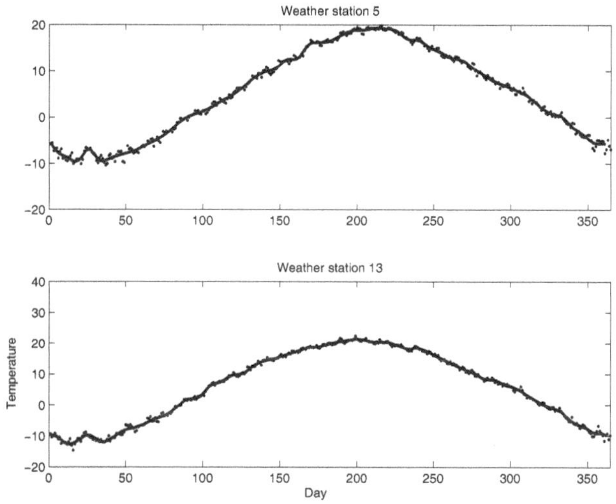

Figure 3.7 *Two reconstructed Canadian temperature curves by the regression spline method described in Figure 3.6. It appears that the reconstructed temperature curves (solid) fit the data (dots) reasonably well although a common sequence of knots is used for all thirty-five temperature curves.*

using a common sequence of knots (3.14) based on the design time points (3.15) for all thirty-five temperature curves. Figure 3.7 displays two reconstructed temperature curves. The raw data are presented as dots. It is seen that the reconstructed temperature curves fit the data reasonably well although a common sequence of knots is used for all thirty-five temperature curves.

If desired, we can also reconstruct the Canadian temperature curves for the Eastern, Western, and Northern weather stations separately using possibly different numbers of knots selected by the GCV rule (3.10).

3.2.5 Smoothing Spline Reconstruction

Assume that the support of all the design time points is $\mathcal{T} = [a, b]$ with $-\infty < a < b < \infty$. The smoothing spline reconstructions of the individual functions $f_i(t), i = 1, 2, \cdots, n$ are defined as the minimizers $\hat{f}_i(t), i = 1, 2, \cdots, n$ of the following penalized least squares criterion:

$$\sum_{i=1}^{n}\sum_{j=1}^{n_i}[y_{ij} - f_i(t_{ij})]^2 + \lambda\sum_{i=1}^{n}\int_a^b \left[f_i^{(k)}(t)\right]^2 dt, \tag{3.18}$$

over the following Sobolev space $\mathcal{W}_2^k[a, b]$:

$$\left\{f : f(t), \cdots, f^{(k)}(t) \text{ abs. continu.}, \int_a^b \{f^{(k)}(t)\}^2\, dt < \infty\right\}, \tag{3.19}$$

where $\lambda > 0$ is a common smoothing parameter for all the individual functions. The smoothing parameter λ controls the size of the roughness penalty and is usually used to trade off the goodness of fit, represented by the first term in (3.18), and the roughness of the resulting curves. The $\hat{f}_i(t), i = 1, 2, \cdots, n$ are known as natural smoothing splines of degree $(2k - 1)$ whose knots are all the distinct design times of the pooled design time points (3.15). Denote the knots of these natural smoothing splines as

$$\tau_1, \tau_2, \cdots, \tau_K. \tag{3.20}$$

Let $\mathbf{f}_i = [f_i(\tau_1), f_i(\tau_2), \cdots, f_i(\tau_K)]^T$ denote the vector containing the values of $f_i(t)$ evaluated at all the knots. For $i = 1, 2, \cdots, n$, set $\mathbf{X}_i = (x_{ijr}) : n_i \times K$ to be the incidence matrix for the ith individual function where

$$x_{ijr} = \begin{cases} 1 & \text{if } t_{ij} = \tau_r, r = 1, 2, \cdots, K; j = 1, 2, \cdots, n_i, \\ 0 & \text{otherwise.} \end{cases}$$

Then we can re-express (3.18) as

$$\sum_{i=1}^{n} \Big\{ \sum_{j=1}^{n_i} \|\mathbf{y}_i - \mathbf{X}_i \mathbf{f}_i\|^2 + \lambda \mathbf{f}_i^T \mathbf{G} \mathbf{f}_i \Big\}, \tag{3.21}$$

where $\mathbf{G}$ is the common roughness matrix for all the individual functions depending on the knots (3.20). We then have the following remark.

Remark 3.4 *For smoothing spline smoothing, reconstructing a whole functional data set using a common sequence of knots and a common smoothing parameter is equivalent to reconstructing each of the individual functions using the common sequence of knots and the common smoothing parameter separately.*

For the given K knots (3.20), when $k = 2$, the roughness matrix $\mathbf{G}$ can be easily calculated using the method given in Green and Silverman (1994, p. 12–13). It is easily seen from the criterion (3.21) that the natural smoothing spline reconstructions of the individual functions can be expressed as

$$\hat{\mathbf{f}}_i = (\mathbf{X}_i^T \mathbf{X}_i + \lambda \mathbf{G})^{-1} \mathbf{X}_i^T \mathbf{y}_i, \ i = 1, 2, \cdots, n. \tag{3.22}$$

Therefore, the fitted response curves can be expressed as

$$\hat{\mathbf{y}}_i = \mathbf{X}_i (\mathbf{X}_i^T \mathbf{X}_i + \lambda \mathbf{G})^{-1} \mathbf{X}_i^T \mathbf{y}_i, \ i = 1, 2, \cdots, n. \tag{3.23}$$

The well-known representer lemma of Wahba (1990) indicates that for any $t \in \mathcal{T}$, for some $\mathbf{a}_i(t), i = 1, 2, \cdots, n$, we have

$$\hat{f}_i(t) = \mathbf{a}_i(t)^T \mathbf{y}_i, \ i = 1, 2, \cdots, n, \tag{3.24}$$

which allows us to evaluate the smoothing spline reconstructions $\hat{f}_i(t), \ i = 1, 2, \cdots, n$ in any desired resolution. The common smoothing parameter λ can

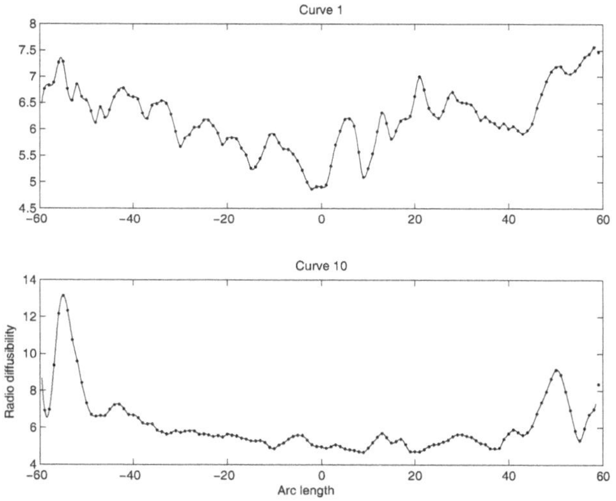

Figure 3.8 *Two reconstructed left-cingulum curves obtained by the cubic smoothing spline method. The optimal smoothing parameter* $\lambda^* = 0.25$, *selected by the GCV rule (3.10), is used. It appears that the reconstructed left-cingulum curves (solid) nearly interpolate the associated data (dots).*

be chosen by minimizing the GCV rule (3.10) with respect to λ. To evaluate the smoothing spline reconstructions $\hat{f}_i(t), i = 1, 2, \cdots, n$ at any desired resolution, alternatively we may use interpolation based on the values of $\hat{f}_i(t)$, $i = 1, 2, \cdots, n$.

Example 3.3 *The left-cingulum data are presented in Section 1.2.6 of Chapter 1. For each left-cingulum curve, there are* 119 *measurements. We now apply the cubic smoothing spline reconstruction method to the left-cingulum data set. The optimal smoothing parameter that minimizes the GCV rule (3.10) is found to be* $\lambda^* = 0.25$. *To show how well the individual left-cingulum curves are fitted, Figure 3.8 displays the two reconstructed left-cingulum curves. The raw data are presented as dots. It is seen that the reconstructed left-cingulum curves (solid) nearly interpolate the associated data (dots).*

3.2.6 P-Spline Reconstruction

For the P-spline reconstruction of a functional data set, we first need to specify a common sequence of knots. To this end, the two knot locating methods presented in Section 2.3.3 of Chapter 2 can be used to select a common sequence of knots $\tau_1, \tau_2, \cdots, \tau_K$ based on the pooled design time points (3.15) for all the individual functions where K should not be too small or too large. When K is too small, the reconstructions will be over-smoothed and when K is too large, the computation is expensive. Using these knots, we can construct the kth-order truncated power basis vector $\mathbf{\Phi}_p(t)$ using (2.14) where $p = K+k+1$. Set $\mathbf{X}_i = [\mathbf{\Phi}_p(t_{i1}), \mathbf{\Phi}_p(t_{i2}), \cdots, \mathbf{\Phi}_p(t_{in_i})]^T$, $i = 1, 2, \cdots, n$. Then the P-spline reconstructions of the individual functions $f_i(t), i = 1, 2, \cdots, n$ are defined as

$$\hat{f}_i(t) = \mathbf{\Phi}_p(t)^T \hat{\boldsymbol{\beta}}_i, \; i = 1, 2, \cdots, n,$$

where $\hat{\boldsymbol{\beta}}_i, i = 1, 2, \cdots, n$ are the minimizers of the following penalized least squares criterion:

$$\sum_{i=1}^{n} \Big\{ \sum_{j=1}^{n_i} \|\mathbf{y}_i - \mathbf{X}_i \boldsymbol{\beta}_i\|^2 + \lambda \boldsymbol{\beta}_i^T \mathbf{G} \boldsymbol{\beta}_i \Big\}, \tag{3.25}$$

where λ is the common smoothing parameter and $\mathbf{G}$ is the common roughness matrix for all the individual functions, specified as

$$\mathbf{G} = \begin{pmatrix} \mathbf{0} & \mathbf{0} \\ \mathbf{0} & \mathbf{I}_K \end{pmatrix},$$

so that the coefficients of the truncated power basis functions $(t - \tau_r)_+^k, r = 1, 2, \cdots, K$ are penalized while the coefficients of the polynomial basis functions $1, t, t^2, \cdots, t^k$ are not penalized. It is easy to find that

$$\hat{\boldsymbol{\beta}}_i = (\mathbf{X}_i^T \mathbf{X}_i + \lambda \mathbf{G})^{-1} \mathbf{X}_i^T \mathbf{y}_i, \; i = 1, 2, \cdots, n.$$

It follows that the P-spline reconstructions of the individual functions can be expressed as

$$\hat{f}_i(t) = \mathbf{\Phi}_p(t)^T (\mathbf{X}_i^T \mathbf{X}_i + \lambda \mathbf{G})^{-1} \mathbf{X}_i^T \mathbf{y}_i = \mathbf{a}_i(t)^T \mathbf{y}_i, \; i = 1, 2, \cdots, n, \tag{3.26}$$

so that in the expression (3.5), we have

$$\mathbf{a}_i(t) = \mathbf{X}_i (\mathbf{X}_i^T \mathbf{X}_i + \lambda \mathbf{G})^{-1} \mathbf{\Phi}_p(t), \; i = 1, 2, \cdots, n. \tag{3.27}$$

Therefore, the fitted response curves can be expressed as

$$\hat{\mathbf{y}}_i = \mathbf{X}_i (\mathbf{X}_i^T \mathbf{X}_i + \lambda \mathbf{G})^{-1} \mathbf{X}_i^T \mathbf{y}_i, \; i = 1, 2, \cdots, n.$$

The common smoothing parameter λ can be chosen by minimizing the GCV rule (3.10) with respect to λ. Therefore, we have the following remark.

Remark 3.5 *For P-spline smoothing, reconstructing a whole functional data set using the same basis vector $\mathbf{\Phi}_p(t)$, the same roughness matrix $\mathbf{G}$, and the same smoothing parameter λ is equivalent to reconstructing each of the individual functions using the same basis vector $\mathbf{\Phi}_p(t)$, the same roughness matrix $\mathbf{G}$, and the same smoothing parameter λ separately.*

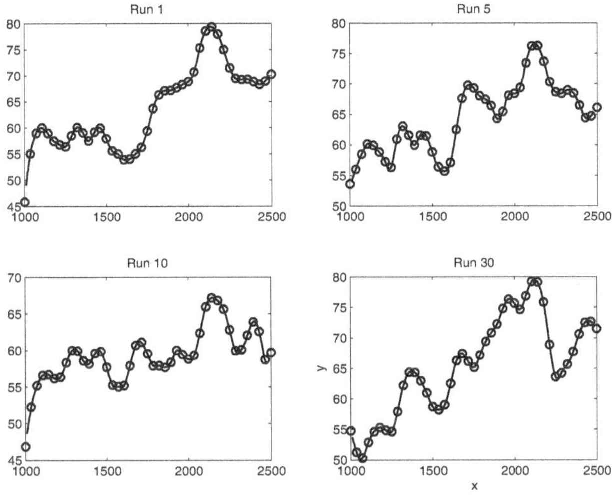

Figure 3.9 *Four reconstructed audible noise curves obtained by the P-spline reconstruction method. It is seen that the reconstructed audible noise curves (solid) nearly interpolate the data (circles). This is possibly due to the fact that the measurement errors of the audible noise data are very small.*

Example 3.4 *We now apply the P-spline reconstruction method to fit the audible noise data set presented in Section 1.2.5 of Chapter 1. We use the quadratic truncated power basis and use all the distinct design time points for all the subjects as knots. That is,* $K = 43$. *The optimal smoothing parameter selected by GCV is* $\lambda^* = 939$. *Figure 3.9 displays four reconstructed audible noise curves obtained by the P-spline reconstruction method. The data are presented as circles. It is seen that the reconstructed audible noise curves nearly interpolate the data. This is possibly due to the fact that the measuremnt errors of the audible noise data are very small.*

3.3 Accuracy of LPK Reconstructions

From the previous section, it is seen that the reconstructions of a functional data set generally fit the data of the individual functions well. In this section, we aim to theoretically investigate how accurate the reconstructions are. That is, we want to know what would happen if we substitute the underlying individual functions with their reconstructions based on the data. As an example, we study the accuracy of the LPK reconstructions (3.12). For convenience, we define the following widely used functionals of a kernel K:

$$\begin{aligned} B_r(K) = \int K(t)t^r dt, \quad V(K) = \int K^2(t)dt, \\ K^{(1)}(t) = \int K(s)K(s+t)ds. \end{aligned} \tag{3.28}$$

For estimating a function instead of its derivatives, Fan and Gijbels (1996) pointed out that even orders are not appealing. Therefore, we assume p is an odd integer; moreover, we denote $\gamma_{k,l}(s,t)$ as the (k,l)-times partial derivative of $\gamma(s,t)$ as

$$\gamma_{k,l}(s,t) = \frac{\partial^{k+l}\gamma(s,t)}{\partial^k s \partial^l t},$$

and use $\mathcal{D} = \{t_{ij}, j = 1,2,\cdots,n_i; i = 1,2,\cdots,n\}$ to collect all the design time points for all the individual functions. Moreover, we use $O_{UP}(1)$ (resp. $o_{UP}(1)$) to denote "bounded (resp. tends to 0) in probability uniformly for any t within the interior of the bounded support $\mathcal{T} = [a,b]$ of the design time points for all the individual functions and all $i = 1,2,\cdots,n$." Finally, the following regularity conditions are imposed.

Condition A

1. The design time points $t_{ij}, j = 1,2,\cdots,n_i; i = 1,2,\cdots,n$, are i.i.d. with the probability density function (pdf) $\pi(t), t \in \mathcal{T}$. For any given t within the interior of $\mathcal{T}$, $\pi'(t)$ exists and is continuous over $\mathcal{T}$.
2. Let s and t be any two interior time points of $\mathcal{T}$. The individual functions $f_i(t), i = 1,2,\cdots,n$, and their mean function $\eta(t)$ have up to $(p+1)$-times continuous derivatives. Their covariance function $\gamma(s,t)$ has up to $(p+1)$-times continuous derivatives for both s and t. The variance function of the measurement errors, $\sigma^2(t)$, is continuous at t.
3. The kernel K is a bounded symmetrical pdf with a bounded support $[-1,1]$.
4. There are two positive constants C and δ such that $n_i \geq Cn^\delta$, for all $i = 1,2,\cdots,n$. As $n \to \infty$, we have $h \to 0$ and $n^\delta h \to \infty$.

The first three conditions in Condition A are regular for LPK smoothing. The last condition requires that the numbers of measurements of all the individual functions be slightly larger than the number of the individual functions. This condition is usually satisfied for functional data with dense designs. Let K^* be the equivalent kernel of the p-order LPK reconstruction (Fan and Gijbels 1996, p. 64). We have the following result.

Theorem 3.1 *Under Condition A, the average mean squared error (MSE) of the p-order LPK reconstructions* $\hat{f}_i(t), i = 1,2,\cdots,n$ *is*

$$\begin{aligned} n^{-1}\sum_{i=1}^n E\Big\{[\hat{f}_i(t) - f_i(t)]^2|\mathcal{D}\Big\} = \Big\{\tfrac{V(K^*)\sigma^2(t)}{\pi(t)}(\tilde{m}h)^{-1} \\ +\tfrac{B_{p+1}^2(K^*)[(\eta^{(p+1)}(t))^2+\gamma_{p+1,p+1}(t,t)]}{(p+1)!^2}h^{2(p+1)}\Big\}[1+o_P(1)], \end{aligned} \quad (3.29)$$

where $\tilde{m} = (n^{-1}\sum_{i=1}^n n_i^{-1})^{-1}$.

Remark 3.6 *It is seen that the quantity $\tilde{m}$ represents an average of the numbers of the measurements, $n_i, i = 1, 2, \cdots, n$, of some kind. Theorem 3.1 indicates that the average MSE of the LPK reconstructions $\hat{f}_i(t), i = 1, 2, \cdots, n$ tends to 0 provided that $h \to 0$ and $\tilde{m}h \to \infty$. Minimizing the average MSE with respect to h shows that the associated optimal bandwidth is*

$$h^* = O_P(\tilde{m}^{-1/(2p+3)}) = O_P(n^{-\delta/(2p+3)}). \tag{3.30}$$

Theorem 3.2 *Under Condition A, for the p-order LPK reconstructions $\hat{f}_i(t), i = 1, 2, \cdots, n$ with the optimal bandwidth $h^* = O(n^{-\delta/(2p+3)})$, we have*

$$\hat{f}_i(t) = f_i(t) + n^{-(p+1)\delta/(2p+3)} O_{UP}(1), \quad i = 1, 2, \cdots, n.$$

Remark 3.7 *Theorem 3.2 implies that the LPK reconstructions $\hat{f}_i(t)$ are asymptotically uniformly little different from the underlying individual functions $f_i(t)$ provided that the number of measurements of each individual function tends to infinity at some reasonable rate with the number of individual functions. We expect that this is true not only for the LPK reconstructions, but also for other three type reconstructions of functional data; see Kraft (2010) for an example.*

3.3.1 Mean and Covariance Function Estimation

It is then natural to estimate the mean function $\eta(t)$ and covariance function $\gamma(s,t)$ by the sample mean and sample covariance functions of the p-order LPK reconstructions $\hat{f}_i(t), i = 1, 2, \cdots, n$:

$$\begin{aligned} \hat{\eta}(t) &= n^{-1}\textstyle\sum_{i=1}^n \hat{f}_i(t), \\ \hat{\gamma}(s,t) &= (n-1)^{-1}\textstyle\sum_{i=1}^n \left\{\hat{f}_i(s) - \hat{\eta}(s)\right\}\left\{\hat{f}_i(t) - \hat{\eta}(t)\right\}. \end{aligned} \tag{3.31}$$

The asymptotic conditional bias, covariance, and variance functions for $\hat{\eta}(t)$ are given below.

Theorem 3.3 *Under Condition A, as $n \to \infty$, the asymptotic conditional bias, covariance, and variance functions of $\hat{\eta}(t)$ are*

$$\begin{aligned} &Bias\{\hat{\eta}(t)|\mathcal{D}\} = \tfrac{B_{p+1}(K^*)\eta^{(p+1)}(t)}{(p+1)!} h^{p+1}[1 + o_{UP}(1)], \\ &Cov\{\hat{\eta}(s), \hat{\eta}(t)|\mathcal{D}\} = \Big\{\gamma(s,t)/n + \tfrac{K^{*(1)}[(s-t)/h]\sigma^2(s)}{\pi(t)}(n\tilde{m}h)^{-1} \\ &\quad + \tfrac{B_{p+1}(K^*)[\gamma_{p+1,0}(s,t)+\gamma_{0,p+1}(s,t)]}{(p+1)!} n^{-1}h^{p+1}\Big\}[1 + o_{UP}(1)], \\ &Var\{\hat{\eta}(t)|\mathcal{D}\} = \Big\{\gamma(t,t)/n + \tfrac{V(K^*)\sigma^2(t)}{\pi(t)}(n\tilde{m}h)^{-1} \\ &\quad \tfrac{B_{p+1}(K^*)[\gamma_{p+1,0}(t,t)+\gamma_{0,p+1}(t,t)]}{(p+1)!} n^{-1}h^{p+1}\Big\}[1 + o_{UP}(1)]. \end{aligned}$$

By Theorem 3.3, we have

$$\text{MSE}\{\hat{\eta}(t)|\mathcal{D}\} = \gamma(t,t)/n + O_{UP}(h^{2(p+1)} + (n\tilde{m}h)^{-1} + n^{-1}h^{p+1}).$$

We then always have $\text{MSE}\{\hat{\eta}(t)|\mathcal{D}\} = \gamma(t,t)/n + o_{UP}(1/n)$, implying a root-$n$ consistency of $\hat{\eta}(t)$, provided

$$\tilde{m}h \to \infty, \; nh^{2(p+1)} \to 0. \tag{3.32}$$

Remark 3.8 *Condition (3.32) is satisfied by any bandwidth $h = O(n^{-\delta^*})$ where $1/[2(p+1)] < \delta^* < \delta$. In particular, provided $\delta > 1 + 1/[2(p+1)]$, it is satisfied by $h^* = O(n^{-\delta/(2p+3)})$, the optimal bandwidth (3.30) for the p-order LPK reconstructions $\hat{f}_i(t), i = 1, 2, \cdots, n$. That is to say, in this case, the p-order LPK reconstruction optimal bandwidth h^* is sufficiently small to guarantee the $\sqrt{n}$-consistency of $\hat{\eta}(t)$. Condition (3.32) is also satisfied by $h^* = O(n^{-(1+\delta)/(2p+3)})$, the optimal bandwidth for $\hat{\eta}(t)$ when $1+1/[2(p+1)] < \delta < 1 + 1/(p+1)$, and by $h^* = O(n^{-\delta/(p+2)})$, the optimal bandwidth for $\hat{\eta}(t)$ when $\delta > 1+1/(p+1)$. Both the bandwidths will not affect the $\sqrt{n}$-consistency of $\hat{\eta}(t)$.*

By pretending all the underlying individual functions $f_i(t)$ were observed, the "ideal" estimators of $\eta(t)$ and $\gamma(s,t)$ are

$$\begin{aligned} \tilde{\eta}(t) &= n^{-1}\textstyle\sum_{i=1}^n f_i(t), \\ \tilde{\gamma}(s,t) &= (n-1)^{-1}\textstyle\sum_{i=1}^n \{f_i(s) - \tilde{\eta}(s)\}\{f_i(t) - \tilde{\eta}(t)\}. \end{aligned} \tag{3.33}$$

Let $\text{GP}(\eta, \gamma)$ denote a Gaussian process with mean function $\eta(t)$ and covariance function $\gamma(s,t)$; see some detailed discussions about Gaussian processes in Chapter 4. Let $v_1(t)$ denote the subject-effect of the first individual function $f_1(t)$ as defined in (3.1). In addition, throughout this book, we let "$\xrightarrow{d}$" denote "convergence in distribution."

Theorem 3.4 *Assume the optimal bandwidth $h^* = O(n^{-\delta/(2p+3)})$ is used for the p-order LPK reconstructions $\hat{f}_i(t), i = 1, 2, \cdots, n$. Then under Condition A, as $n \to \infty$, we have*

$$\hat{\eta}(t) = \tilde{\eta}(t) + n^{-\frac{(p+1)\delta}{2p+3}} O_{UP}(1), \quad \hat{\gamma}(s,t) = \tilde{\gamma}(s,t) + n^{-\frac{(p+1)\delta}{2p+3}} O_{UP}(1). \tag{3.34}$$

In addition, assume that $\delta > 1 + 1/[2(p+1)]$. Then as $n \to \infty$, we have

$$\sqrt{n}\{\hat{\eta}(t) - \eta(t)\} \xrightarrow{d} GP(0, \gamma), \quad \sqrt{n}\{\hat{\gamma}(s,t) - \gamma(s,t)\} \xrightarrow{d} GP(0, \varpi), \tag{3.35}$$

where

$$\begin{aligned} \varpi\{(s_1,t_1),(s_2,t_2)\} &= E\{v_1(s_1)v_1(t_1)v_1(s_2)v_1(t_2)\} \\ &\quad -\gamma(s_1,t_1)\gamma(s_2,t_2). \end{aligned} \tag{3.36}$$

When the subject-effect process $v_1(t)$ is Gaussian,

$$\varpi\{(s_1,t_1),(s_2,t_2)\} = \gamma(s_1,t_2)\gamma(s_2,t_1) + \gamma(s_1,s_2)\gamma(t_1,t_2).$$

Remark 3.9 *Theorem 3.4 indicates that under some mild conditions, the LPK-reconstructions based estimators (3.31) are asymptotically identical to the "ideal" estimators (3.33). The required key condition is $\delta > 1+1/[2(p+1)]$. It follows that, to make the measurement errors ignorable by the LPK reconstruction, we need the numbers of measurements, $n_i, i = 1, 2, \cdots, n$, for all the individual functions, to tend to infinity slightly faster than the number of the individual functions, n. This condition is approximately satisfied by many functional data sets, for example, the audible noise data, the Canadian temperature data, and the left-cingulum data introduced in Section 1.2 of Chapter 1.*

3.3.2 Noise Variance Function Estimation

The noise variance function $\sigma^2(t)$ measures the variation of the measurement errors ϵ_{ij} of the model (3.3). Following Fan and Yao (1998), we can construct the $\tilde{p}$-order LPK estimator of $\sigma^2(t)$ based on the p-order LPK residuals $\hat{\epsilon}_{ij} = y_{ij} - \hat{f}_i(t_{ij})$ although the setting here is more complicated. As expected, the resulting $\tilde{p}$-order LPK estimator of $\sigma^2(t)$ will be consistent but its convergence rate will be affected by that of the p-order LPK reconstructions $\hat{f}_i(t), i = 1, 2, \cdots, n$.

As an illustration, let us consider the simplest LPK estimator. That is, we consider the following kernel estimator for $\sigma^2(t)$ based on $\hat{\epsilon}_{ij}$:

$$\hat{\sigma}^2(t) = \frac{\sum_{i=1}^{n}\sum_{j=1}^{n_i} H_b(t_{ij} - t)\hat{\epsilon}_{ij}^2}{\sum_{i=1}^{n}\sum_{j=1}^{n_i} H_b(t_{ij} - t)}, \tag{3.37}$$

where $H_b(\cdot) = H(\cdot/b)/b$ with the kernel function H and the bandwidth b. Denote $N = \sum_{i=1}^{n} n_i$ to be the total number of the measurements for all the subjects.

Pretending $\hat{\epsilon}_{ij} \equiv \epsilon_{ij}$, by standard kernel estimation theory (Wand and Jones 1995, Fan and Gijbels 1996, among others), the optimal bandwidth of $\hat{\sigma}^2(t)$ is $b = O_P(N^{-1/5})$ and the associated convergence rate of $\hat{\sigma}^2(t)$ is $O_P(N^{-2/5})$. However, this convergence rate will be affected by the convergence rate of the p-order LPK reconstructions $\hat{f}_i(t), i = 1, 2, \cdots, n$ as under Condition A and by Theorem 3.2, we actually only have $\hat{\epsilon}_{ij} = \epsilon_{ij} + n^{-(p+1)\delta/(2p+3)}O_{UP}(1)$. Let $\nu_1(t) = \mathrm{E}[\epsilon_i^2(t)] = \sigma^2(t)$ and $\nu_2(t) = \mathrm{Var}[\epsilon_i^2(t)]$.

Theorem 3.5 *Assume Condition A is satisfied and the p-order LPK reconstructions $\hat{f}_i(t)$ use a bandwidth $h = O(n^{-\delta/(2p+3)})$. In addition, assume $\nu_1'(t)$ and $\nu_2'(t)$ exist and are continuous at $t \in \mathcal{T}$, and the kernel estimator $\hat{\sigma}^2(t)$ uses a bandwidth $b = O(N^{-1/5})$. Then we have*

$$\hat{\sigma}^2(t) = \sigma^2(t) + O_{UP}\left(n^{-2(1+\delta)/5} + n^{-(p+1)\delta/(2p+3)}\right). \tag{3.38}$$

Remark 3.10 *By the above theorem, it is seen that when $\delta < 2(2p+3)/(p-$

1), the second term in $O_{UP}(\cdot)$ dominates the first term; and in particular, when $p = 1$, we have $\hat{\sigma}^2(t) = \sigma^2(t) + O_{UP}(n^{-2\delta/5})$. In this case, the optimal convergence rate of $\hat{\sigma}^2(t)$ is not attainable. It is attainable only when $\delta > 2(2p+3)/(p-1)$ so that the first term in $O_{UP}(\cdot)$ in (3.38) dominates the second term. This is the case only when $p \geq 3$. When $p = 3$, $\delta > 9$ is required; and when $p = 2k+1 \to \infty$, $\delta > 4$ is required. Therefore, it is usually difficult to make the convergence rate of $\hat{\sigma}^2(t)$ unaffected by the convergence rate of the p-order LPK reconstructions $\hat{f}_i(t)$.

3.3.3 Effect of LPK Smoothing

Suppose that we can "observe" the sample curves $y_i(t), i = 1, 2, \cdots, n$ at any time point t. It is then natural to estimate $\eta(t)$ and $\gamma(s,t)$ by the sample mean and sample covariance functions of $y_i(t), i = 1, 2, \cdots, n$:

$$\begin{aligned} \bar{y}(t) &= n^{-1}\textstyle\sum_{i=1}^n y_i(t), \\ \hat{\gamma}_0(s,t) &= (n-1)^{-1}\textstyle\sum_{i=1}^n \{y_i(s) - \bar{y}(s)\}\{y_i(t) - \bar{y}(t)\}, \end{aligned}$$

without involving any smoothing. Notice that $\mathrm{E}[y_i(t)] = \eta(t)$ and $\mathrm{Cov}[y_i(s), y_i(t)] = \gamma_0(s,t) = \gamma(s,t) + \sigma^2(t)1_{\{s=t\}}$, containing the variation of the measurement errors, we have $\mathrm{E}\bar{y}(t) = \eta(t)$ and $\mathrm{E}\hat{\gamma}_0(s,t) = \gamma_0(s,t)$. Moreover, by the central limit theorems for i.i.d. stochastic processes, Theorems 4.12 and 4.13 in Chapter 4, we have

Theorem 3.6 *As $n \to \infty$, we have $\sqrt{n}\{\bar{y}(t) - \eta(t)\} \xrightarrow{d} GP(0, \gamma_0)$, and $\sqrt{n}\{\hat{\gamma}_0(s,t) - \gamma_0(s,t)\} \xrightarrow{d} GP(0, \varpi_0)$, where*

$$\varpi_0\{(s_1,t_1),(s_2,t_2)\} = E\{\tilde{v}_1(s_1)\tilde{v}_1(t_1)\tilde{v}_1(s_2)\tilde{v}_1(t_2)\} - \gamma_0(s_1,t_1)\gamma_0(s_2,t_2),$$

with $\tilde{v}_1(t) = v_1(t) + \epsilon_1(t)$.

Remark 3.11 *We can now discuss the effect of LPK smoothing by comparing the estimators $\hat{\eta}(t)$ and $\hat{\gamma}(s,t)$ (with LPK smoothing) against the estimators $\bar{y}(t)$ and $\hat{\gamma}_0(s,t)$ (with no smoothing) using Theorems 3.5 and 3.6. First of all, we notice that $\hat{\eta}(t)$ is asymptotically more efficient than $\bar{y}(t)$ because we always have $\gamma(s,t) \leq \gamma_0(s,t)$. Second, we notice that $\hat{\gamma}(s,t)$ is asymptotically less biased than $\hat{\gamma}_0(s,t)$ in the sense of estimating $\gamma(s,t)$. These results are reasonable as the estimators $\hat{\eta}(t)$ and $\hat{\gamma}(s,t)$ use more information than the estimators $\bar{y}(t)$ and $\hat{\gamma}_0(s,t)$. Of course, we should keep in mind that the effect of LPK smoothing is second order because under some mild conditions, all the estimators are $\sqrt{n}$-consistent.*

3.3.4 A Simulation Study

In this subsection we aim to investigate the effect of the bandwidth selected by the GCV rule (3.10) on the average MSE (3.29) of the p-order LPK reconstructions $\hat{f}_i(t), i = 1, 2, \cdots, n$ and the MSE of the mean function estimator $\hat{\eta}(t)$ by a simulation study.

We generated simulation samples from the following model:

$$\begin{aligned} y_i(t) &= \eta(t) + v_i(t) + \epsilon_i(t), \\ \eta(t) &= a_0 + a_1\phi_1(t) + a_2\phi_2(t), \\ v_i(t) &= b_{i0} + b_{i1}\psi_1(t) + b_{i2}\psi_2(t), \\ \mathbf{b}_i &= [b_{i0}, b_{i1}, b_{i2}]^T \sim N[0, \text{diag}(\sigma_0^2, \sigma_1^2, \sigma_2^2)], \\ \epsilon_i(t) &\sim N[0, \sigma_\epsilon^2(1+t)], \quad i = 1, 2, \cdots, n, \end{aligned}$$

where n is the number of subjects, and $\mathbf{b}_i$ and $\epsilon_i(t)$ are independent. The scheduled design time points are $t_j = j/(m+1), j = 1, 2, \cdots, m$. To obtain an imbalanced design that is more realistic, we randomly removed some responses on a subject at a rate r_{miss} so that on average there are about $m(1 - r_{miss})$ measurements on a subject, and $nm(1 - r_{miss})$ measurements in a whole simulated sample. For simplicity, in this simulation, the parameters we actually used were $[a_0, a_1, a_2] = [1, 3.2, 4.5], [\sigma_0^2, \sigma_1^2, \sigma_2^2, \sigma_\epsilon^2] = [1, 2, 4, .1]$, $\phi_1(t) = \psi_1(t) = \cos(2\pi t), \phi_2(t) = \psi_2(t) = \sin(2\pi t), r_{miss} = 10\%$, $m = 100$, and $n = 20, 30$, and 40.

For a simulated sample, the p-order LPK reconstructions $\hat{f}_i(t)$ were obtained using a local linear smoother (Fan 1992) with the well-known Gaussian kernel $K(t) = \exp(-\frac{t^2}{2})/\sqrt{2\pi}$. We considered five bandwidth choices: $0.5h^*, 0.75h^*, h^*, 1.25h^*$, and $1.5h^*$, where h^* is the bandwidth selected by the GCV rule (3.10). For a simulated sample, the average MSE for $\hat{f}_i(t)$ and the MSE for the mean function estimator $\hat{\eta}(t)$ were computed respectively as

$$\begin{aligned} \text{MSE}_f &= (nM)^{-1}\textstyle\sum_{i=1}^n \sum_{j=1}^M \{\hat{f}_i(\tau_j) - f_i(\tau_j)\}^2, \\ \text{MSE}_\eta &= M^{-1}\textstyle\sum_{j=1}^M \{\hat{\eta}(\tau_j) - \eta(\tau_j)\}^2, \end{aligned}$$

where $\tau_1, \cdots, \tau_M$ are M time points equally spaced in $[0, 1]$, for some M large. In this simulation study, we took $M = m = 100$ for simplicity.

Figure 3.10 presents the simulation results for $n = 20$. The boxplots were based on $1,000$ simulated samples. From top to bottom, panels are respectively for GCV, MSE_f and MSE_η. In each of the panels, the first 5 boxplots are associated with the 5 bandwidth choices: $.5h^*, .75h^*, h^*, 1.25h^*$ and $1.5h^*$ respectively. The sixth boxplot in the MSE_η panel is associated with the "ideal" estimator $\tilde{\eta}(t)$; see (3.33) for its definition. From Figure 3.10, we may conclude that

- Overall, the GCV rule (3.10) performed well in the sense of choosing the right bandwidths to minimize the average MSE (3.29).
- Moderately smaller or larger bandwidths than h^* do not affect the MSE_η too much.
- The MSE_η based on $\hat{\eta}(t)$ and those based on the "ideal" estimator $\tilde{\eta}(t)$ are nearly the same.

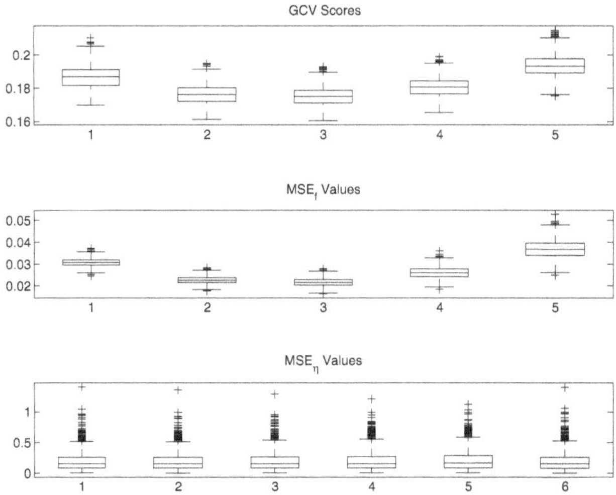

Figure 3.10 *Simulation results about the LPK reconstruction for functional data ($n = 20$). From top to bottom, panels are respectively for GCV, MSE_f and MSE_η. In each of the panels, the first 5 boxplots are associated with the 5 bandwidth choices: $0.5h^*, 0.75h^*, h^*, 1.25h^*$, and $1.5h^*$, where h^* is the GCV bandwidth. The sixth boxplot in the MSE_η panel is associated with the "ideal" estimator $\tilde{\eta}(t)$.*

The same conclusions can be made by checking the simulation results for $n = 30$ and 40 (not shown). They are consistent with the results stated in Theorems 3.1, 3.2, and 3.3.

3.4 Accuracy of LPK Reconstruction in FLMs

In the model (3.1), the only covariate for the mean function $\eta(t)$ is time. In many applications, $\eta(t)$ may also depend on some time-independent covariates and can be written as $\eta(t; \mathbf{x}) = \mathbf{x}^T \boldsymbol{\beta}(t)$, where the covariate vector $\mathbf{x} = [x_1, \cdots, x_q]^T$ and the unknown but smooth coefficient function vector $\boldsymbol{\beta}(t) = [\beta_1(t), \cdots, \beta_q(t)]^T$. A replacement of $\eta(t)$ by $\eta(t; \mathbf{x}_i) = \mathbf{x}_i^T \boldsymbol{\beta}(t)$ in (3.1) leads to the following so-called functional linear model (FLM):

$$y_i(t) = \mathbf{x}_i^T \boldsymbol{\beta}(t) + v_i(t) + \epsilon_i(t), \quad i = 1, 2, \cdots, n, \tag{3.39}$$

where $y_i(t), v_i(t)$, and $\epsilon_i(t)$ are the same as those defined in (3.1). In many applications, we may have $x_1 \equiv 1$ so that the first coefficient function $\beta_1(t)$ in $\boldsymbol{\beta}(t)$ represents an intercept function.

The ignorability of the substitution effect is also applicable to the LPK reconstructions $\hat{f}_i(t)$ of the individual functions $f_i(t) = \mathbf{x}_i^T \boldsymbol{\beta}(t) + v_i(t)$ of the above FLM (3.39). Based on this, in this section, we construct the estimators $\hat{\boldsymbol{\beta}}(t)$ and $\hat{\gamma}(s, t)$ and investigate their asymptotics; in particular, we show that

$\hat{\boldsymbol{\beta}}(t)$ is $\sqrt{n}$-consistent and asymptotically Gaussian. In addition, as an illustrative example, we study the effect of the LPK reconstruction on a global test statistic T_n, which is a L^2-norm-based test, to test a general linear hypothesis testing (GLHT) problem about the covariate effects $\boldsymbol{\beta}(t)$. Its asymptotic random expression is derived.

3.4.1 *Coefficient Function Estimation*

Notice that Theorem 3.2 is also applicable to the p-order LPK reconstructions $\hat{f}_i(t)$ of the underlying individual functions $f_i(t) = \mathbf{x}_i^T\boldsymbol{\beta}(t) + v_i(t), i = 1, 2, \cdots, n$ of the FLM (3.39). Let $\hat{\mathbf{f}}(t) = [\hat{f}_1(t), \cdots, \hat{f}_n(t)]^T$ and $\mathbf{X} = [\mathbf{x}_1, \cdots, \mathbf{x}_n]^T$. Here, we assume $\mathbf{X}$ has full rank. Then the least-squares estimator of $\boldsymbol{\beta}(t)$ is

$$\hat{\boldsymbol{\beta}}(t) = \left\{\sum_{i=1}^{n} \mathbf{x}_i\mathbf{x}_i^T\right\}^{-1} \sum_{i=1}^{n} \mathbf{x}_i\hat{f}_i(t) = (\mathbf{X}^T\mathbf{X})^{-1}\mathbf{X}^T\hat{\mathbf{f}}(t), \tag{3.40}$$

which minimizes $Q(\boldsymbol{\beta}) = n^{-1}\sum_{i=1}^{n}\int_{\mathcal{T}}[\hat{f}_i(t) - \mathbf{x}_i^T\boldsymbol{\beta}(t)]^2dt$. It follows that the subject-effects $v_i(t)$ can be estimated by $\hat{v}_i(t) = \hat{f}_i(t) - \mathbf{x}_i^T\hat{\boldsymbol{\beta}}(t)$ and their covariance function $\gamma(s,t)$ can be estimated by

$$\hat{\gamma}(s,t) = (n-q)^{-1}\sum_{i=1}^{n} \hat{v}_i(s)\hat{v}_i(t) = (n-q)^{-1}\hat{\mathbf{v}}(s)^T\hat{\mathbf{v}}(t), \tag{3.41}$$

where $\hat{\mathbf{v}}(t) = [\hat{v}_1(t), \hat{v}_2(t), \cdots, \hat{v}_n(t)]^T = \hat{\mathbf{f}}(t) - \mathbf{X}(\mathbf{X}^T\mathbf{X})^{-1}\mathbf{X}^T\hat{\mathbf{f}}(t) = (\mathbf{I}_n - \mathbf{P}_{\mathbf{X}})\hat{\mathbf{f}}(t)$, and $\mathbf{P}_{\mathbf{X}} = \mathbf{X}(\mathbf{X}^T\mathbf{X})^{-1}\mathbf{X}^T$ is a projection matrix with $\mathbf{P}_{\mathbf{X}}^T = \mathbf{P}_{\mathbf{X}}, \mathbf{P}_{\mathbf{X}}^2 = \mathbf{P}_{\mathbf{X}}$, and $\text{tr}(\mathbf{P}_{\mathbf{X}}) = q$.

Pretending that $f_i(t), i = 1, 2, \cdots, n$ were known, the "ideal" estimators of $\boldsymbol{\beta}(t)$ and $\gamma(s,t)$ are

$$\tilde{\boldsymbol{\beta}}(t) = (\mathbf{X}^T\mathbf{X})^{-1}\mathbf{X}^T\mathbf{f}(t), \quad \tilde{\gamma}(s,t) = (n-q)^{-1}\tilde{\mathbf{v}}(s)^T\tilde{\mathbf{v}}(t), \tag{3.42}$$

where $\mathbf{f}(t) = [f_1(t), \cdots, f_n(t)]^T$ and $\tilde{\mathbf{v}}(t) = (\mathbf{I}_n - \mathbf{P}_{\mathbf{X}})\mathbf{f}(t)$. It is easy to show that $\text{E}\tilde{\boldsymbol{\beta}}(t) = \boldsymbol{\beta}(t)$ and $\text{E}\tilde{\gamma}(s,t) = \gamma(s,t)$. For further investigation, we impose the following conditions:

Condition B

1. The covariate vectors $\mathbf{x}_i, i = 1, 2, \cdots, n$ are i.i.d. with finite and invertible second moment $\text{E}\mathbf{x}_1\mathbf{x}_1^T = \boldsymbol{\Omega}$. Moreover, they are uniformly bounded in probability. That is, $\mathbf{x}_i = O_{UP}(1)$.
2. The subject-effects $v_i(t)$ are uniformly bounded in probability. That is, $\mathbf{v}_i(t) = O_{UP}(1)$.

Theorem 3.7 *Assume Conditions A and B are satisfied, and the p-order*

LPK reconstructions $\hat{f}_i(t)$ use a bandwidth $h = O(n^{-\delta/(2p+3)})$. Then as $n \to \infty$, we have

$$\hat{\boldsymbol{\beta}}(t) = \tilde{\boldsymbol{\beta}}(t) + n^{-\frac{(p+1)\delta}{2p+3}} O_{UP}(1), \quad \hat{\gamma}(s,t) = \tilde{\gamma}(s,t) + n^{-\frac{(p+1)\delta}{2p+3}} O_{UP}(1). \quad (3.43)$$

In addition, assume $\delta > 1 + 1/[2(p+1)]$. Then as $n \to \infty$, we have

$$\sqrt{n}\left\{\hat{\boldsymbol{\beta}}(t) - \boldsymbol{\beta}(t)\right\} \xrightarrow{d} GP_q(0, \gamma_\beta), \quad (3.44)$$

where $\gamma_\beta(s,t) = \gamma(s,t)\boldsymbol{\Omega}^{-1}$ and $GP_q(\eta, \gamma)$ denotes a q-dimensional Gaussian process with mean function $\eta(t)$ and covariance function $\gamma(s,t)$.

Remark 3.12 *Theorem 3.7 implies that under the given conditions, the proposed estimators $\hat{\boldsymbol{\beta}}(t)$ and $\hat{\gamma}(s,t)$ are asymptotically identical to the "ideal" estimators $\tilde{\boldsymbol{\beta}}(t)$ and $\tilde{\gamma}(s,t)$ respectively. Therefore, in FDA, it seems reasonable to directly assume the underlying individual functions are "observed" as done in Ramsay and Silverman (2002, 2005). Alternatively speaking, statistical inferences for functional data can be based on the reconstructions of the observed functional data without loss of too much information. In this book, this strategy will be used from the next chapter onward.*

3.4.2 Significance Tests of Covariate Effects

Consider the following GLHT problem:

$$H_0 : \mathbf{C}\boldsymbol{\beta}(t) \equiv \mathbf{c}(t), \quad \text{versus} \quad H_1 : \mathbf{C}\boldsymbol{\beta}(t) \neq \mathbf{c}(t) \text{ for some } t \in \mathcal{T}, \quad (3.45)$$

where $\mathcal{T} = [a,b]$, $\mathbf{C}$ is a given $k \times q$ full rank matrix, and $\mathbf{c}(t) = [c_1(t), \cdots, c_k(t)]^T$ a given vector of functions, often specified as $\mathbf{0}$. The above GLHT problem is very general. For example, in order to check the significance of the rth covariate effect, one takes $\mathbf{C} = \mathbf{e}_{r,q}^T$ and $\mathbf{c}(t) = \mathbf{0}$; in order to check if the first two coefficient functions are the same, say, $\beta_1(t) = \beta_2(t)$, one takes $\mathbf{C} = (\mathbf{e}_{1,q} - \mathbf{e}_{2,q})^T$ and $\mathbf{c}(t) = \mathbf{0}$.

It is natural to estimate $\mathbf{C}\boldsymbol{\beta}(t)$ by $\mathbf{C}\hat{\boldsymbol{\beta}}(t)$. By Theorem 3.7, as $n \to \infty$, we have

$$\sqrt{n}[\mathbf{C}\hat{\boldsymbol{\beta}}(t) - \mathbf{c}(t)] - \boldsymbol{\eta}_C(t) \xrightarrow{d} \text{GP}_k(\mathbf{0}, \gamma_C), \quad (3.46)$$

where $\boldsymbol{\eta}_C(t) = \sqrt{n}[\mathbf{C}\boldsymbol{\beta}(t) - \mathbf{c}(t)]$ and $\gamma_C(s,t) = \gamma(s,t)\mathbf{C}\Omega^{-1}\mathbf{C}^T$. Based on the above result, pointwise t- and F-tests can be easily conducted. We here aim to study the effect of the LPK reconstructions on the so-called L^2-norm-based test (Zhang and Chen 2007) based on the following global test statistic:

$$T_n = \int_a^b [\mathbf{C}\hat{\boldsymbol{\beta}}(t) - \mathbf{c}(t)]^T [\mathbf{C}(\mathbf{X}^T\mathbf{X})^{-1}\mathbf{C}^T]^{-1} [\mathbf{C}\hat{\boldsymbol{\beta}}(t) - \mathbf{c}(t)] dt. \quad (3.47)$$

Let $\tilde{T}_n$ be the "ideal" global test statistic, obtained by replacing $\hat{\boldsymbol{\beta}}(t)$ by the "ideal" estimator $\tilde{\boldsymbol{\beta}}(t)$ as defined in (3.42).

To derive the asymptotic random expression of T_n, we assume $\gamma(s,t)$ has a finite trace. That is, we assume that $\mathrm{tr}(\gamma) = \int_a^b \gamma(t,t)dt < \infty$ so that it has the following singular value decomposition:

$$\gamma(s,t) = \sum_{r=1}^{m} \lambda_r \phi_r(s)\phi_r(t),\ s,t \in \mathcal{T}, \tag{3.48}$$

where $\lambda_1 \geq \lambda_2 \geq \cdots \geq \lambda_m > 0$ are all the positive eigenvalues for some $m \leq \infty$ and $\phi_1(t), \phi_2(t), \cdots, \phi_m(t)$ are the associated orthonormal eigenfunctions of $\gamma(s,t)$. Let $X \stackrel{d}{=} Y$ denote that the random variables X and Y have the same distribution, χ_r^2 denote a χ^2-distribution with r degrees of freedom, and $\|\cdot\|$ the usual L^2-norm.

Theorem 3.8 *Under Conditions A, B, and the null hypothesis, as $n \to \infty$, we have*

$$T_n = \tilde{T}_n + n^{1/2-(p+1)\delta/(2p+3)} O_P(1). \tag{3.49}$$

In addition, assume $\delta > 1 + 1/[2(p+1)]$ and $\gamma(s,t)$ has a finite trace so that it has a singular value decomposition (3.48). Then as $n \to \infty$, we have

$$T_n \stackrel{d}{\to} \sum_{r=1}^{m} \lambda_r A_r, \quad A_r \stackrel{i.i.d.}{\sim} \chi_k^2. \tag{3.50}$$

Remark 3.13 *Theorem 3.8 suggests that the distribution of T_n is asymptotically the same as that of a χ^2-type mixture. Methods for approximating the distribution of a χ^2-type mixture will be discussed in Section 4.3 of Chapter 4. Some of those methods are due to Zhang and Chen (2007) and are strongly related to Buckley and Eagleson (1988) and Zhang (2005).*

In fact, Zhang and Chen (2007) proposed three methods to approximate the null distribution of T_n: χ^2-approximation, direct simulation, and nonparametric bootstrapping. In the first two methods, the null distribution of T_n is approximated by that of the χ^2-type mixture $S = \sum_{r=1}^m \hat{\lambda}_r A_r$, where $A_r \sim \chi_k^2$ and $\hat{\lambda}_r$ are the eigenvalues of $\hat{\gamma}(s,t)$. In the χ^2-approximation method, the distribution of S is approximated by that of a random variable $R = \alpha\chi_d^2 + \beta$ by matching the first three cumulants of R and S to determine the unknown parameters α, d, and β (Buckley and Eagleson 1988, Zhang 2005). In the direct simulation method, the sampling distribution of S is computed based on a sample of S obtained by repeatedly generating $(A_1, A_2, \cdots, A_m)$. The nonparametric bootstrapping method is slightly more complicated. In the nonparametric bootstrapping method, a sample of subject-effects $v_i^*(t), i = 1,2,\cdots,n$ are generated from the estimated subject-effects $\hat{v}_i(t), i = 1,2,\cdots,n$ and then construct a bootstrapping sample: $f_i^*(t) = \mathbf{x}_i^T\hat{\boldsymbol{\beta}}(t) + v_i^*(t), i = 1,2,\cdots,n$. Let $\hat{\boldsymbol{\beta}}^*(t)$ be the bootstrapping estimator of $\boldsymbol{\beta}(t)$ based on the above bootstrapping sample. We then use it to compute

$$T_n^* = \int_a^b [\hat{\boldsymbol{\beta}}^*(t) - \hat{\boldsymbol{\beta}}(t)]^T \mathbf{C}^T [\mathbf{C}(\mathbf{X}^T\mathbf{X})^{-1}\mathbf{C}^T]^{-1}\mathbf{C}[\hat{\boldsymbol{\beta}}^*(t) - \hat{\boldsymbol{\beta}}(t)]dt.$$

Then the bootstrapping null distribution of T_n is obtained by the sampling distribution of T_n^* by B replications of the above bootstrapping process for some B large, say, $B = 10,000$.

3.4.3 A Real Data Example

Example 3.5 *The real functional data set we used here is the progesterone data set introduced in some detail in Section 1.2.1 of Chapter 1. We modeled the progesterone data set by the FLM (3.39) with the covariates*

$$\mathbf{x}_i = \begin{cases} [1,0]^T, & \text{if curve } i \text{ is nonconceptive}, \\ [0,1]^T, & \text{if curve } i \text{ is conceptive}, \end{cases} \quad i = 1, 2, \cdots, 91,$$

and the coefficient function vector $\boldsymbol{\beta}(t) = [\beta_1(t), \beta_2(t)]^T$, where $\beta_1(t)$ and $\beta_2(t)$ are the nonconceptive and conceptive covariate effect (group mean) functions, respectively.

The progesterone curves were first reconstructed by the LPK reconstruction method discussed in Section 3.2.3 using local linear smoothing with the well-known Gaussian kernel and the bandwidth $h^* = 1.40$ selected by the GCV rule (3.10); see Figure 3.1 for the GCV score against $\log_{10}(h)$. Figure 3.2 presents the nonconceptive (upper panel) and conceptive (lower panel) progesterone curves. Figure 3.11 displays the estimated nonconceptive (solid) and conceptive (dashed) covariate effect functions with their 95% pointwise confidence bands (nonconceptive, solid; conceptive, dashed).

From Figure 3.2, it is seen that before Day 8, the nonconceptive and conceptive curves behaved similarly but they then moved in different directions; so did their mean functions as indicated in Figure 3.11. Based on the 95% pointwise confidence bands, we can conclude that the nonconceptive and conceptive mean function differences over the intervals $[a,b] = [-8,15]$, $[-8,8]$, $[0,8]$, or $[8,15]$ will be "significant," "not significant," "not significant," and "highly significant." These conclusions can be made more clear by testing the hypothesis testing problem (3.45) with $\mathbf{C} = [1,-1]$, $\mathbf{c}(t) = [0,0]^T$, and $t \in \mathcal{T} = [a,b]$ using the test statistic T_n (3.47).

Table 3.1 shows the results of the above significance tests for the progesterone data. For each choice of the interval $[a,b]$, we used three different bandwidth choices: $h^*/2$, h^*, and $1.5h^*$, where $h^* = 1.40$ was selected by the GCV rule (3.10). For each bandwidth choice, the associated test statistics T_n were computed using (3.47) with each curve evaluated at 1,000 locations that are equally spaced over $[-8,15]$. For each T_n, we computed its P-value using the χ^2-approximation, direct simulation, and nonparametric bootstrapping

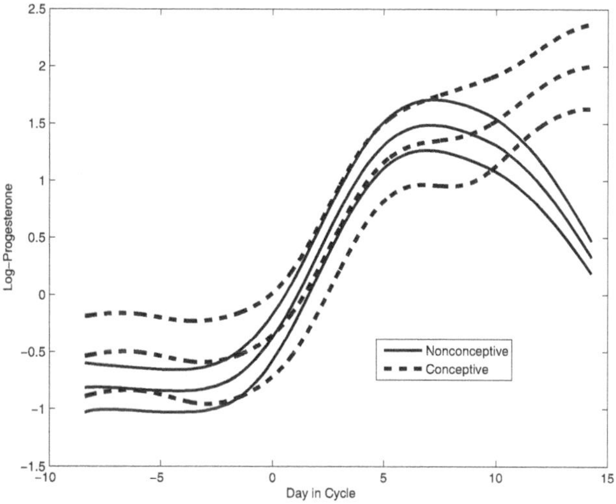

Figure 3.11 *Estimated nonconceptive (solid) and conceptive (dashed) mean functions with 95% pointwise confidence bands (nonconceptive, solid; conceptive, dashed). It appears that the mean function of the nonconceptive progesterone curves is comparable with that of the conceptive progesterone curves before Day 8 but they are different from each other after Day 8.*

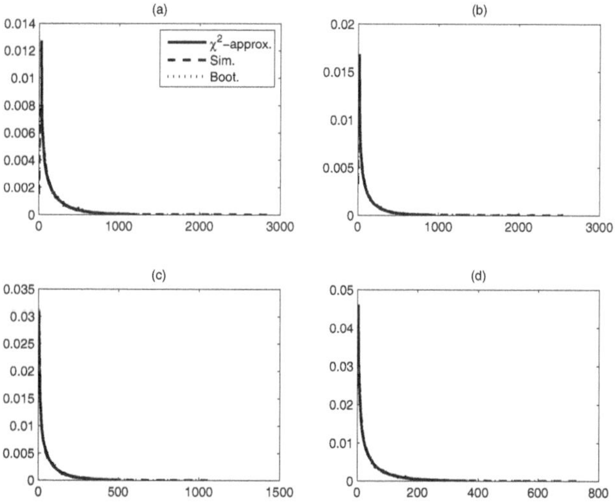

Figure 3.12 *Null distribution approximations (χ^2-approximation, solid; direct simulation, dashed; nonparametric bootstrapping, dotted) of T_n (3.47) when $h = h^* =$ 1.40 over various periods: (a) $[a, b] = [-8, 15]$, (b) $[a, b] = [-8, 8]$, (c) $[a, b] = [0, 8]$, and (d) $[a, b] = [8, 15]$.*

Table 3.1 *Test results for comparing the two group mean functions of the progesterone data where $h^* = 1.40$ was selected by the GCV rule (3.10) and the direct simulation and bootstrapping P-values were computed based on 50,000 and 10,000 replications, respectively.*

			P-Value		
$[a,b]$	h	T_n	χ^2-Approx.	Direct Sim.	Nonpara. Boots.
$[-8,15]$	$h^*/2$	2.077e4	0.0138	0.0136	0.0126
	h^*	2.033e4	0.0133	0.0147	0.0117
	$1.5h^*$	2.057e4	0.0116	0.0117	0.0103
$[-8,8]$	$h^*/2$	3.066e3	0.3041	0.3067	0.2991
	h^*	2.670e3	0.3257	0.3267	0.3313
	$1.5h^*$	2.355e3	0.3503	0.3519	0.3539
[0,8]	$h^*/2$	7.656e2	0.5084	0.5167	0.5138
	h^*	5.768e2	0.5497	0.5594	0.5590
	$1.5h^*$	3.873e2	0.6201	0.6244	0.6309
[8,15]	$h^*/2$	1.770e4	0.0001	0.0001	0.0001
	h^*	1.766e4	0.0000	0.0001	0.0000
	$1.5h^*$	1.822e4	0.0000	0.0000	0.0000

methods that were described briefly in Section 3.4.2. Figure 3.12 displays the null probability density function (pdf) approximations obtained using the three methods. It seems that all three approximations perform reasonably well except at the left boundary where the χ^2-approximations seem problematic. Nevertheless, from this table, we can conclude that the mean function differences over the intervals $[a,b] = [-8,15]$, $[-8,8]$, $[0,8]$ and $[8,15]$ are, respectively, "very significant(P-value < 0.015)," "not significant (P-value > 0.29)," "not significant (P-value > 0.50)," and "highly significant (P-value ≤ 0.0001)." Notice that the test over $[-8,15]$ is less significant (in the sense of P-value) than the test over $[8,15]$. This is reasonable because the test over $[-8,8]$ is not significant, which reduces the significance of the test over $[-8,15]$.

3.5 Technical Proofs

In this section, we outline the technical proofs of some of the asymptotic results presented in this chapter. For applied statisticians, these technical details may be skipped.

Before we proceed, we list some useful lemmas. Proof of Lemma 3.1 can be found in Fan and Gijbels (1996, p. 64). Notice that under Condition A 4, "$n \to \infty$" implies that "$n_i \to \infty$".

Lemma 3.1 *Under Condition A, as $n \to \infty$, we have*

$$K_h^{n_i}(t_{ij} - t) = \frac{1}{n_i \pi(t)} K_h^*(t_{ij} - t)[1 + o_P(1)],$$

where $K^*(\cdot)$ is the LPK equivalent kernel (Fan and Gijbels 1996, p. 64).

Lemma 3.2 *We always have*

$$\sum_{j=1}^{n_i} K_h^{n_i}(t_{ij}-t)(t_{ij}-t)^r = \begin{cases} 1, & \text{when } r=0, \\ 0, & \text{otherwise.} \end{cases}$$

In addition, under Condition A, as $n \to \infty$, we have

$$\sum_{j=1}^{n_i} K_h^{n_i}(t_{ij}-t)(t_{ij}-t)^{p+1} = \frac{B_{p+1}(K^*)h^{p+1}}{\pi(t)}[1+o_P(1)],$$
$$\sum_{j=1}^{n_i} \{K_h^{n_i}(t_{ij}-t)\}^2 = \frac{V(K^*)}{\pi(t)}(n_i h)^{-1}[1+o_P(1)],$$
$$\sum_{j=1}^{n_i} K_h^{n_i}(t_{ij}-s)K_h^{n_i}(t_{ij}-t) = \frac{K^{*(1)}((s-t)/h)}{\pi(t)}(n_i h)^{-1}[1+o_P(1)],$$

where $B_r(\cdot)$ and $V(\cdot)$ are defined in (3.28).

Let $r_i(t) = \hat{f}_i(t) - f_i(t), i = 1, 2, \cdots, n$, where $\hat{f}_i(t)$ are the p-order LPK reconstructions of $f_i(t)$ given in Section 3.2.3. Let $\bar{r}(t) = n^{-1}\sum_{i=1}^n r_i(t)$ and $\bar{f}(t) = n^{-1}\sum_{i=1}^n f_i(t)$. Using Lemmas 3.1 and 3.2, we can show the following useful lemma.

Lemma 3.3 *Under Condition A, as $n \to \infty$, we have*

$$\begin{aligned} E\{r_i(t)|\mathcal{D}\} &= \tfrac{B_{p+1}(K^*)\eta^{(p+1)}(t)}{(p+1)!}h^{p+1}[1+o_P(1)], \\ Cov\{r_i(s), r_i(t)|\mathcal{D}\} &= \Big\{\tfrac{K^{*(1)}((s-t)/h)}{\pi(t)}(n_i h)^{-1} \\ &\quad + \tfrac{B_{p+1}^2(K^*)\gamma_{p+1,p+1}(s,t)}{(p+1)!^2}h^{2(p+1)}\Big\}[1+o_P(1)], \\ Cov\{r_i(s), f_i(t)|\mathcal{D}\} &= \tfrac{B_{p+1}(K^*)\gamma_{p+1,0}(s,t)}{(p+1)!}h^{p+1}[1+o_P(1)]. \end{aligned}$$

Proof of Lemma 3.3 By (3.12) and Lemma 3.1, we have

$$r_i(t) = \sum_{j=1}^{n_i} K_h^{n_i}(t_{ij}-t)\epsilon_{ij} + \sum_{j=1}^{n_i} K_h^{n_i}(t_{ij}-t)\left\{f_i(t_{ij}) - f_i(t)\right\}.$$

It follows that

$$\mathrm{E}(r_i(t)|\mathcal{D}) = \sum_{j=1}^{n_i} K_h^{n_i}(t_{ij}-t)\left\{\eta(t_{ij}) - \eta(t)\right\}.$$

Applying Taylor expansion and Lemmas 3.1 and 3.2, we have

$$\begin{aligned} \mathrm{E}(r_i(t)|\mathcal{D}) &= \textstyle\sum_{j=1}^{n_i} K_h^{n_i}(t_{ij}-t)\left\{\sum_{l=1}^{p+1}\eta^{(l)}(t)\frac{(t_{ij}-t)^l}{l!}\right. \\ &\quad \left. + o[(t_{ij}-t)^{p+1}]\right\} \\ &= \frac{B_{p+1}(K^*)\eta^{(p+1)}(t)}{(p+1)!}h^{p+1}[1+o_P(1)], \end{aligned} \tag{3.51}$$

Similarly, by the independence of $f_i(t)$ and $\epsilon_i(t)$, we have

$$\begin{aligned} &\mathrm{Cov}(r_i(s), r_i(t)|\mathcal{D}) = \textstyle\sum_{j=1}^{n_i} K_h^{n_i}(t_{ij}-s)K_h^{n_i}(t_{ij}-t)\sigma^2(t_{ij}) \\ &\quad + \textstyle\sum_{j=1}^{n_i}\sum_{l=1}^{n_i} K_h^{n_i}(t_{ij}-s)K_h^{n_i}(t_{il}-t) \\ &\quad \times\{\gamma(t_{ij},t_{il}) - \gamma(t_{ij},t) - \gamma(s,t_{il}) + \gamma(s,t)\} \\ &= \left\{\frac{K^{*(1)}[(s-t)/h]\sigma^2(s)}{\pi(t)}(n_ih)^{-1} + \frac{B_{p+1}^2(K^*)\gamma_{p+1,p+1}(s,t)}{(p+1)!^2}h^{2(p+1)}\right\} \\ &\quad \times[1+o_P(1)]. \end{aligned}$$

In particular, letting $s = t$, we obtain

$$\begin{aligned} \mathrm{Var}(r_i(t)|\mathcal{D}) &= \left\{\frac{V(K^*)\sigma^2(t)}{\pi(t)}(n_ih)^{-1}\right. \\ &\quad \left. + \frac{B_{p+1}^2(K^*)\gamma_{p+1,p+1}(t,t)}{(p+1)!^2}h^{2(p+1)}\right\}[1+o_P(1)], \end{aligned} \tag{3.52}$$

as desired. Lemma 3.3 is proved.

Direct application of Lemma 3.3 leads to Lemma 3.4.

Lemma 3.4 *Under Condition A, as $n \to \infty$, we have*

$$\begin{aligned} &E(\bar{r}(t)|\mathcal{D}) = \frac{B_{p+1}(K^*)\eta^{(p+1)}(t)}{(p+1)!}h^{p+1}[1+o_P(1)], \\ &Cov(\bar{r}(s),\bar{r}(t)|\mathcal{D}) = n^{-1}\left\{\frac{K^{*(1)}((s-t)/h)}{\pi(t)}(\tilde{m}h)^{-1}\right. \\ &\quad \left. + \frac{B_{p+1}^2(K^*)\gamma_{p+1,p+1}(s,t)}{(p+1)!^2}h^{2(p+1)}\right\}[1+o_P(1)], \\ &Cov(\bar{r}(s),\bar{f}(t)|\mathcal{D}) = \frac{B_{p+1}(K^*)\gamma_{p+1,0}(s,t)}{(p+1)!}n^{-1}h^{p+1}[1+o_P(1)], \end{aligned}$$

where $\tilde{m} = (n^{-1}\sum_{i=1}^n n_i^{-1})^{-1}$ as defined in Theorem 3.2.

Proof of Theorem 3.1 For each $i = 1, 2, \cdots, n$, by (3.51) and (3.52), we have

$$\begin{aligned} &\mathrm{E}\left\{[\hat{f}_i(t) - f_i(t)]^2|\mathcal{D}\right\} = \mathrm{E}\left\{r_i^2(t)|\mathcal{D}\right\} \\ &= \mathrm{Var}\Big(r_i(t)|\mathcal{D}\Big) + \Big\{\mathrm{E}(r_i(t)|\mathcal{D})\Big\}^2 \\ &= \left\{\frac{V(K^*)\sigma^2(t)}{\pi(t)}(n_ih)^{-1}\right. \\ &\quad \left. + \frac{B_{p+1}^2(K^*)[(\eta^{(p+1)}(t))^2 + \gamma_{p+1,p+1}(t,t)]}{(p+1)!^2}h^{2(p+1)}\right\}[1+o_P(1)]. \end{aligned} \tag{3.53}$$

Theorem 3.1 then follows directly.

Proof of Theorem 3.2 Under Condition A, the coefficients of $h^{2(p+1)}$ and $(n_i h)^{-1}$ in expression (3.53) are uniformly bounded over the finite interval $\mathcal{T} = [a, b]$. Moreover, as $n_i \geq Cn^{\delta}$ and $h = O(n^{-\delta/(2p+3)})$, we have $O(h^{2(p+1)}) = O((n_i h)^{-1}) = O(n^{-2(p+1)\delta/(2p+3)}) = n^{-2(p+1)\delta/(2p+3)}O(1)$. Thus, by (3.51) and (3.52), we have

$$\begin{aligned} & \mathrm{E}\{r_i^2(t)|\mathcal{D}\} = \mathrm{E}^2(r_i(t)|\mathcal{D}) + \mathrm{Var}(r_i(t)|\mathcal{D}) \\ = \; & O_{UP}[h^{2(p+1)} + (n_i h)^{-1}] = n^{-2(p+1)\delta/(2p+3)}O_{UP}(1). \end{aligned}$$

Therefore, $\hat{f}_i(t) = f_i(t) + n^{-(p+1)\delta/(2p+3)}O_{UP}(1)$. Theorem 3.2 is then proved.

Proof of Theorem 3.3 First of all, notice that $\hat{\eta}(t) = n^{-1}\sum_{i=1}^n \hat{f}_i(t) = \bar{f}(t) + \bar{r}(t)$. It follows that $\mathrm{Bias}(\hat{\eta}(t)|\mathcal{D}) = \mathrm{E}(\bar{r}(t)|\mathcal{D})$, $\mathrm{Cov}(\hat{\eta}(s), \hat{\eta}(t)|\mathcal{D}) = \mathrm{Cov}(\bar{f}(s), \bar{f}(t)) + \mathrm{Cov}(\bar{f}(s), \bar{r}(t)) + \mathrm{Cov}(\bar{r}(s), \bar{f}(t)) + \mathrm{Cov}(\bar{r}(s), \bar{r}(t))$. The results of Theorem 3.3 follow directly from Lemma 3.4.

Proof of Theorem 3.4 As $\hat{\eta}(t) = \bar{f}(t) + \bar{r}(t) = \tilde{\eta}(t) + \bar{r}(t)$, in order to show the first expression in (3.34), it is sufficient to prove that $\mathrm{E}\{\bar{r}^2(t)|\mathcal{D}\} = n^{-2(p+1)\delta/(2p+3)}O_{UP}(1)$. This result follows directly from $\mathrm{E}\{\bar{r}^2(t)|\mathcal{D}\} = \{\mathrm{E}(\bar{r}(t)|\mathcal{D})\}^2 + \mathrm{Var}(\bar{r}(t)|\mathcal{D})$ and Lemma 3.4. To show the second expression in (3.34), notice that the covariance estimator $\hat{\gamma}(s,t)$ can be expressed as

$$\begin{aligned} \hat{\gamma}(s,t) &= \frac{1}{n}\sum_{i=1}^n \{f_i(s) - \bar{f}(s)\}\{f_i(t) - \bar{f}(t)\} \\ &\quad + \frac{1}{n}\sum_{i=1}^n \{f_i(s) - \bar{f}(s)\}\{r_i(t) - \bar{r}(t)\} \\ &\quad + \frac{1}{n}\sum_{i=1}^n \{r_i(s) - \bar{r}(s)\}\{f_i(t) - \bar{f}(t)\} \\ &\quad + \frac{1}{n}\sum_{i=1}^n \{r_i(s) - \bar{r}(s)\}\{r_i(t) - \bar{r}(t)\} \\ &\equiv \tilde{\gamma}(s,t) + I_1 + I_2 + I_3, \end{aligned}$$

where $r_i(t) = \hat{f}_i(t) - f_i(t), i = 1, 2, \cdots, n$ are independent, and asymptotically have the same variance. By the law of large numbers and by Lemma 3.3, we have

$$\begin{aligned} I_1 &= \mathrm{E}\left\{n^{-1}\sum_{i=1}^n E\left[(f_i(s) - \bar{f}(s))(r_i(t) - \bar{r}(t))|\mathcal{D})\right]\right\} O_P(1) \\ &= \mathrm{E}\{\mathrm{Cov}(f_1(s), r_1(t)|\mathcal{D})\}\, O_P(1) \\ &= n^{-(p+1)\delta/(2p+3)}O_{UP}(1). \end{aligned}$$

Similarly, we can show that

$$I_2 = n^{-(p+1)\delta/(2p+3)}O_{UP}(1), \quad \text{and} \quad I_3 = n^{-2(p+1)\delta/(2p+3)}O_{UP}(1).$$

The second expression in (3.34) then follows. When $\delta > 1 + 1/[2(p+1)]$, we have

$$n^{1/2}\{\tilde{\eta}(t) - \hat{\eta}(t)\} = o_{UP}(1), \quad n^{1/2}\{\tilde{\gamma}(s,t) - \hat{\gamma}(s,t)\} = o_{UP}(1).$$

By the definition of $\tilde{\eta}(t)$ and $\tilde{\gamma}(s,t)$, we have

$$\tilde{\eta}(t) = \eta(t) + \bar{v}(t), \quad \tilde{\gamma}(s,t) = n^{-1}\sum_{i=1}^{n} v_i(s)v_i(t) - \bar{v}(s)\bar{v}(t).$$

By he central limit theorem for i.i.d. stochastic processes (see Theorems 4.12 and 4.13 in Chapter 4), it is easy to show that as $n \to \infty$, we have

$$n^{1/2}\{\tilde{\eta}(t) - \eta(t)\} \xrightarrow{d} \text{GP}(0,\gamma), \quad n^{1/2}\{\tilde{\gamma}(s,t) - \gamma(s,t)\} \xrightarrow{d} \text{GP}(0,\varpi),$$

where

$$\begin{aligned}\varpi\{(s_1,t_1),(s_2,t_2)\} &= \text{Cov}\{v_1(s_1)v_1(t_1), v_1(s_2)v_1(t_2)\}\\ &= \text{E}\{v_1(s_1)v_1(t_1)v_1(s_2)v_1(t_2)\} - \gamma(s_1,t_1)\gamma(s_2,t_2).\end{aligned}$$

In particular, when $v(t)$ is a Gaussian process, we have

$$\begin{aligned}\text{E}\{v_1(s_1)v_1(t_1)v_1(s_2)v_1(t_2)\} &= \gamma(s_1,t_1)\gamma(s_2,t_2)\\ &+\gamma(s_1,t_2)\gamma(s_2,t_1) + \gamma(s_1,s_2)\gamma(t_1,t_2).\end{aligned}$$

Thus, $\varpi\{(s_1,t_1),(s_2,t_2)\} = \gamma(s_1,t_2)\gamma(s_2,t_1) + \gamma(s_1,s_2)\gamma(t_1,t_2)$. The proof of Theorem 3.4 is finished.

Proof of Theorem 3.5 Under Condition A and by Theorem 3.2, we have $\hat{f}_i(t_{ij}) = f_i(t_{ij}) + n^{-(p+1)\delta/(2p+3)}O_{UP}(1)$. It follows that

$$\begin{aligned}\hat{\epsilon}_{ij}^2 &= \left\{y_{ij} - \hat{f}_i(t_{ij})\right\}^2 = \left\{\epsilon_{ij} + n^{-(p+1)\delta/(2p+3)}O_{UP}(1)\right\}^2\\ &= \epsilon_{ij}^2 + 2n^{-(p+1)\delta/(2p+3)}\epsilon_{ij}O_{UP}(1) + n^{-2(p+1)\delta/(2p+3)}O_{UP}(1).\end{aligned}$$

Plugging this into (3.37) with $b = O(N^{-1/5})$, we have $\hat{\sigma}^2(t) = I_1 + I_2 + I_3$, where under the given conditions and by standard kernel estimation theory,

$$\begin{aligned}I_1 &= \frac{\sum_{i=1}^n\sum_{j=1}^{n_i} H_b(t_{ij}-t)\epsilon_{ij}^2}{\sum_{i=1}^n\sum_{j=1}^{n_i} H_b(t_{ij}-t)} = \sigma^2(t) + N^{-2/5}O_{UP}(1),\\ I_2 &= 2\frac{\sum_{i=1}^n\sum_{j=1}^{n_i} H_b(t_{ij}-t)n^{-(p+1)\delta/(2p+3)}\epsilon_{ij}O_{UP}(1)}{\sum_{i=1}^n\sum_{j=1}^{n_i} H_b(t_{ij}-t)}\\ &= n^{-(p+1)\delta/(2p+3)}O_{UP}(1),\\ I_3 &= \frac{\sum_{i=1}^n\sum_{j=1}^{n_i} H_b(t_{ij}-t)n^{-2(p+1)\delta/(2p+3)}O_{UP}(1)}{\sum_{i=1}^n\sum_{j=1}^{n_i} H_b(t_{ij}-t)}\\ &= n^{-2(p+1)\delta/(2p+3)}O_{UP}(1).\end{aligned}$$

Under Condition A4, $n_i \geq Cn^\delta$. This implies that $N = \sum_{i=1}^n n_i > Cn^{1+\delta}$. Thus $N^{-2/5} = O(n^{-2(1+\delta)/5})$. It follows that $\hat{\sigma}^2(t) = \sigma^2(t) + O_{UP}(n^{-2(1+\delta)/5} + n^{-(p+1)\delta/(2p+3)})$, as desired. The proof of the theorem is completed.

Proof of Theorem 3.6 The theorem follows from Theorems 4.12 and 4.13 of Chapter 4. See also Laha and Rohatgi (1979, p. 474) and van der Vaart and Wellner (1996, p. 50–51).

Proof of Theorem 3.7 Under the conditions of Theorem 3.2, we have $|r_i(t)| = |\hat{f}_i(t) - f_i(t)| \leq n^{-(p+1)\delta/(2p+3)}C$ for some $C > 0$ for all i and t. Let $\Delta(t) = [\Delta_1(t), \cdots, \Delta_q(t)]^T = (\mathbf{X}^T\mathbf{X})^{-1}\mathbf{X}^T(\hat{\mathbf{f}}(t) - \mathbf{f}(t))$. Then for $r = 1, 2, \cdots, q$, we have

$$\begin{aligned}|\Delta_r(t)| &= |n^{-1}\sum_{i=1}^n \mathbf{e}_{r,q}^T(n^{-1}\sum_{j=1}^n \mathbf{x}_j\mathbf{x}_j^T)^{-1}\mathbf{x}_i r_i(t)| \\ &\leq n^{-1}\sum_{i=1}^n |\mathbf{e}_{r,q}^T(n^{-1}\sum_{j=1}^n \mathbf{x}_j\mathbf{x}_j^T)^{-1}\mathbf{x}_i||r_i(t)| \\ &\leq Cn^{-(p+1)\delta/(2p+3)}\mathrm{E}|\mathbf{e}_{r,q}^T\Omega^{-1}\mathbf{x}_1|[1 + o_p(1)].\end{aligned}$$

It follows that $\Delta(t) = n^{-(p+1)\delta/(2p+3)}O_{UP}(1)$. The first expression in (3.43) follows directly from the fact that $\hat{\boldsymbol{\beta}}(t) - \tilde{\boldsymbol{\beta}}(t) = \Delta(t)$.

To show the second expression in (3.43), notice that $\hat{v}_i(t) = \tilde{v}_i(t) + r_i(t) + \mathbf{x}_i^T[\hat{\boldsymbol{\beta}}(t) - \tilde{\boldsymbol{\beta}}(t)] = \tilde{v}_i(t) + n^{-(p+1)\delta/(2p+3)}O_{UP}(1)$ because under the given conditions, we have $\mathbf{x}_i = O_{UP}(1)$, $r_i(t) = n^{-(p+1)\delta/(2p+3)}O_{UP}(1)$, and $\hat{\boldsymbol{\beta}}(t) - \tilde{\boldsymbol{\beta}}(t) = n^{-(p+1)\delta/(2p+3)}O_{UP}(1)$. Further, by Condition B, we have $v_i(t) = O_{UP}(1)$; therefore $\hat{v}_i(s)\hat{v}_i(t) = \tilde{v}_i(s)\tilde{v}_i(t) + n^{-(p+1)\delta/(2p+3)}O_{UP}(1)$. The second expression in (3.43) follows immediately.

When $\delta > 1 + 1/[2(p+1)]$, we have $(p+1)\delta/(2p+3) > 1/2$. Therefore, $\sqrt{n}[\hat{\boldsymbol{\beta}}(t) - \tilde{\boldsymbol{\beta}}(t)] = n^{1/2-(p+1)\delta/(2p+3)}O_{UP}(1) = o_{UP}(1)$. Moreover, it is easy to show that as $n \to \infty$, we have $\sqrt{n}[\tilde{\boldsymbol{\beta}}(t) - \boldsymbol{\beta}(t)] \xrightarrow{d} \mathrm{GP}_q(0, \gamma_\beta)$, where $\gamma_\beta(s,t) = \gamma(s,t)\boldsymbol{\Omega}^{-1}$. The result in (3.44) follows immediately. The proof of the theorem is completed.

Proof of Theorem 3.8 Let $\hat{\mathbf{w}}(t) = [\mathbf{C}(\mathbf{X}^T\mathbf{X})^{-1}\mathbf{C}^T]^{-1/2}[\mathbf{C}\hat{\boldsymbol{\beta}}(t) - \mathbf{c}(t)]$ and $\tilde{\mathbf{w}}(t)$ is similarly defined but now using $\tilde{\boldsymbol{\beta}}(t)$ instead of $\hat{\boldsymbol{\beta}}(t)$. Then $T_n = \int_a^b \|\hat{\mathbf{w}}(t)\|^2 dt$ and $\tilde{T}_n = \int_a^b \|\tilde{\mathbf{w}}(t)\|^2 dt$.

Let $\Delta(t) = \hat{\mathbf{w}}(t) - \tilde{\mathbf{w}}(t) = [\mathbf{C}(\mathbf{X}^T\mathbf{X})^{-1}\mathbf{C}^T]^{-1/2}\mathbf{C}[\hat{\boldsymbol{\beta}}(t) - \tilde{\boldsymbol{\beta}}(t)]$. Then under the given conditions and by Theorem 3.7, we can show that $\Delta(t) = n^{1/2-(p+1)\delta/(2p+3)}O_{UP}(1)$. It follows that $\hat{\mathbf{w}}(t) = \tilde{\mathbf{w}}(t) + n^{1/2-(p+1)\delta/(2p+3)}O_{UP}(1)$ and hence $T_n = \tilde{T}_n + 2\int_a^b \tilde{\mathbf{w}}(t)^T\Delta(t)dt + \int_a^b \|\Delta(t)\|^2 dt = \tilde{T}_n + n^{1/2-(p+1)\delta/(2p+3)}O_p(1)$ as desired.

When $\delta > 1 + 1/[2(p+1)]$, we have $T_n = \tilde{T}_n + o_P(1)$ as $n \to \infty$. Thus, to show (3.50), it is sufficient to show that as $n \to \infty$, we have $\tilde{T}_n \xrightarrow{d} \sum_{r=1}^m \lambda_r A_r$. Using (3.46), under the null hypothesis, it is easy to show that $n \to \infty$, we have $\tilde{\mathbf{w}}(t) \xrightarrow{d} \mathrm{GP}_k(0, \gamma_{\tilde{w}})$, where $\gamma_{\tilde{w}}(s,t) = \gamma(s,t)\mathbf{I}_k$. It follows that the k components of $\tilde{\mathbf{w}}(t)$ are independent of each other, and as $n \to \infty$, the lth component $\tilde{w}_l(t) \xrightarrow{d} \mathrm{GP}(0, \gamma)$. As $\gamma(s,t)$ has the singular value decomposition (3.48), we have $\tilde{w}_l(t) = \sum_{r=1}^m \xi_{lr}\phi_r(t)$ where as $n \to \infty$, we have

$$\xi_{lr} = \int_a^b \tilde{w}_l(t)\phi_r(t)dt \xrightarrow{d} N(0, \lambda_r),\ r = 1, 2, \cdots, m. \tag{3.54}$$

It follows that $\tilde{T}_n = \int_a^b \|\tilde{\mathbf{w}}(t)\|^2 dt = \sum_{l=1}^k \int_a^b \tilde{w}_l^2(t)dt = \sum_{l=1}^k \sum_{r=1}^m \xi_{lr}^2$ because the eigenfunctions $\phi_r(t)$ are orthonormal over $\mathcal{T} = [a, b]$. By (3.54), we have $\sum_{l=1}^k \xi_{lr}^2 \stackrel{d}{=} \lambda_r A_r$ where $A_r \sim \chi_k^2$. It follows that as $n \to \infty$, we have $\tilde{T}_n \xrightarrow{d} \sum_{r=1}^m \lambda_r A_r$ as desired. The proof of the theorem is completed.

3.6 Concluding Remarks and Bibliographical Notes

In this chapter, we studied reconstruction of functional data by the four smoothing techniques described in Chapter 2. The GCV rule of Zhang and Chen (2007) is adopted for bandwidth or smoothing parameter selection. It is noticed that by properly choosing the reconstruction setup, reconstructing a functional data set is equivalent to reconstructing the individual functions separately in a similar manner. It is shown that under some mild conditions, the effect of replacing the underlying individual functions with their LPK reconstructions is asymptotically ignorable. This result is also true for smoothing spline reconstructions of functional data as shown by Krafty (2010). We believe that this is also valid for regression spline and P-spline reconstructions of functional data. Therefore, we can use the reconstructed individual functions as "observed" continuous functions to conduct statistical inferences about functional data without loss of much information provided that the minimum number of measurements per subject tends to ∞ slightly faster than the number of subjects. This allows the "smoothing first, then estimation or inference" method adopted in this book.

Reconstruction of a functional data set is also known as "presmoothing functional data" (Martínez-Calvo 2008). As mentioned in this chapter, presmoothing functional data aims to reconstruct the individual functions, to remove as much measurement errors as possible, and to allow evaluating the individual functions at any prespecified resolution. It has wide applications and has caught more and more attentions recently. For example, Hitchcock, Casella, and Booth (2006) and Hitchcock, Booth, and Casella (2007) showed by theoretical analysis and intensive simulations that presmoothing helps to produce more accurate clustering for functional data. Most recently, Ferraty, González-Manteiga, Martnez-Calvo, and Vieu (2012) studied the properties of presmoothing in functional linear regression models with scalar responses and functional covariates. More studies in this area are expected and warranted.

3.7 Exercises

1. Assume that all the subjects in a functional data set have the same design time points:

$$t_{ij} = t_j, j = 1, 2, \cdots, m; i = 1, 2, \cdots, n. \tag{3.55}$$

   (a) Show that $\mathbf{a}_i(t) = \mathbf{a}_1(t), i = 2, \cdots, n$, where $\mathbf{a}_i(t), i = 1, 2, \cdots, n$ are defined in (3.13).
   (b) Show that $\hat{\eta}(t) = \mathbf{a}_1(t)^T \bar{\mathbf{y}}$, where $\hat{\eta}(t)$ is defined in (3.31) and $\bar{\mathbf{y}} = n^{-1} \sum_{i=1}^n \mathbf{y}_i$ with $\mathbf{y}_i = [y_{i1}, y_{i2}, \cdots, y_{im}]^T$.

2. Under the assumption (3.55) and Condition A, show that for the local linear kernel reconstructions $\hat{f}_i(t), i = 1, 2, \cdots, n$, we have

$$\text{Cov}(\hat{\eta}(s), \hat{\eta}(t)|t_1, t_2, \cdots, t_m) = \gamma(s,t)/n + o_{UP}(1/n),$$

   provided that $mh \to \infty$ and $nh^4 \to 0$ as $n \to \infty$.

3. Under the assumption (3.55), do the following problems.
   (a) Show that $\mathbf{a}_i(t) = \mathbf{a}_1(t), i = 2, \cdots, n$, where $\mathbf{a}_i(t), i = 1, 2, \cdots, n$ are defined in (3.27).
   (b) Show that $\hat{\eta}(t) = \mathbf{a}_1(t)^T \bar{\mathbf{y}}$, where $\bar{\mathbf{y}}$ is defined in Exercise 1 and $\hat{\eta}(t)$ is the sample mean function of the P-spline reconstructions $\hat{f}_i(t), i = 1, 2, \cdots, n$.
   (c) Comment if the same property holds for the regression spline reconstruction method.

4. Under the assumption (3.55), let $\hat{\mathbf{\Sigma}}_0$ and $\hat{\mathbf{\Sigma}}$ be the sample covariance matrix of $\mathbf{y}_i, i = 1, 2, \cdots, n$ and the sample covariance matrix of the cubic smoothing spline reconstructions $\hat{\mathbf{y}}_i, i = 1, 2, \cdots, n$ as defined in (3.23). Show that $\hat{\mathbf{\Sigma}} = \mathbf{C}\hat{\mathbf{\Sigma}}_0\mathbf{C}^T$ where $\mathbf{C}$ is some constant matrix, depending on the smoothing parameter λ and the roughness matrix $\mathbf{G}$ only.

5. Show that the optimal bandwidth h^* (3.30) for all the individual function reconstructions is generally larger than the optimal bandwidth for the estimator $\hat{\eta}(t)$ of the population mean function $\eta(t)$. Explain intuitively why this is reasonable.

Chapter 4

Stochastic Processes

4.1 Introduction

In the previous chapter, we studied how to reconstruct the individual functions from observed discrete functional data so that the reconstructed individual functions of a functional data set can be approximately modeled as i.i.d realizations of an underlying stochastic process. In this chapter, we introduce two important stochastic processes, namely, Gaussian process and Wishart process. These two processes play a central role in this book. Some of their important properties are investigated in Section 4.2 where we show that the squared L^2-norm of a Gaussian process is a χ^2-type mixture, and the trace of a Wishart process is also a χ^2-type mixture. In Section 4.3, methods for approximating the distribution of a χ^2-type mixture are introduced. The ratio of two independent χ^2-type mixtures is called an F-type mixture. Some methods for approximating the distribution of an F-type mixture are given in Section 4.4. These methods are useful for statistical inferences about functional data. As applications of these methods, a one-sample problem for functional data is introduced in Section 4.5 where some basic inference techniques for functional data are introduced. Some remarks and bibliographical notes are given in Section 4.7. The chapter concludes with technical proofs of some main results in Section 4.6 and some exercises in Section 4.8.

4.2 Stochastic Processes

Throughout this book, let $\mathcal{T}$ be a finite interval and we use $||f||$ to denote the L^2-norm of a function $f(t), t \in \mathcal{T}$:

$$||f|| = \left[\int_{\mathcal{T}} f^2(t)dt\right]^{1/2}.$$

If $||f|| < \infty$, we say that $f(t), t \in \mathcal{T}$ is a squared integrable function. In this case, we write $f(t) \in \mathcal{L}^2(\mathcal{T})$, where $\mathcal{L}^2(\mathcal{T})$ denotes the Hilbert space formed by all the squared integrable functions over $\mathcal{T}$ and the associated inner-product is defined as

$$< f, g >= \int_{\mathcal{T}} f(t)g(t)dt, \quad f(t), g(t) \in \mathcal{L}^2(\mathcal{T}).$$

A stochastic process or random process may be defined as a sequence of random variables or random vectors over time or space. A stochastic process over time is usually known as a time series. Gaussian processes and Wishart processes are two very useful stochastic processes that we will introduce below.

Let $y(t), t \in \mathcal{T}$ be a stochastic process having mean function $\eta(t), t \in \mathcal{T}$ and covariance function $\gamma(s,t), s, t \in \mathcal{T}$, where $\mathcal{T}$ is a compact support of t. We write $y(t) \sim \text{SP}(\eta, \gamma)$ for simplicity. When $\gamma(s,t)$ has finite trace: $\text{tr}(\gamma) = \int_{\mathcal{T}} \gamma(t,t)dt < \infty$, it has the following singular value decomposition (SVD) (Wahba 1990, p. 3):

$$\gamma(s,t) = \sum_{r=1}^{m} \lambda_r \phi_r(s)\phi_r(t), \tag{4.1}$$

where $\lambda_1, \lambda_2, \cdots, \lambda_m$ are all the decreasingly ordered positive eigenvalues of $\gamma(s,t)$, $\phi_1(t), \phi_2(t), \cdots, \phi_m(t)$ are the associated orthonormal eigenfunctions of $\gamma(s,t)$ such that

$$\int_{\mathcal{T}} \phi_r^2(t)dt = 1, \ \int_{\mathcal{T}} \phi_r(t)\phi_l(t)dt = 0, \ r \neq l,$$

and m is the smallest integer such that when $r > m$, $\lambda_r = 0$. Notice that $m = \infty$ when all the eigenvalues are positive. Then $y(t)$ has the following Karhunen-Loéve expansion:

$$y(t) = \eta(t) + \sum_{r=1}^{m} \xi_r \phi_r(t), \tag{4.2}$$

where $\xi_r = < y, \phi_r >, r = 1, 2, \cdots, m$ are uncorrelated, $\text{E}\xi_r = 0$, and $\text{E}\xi_r^2 = \lambda_r$. The SVD of a covariance function and the Karhunen-Loéve expansion of a random function play a central role in FDA.

To investigate the properties of $\gamma(s,t)$, we define the product of two bivariate functions $f_1(s,t)$ and $f_2(s,t)$ as

$$(f_1 \otimes f_2)(s,t) = \int_{\mathcal{T}} f_1(s,u) f_2(u,t)du,$$

and generally define the product of k bivariate functions $f_i(s,t), i = 1, 2, \cdots, k$ as

$$(\otimes_{i=1}^{k} f_i)(s,t) = \int_{\mathcal{T}} (\otimes_{i=1}^{k-1} f_i)(s,u) f_k(u,t)du. \tag{4.3}$$

When $f_i(s,t), i = 1, 2, \cdots, k$ are all equal to $f(s,t)$, the kth power of $f(s,t)$ can then be expressed as

$$f^{\otimes k}(s,t) = \int_{\mathcal{T}} \cdots \int_{\mathcal{T}} f(s,u_1) f(u_1,u_2) \cdots f(u_{k-1},t) du_1 \cdots du_{k-1}. \tag{4.4}$$

In particular, the product of any two bivariate functions $f(s,t)$ and $g(s,t)$ and the square of $f(s,t)$ are

$$\begin{aligned} (f \otimes g)(s,t) &= \int_{\mathcal{T}} f(s,u)g(u,t)du, \\ f^{\otimes 2}(s,t) &= \int_{\mathcal{T}} f(s,u)f(u,t)du. \end{aligned} \tag{4.5}$$

As the covariance function $\gamma(s,t)$ is symmetric with respect to s and t, we have $\text{tr}(\gamma^{\otimes 2}) = \int_{\mathcal{T}}\int_{\mathcal{T}} \gamma^2(s,t)dsdt$. By the SVD (4.1) of $\gamma(s,t)$, we can show that

$$\text{tr}(\gamma) = \sum_{r=1}^{m} \lambda_r, \ \text{tr}(\gamma^{\otimes 2}) = \sum_{r=1}^{m} \lambda_r^2. \tag{4.6}$$

The above two expressions are very useful. In fact, we have a more general result as stated below.

Theorem 4.1 *If $tr(\gamma) < \infty$, then for any finite integer $k = 1, 2, \cdots$, we have $tr(\gamma^{\otimes k}) = \sum_{r=1}^{m} \lambda_r^k$ where $\lambda_r, r = 1, 2, \cdots, m$ are all the m positive eigenvalues of $\gamma(s,t)$.*

When $y(t) \sim \text{SP}(\eta, \gamma)$ with $\eta(t) \in \mathcal{L}^2(\mathcal{T})$ and $\text{tr}(\gamma) < \infty$, it is easy to show that the squared L^2-norm of $y(t)$ is a random variable with

$$\text{E}\|y\|^2 = \|\eta\|^2 + \text{tr}(\gamma) < \infty. \tag{4.7}$$

For a general stochastic process $y(t) \sim \text{SP}(\eta, \gamma)$, it is not easy to find the variance of $\|y\|^2$.

The methodologies for estimating the eigenvalues and eigenfunctions of the underlying covariance function of a functional data set are known as principal components analysis (PCA) of functional data. Ramsay (1982) and Besse and Ramsay (1986) first addressed this problem for functional data. Further work includes Ramsay and Dalzell (1991), Rice and Silverman (1991), Silverman (1995, 1996), Huang, Shen, and Buja (2008), among others. Properties of functional PCA are established in Hall and Hosseini-Nasab (2006) and Hall, Müller, and Wang (2006). A comprehensive account of PCA for functional data can be found in Ramsay and Silverman (2005, Chapters 8–10). Because we use the eigenvalues and eigenfunctions of a covariance function indirectly throughout this book, we decided not to further discuss this topic in this book.

4.2.1 Gaussian Processes

Gaussian processes are special stochastic processes. In multivariate data analysis (MDA), Gaussian distributions play a central role. In FDA, Gaussian processes play a similar role as Gaussian distributions in MDA.

A process $y(t), t \in \mathcal{T}$ is Gaussian with mean function $\eta(t), t \in \mathcal{T}$ and covariance function $\gamma(s,t), s,t \in \mathcal{T}$, denoted as $\text{GP}(\eta, \gamma)$, if and only if for any p time points, $t_j, j = 1, 2, \cdots, p$, the random vector $[y(t_1), \cdots, y(t_p)]^T$ follows a multivariate normal distribution $N_p(\boldsymbol{\eta}, \boldsymbol{\Gamma})$, where $\boldsymbol{\eta} = [\eta(t_1), \cdots, \eta(t_p)]^T$ and $\boldsymbol{\Gamma} = (\gamma(t_i, t_j)) : p \times p$. Throughout this book, $X \stackrel{d}{=} Y$ denotes that X and Y have the same distribution, and $f(t) \equiv 0, t \in \mathcal{T}$ denotes $f(t) = 0$ for all $t \in \mathcal{T}$.

Theorem 4.2 *If $y(t) \sim GP(\eta, \gamma), t \in \mathcal{T}$ with $\eta(t) \in \mathcal{L}^2(\mathcal{T})$ and $tr(\gamma) < \infty$, then the squared L^2-norm of $y(t)$ can be expressed as*

$$\|y\|^2 \stackrel{d}{=} \sum_{r=1}^{m} \lambda_r A_r + \sum_{r=m+1}^{\infty} \delta_r^2, \ A_r \sim \chi_1^2\left(\lambda_r^{-1}\delta_r^2\right) \text{ independent}, \tag{4.8}$$

where $\lambda_r, r = 1, 2, \cdots, \infty$ are the decreasing-ordered eigenvalues of $\gamma(s,t)$, $\phi_r(t), r = 1, 2, \cdots, \infty$ are the associated eigenfunctions, $\delta_r = \int_{\mathcal{T}} \eta(t)\phi_r(t)dt, r = 1, 2, \cdots, \infty$, and m is the number of all the positive eigenvalues so that $\lambda_m > 0$ and $\lambda_r = 0, r > m$. In particular, when $\eta(t) \equiv 0$, we have $\delta_r = 0, r = 1, 2, \cdots, \infty$ so that we have

$$\|y\|^2 \stackrel{d}{=} \sum_{r=1}^{m} \lambda_r A_r, \ \ A_r \stackrel{i.i.d.}{\sim} \chi_1^2, \tag{4.9}$$

which is a central χ^2-type mixture.

Theorem 4.2 plays an import role in this book. It says that the squared L^2-norm of a Gaussian process is a χ^2-type mixture plus a constant. Therefore, its distribution can be approximated properly using the methods presented in Section 4.3. We will apply this theorem frequently in later chapters. In particular, by Theorem 4.2, when $y(t) \sim \text{GP}(\eta, \gamma)$, we have

$$\text{E}\|y\|^2 = \|\eta\|^2 + \text{tr}(\gamma), \ \text{Var}(\|y\|^2) = 2\text{tr}(\gamma^{\otimes 2}) + 4\sum_{r=1}^{m} \lambda_r \delta_r^2. \tag{4.10}$$

The following theorem is useful in Chapter 10 on tests of equality of several covariance functions. It gives a simple formula for computing the fourth moments of a Gaussian process.

Theorem 4.3 *If $y(t) \sim GP(0, \gamma), t \in \mathcal{T}$ with $tr(\gamma) < \infty$, then for any $t_1, t_2, t_3, t_4 \in \mathcal{T}$, we have $E[y(t_1)y(t_2)y(t_3)y(t_4)] = \gamma(t_1, t_2)\gamma(t_3, t_4) + \gamma(t_1, t_3)\gamma(t_2, t_4) + \gamma(t_1, t_4)\gamma(t_2, t_3)$.*

4.2.2 Wishart Processes

Wishart processes are natural generalizations of Wishart random matrices. Throughout, we use $\text{WP}(n, \gamma)$ to denote a Wishart process with n degrees of freedom and a covariance function $\gamma(s,t)$. A general Wishart process $W(s,t) \sim \text{WP}(n, \gamma)$ can be written as

$$W(s,t) = \sum_{i=1}^{n} W_i(s,t) = \sum_{i=1}^{n} v_i(s)v_i(t), \tag{4.11}$$

where $W_i(s,t) = v_i(s)v_i(t), i = 1, 2, \cdots, n, \stackrel{i.i.d.}{\sim} \text{WP}(1, \gamma)$ and $v_i(t), i = 1, 2, \cdots, n, \stackrel{i.i.d.}{\sim} \text{GP}(0, \gamma)$. By the definition, we have the following obvious but useful result.

Theorem 4.4 *Let $W_i(s,t) \sim WP(n_i, \gamma)$, $i = 1, 2, \cdots, k$. Then we have*

$$W_1(s,t) + W_2(s,t) + \cdots W_k(s,t) \sim WP(n_1 + n_2 + \cdots + n_k, \gamma).$$

The above theorem says that the sum of several independent Wishart processes with a common covariance function is also a Wishart process with the degree of freedom being the sum of the degrees of freedom of the individual Wishart processes and the covariance function being unchanged. This property is a natural extension of the property of "the sum of several independent χ^2 random variables is also a χ^2 random variable" and it is useful in analysis of variance for functional data.

Notice that when $\gamma(s,t)$ has finite trace, it has the SVD (4.1). By (4.2), the $v_i(t)$'s in (4.11) have the Karhunen-Loéve expansions:

$$v_i(t) = \sum_{r=1}^{m} \xi_{ir}\phi_r(t), i = 1, \cdots, n, \tag{4.12}$$

where $\xi_{ir} =< v_i, \phi_r >= \int_{\mathcal{T}} v_i(t)\phi_r(t)dt$ are independent, $\xi_{ir} \sim N(0, \lambda_r)$, and $\lambda_r, \phi_r(t)$ are the rth eigenvalue and eigenfunction of $\gamma(s,t)$. Using (4.12), we have the following useful theorem.

Theorem 4.5 *Assume $W(s,t) \sim WP(n, \gamma)$ with $tr(\gamma) < \infty$. Then we have*

(a) $EW(s,t) = n\gamma(s,t)$;

(b) $tr(W) \stackrel{d}{=} \sum_{r=1}^{m} \lambda_r A_r$, $A_r \stackrel{i.i.d.}{\sim} \chi_n^2$;

(c) $Etr(W) = ntr(\gamma)$, $Etr^2(W) = 2ntr(\gamma^{\otimes 2}) + n^2 tr^2(\gamma)$, & $Var(tr(W)) = 2ntr(\gamma^{\otimes 2})$;

(d) $Etr(W^{\otimes 2}) = n(n+1)tr(\gamma^{\otimes 2}) + ntr^2(\gamma)$.

Part (a) of Theorem 4.5 gives the mean function of a Wishart process. Part (b) shows that the trace of a Wishart process is a χ^2-type mixture. This result will be used frequently in this book. Part (c) gives the first, second moments and variance of the trace of a Wishart process. Direct application of Theorem 4.5 leads to the following useful result about the unbiased estimators of $\gamma(s,t)$, $\text{tr}(\gamma)$, $\text{tr}^2(\gamma)$, and $\text{tr}(\gamma^{\otimes 2})$. This result will be used frequently in later chapters.

Theorem 4.6 *Assume $W(s,t) \sim WP(n, \gamma)$ with $n > 1$ and $tr(\gamma) < \infty$. Set $\hat{\gamma}(s,t) = W(s,t)/n$. Then $\hat{\gamma}(s,t)$ and $tr(\hat{\gamma})$ are the unbiased estimators of $\gamma(s,t)$ and $tr(\gamma)$, respectively. Moreover, the unbiased estimators of $tr(\gamma^{\otimes 2})$ and $tr^2(\gamma)$, respectively, are*

$$\frac{n^2}{(n-1)(n+2)}\left[tr(\hat{\gamma}^{\otimes 2}) - \tfrac{1}{n}tr^2(\hat{\gamma})\right] \quad \text{and}$$
$$\frac{n(n+1)}{(n-1)(n+2)}\left[tr^2(\hat{\gamma}) - \tfrac{2}{n+1}tr(\hat{\gamma}^{\otimes 2})\right].$$

Remark 4.1 *It is seen from Theorem 4.6 that* $tr^2(\hat{\gamma})$ *and* $tr(\hat{\gamma}^{\otimes 2})$ *are biased estimates of* $tr^2(\gamma)$ *and* $tr(\gamma^{\otimes 2})$*, respectively, but they are asymptotically unbiased provided that* $tr(\gamma) < \infty$*. This latter condition is assumed throughout the book.*

4.2.3 Linear Forms of Stochastic Processes

Let $y_1(t), y_2(t), \cdots, y_n(t) \in \mathcal{L}^2(\mathcal{T})$ be stochastic processes. Then $\mathbf{y}(t) = [y_1(t), y_2(t), \cdots, y_n(t)]^T$ is an n-dimensional stochastic process. We shall encounter such a multidimensional stochastic process frequently in this book. When $y_i(t) \sim \text{SP}(\eta_i, \gamma_i), i = 1, 2, \cdots, n$ are independent, then throughout this book, we write

$$\mathbf{y}(t) \sim \text{SP}_n(\boldsymbol{\eta}, \boldsymbol{\Gamma}),$$
$$\boldsymbol{\eta}(t) = [\eta_1(t), \cdots, \eta_n(t)]^T, \;\; \boldsymbol{\Gamma}(s,t) = \text{diag}(\gamma_1(s,t), \gamma_2(s,t), \cdots, \gamma_n(s,t)).$$

In particular, when $y_1(t), \cdots, y_n(t) \stackrel{i.i.d.}{\sim} \text{SP}(\eta, \gamma)$, we have

$$\mathbf{y}(t) \sim \text{SP}_n(\boldsymbol{\eta}, \boldsymbol{\Gamma}), \;\; \boldsymbol{\eta}(t) = \eta(t)\mathbf{1}_n, \;\; \boldsymbol{\Gamma}(s,t) = \gamma(s,t)\mathbf{I}_n,$$

where again $\mathbf{1}_n$ is the n-dimensional vector of ones and $\mathbf{I}_n$ is the identity matrix of size $n \times n$. In this case, for simplicity, we can write $\mathbf{y}(t) \sim \text{SP}_n(\eta\mathbf{1}_n, \gamma\mathbf{I}_n)$. Similarly, throughout this book, we write $\text{GP}_n(\boldsymbol{\eta}, \boldsymbol{\Gamma})$ for an n-dimensional Gaussian process.

Let $\mathbf{C} : q \times n$ be a constant matrix. Then $\mathbf{C}\mathbf{y}(t)$ is called a linear form of $\mathbf{y}(t)$. It is easy to show that

- When $\mathbf{y}(t) \sim \text{SP}_n(\boldsymbol{\eta}, \boldsymbol{\Gamma})$, we have $\mathbf{C}\mathbf{y}(t) \sim \text{SP}_q(\mathbf{C}\boldsymbol{\eta}, \mathbf{C}\boldsymbol{\Gamma}\mathbf{C}^T)$.
- When $\mathbf{y}(t) \sim \text{GP}_n(\boldsymbol{\eta}, \boldsymbol{\Gamma})$, we have $\mathbf{C}\mathbf{y}(t) \sim \text{GP}_q(\mathbf{C}\boldsymbol{\eta}, \mathbf{C}\boldsymbol{\Gamma}\mathbf{C}^T)$.

In particular, when $\boldsymbol{\Gamma}(s,t) = \gamma(s,t)\mathbf{I}_n$, we have $\mathbf{C}\boldsymbol{\Gamma}(s,t)\mathbf{C}^T = \gamma(s,t)\mathbf{C}\mathbf{C}^T$. Furthermore, let $c(t), t \in \mathcal{T}$ be a squared integrable function. Let $y(t) \sim \text{SP}(\eta, \gamma)$ with $\eta(t) \in \mathcal{L}^2(\mathcal{T})$ and $\text{tr}(\gamma) < \infty$. Set $\xi = < c, y > = \int_{\mathcal{T}} c(t)y(t)dt$. Then

$$\mu_\xi = \text{E}\xi = \int_{\mathcal{T}} c(t)\eta(t)dt, \;\; \sigma_\xi^2 = \text{Var}(\xi) = \int_{\mathcal{T}}\int_{\mathcal{T}} c(s)\gamma(s,t)c(t)dsdt.$$

Moreover, when $\mathbf{y}(t) \sim \text{GP}_n(\boldsymbol{\eta}, \gamma\mathbf{I}_n)$, we have $\xi \sim N(\mu_\xi, \sigma_\xi^2)$.

4.2.4 Quadratic Forms of Stochastic Processes

Let $\mathbf{y}(t)$ be an n-dimensional stochastic process and $\mathbf{A}$ be a symmetrical real matrix of size $n \times n$. We define the following quadratic form of $\mathbf{y}(t)$ as

$$q(s,t) = \mathbf{y}(s)^T\mathbf{A}\mathbf{y}(t), \;\; s, t \in \mathcal{T}.$$

Notice that $\mathbf{A}$ has the following SVD:

$$\mathbf{A} = \mathbf{U}\mathbf{D}\mathbf{U}^T, \tag{4.13}$$

where $\mathbf{U} = [\mathbf{u}_1, \mathbf{u}_2, \cdots, \mathbf{u}_n]$ is an orthonormal matrix with columns being the eigenvectors of $\mathbf{A}$ and $\mathbf{D} = \text{diag}(d_1, d_2, \cdots, d_n)$ is a diagonal matrix with diagonal entries being the eigenvalues of $\mathbf{A}$. Then when $\mathbf{y}(t) \sim \text{SP}_n(\boldsymbol{\eta}, \gamma\mathbf{I}_n)$, we have

$$q(s,t) \stackrel{d}{=} \mathbf{z}(s)^T\mathbf{D}\mathbf{z}(t) = \sum_{i=1}^n d_i z_i(s) z_i(t), \tag{4.14}$$

where $\mathbf{z}(t) = \mathbf{U}^T\mathbf{y}(t) \sim \text{SP}_n(\boldsymbol{\eta}_z, \gamma\mathbf{I}_n)$ with $\boldsymbol{\eta}_z(t) = \mathbf{U}^T\boldsymbol{\eta}(t)$. Notice that the entries of $\mathbf{z}(t)$ are uncorrelated. In particular, when $\mathbf{y}(t)$ is a multidimensional Gaussian process whose components are independent, we have the following important result.

Theorem 4.7 *Assume $\mathbf{y}(t) \sim GP_n(\boldsymbol{\eta}, \gamma\mathbf{I}_n)$ and $\mathbf{A} : n \times n$ is a symmetric real matrix, having the SVD (4.13). Then the quadratic form $q(s,t) = \mathbf{y}(s)^T\mathbf{A}\mathbf{y}(t)$ has the following random expression:*

$$q(s,t) \stackrel{d}{=} \sum_{i=1}^n d_i z_i(s) z_i(t), \tag{4.15}$$

where $z_1(t), \cdots, z_n(t)$ are the entries of $\mathbf{z}(t) = \mathbf{U}^T\mathbf{y}(t) \sim GP_n(\mathbf{U}^T\boldsymbol{\eta}, \gamma\mathbf{I}_n)$. In particular, when $\boldsymbol{\eta}(t) \equiv 0$, $q(s,t)$ is a Wishart mixture. We then have the following random expression:

$$q(s,t) \stackrel{d}{=} \sum_{i=1}^n d_i W_i(s,t), \tag{4.16}$$

where $W_i(s,t), i = 1, 2, \cdots, n \stackrel{i.i.d.}{\sim} WP(1, \gamma)$.

The above theorem says that "a quadratic form of a multidimensional Gaussian process is a mixture of a few Wishart processes." An application of Theorem 4.5 leads to the fact that the trace of a quadratic form of a multidimensional Gaussian process is a χ^2-type mixture. This result will be implicitly used frequently in this book.

When $\mathbf{A}$ is an idempotent matrix of rank k such that the SVD (4.13) can be further written as

$$\mathbf{A} = \mathbf{U}\begin{pmatrix} \mathbf{I}_k & \mathbf{0} \\ \mathbf{0} & \mathbf{0} \end{pmatrix}\mathbf{U}^T, \tag{4.17}$$

we have the following useful result.

Theorem 4.8 *Assume $\mathbf{y}(t) \sim GP_n(\boldsymbol{\eta}, \gamma\mathbf{I}_n)$ and $\mathbf{A} : n \times n$ is an idempotent matrix of rank k, having the SVD (4.17). Then the quadratic form $q(s,t) = \mathbf{y}(s)^T\mathbf{A}\mathbf{y}(t)$ has the following random expression:*

$$q(s,t) \stackrel{d}{=} \sum_{i=1}^k z_i(s) z_i(t), \tag{4.18}$$

where $z_1(t), \cdots, z_n(t)$ *are the entries of* $\mathbf{z}(t) = \mathbf{U}^T\mathbf{y}(t) \sim GP_n(\mathbf{U}^T\boldsymbol{\eta}, \gamma\mathbf{I}_n)$*. In particular, when* $\boldsymbol{\eta}(t) \equiv 0$,

$$q(s,t) \sim WP(k, \gamma). \tag{4.19}$$

The above theorem says that "a quadratic form of a multidimensional Gaussian process with an idempotent matrix is a Wishart process." This property may be regarded as an extension of the property that "a quadratic form of a multidimensional normal random variable with an idempotent matrix is a χ^2 random variable." The following theorem states the sufficient and necessary conditions when a few quadratic forms of a multidimensional Gaussian process are independent. This theorem can be regarded as an extension of the well-known Cochran's theorem for quadratic forms in multivariate data analysis.

Theorem 4.9 *Assume* $\mathbf{y}(t) \sim GP_n(\boldsymbol{\eta}, \gamma\mathbf{I}_n)$ *and* $\mathbf{A}_i : n \times n, i = 1, 2, \cdots, k$ *are symmetric real matrices. Then the quadratic forms* $q_i(s,t) = \mathbf{y}(s)^T\mathbf{A}_i\mathbf{y}(t), i = 1, 2, \cdots, k$ *are independent of each other if and only if* $\mathbf{A}_i\mathbf{A}_j = \mathbf{0}, 1 \le i < j \le k$.

The following theorem gives the random expression of the trace of a quadratic form of a multidimensional Gaussian process whose components have a common covariance function. This theorem plays a central role in this book for studying various tests for functional data analysis. It will be used frequently in later chapters.

Theorem 4.10 *Let* $\mathbf{y}(t), t \in \mathcal{T} \sim GP_n(\boldsymbol{\eta}, \gamma\mathbf{I}_n)$ *with* $\boldsymbol{\eta}(t) = [\eta_1(t), \cdots, \eta_n(t)]^T$ *and* $\eta_1(t), \cdots, \eta_n(t) \in \mathcal{L}^2(\mathcal{T})$ *and* $tr(\gamma) < \infty$*. Set* $T = tr(q)$*, where* $q(s,t) = \mathbf{y}(s)^T\mathbf{A}\mathbf{y}(t)$ *is a quadratic form of* $\mathbf{y}(t)$ *and* $\mathbf{A}$ *is an idempotent matrix of rank* k*, having the SVD (4.17). Then we have*

$$\begin{array}{rcl} T & \stackrel{d}{=} & \int_{\mathcal{T}} \|\mathbf{w}(t)\|^2 dt = \sum_{i=1}^k \int_{\mathcal{T}} w_i^2(t) dt \\ & \stackrel{d}{=} & \sum_{r=1}^m \lambda_r A_r + \sum_{r=m+1}^\infty \delta_r^2, \\ & & A_r \sim \chi_k^2\left(\lambda_r^{-1}\delta_r^2\right) \ \ independent, \end{array} \tag{4.20}$$

where $\mathbf{w}(t) = [w_1(t), \cdots, w_k(t)]^T \sim GP_k(\boldsymbol{\eta}_w, \gamma\mathbf{I}_k)$ *with* $\boldsymbol{\eta}_w(t) = [\mathbf{I}_k, \mathbf{0}]\mathbf{U}^T\boldsymbol{\eta}(t)$*,* $\lambda_r, r = 1, 2, \cdots, \infty$ *are the eigenvalues of* $\gamma(s,t)$ *with only the first* m *eigenvalues being positive,* $\phi_r(t), r = 1, 2, \cdots, \infty$ *are the associated eigenfunctions, and* $\delta_r^2 = \| \int_{\mathcal{T}} \boldsymbol{\eta}_w(t)\phi_r(t)dt\|^2, r = 1, 2, \cdots, \infty$*. In particular, when* $\boldsymbol{\eta}(t) \equiv 0$*, we have* $\delta_r^2 = 0, r = 1, 2, \cdots, \infty$ *so that we have*

$$T \stackrel{d}{=} \sum_{r=1}^m \lambda_r A_r, \ \ A_r \stackrel{i.i.d.}{\sim} \chi_k^2. \tag{4.21}$$

When the components in a multidimensional Gaussian process do not have the same covariance function, the above theorem is no longer valid. However, we still can derive the mean and variance of the trace of a quadratic form of a multidimensional Gaussian process whose components have different covariance functions, as stated in the following theorem. This theorem is useful in Chapter 9 when we study a k-sample Behrens-Fisher problem for functional data.

Theorem 4.11 *Let $y_i(t), i = 1, 2, \cdots, n$ be independent Gaussian processes with $y_i(t) \sim GP(0, \gamma_i)$ and $tr(\gamma_i) < \infty, i = 1, 2, \cdots, n$. Let $\mathbf{A} = (a_{ij}) : n \times n$ be a symmetric matrix and $q(s,t) = \mathbf{y}(s)^T\mathbf{A}\mathbf{y}(t)$ be a quadratic form of $\mathbf{y}(t) = [y_1(t), \cdots, y_n(t)]^T$. Set $T = tr(q)$. Then*

$$E(T) = \sum_{i=1}^{n} a_{ii} tr(\gamma_i), \text{ and } Var(T) = 2\sum_{i=1}^{n}\sum_{j=1}^{n} a_{ij}^2 tr(\gamma_i \otimes \gamma_j),$$

where $(\gamma_i \otimes \gamma_j)(s,t) = \int_{\mathcal{T}} \gamma_i(s,u)\gamma_j(u,t)du$.

By the above theorem, when $\gamma_1(s,t) \equiv \gamma_2(s,t) \equiv \cdots \equiv \gamma_n(s,t) \equiv \gamma(s,t)$, we have $\text{E}(T) = \text{tr}(\mathbf{A})\text{tr}(\gamma)$ and $\text{Var}(T) = 2\text{tr}(\mathbf{A}^2)\text{tr}(\gamma^{\otimes 2})$.

4.2.5 Central Limit Theorems for Stochastic Processes

Let $\mathcal{R}^p$ denote the p-dimensional Euclidean space and let $\mathcal{T} \subset \mathcal{R}^p$ be a finite compact interval. Let $y_i(\mathbf{t}), i = 1, 2, \cdots, n$ denote n realizations of a stochastic process $y(\mathbf{t})$ defined on $\mathcal{T}$.

Theorem 4.12 *If $y_1(\mathbf{t}), y_2(\mathbf{t}), \cdots, y_n(\mathbf{t}) \overset{i.i.d.}{\sim} SP(\eta, \gamma)$ such that $E\|y_1\|^2 < \infty$ where $\mathbf{t} \in \mathcal{T}$, then as $n \to \infty$, we have*

$$\sqrt{n}[\bar{y}(\mathbf{t}) - \eta(\mathbf{t})] \xrightarrow{d} GP(0, \gamma),$$

where $\bar{y}(\mathbf{t}) = n^{-1}\sum_{i=1}^{n} y_i(\mathbf{t})$ is the usual sample mean function of $y_i(\mathbf{t}), i = 1, 2, \cdots, n$.

Theorem 4.12 is the central limit theorem for stochastic processes (random elements taking values in a Hilbert space); see Laha and Rohatgi (1979, p. 474). It is also a direct result of a theorem given by van der Vaart and Wellner (1996, p. 50–51). Notice that as $\text{E}\|y_1\|^2 = \|\eta\|^2 + \text{tr}(\gamma)$, the condition $\text{E}\|y_1\|^2 < \infty$ if and only if $\eta(\mathbf{t}) \in \mathcal{L}^2(\mathcal{T})$ and $\text{tr}(\gamma) < \infty$. A direct application of Theorem 4.12 leads to the following result.

Theorem 4.13 *Suppose that $y_1(t), y_2(t), \cdots, y_n(t) \overset{i.i.d.}{\sim} SP(0, \gamma)$ such that $E\|y_1\|^4 < \infty$. Set $\bar{q}(s,t) = n^{-1}\sum_{i=1}^{n} q_i(s,t)$, where $q_i(s,t) = y_i(s)y_i(t), s, t \in \mathcal{T}, i = 1, 2, \cdots, n$. Then as $n \to \infty$, we have*

$$\sqrt{n}[\bar{q}(s,t) - \gamma(s,t)] \xrightarrow{d} GP(0, \varpi),$$

where

$$\begin{aligned}\varpi\left[(s_1,t_1),(s_2,t_2)\right] &= Cov(q_1(s_1,t_1),q_1(s_2,t_2))\\ &= Ey_1(s_1)y_1(t_1)y_1(s_2)y_1(t_2)-\gamma(s_1,t_1)\gamma(s_2,t_2).\end{aligned}$$

Notice that the above theorem is the central limit theorem of the quadratic forms $q_i(s,t) = y_i(s)y_i(t), i = 1,2,\cdots,n$. Thus, it requires much stronger condition $\mathrm{E}\|y_1\|^4 < \infty$ than that required by Theorem 4.12. Notice also that the above theorem is important to study the asymptotic property of sample covariance functions. It will be used frequently in later chapters.

4.3 χ^2-Type Mixtures

From the previous section, it is seen that when $y(t) \sim \mathrm{GP}(0,\gamma)$, $\|y\|^2$ is a χ^2-type mixture. In addition, when $W(s,t) \sim \mathrm{WP}(n,\gamma)$, $\mathrm{tr}(W)$ is a χ^2-type mixture. In this section, we shall study how to approximate the distribution of a χ^2-type mixture. A general χ^2-type mixture can be defined as

$$T = \sum_{r=1}^{q} c_r A_r, \quad A_r \sim \chi^2_{d_r}(u_r^2) \ \text{ independent}, \tag{4.22}$$

where $c_r, r = 1,2,\cdots,q$, are nonzero real coefficients, and $u_r^2, r = 1,2,\cdots,q$, the noncentral parameters of the χ^2-variates $A_r, r = 1,2,\cdots,q$. When all the $u_r^2 = 0$, T is called a central χ^2-type mixture; otherwise noncentral.

Example 4.1 *We here define a central χ^2-type mixture and a noncentral χ^2-type mixture that will be used to illustrate the methodologies described in this section. The central χ^2-type mixture is defined as*

$$\begin{aligned} T_a &= 1.39A_1 + 7.09A_2 + 41.6A_3 + 21.09A_4 + 4.5A_5\\ &\quad +11.41A_6 + 1.55A_7 + 10.13A_8 + 21.86A_9 + 27.99A_{10},\end{aligned} \tag{4.23}$$

where $A_r, r = 1,2,\cdots,10$ are independent and

$$A_1, A_2, A_4, A_6 \sim \chi^2_1, \ \textit{while } A_3, A_5, A_7, A_8, A_9, A_{10} \sim \chi^2_2.$$

The noncentral χ^2-type mixture with a few negative coefficients is defined as

$$\begin{aligned} T_b &= 2.73A_1 + 5.90A_2 - 7.39A_3 + 5.79A_4 - 8.85A_5\\ &\quad +4.5A_6 + 2.47A_7 - 2.73A_8 - 1.71A_9 - 19.75A_{10},\end{aligned} \tag{4.24}$$

where $A_r, r = 1,2,\cdots,10$ are independent and

$$\begin{array}{llll} A_1 \sim \chi^2_1(.62), & A_2 \sim \chi^2_2(.31), & A_3 \sim \chi^2_2(.2), & \\ A_4 \sim \chi^2_3(1.87), & A_5 \sim \chi^2_2(1.59), & A_6 \sim \chi^2_1(1.11), & \\ A_7 \sim \chi^2_1(.54), & A_8 \sim \chi^2_3(.85), & A_9 \sim \chi^2_3(1.11), & A_{10} \sim \chi^2_2(1.74). \end{array}$$

As mentioned earlier, in this section, we aim to describe various methods for approximating the distribution of a χ^2-type mixture. Note that the distribution of a χ^2-type mixture is of interest not only in the current context of analysis of variances for functional data, but also in nonparametric goodness-of-fit tests as discussed in Zhang (2005) and in the classical analysis of variance (Satterthwaite 1946) among other areas of statistics. However, except for a few special cases, the exact distribution of a χ^2-type mixture (4.22) is in general not tractable, especially when q is large, say $q > 100$.

4.3.1 Cumulants

To develop some methods for approximating the distribution of a χ^2-type mixture, we need the concept of cumulants of a random variable. Let X be a random variable and its characteristic function is denoted as $\psi_X(t)$. Assume $\log(\psi_X(t))$ has the following expansion:

$$\log(\psi_X(t)) = \sum_{l=1}^{\infty} \mathcal{K}_l(X)\frac{(it)^l}{l!}.$$

The constants $\mathcal{K}_l(X), l = 1, 2, \cdots$ are known as the cumulants of X (Muirhead 1982, p. 40). In particular, the first four cumulants are

$$\begin{array}{ll} \mathcal{K}_1(X) = \mathrm{E}(X), & \mathcal{K}_2(X) = \mathrm{Var}(X), \\ \mathcal{K}_3(X) = \mathrm{E}(X - \mathrm{E}X)^3, & \mathcal{K}_4(X) = \mathrm{E}(X - \mathrm{E}X)^4 - 3\mathrm{Var}^2(X). \end{array} \tag{4.25}$$

Using these, the skewness and kurtosis of X can be expressed as

$$sk(T) = \mathcal{K}_3(X)/\mathcal{K}_2^{3/2}(X), \; ku(T) = \mathcal{K}_4(X)/\mathcal{K}_2^2(X), \tag{4.26}$$

respectively. When the distribution of X is hard to obtain, we can approximate the distribution of X using that of another random variable R by matching the first two or three cumulants of X and R when X and R have very similar distributions in shape and when the distribution of R can be easily obtained. The approximation distribution of X is then easily obtained as

$$P(X \leq x) \approx P(R \leq x).$$

Simple algebra yields that the cumulants of the χ^2 type mixture T (4.22) are given by

$$\mathcal{K}_l(T) = 2^{l-1}(l-1)! \sum_{r=1}^{q} c_r^l (d_r + l u_r^2), \; l = 1, 2, \cdots. \tag{4.27}$$

See Zhang (2005) for details. In particular, the first 4 cumulants of T are

$$\begin{array}{rcl} \mathcal{K}_1(T) & = & \sum_{r=1}^q c_r(d_r + u_r^2), \\ \mathcal{K}_2(T) & = & 2\sum_{r=1}^q c_r^2(d_r + 2u_r^2), \\ \mathcal{K}_3(T) & = & 8\sum_{r=1}^q c_r^3(d_r + 3u_r^2), \\ \mathcal{K}_4(T) & = & 48\sum_{r=1}^q c_r^4(d_r + 4u_r^2). \end{array} \tag{4.28}$$

Example 4.2 *Using the above formulas, for the central χ^2-type mixture T_a defined in (4.23), we have*

$$\mathcal{K}_1(T_a) = 25.624, \quad \mathcal{K}_2(T_a) = 13724,$$
$$\mathcal{K}_3(T_a) = 1.7778e6, \quad \mathcal{K}_4(T_a) = 3.7983e8,$$

where and throughout, we use 1.7778e6 *and* 1.7778e − 8 *to denote* 1.7778×10^6 *and* 1.7778×10^{-8}*, respectively. Using (4.26), the associated skewness and kurtosis of T_a are $sk(T_a) = 1.1059$ and $ku(T_a) = 2.017$, indicating that T_a is not normally distributed. For the noncentral χ^2-type mixture T_b defined in (4.24), we have*

$$\mathcal{K}_1(T_b) = -79.885, \quad \mathcal{K}_2(T_b) = 6272.7,$$
$$\mathcal{K}_3(T_b) = -4.6994e5, \quad \mathcal{K}_4(T_b) = 6.9195e7.$$

The associated skewness and kurtosis of T_b are $sk(T_b) = 0.9459$ and $ku(T_b) =$ 1.7586, indicating that T_b is also not normally distributed.

4.3.2 Distribution Approximation

In general, the exact distribution of the χ^2-type mixture T (4.22) is not tractable. However, various approximation methods described below can be used to approximate the distribution of T.

Direct Simulation This is the most direct method to approximate the distribution of T by simulation. For a given large number, $N = 10,000$, say, we first generate $A_{ir} \sim \chi^2_{d_r}(u_r^2), r = 1, 2, \cdots, q$ for $i = 1, 2, \cdots, N$. We then obtain a sample of T as

$$T_i = \sum_{r=1}^{q} c_r A_{ir}, i = 1, 2, \cdots, N.$$

Based on this sample, it is easy to approximately compute the critical values of T or to approximately compute the P-value of T. To see how the distribution of T looks, one can draw the histogram of T based on the above sample. In addition, one can also estimate the probability density function (pdf) of T using a nonparametric method, for example, the kernel density estimator (KDE) described in Wand and Jones (1995).

Example 4.3 *Figure 4.1 displays the histograms of the χ^2-type mixtures T_a (upper panel) and T_b (lower panel) superimposed by their associated KDEs multiplied by proper constants. The direct simulation method with the number of replicates $N = 10,000$ is used. The required constant for superimposing a KDE on a histogram is the product of the sample size, N, and the bar width between any two neighboring bars of the histogram. It is seen that the distributions of T_a and T_b are skewed to the right and to the left, respectively, and that the KDEs of T_a and T_b work well in approximating the underlying pdfs of T_a and T_b. Therefore, we can use the KDEs of T_a and T_b as the benchmarks for comparing various methods.*

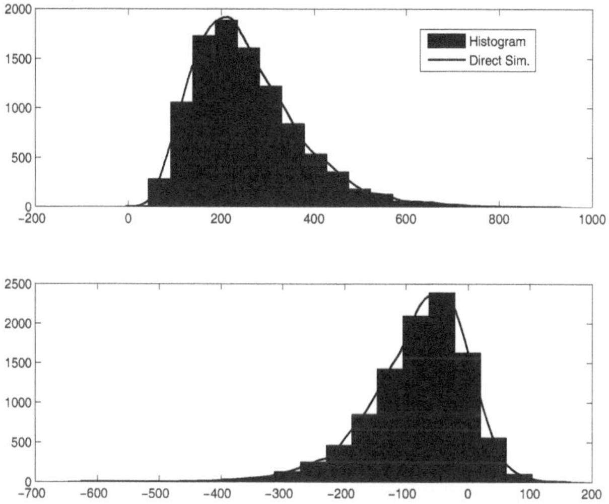

Figure 4.1 *Histograms of the central χ^2-type mixture T_a (upper panel) and the non-central χ^2-type mixture T_b (lower panel), superimposed by their associated KDEs multiplied by proper constants. The direct simulation method with the number of replicates $N = 10,000$ is used. The required constant for superimposing a KDE on a histogram is the product of the sample size, N, and the bar width between any two neighboring bars of the histogram. It appears that the KDEs of T_a and T_b work well in approximating the underlying pdfs of T_a and T_b.*

Remark 4.2 *The direct simulation method is simple to understand and easy to implement. However, like any simulation-based approaches, the direct simulation method is time consuming, especially when q is large. Moreover, its accuracy is limited by $O_p(N^{-1/2})$. This motivates the development of other methods.*

Normal Approximation *We now consider the normal approximation method, which uses the fact that under some conditions, the distribution of the χ^2-type mixture T (4.22) is asymptotically normal. In practice, when the skewness and kurtosis of T are close to 0, the distribution of T can be well approximated by that of $R \sim N(\mu, \sigma^2)$ by matching the first two cumulants or moments of T and R. The resulting parameters are*

$$\mu = \mathcal{K}_1(T),\ \sigma^2 = \mathcal{K}_2(T).$$

Remark 4.3 *As the distribution of a χ^2-type mixture is generally skewed while the normal pdfs are always symmetrical, the normal approximation method will in general not work well.*

Example 4.4 *Consider the normal approximation of the distributions of T_a*

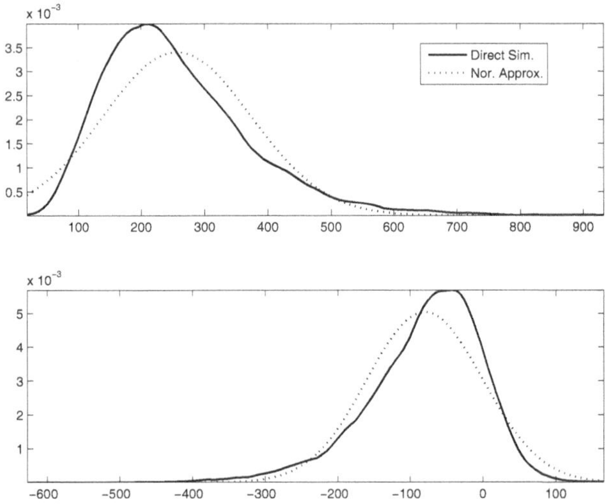

Figure 4.2 *Estimated pdfs (dashed curves) of T_a (upper panel) and T_b (lower panel) by the normal approximation method. The superimposed curves (solid) are the KDEs of T_a (upper panel) and T_b (lower panel). It appears that the normal approximation method does not work well in approximating the distributions of T_a and T_b.*

and T_b. Using the cumulants given in Example 4.2 for T_a, the values of μ and σ^2 are 25.624 *and* 13, 724, *respectively; and for T_b, they are* −79.885 *and* 6, 272.7, *respectively. Based on these values, we computed and displayed the approximate normal pdfs (dashed curves) of T_a (upper panel) and T_b (lower panel) in Figure 4.2, superimposed by their associated KDEs (solid curves). It is observed that the normal approximation method indeed does not work well in approximating the distributions of T_a and T_b.*

Two-Cumulant Matched χ^2-Approximation This method dates back to Satterthwaite (1946) and Welch (1947). It is often referred to as the Welch-Satterthwaite χ^2-approximation. As the distribution of the χ^2-type mixture T (4.22) is usually skewed, it is natural to approximate its distribution using that of a random variable of form $R \sim \beta\chi_d^2$, whose distribution is known to be skewed. From this point of view, the two-cumulant matched χ^2-approximation should be generally better than the normal approximation described above for a χ^2-type mixture with positive coefficients. The parameters β and d are determined by matching the first two-cumulants or moments of T and R. As R is a special χ^2-type mixture, by (4.27) we have $\mathcal{K}_1(R) = \beta d$ and $\mathcal{K}_2(R) = 2\beta^2 d$. Matching the first two-cumulants of T and R leads to

$$\beta = \frac{\mathcal{K}_2(T)}{2\mathcal{K}_1(T)}, \; d = \frac{2\mathcal{K}_1^2(T)}{\mathcal{K}_2(T)}. \tag{4.29}$$

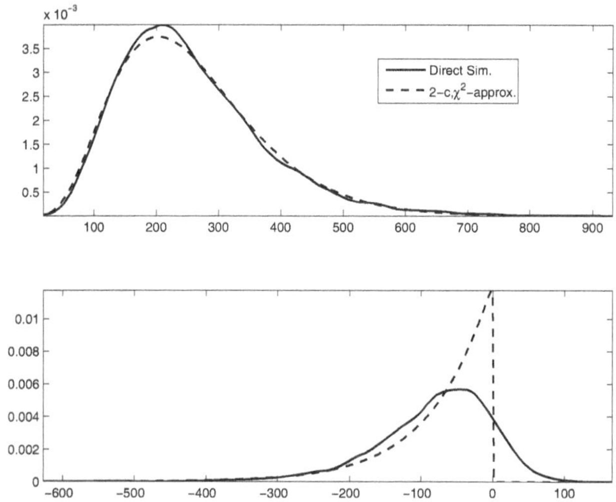

Figure 4.3 *Estimated pdfs (dashed curves) of T_a (upper panel) and T_b (lower panel) obtained by the two-cumulant matched χ^2-approximation method. The superimposed curves (solid) are the KDEs of T_a (upper panel) and T_b (lower panel). It appears that the two-cumulant matched χ^2-approximation works well for T_a but does not work for T_b.*

Remark 4.4 *When the coefficients of a χ^2-type mixture are all positive, the two-cumulant matched χ^2-approximation method will generally work well, especially when d is reasonably large. When some of the coefficients of the χ^2-type mixture T (4.22) are negative, however, it no longer works well. This is because in this case, the values of T can be positive or negative, while $R \sim \beta\chi_d^2$ can only take on positive (when $\beta > 0$) or negative (when $\beta < 0$) values.*

Example 4.5 *For the χ^2-type mixture T_a (4.23), applying the formulas in (4.29) leads to $\beta = 26.777$ and $d = 9.5693$. The associated pdf of $R \sim 26.777\chi^2_{9.5693}$ is depicted as a dashed curve in the upper panel of Figure 4.3, superimposed by the KDE of T_a (solid curve). It is seen that the two pdfs are close to each other, showing that the two-cumulant matched χ^2-approximation method indeed works well in this case. However, for the χ^2-type mixture T_b (4.24), applying (4.29) leads to $\beta = -39.261$ and $d = 2.0347$ so that $R \sim -39.261\chi^2_{2.0347}$ takes only negative values. In this case, the two-cumulant matched χ^2-approximation method does not work well, as shown in the lower panel of Figure 4.3 where the pdf of R is depicted as a dashed curve while the KDE of T_b is depicted as a solid curve.*

Three-Cumulant Matched χ^2-Approximation To overcome the above problem and to get a better approximation of the pdf of T (4.22), Zhang (2005) proposed to approximate the distribution of T by that of a random

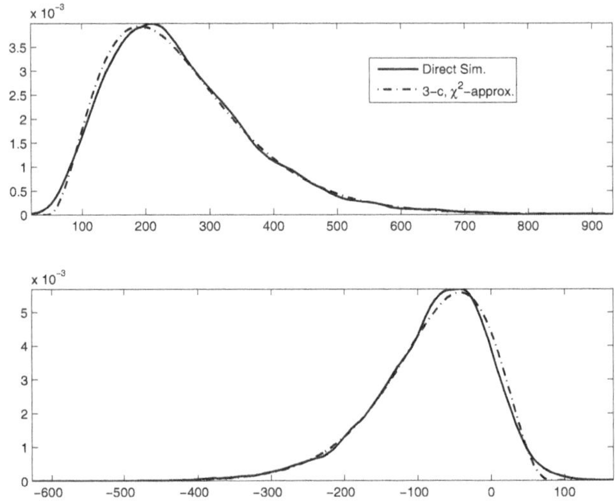

Figure 4.4 *Estimated pdfs (dashed curves) of T_a (upper panel) and T_b (lower panel) obtained by the three-cumulant matched χ^2-approximation method. The superimposed curves (solid) are the KDEs of T_a (upper panel) and T_b (lower panel). It appears that the three-cumulant matched χ^2-approximation method works for both T_a and T_b.*

variable of form $R \sim \beta\chi_d^2 + \beta_0$, by matching the first three cumulants of T and R, generalizing the work of Buckley and Eagleson (1988). It is easy to see that

$$\mathcal{K}_1(R) = \beta d + \beta_0, \quad \mathcal{K}_l(R) = 2^{l-1}(l-1)!\beta^l d, \quad l = 2, 3, \cdots. \tag{4.30}$$

Matching the first three cumulants of T and R leads to

$$\beta = \frac{\mathcal{K}_3(T)}{4\mathcal{K}_2(T)}, \ d = \frac{8\mathcal{K}_2^3(T)}{\mathcal{K}_3^2(T)}, \ \beta_0 = \mathcal{K}_1(T) - \frac{2\mathcal{K}_2^2(T)}{\mathcal{K}_3(T)}. \tag{4.31}$$

Remark 4.5 *It is easy to show that the estimated d in (4.29) [and in (4.31) as well] is positive. However, it is often not an integer, so that the χ_d^2-distribution is actually a gamma distribution with shape parameter $d/2$ and scale parameter $1/2$. Nevertheless, the distribution of χ_d^2 can be easily evaluated in popular statistical software such as MATLAB®, R, and S-PLUS, among others.*

Example 4.6 *For the χ^2-type mixture T_a, the resulting parameters are $\beta = 32.388, d = 6.5412$, and $\beta_0 = 44.386$. While for the χ^2-type mixture T_b, the resulting parameters are $\beta = -18.729, d = 8.9408$, and $\beta_0 = 87.571$. Figure 4.4 displays the estimated pdfs (dashed curves) of T_a and T_b in the upper panel and the lower panel, respectively, obtained by the three-cumulant matched χ^2-approximation method. Again, the superimposed solid curves are the associated*

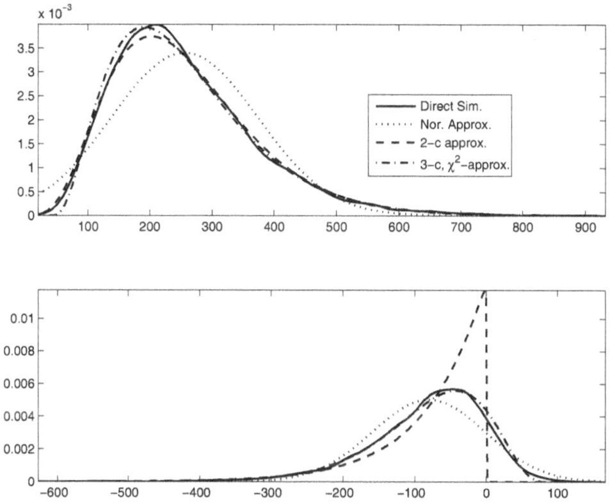

Figure 4.5 *Estimated pdfs of T_a (upper panel) and T_b (lower panel) by the direct simulation (solid), normal approximation (dot-dashed), two-cumulant matched χ^2-approximation (dashed), and three-cumulant matched χ^3-approximation methods (thick dashed). It appears that for T_a, the two-cumulant matched χ^2-approximation method slightly outperforms the three-cumulant matched χ^2-approximation method while for T_b, the two-cumulant matched χ^2-approximation method does not work but the three-cumulant matched χ^2-approximation method works very well.*

KDEs of T_a (upper panel) and T_b (lower panel). It is seen that for both cases, the three-cumulant matched χ^2-approximation method works rather well.

Example 4.7 *To give an overall comparison of the four methods, we superimpose the estimated pdfs of T_a and T_b in the upper and lower panels of Figure 4.5. It is observed that for T_a, the two-cumulant matched χ^2-approximation method (dashed curve) slightly outperforms the three-cumulant matched χ^2-approximation method (thick dashed curve) while the normal approximation method (dot-dashed curve) performs worse than the other two methods. For T_b, which has a few negative coefficients, however, the two-cumulant matched χ^2-approximation method (dashed curve) performs much worse than the normal approximation method (dot-dashed curve) while the three-cumulant matched χ^2-approximation method (thick dashed curve) performs best.*

Remark 4.6 *Some comments on the above approximation methods may be worthwhile. When q is relatively small, the direct simulation method can be used, and when q is very large, especially in simulation studies where thousands of replicates are conducted, this method can be very time consuming. When* *the χ^2-type mixture T (4.22) is approximately normally distributed, the normal approximation is advantageous as normal distributions have many good properties and are widely used. When all the coefficients $c_r, r = 1, 2, \cdots, q$, of T are nonnegative, the two-cumulant matched χ^2-approximation method is a nice choice as it can be simply conducted using the usual χ^2-distribution with acceptable accuracy. This method will be used frequently in this book due to its simplicity and accuracy. When some of the coefficients $c_r, r = 1, 2, \cdots, q$ are negative, the three-cumulant matched χ^2-approximation method is advantageous as in this case the two-cumulant matched χ^2-approximation method may not work. Zhang (2005), theoretically and by simulations, showed that the three-cumulant matched χ^2-approximation method generally outperforms the normal approximation method. Zhang (2005) also noted that the three-cumulant matched χ^2-approximation method does not perform at the left (right) boundary region as well as at the right (left) boundary region when the estimated $\beta > 0$ (< 0).*

4.4 F-Type Mixtures

In statistical inferences for functional data, one may encounter the ratio of two independent χ^2-type mixtures (Shen and Faraway 2004, Zhang 2011a), which is known as an F-type mixture. In this section we consider how to approximate the distribution of a general F-type mixture, which is defined as

$$F = \frac{\sum_{r=1}^{q_1} c_r A_r}{\sum_{s=1}^{q_2} h_s B_s}, \quad A_r \sim \chi^2_{a_r}(u_r^2), \quad B_s \sim \chi^2_{b_s}(v_s^2), \tag{4.32}$$

where A_r, B_s are all independent of each other; the coefficients h_s of the denominator are usually assumed to be nonnegative; the degrees of freedom a_r, b_s are all positive integers; and the noncentral parameters u_r^2, v_s^2 are nonnegative. In practice, the denominator's noncentral parameters $v_s, s = 1, 2, \cdots, q_2$ are often taken as 0. For simplicity, here and throughout, we denote the numerator and the denominator of F as A and B, respectively. That is,

$$A = \sum_{r=1}^{q_1} c_r A_r, \ B = \sum_{s=1}^{q_2} h_s B_s. \tag{4.33}$$

Example 4.8 *We here define two F-type mixtures. The first one is the following central F-type mixture:*

$$\begin{aligned} F_a &= [6.5A_1 + 4.81A_2 + 5.95A_3 + 9.26A_4 + 4.83A_5]/B_a, \\ B_a &= 6.05B_1 + 1.31B_2 + 9.72B_3 + 9.62B_4 + 6.97B_5 \\ &\quad +7.21B_6 + 2.21B_7 + 5.80B_8 + 4.26B_9 + 5.35B_{10}, \end{aligned} \tag{4.34}$$

where $A_r, B_s, r = 1, 2, \cdots, 5; s = 1, 2, \cdots, 10$ are independent and

$$\begin{aligned} &A_2, A_3, A_4, A_5 \sim \chi^2_1, \ A_1 \sim \chi^2_3, \\ &B_2, B_7, B_8 \sim \chi^2_1, \ B_5 \sim \chi^2_2, \\ &B_3, B_4, B_6, B_9 \sim \chi^2_3, \ B_{10} \sim \chi^2_4, \ B_1 \sim \chi^2_5. \end{aligned}$$

The second one is the following noncentral F-type mixture:

$$\begin{aligned} F_b &= [8.4A_1 + 3.93A_2 + 7.53A_3 + 10.88A_4 + 10.87A_5]/B_b, \\ B_b &= 0.53B_1 + 6.87B_2 + 2.39B_3 + 5.15B_4 + 1.63B_5 \\ &\quad +0.08B_6 + .39B_7 + 1.72B_8 + 8.97B_9 + 9.4B_{10}, \end{aligned} \tag{4.35}$$

where $A_r, B_s, r = 1, 2, \cdots, 5; s = 1, 2, \cdots, 10$ *are independent and*

$$\begin{array}{llll} A_1 \sim \chi_1^2(9.44), & A_2 \sim \chi_3^2(8.95), & A_3 \sim \chi_1^2(.03), & A_4 \sim \chi_2^2(6.68), \\ A_5 \sim \chi_2^2(9.12), & B_3 \sim \chi_1^2, & B_2, B_{10} \sim \chi_2^2, & B_8, B_9 \sim \chi_3^2, \\ B_4, B_7 \sim \chi_4^2, & B_6 \sim \chi_5^2, & B_1, B_5 \sim \chi_{10}^2. & \end{array}$$

The two F-type mixtures defined in Example 4.8 will be used to illustrate various methods described below for approximating the distribution of a general F-type mixture.

4.4.1 Distribution Approximation

In general, the exact distribution of the F-type mixture F (4.32) is not tractable. However, the following approximation methods may be used to approximate the distribution of F. They are associated with the four methods for approximating the distribution of a χ^2-type mixture presented in the previous subsection.

Direct Simulation The key idea of this method is along the same lines as those of the direct simulation method for approximating the distribution of a χ^2-type mixture. That is, for a given large number, $N = 10,000$, say, we first generate $A_{ir} \sim \chi_{a_r}^2(u_r^2), r = 1, 2, \cdots, q_1$ and $B_{is} \sim \chi_{b_s}^2(v_s^2), s = 1, 2, \cdots, q_2$ for $i = 1, 2, \cdots, N$. We then obtain a sample of F (4.32) as

$$F_i = \frac{\sum_{r=1}^{q_1} c_r A_{ir}}{\sum_{s=1}^{q_2} h_s B_{is}}, i = 1, 2, \cdots, N.$$

Based on this sample, we can approximately compute the critical values of F or the P-value of F. To see how the distribution of F looks, we can construct the histogram of F based on the above sample. Alternatively, we can estimate the pdf of F using KDE (Wand and Jones 1995).

Example 4.9 *Using the above direct simulation method with the number of replicates* $N = 10,000$, *we now approximate the distributions of the F-type mixtures* F_a *(4.34) and* F_b *(4.35) defined in Example 4.8. Figure 4.6 displays the histograms of* F_a *(upper panel) and* F_b *(lower panel) superimposed on their associated KDEs multiplied by proper constants. It is seen that the KDEs of* F_a *and* F_b *work reasonably well in approximating the underlying pdfs of* F_a *and* F_b. *Therefore, we can use the KDEs of* F_a *and* F_b *as the benchmarks for comparisons.*

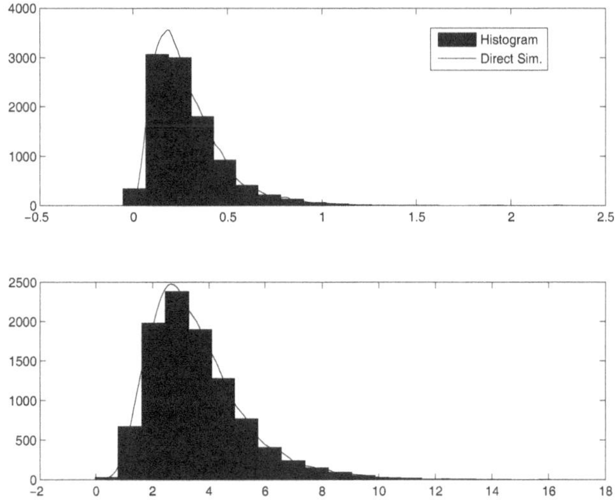

Figure 4.6 *Histograms of the central F-type mixture F_a (upper panel) and the non-central F-type mixture F_b (lower panel) superimposed by their associated KDEs multiplied by proper constants. The direct simulation method with the number of replicates $N = 10,000$ is used. It appears that the KDEs of F_a and F_b work reasonably well in approximating the underlying pdfs of F_a and F_b, respectively.*

Remark 4.7 *Although the direct simulation method is simple to implement, it is time consuming, especially when q_1, q_2, and N are large. Like all simulation-based approaches, the accuracy of the direct simulation is limited by the order $O_p(N^{-1/2})$.*

Normal Approximation The normal approximation method is applied when the numerator A and the denominator B of $F = A/B$ are asymptotically normal. In this case, F is also asymptotically normal. Then the distribution of F can be approximated by the normal distribution with mean and variance, respectively, as

$$\mu_F = \frac{\mathcal{K}_1(A)}{\mathcal{K}_1(B)}, \ \sigma_F^2 = \frac{\mathcal{K}_1^2(A)}{\mathcal{K}_1^2(B)}\left[\frac{\mathcal{K}_2(A)}{\mathcal{K}_1^2(A)} + \frac{\mathcal{K}_2(B)}{\mathcal{K}_1^2(B)}\right].$$

Example 4.10 *For the F-type mixture F_a (4.34), the associated μ_F and σ_F^2 are* 0.265 *and* 0.027. *For the F-type mixture F_b(4.35), the associated μ_F and σ_F^2 are* 3.22 *and* 2.08. *These quantities allow us to obtain the approximate pdfs of F_a and F_b by the normal approximation method. Figure 4.7 displays the estimated pdfs (dashed curves) of F_a (upper panel) and F_b (lower panel) by the normal approximation method, superimposed by their associated KDEs*

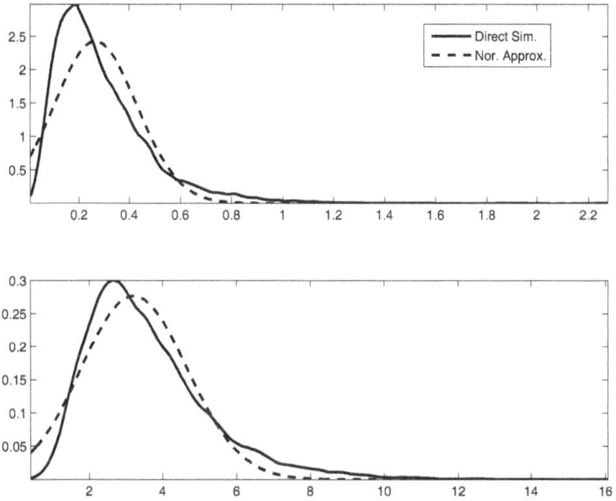

Figure 4.7 *Estimated pdfs (dashed curves) of F_a (upper panel) and F_b (lower panel) obtained by the normal approximation method. The superimposed curves (solid) are the KDEs of F_a (upper panel) and F_b (lower panel). It appears that the normal approximation method does not work very well for approximating the distributions of F_a and F_b.*

(solid curves). It is seen that the normal approximation method does not work very well for approximating the distributions of F_a and F_b.

Two-Cumulant Matched F-Approximation Note that $F = A/B$, with A and B being the χ^2-type mixtures. In the previous section we showed that when the coefficients of a χ^2-type mixture are all positive, the two-cumulant matched χ^2-approximation method works well for approximating its distribution. As the coefficients of A and B are usually assumed to be nonnegative, their distributions can be well approximated by the two-cumulant matched χ^2-approximation method. When we approximate the distributions of A and B by the two-cumulant matched χ^2-approximation method, we approximate A and B by $\beta_1\chi^2_{d_1}$ and $\beta_2\chi^2_{d_2}$, respectively. By (4.29), the associated parameters are

$$\begin{aligned} \beta_1 &= \frac{\mathcal{K}_2(A)}{2\mathcal{K}_1(A)},\ d_1 = \frac{2\mathcal{K}_1^2(A)}{\mathcal{K}_2(A)}, \\ \beta_2 &= \frac{\mathcal{K}_2(B)}{2\mathcal{K}_1(B)},\ d_2 = \frac{2\mathcal{K}_1^2(B)}{\mathcal{K}_2(B)}. \end{aligned} \tag{4.36}$$

In this way, we have

$$F \sim \frac{\beta_1 d_1}{\beta_2 d_2} F_{d_1,d_2} \text{ approximately.} \tag{4.37}$$

In particular, when A and B have the same expectation, that is, $\mathrm{E}(A) = \mathrm{E}(B)$,

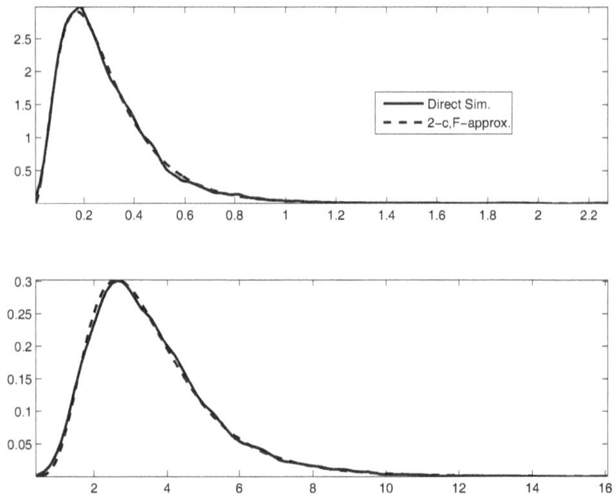

Figure 4.8 *Estimated pdfs (dashed curves) of F_a (upper panel) and F_b (lower panel) obtained by the two-cumulant matched F-approximation method. The superimposed curves (solid) are the KDEs of F_a (upper panel) and F_b (lower panel). It appears that the two-cumulant matched F-approximation method works rather well in approximating the distributions of F_a and F_b.*

we have $\beta_1 d_1 = \beta_2 d_2$. This leads to $F \sim F_{d_1,d_2}$ approximately. In this case, there is no need to compute β_1 and β_2 as in (4.36). The above method is similar to the one investigated by Cleveland and Devlin (1988) in the context of nonparametric goodness of fit tests.

Example 4.11 *We now consider applying the above method to the central F-type mixture F_a (4.34) and the noncentral F-type mixture F_b (4.35). Applying (4.36) to F_a leads to $\beta_1 = 6.637, d_1 = 6.681, \beta_2 = 7.207, d_2 = 23.215$. The estimated pdf of F_a by the two-cumulant matched F-approximation method is depicted as a dashed curve in the upper panel of Figure 4.8, superimposed by the KDE of F_a (solid curve). Similarly, applying (4.36) to F_b leads to $\beta_1 = 16.759, d_1 = 21.337, \beta_2 = 5.966, d_2 = 18.632$. The estimated pdf of F_b by the two-cumulant matched F-approximation method is depicted as a dashed curve in the lower panel of Figure 4.8, superimposed by the KDE of F_b (solid curve). It is seen that in both cases, the estimated pdfs of F_a and F_b match the associated KDEs rather well, indicating that the two-cumulant matched F-approximation method indeed works rather well.*

Three-Cumulant Matched F-Approximation The basic idea of this method is similar to that of the above two-cumulant matched F-approximation method but now we approximate A and B by the three-cumulant matched χ^2-approximation method. That is, we approximate A and

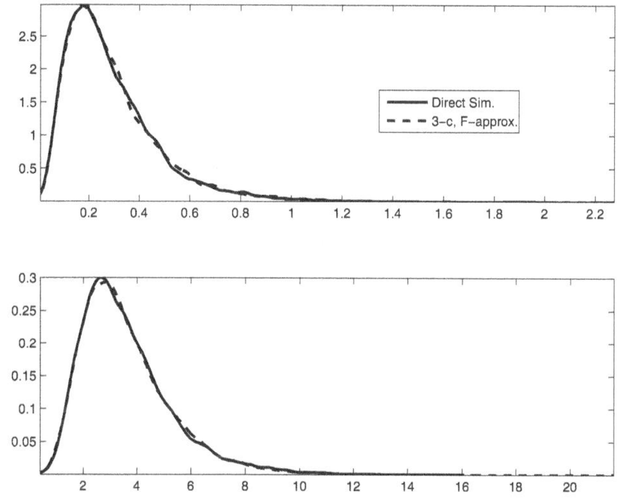

Figure 4.9 *Estimated pdfs (dashed curves) of F_a (upper panel) and F_b (lower panel) obtained by the three-cumulant matched F-approximation method. The superimposed curves (solid) are the KDEs of F_a (upper panel) and F_b (lower panel). It appears that the three-cumulant matched F-approximation method works rather well in approximating the distributions of F_a and F_b.*

B by $\beta_1\chi^2_{d_1} + \beta_{10}$ and $\beta_2\chi^2_{d_2} + \beta_{20}$ by matching the first three cumulants. By (4.31), the resulting parameters are

$$\beta_1 = \frac{\mathcal{K}_3(A)}{4\mathcal{K}_2(A)}, \quad d_1 = \frac{8\mathcal{K}_2^3(A)}{\mathcal{K}_3^2(A)}, \quad \beta_{10} = \mathcal{K}_1(A) - \frac{2\mathcal{K}_2^2(A)}{\mathcal{K}_3(A)},$$
$$\beta_2 = \frac{\mathcal{K}_3(B)}{4\mathcal{K}_2(B)}, \quad d_2 = \frac{8\mathcal{K}_2^3(B)}{\mathcal{K}_3^2(B)}, \quad \beta_{20} = \mathcal{K}_1(B) - \frac{2\mathcal{K}_2^2(B)}{\mathcal{K}_3(B)}.$$

Then we have

$$F \sim \frac{\beta_1\chi^2_{d_1} + \beta_{10}}{\beta_2\chi^2_{d_2} + \beta_{20}} \text{ approximately.} \tag{4.38}$$

The distribution of F can then be simulated.

Example 4.12 *Applying the above method to F_a leads to $\beta_1 = 6.973, d_1 = 6.053, \beta_{10} = 2.136, \beta_2 = 7.778, d_2 = 19.930$, and $\beta_{20} = 12.289$. A sample of size $N = 10,000$ was then simulated from (4.38). The associated KDE of F_a by the three-cumulant matched F-approximation method is depicted as a dashed curve in the upper panel of Figure 4.9, superimposed by the KDE (solid curve) yielded from the direct simulation method. Similarly, the estimated pdf of F_b by the three-cumulant matched F-approximation method is depicted as a dashed curve in the lower panel of Figure 4.9, superimposed by the KDE (solid curve) yielded from the direct simulation method.*

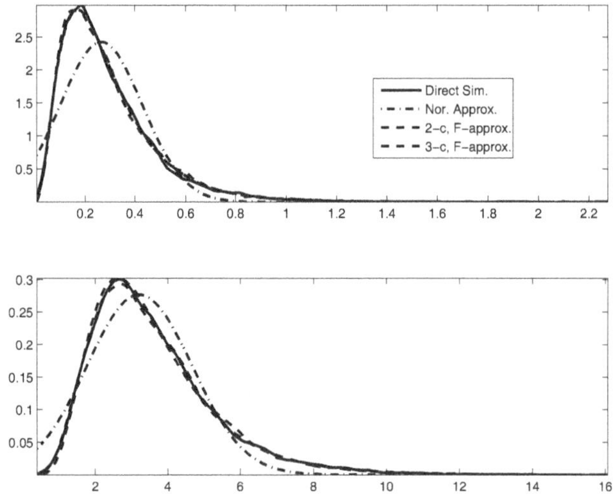

Figure 4.10 *Estimated pdfs of F_a (upper panel) and F_b (lower panel) by the direct simulation (solid), normal approximation (dot-dashed), two-cumulant matched F-approximation (dashed), and three-cumulant matched F-approximation (thick dashed) methods. It appears that the direct simulation, two-cumulant matched F-approximation, and three-cumulant matched F-approximation methods for both F_a and F_b are comparable and they outperform the normal approximation method.*

Example 4.13 *To give an overall comparison of the four methods, we superimpose the estimated pdfs of F_a and F_b in the upper and lower panels of Figure 4.10. It is observed that the direct simulation (solid), two-cumulant matched F-approximation (dashed), and three-cumulant matched F-approximation (thick dashed) methods for both F_a and F_b are comparable and perform better than the normal approximation (dotted) method. This is not a surprise from the definitions of F_a and F_b given in (4.34) and (4.35).*

Remark 4.8 *We now provide some comments on the four methods for approximating the distribution of an F-type mixture. Keep in mind that all the coefficients of the denominator B are assumed to be nonnegative. When q_1 and q_2 are relatively small, the direct simulation method may be used. This method is relatively time consuming when both q_1 and q_2 are very large. When F is asymptotically normally distributed, the normal approximation method is advantageous. When all the coefficients $c_r, r = 1, 2, \cdots, q_1$ of A are nonnegative, the two-cumulant matched F-approximation method is a nice choice as it can be simply conducted using the usual F-distribution with good accuracy and without using simulation. This method will be used frequently in this book due to its simplicity and accuracy. When some of the coefficients $c_r, r = 1, 2, \cdots, q_1$ are negative, the three-cumulant matched F-approximation method is advan-*

tageous as in this case it is more accurate than the two-cumumlant matched F-approximation method.

4.5 One-Sample Problem for Functional Data

To illustrate the methods described in the previous sections for statistical inference for functional data, in this section, we introduce a one-sample problem for functional data. Some basic inference techniques such as pointwise, L^2-norm-based, F-type, and bootstrap tests, based on the concepts of stochastic processes, χ^2-type mixtures, and F-type mixtures among others, are introduced for the one-sample problem for functional data. Almost all inference techniques for more complicated designs and functional linear models with functional responses developed in later chapters can be regarded as generalizations of these basic inference techniques.

Example 4.14 *Figure 4.11 (a) displays the reconstructed functions of the conceptive progesterone data introduced in Section 1.2.1 of Chapter 1. Their sample mean and covariance functions were computed using the formulas (4.42) given below and are displayed in panels (b) and (c), respectively. From panel (b), it is observed that before the ovulation day (Day 0), the sample mean function of the logarithm of the conceptive progesterone data is near a constant* -0.50. *But after the ovulation day, the sample mean function is no longer a constant; it increases over time. It is then of interest to test the following one-sample problem:*

$$\begin{array}{lll} & H_0 & : \eta(t) \equiv -0.50, \ t \in [a,b], \\ \text{versus} & H_1 & : \eta(t) \neq -0.50, \ \textit{for some } t \in [a,b], \end{array} \tag{4.39}$$

where $[a,b]$ *is any time period of interest. When* $[a,b] = [-8,0], [0,15]$ *and* $[-8,15]$, *we are interested in testing if the underlying mean function of the logarithm of the conceptive progesterone data is a constant* -0.50 *before the ovulation day, after the ovulation day, and over the whole observation period, respectively.*

A general one-sample problem for functional data can be described as follows. Suppose we have a functional sample

$$y_1(t), \cdots, y_n(t) \stackrel{i.i.d.}{\sim} \text{SP}(\eta, \gamma), \tag{4.40}$$

and we wish to test the following hypothesis testing problem:

$$H_0 : \eta(t) \equiv \eta_0(t), \ t \in \mathcal{T}, \quad \text{versus} \quad H_1 : \eta(t) \neq \eta_0(t), \ \text{for some } t \in \mathcal{T}, \tag{4.41}$$

where $\eta_0(t)$ is some known function that is prespecified based on related physical theories, past experiences, or past experimental results. In many situations, $\eta_0(t)$ is specified as 0 to test if the sample is purely noise or if there is some time-effect over $\mathcal{T}$.

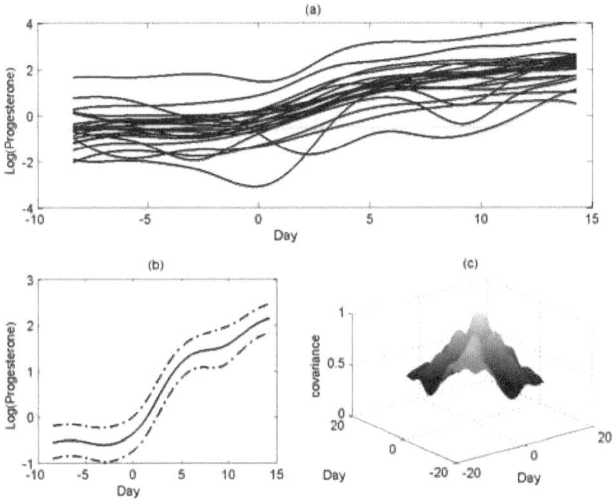

Figure 4.11 *The conceptive progesterone data example with (a) reconstructive individual curves, (b) sample mean function with its 95% pointwise confidence bands, and (c) sample covariance function.*

Based on the sample (4.40), the unbiased estimators of $\eta(t)$ and $\gamma(s,t)$, that is, the sample mean and covariance functions, are respectively

$$\begin{aligned} \hat{\eta}(t) &= \bar{y}(t) = n^{-1}\sum_{i=1}^{n} y_i(t), \\ \hat{\gamma}(s,t) &= (n-1)^{-1}\sum_{i=1}^{n}[y_i(s)-\hat{\eta}(s)][y_i(t)-\hat{\eta}(t)]. \end{aligned} \tag{4.42}$$

To describe various testing procedures for the one-sample problem (4.41), we need to investigate the properties of $\hat{\eta}(t)$ and $\hat{\gamma}(s,t)$. To this end, we list the following assumptions:

One-Sample Problem Assumptions (OS)

1. The functional sample (4.40) is with $\eta(t) \in \mathcal{L}^2(\mathcal{T})$ and $\text{tr}(\gamma) < \infty$.
2. The functional sample (4.40) is Gaussian.
3. The subject-effect function $v_1(t)$ satisfies $\text{E}\|v_1\|^4 = \text{E}\left[\int_{\mathcal{T}} v_1^2(t)dt\right]^2 < \infty$.
4. The maximum variance $\rho = \max_{t\in\mathcal{T}} \gamma(t,t) < \infty$.
5. The expectation $\text{E}[v_1^2(s)v_1^2(t)]$ is uniformly bounded. That is, for any $(s,t) \in \mathcal{T}^2$, we have $\text{E}[v_1^2(s)v_1^2(t)] < C < \infty$, where C is some constant independent of any $(s,t) \in \mathcal{T}^2$.

Assumptions OS1 and OS2 are regular. We impose Assumptions OS3 and OS4 for easily investigating the asymptotical properties of $\hat{\gamma}(s,t)$ for non-

Gaussian data. Assumptions OS2 and OS4 imply Assumption OS5. Assumption OS5 is satisfied when $v_1(t)$ is uniformly bounded in probability over $\mathcal{T}$ which is a finite interval. We first of all have the following simple result.

Theorem 4.14 *Under Assumptions OS1 and OS2, we have*

$$\sqrt{n}\,\{\hat{\eta}(t)-\eta(t)\}\sim GP(0,\gamma),\quad (n-1)\hat{\gamma}(s,t)\sim WP(n-1,\gamma).$$

The above theorem indicates that under the Gaussian assumption, the sample mean function is a Gaussian process and a sample covariance function is proportional to a Wishart process. This result will be used for constructing various tests for the one-sample problem (4.41) when the Gaussian assumption OS2 is satisfied. When the functional sample (4.40) is not Gaussian but the sample size n is large, by the central limit theorem of i.i.d. stochastic processes, Theorem 4.12, we have the following two theorems. They show that both the sample mean and covariance functions are asymptotically Gaussian. These two theorems will be used to construct various tests for the one-sample problem (4.41) when the functional data are not Gaussian but have a large sample size.

Theorem 4.15 *Under Assumption OS1, as $n\to\infty$, we have*

$$\sqrt{n}\,\{\hat{\eta}(t)-\eta(t)\}\stackrel{d}{\to} GP(0,\gamma),$$

where and throughout, "$\stackrel{d}{\to}$" denotes "convergence in distribution."

Theorem 4.16 *Under Assumptions OS1, OS3, and OS4, as $n\to\infty$, we have*

$$\sqrt{n}\,\{\hat{\gamma}(s,t)-\gamma(s,t)\}\stackrel{d}{\to} GP(0,\varpi),$$

where

$$\varpi\{(s_1,t_1),(s_2,t_2)\}=E\{v_1(s_1)v_1(t_1)v_1(s_2)v_1(t_2)\}-\gamma(s_1,t_1)\gamma(s_2,t_2).$$

To test (4.41), set $\Delta(t)=\sqrt{n}[\hat{\eta}(t)-\eta_0(t)]$ as the pivotal test function.

4.5.1 Pointwise Tests

Pointwise tests aim to test the null hypothesis in the one-sample problem (4.41) at each time point $t\in\mathcal{T}$. For any fixed t, the sub-problem is

$$H_{0t}:\eta(t)=\eta_0(t),\quad \text{versus}\quad H_{1t}:\eta(t)\neq\eta_0(t). \tag{4.43}$$

Based on the estimators (4.42), the pivotal test statistic for the above local hypothesis testing problem is

$$z(t)=\frac{\Delta(t)}{\sqrt{\hat{\gamma}(t,t)}}=\frac{\sqrt{n}[\bar{y}(t)-\eta_0(t)]}{\sqrt{\hat{\gamma}(t,t)}}. \tag{4.44}$$

When the functional sample (4.40) is Gaussian, by Theorem 4.14, for each $t \in \mathcal{T}$, we have

$$z(t) \sim t_{n-1}. \tag{4.45}$$

The pointwise t-test is conducted by rejecting H_{0t} whenever $|z(t)| > t_{n-1}(1-\alpha/2)$, where $t_{n-1}(1-\alpha/2)$ denotes the $100(1-\alpha/2)$ percentile of the t-distribution with $n-1$ degrees of freedom. Alternatively, we can construct the $100(1-\alpha)\%$ pointwise confidence bands as

$$\bar{y}(t) \pm t_{n-1}(1-\alpha/2)\sqrt{\hat{\gamma}(t,t)/n},\ t \in \mathcal{T}. \tag{4.46}$$

The pointwise t-test aims to conduct a t-test for (4.41) at each time point t based on the distribution (4.45). When the Gaussian assumption is not valid, for large samples, one can use the pointwise z-test. Notice that for any fixed $t \in \mathcal{T}$, as $n \to \infty$, by Theorem 4.15, we have

$$z(t) \xrightarrow{d} N(0,1), t \in \mathcal{T}. \tag{4.47}$$

The pointwise z-test is conducted by rejecting H_{0t} whenever $|z(t)| > z_{1-\alpha/2}$, where $z_{1-\alpha/2}$ denotes the $100(1-\alpha/2)$-percentile. Alternatively, one can construct the $100(1-\alpha)\%$ asymptotic pointwise confidence bands as

$$\bar{y}(t) \pm z_{1-\alpha/2}\sqrt{\hat{\gamma}(t,t)/n},\ t \in \mathcal{T}. \tag{4.48}$$

When the functional sample (4.40) is not Gaussian and n is small, then a nonparametric bootstrap approach can be used to bootstrap the critical values. Let

$$y_i^*(t), i = 1, 2, \cdots, n, \tag{4.49}$$

be a bootstrap sample randomly generated from (4.40). Then we can construct the sample mean and covariance functions as $\bar{y}^*(t)$ and $\hat{\gamma}^*(s,t)$, computed as in (4.42) but based on the bootstrap sample. The bootstrap test statistic $z^*(t)$ is calculated as in (4.44) but we replace $\bar{y}(t), \hat{\gamma}(t,t)$, and $\eta_0(t)$ by $\bar{y}^*(t), \hat{\gamma}^*(t,t)$, and $\bar{y}(t)$, respectively. This is because the bootstrap sample has the known mean function $\bar{y}(t)$. Repeat the above bootstrapping process a large number of times, calculate its $100(1-\alpha/2)$-percentile, and then conduct the pointwise test or construct the pointwise bootstrap confidence bands accordingly.

Remark 4.9 *The significance level α for the pointwise tests considered in this book is the usual significance level of a pointwise test restricted at a given time point $t \in \mathcal{T}$. That is, for simplicity, the associated P-values of these pointwise t- and z-tests were not corrected using some multiple comparison methods (for example, Cox and Lee 2008).*

Example 4.15 *Figure 4.12 shows the pointwise t-test, z-test, and bootstrap test for the conceptive progesterone data. Panel (a) displays the case when*

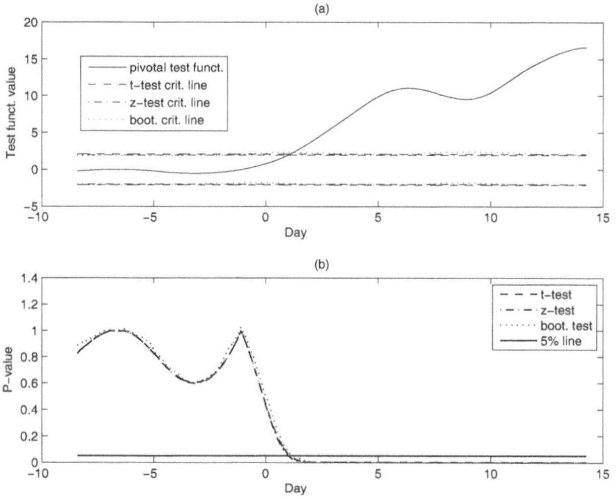

Figure 4.12 *Pointwise t-test, z-test, and bootstrap test for the conceptive progesterone data: (a) pivotal test statistic function (solid) with t-test (dashed), z-test (dot-dashed), and bootstrap test (dotted) critical lines; and (b) P-values of the pointwise t-test (dashed), z-test(dot-dashed), and bootstrap test (dotted) with the 5% significance level line (solid).*

the pivotal test function (solid) is used, together with the 5% upper and lower critical lines of the pointwise t-test (dashed), z-test (dot-dashed), and bootstrap test (dotted). It is seen that the critical lines of the three pointwise tests are comparable and the pivotal test function crosses the critical lines at some day near the ovulation day (Day 0), showing that the pointwise tests are not significant before that day but they are significant after that day. Panel (b) displays the case when the P-values of the pointwise t-test (dashed), z-test (dot-dashed), and bootstrap test (dotted) are used, together with the 5% significance level line (solid). It is seen that the P-values of the three pointwise tests are comparable and the P-values of all the three pointwise tests are larger than 5% at some day near the ovulation day and they are much less than 5% in the remaining days, showing that the pointwise tests are not significant before that day and they are significant after that day.

4.5.2 L^2-Norm-Based Test

In the previous subsection, we presented some pointwise tests. In practice, one may prefer some global tests as a global test will give one a single summary of the evidence about the null hypothesis based on the given sample. In this subsection, we study the L^2-norm-based test which is based on the squared L^2-norm of the pivotal test function $\Delta(t)$ defined in the previous subsection.

Under H_0, when the Gaussian assumption is valid, by Theorem 4.14, the pivotal test function $\Delta(t) \sim \text{GP}(0, \gamma)$ and when the Gaussian assumption is not valid but n is large, by Theorem 4.15, we have $\Delta(t) \xrightarrow{d} \text{GP}(0, \gamma)$. The L^2-norm-based test uses the squared L^2-norm of $\Delta(t)$ as its test statistic:

$$T_n = \|\Delta\|^2 = n \int_{\mathcal{T}} [\bar{y}(t) - \eta_0(t)]^2 dt. \tag{4.50}$$

It is easy to see that T_n will be small under the null hypothesis and it will be large under the alternatives. By Theorem 4.2 and under H_0,

$$T_n \stackrel{d}{=} \sum_{r=1}^{m} \lambda_r A_r, \ A_r \stackrel{i.i.d.}{\sim} \chi_1^2,$$

which is valid when the Gaussian assumption holds or is asymptotically valid when n is large.

The null distribution of T_n can then be approximated using the methods described in Section 4.3. For example, by the Welch-Satterthwaite χ^2-approximation, we have

$$T_n \sim \beta \chi_d^2 \text{ approximately, } \quad \text{where } \beta = \frac{\text{tr}(\gamma^{\otimes 2})}{\text{tr}(\gamma)}, \ d = \kappa = \frac{\text{tr}^2(\gamma)}{\text{tr}(\gamma^{\otimes 2})}. \tag{4.51}$$

Remark 4.10 *The parameters β and κ depend on the underlying covariance function $\gamma(s,t)$ only. They not only depend on the variances $\gamma(t,t), t \in \mathcal{T}$ via tr(γ), but also depend on the covariances $\gamma(s,t)$, $s \neq t$, $s,t \in \mathcal{T}$ by tr($\gamma^{\otimes 2}$). In this sense, the L^2-norm-based test and the F-type test defined in the next subsection do partially take into account the dependence of functional data. The parameter β is proportional to the scale of the functional data while the parameter κ is scale invariant. Later in this book we shall see that the approximate degrees of freedom of many test statistics for functional data are proportional to κ. That is why Shen and Faraway (2004) called κ the degrees of freedom adjustment factor.*

In practice, the parameters β and κ must be estimated based on the functional data (4.40). A natural way to do this is to replace $\gamma(s,t)$ in $\text{tr}(\gamma), \text{tr}^2(\gamma)$ and $tr(\gamma^{\otimes 2})$ with its unbiased estimator $\hat{\gamma}(s,t)$ so that

$$\hat{\beta} = \frac{\text{tr}(\hat{\gamma}^{\otimes 2})}{\text{tr}(\hat{\gamma})}, \ \hat{d} = \hat{\kappa} = \frac{\text{tr}^2(\hat{\gamma})}{\text{tr}(\hat{\gamma}^{\otimes 2})}, \tag{4.52}$$

where $\hat{\gamma}(s,t)$ is given in (4.42) based on the sample (4.40). In this case, we have

$$T_n \sim \hat{\beta} \chi_{\hat{d}}^2 \text{ approximately.} \tag{4.53}$$

The above method is usually known as the naive method (Zhang 2011a), which often works well but is biased as both $\text{tr}^2(\hat{\gamma})$ and $\text{tr}(\hat{\gamma}^{\otimes 2})$ are biased for $\text{tr}^2(\gamma)$

Table 4.1 *Testing (4.39) for the conceptive progesterone data by the L^2-norm-based test.*

Method	$[a,b]$	T_n	$\hat{\beta}$	$\hat{d}=\hat{\kappa}$	P-value
Naive	$[-8,0]$	29.96	235.26	1.13	0.769
	$[0,15]$	40,767	369.85	1.26	0
	$[-8,15]$	40,797	458.72	1.59	0
Bias-reduced	$[-8,0]$	29.96	213.38	1.15	0.763
	$[0,15]$	40,767	333.32	1.30	0
	$[-8,15]$	40,797	406.36	1.71	0

Note: The resolution number of an individual function is $M = 1,000$. The P-values by the naive method are nearly the same as those by the bias-reduced method.

and $\text{tr}(\gamma^{\otimes 2})$. By Theorem 4.6, we can obtain the unbiased estimators of $\text{tr}^2(\gamma)$ and $\text{tr}(\gamma^{\otimes 2})$, respectively, as

$$\begin{aligned}\widehat{\text{tr}^2(\gamma)} &= \frac{(n-1)n}{(n-2)(n+1)}\left[\text{tr}^2(\hat{\gamma}) - \frac{2\text{tr}(\hat{\gamma}^{\otimes 2})}{n}\right],\\ \widehat{\text{tr}(\gamma^{\otimes 2})} &= \frac{(n-1)^2}{(n-2)(n+1)}\left[\text{tr}(\hat{\gamma}^{\otimes 2}) - \frac{\text{tr}^2(\hat{\gamma})}{n-1}\right].\end{aligned} \tag{4.54}$$

Replacing $\text{tr}^2(\gamma)$ and $\text{tr}(\gamma^{\otimes 2})$ in (4.51) by their unbiased estimator defined above results in the so-called bias-reduced method for estimating the parameters β and d:

$$\hat{\beta} = \frac{\widehat{\text{tr}(\gamma^{\otimes 2})}}{\text{tr}(\hat{\gamma})}, \quad \hat{d} = \hat{\kappa} = \frac{\widehat{\text{tr}^2(\gamma)}}{\widehat{\text{tr}(\gamma^{\otimes 2})}}. \tag{4.55}$$

Throughout this book "$\xrightarrow{P}$" denotes "convergence in probability." We can show that with the sample size n growing to infinity, the estimators $\hat{\beta}$ and $\hat{\kappa}$ are consistent in the sense of the following theorem.

Theorem 4.17 *Under Assumptions OS1 and OS3 through OS5, as $n \to \infty$, we have $tr(\hat{\gamma}) \xrightarrow{P} tr(\gamma)$ and $tr(\hat{\gamma}^{\otimes 2}) \xrightarrow{P} tr(\gamma^{\otimes 2})$. Furthermore, as $n \to \infty$, we have $\hat{\beta} \xrightarrow{P} \beta$ and $\hat{\kappa} \xrightarrow{P} \kappa$, where $\hat{\beta}$ and $\hat{\kappa}$ are the naive or bias-reduced estimators of β and κ.*

The one-sample test (4.41) can then be conducted as follows. We reject the null hypothesis in (4.41) whenever $T_n > \hat{\beta}\chi^2_{\hat{d}}(1-\alpha)$ for a given significance level α where $\hat{\beta}$ and $\hat{d} = \hat{\kappa}$ are obtained with either the naive method or the bias-reduced method. Alternatively, the one-sample test (4.41) can be conducted by computing the P-value $P(\chi^2_{\hat{d}} \geq T_n/\hat{\beta})$.

Example 4.16 *Table 4.1 displays the L^2-norm-based test for the one-sample problem (4.39) with the conceptive progesterone data. The functions were discretized with the resolution number $M = 1,000$ as described in Section 4.5.5.*

That is, each function is evaluated at $1,000$ *design time points uniformly scattered over the whole data range* $[-8, 15]$, *and the quantities* $T_n, \hat{\beta}$, *and* $\hat{d}$ *are numerically computed as in Section 4.5.5 but ignoring the constant term* $v(\mathcal{T})/M$ *in both the approximate expressions of* T_n *and* $\hat{\beta}$ *for ease computation. It is seen that both the naive and bias-reduced methods produced very similar results, and the null hypothesis (4.39) is not significant before ovulation day, suggesting that the average of the logarithm of the conceptive progesterone data is very likely a constant* -0.50 *before ovulation day while after ovulation day, the test results indicate that it is unlikely to be the constant* -0.50.

4.5.3 F-Type Test

In the previous subsection we presented the L^2-norm-based test, which does not take the variation of the sample covariance function $\hat{\gamma}(s, t)$ into account but works even when the normality assumption is not valid. Under the Gaussian assumption, we can take this into account partially by the so-called F-type test defined below. The F-type test has two advantages. First of all, it is scale invariant. That is, the F-type test is invariant when the functional data are multiplied by a nonzero real number. Second, the null distribution of the F-type test can be well approximated by a usual F-distribution with degrees of freedom proportional to the degrees of freedom adjustment factor κ defined in (4.51) that depends on the covariance function only. This allows easy and fast implementation of the F-type test provided κ is properly estimated.

For the one-sample problem (4.41), recall that the L^2-norm-based test is based on $\|\Delta\|^2$ with the pivotal test function $\Delta(t) = \sqrt{n}\,[\bar{y}(t) - \eta_0(t)]$. Under H_0 and the Gaussian assumption, $\Delta(t) \sim \text{GP}(0, \gamma)$ and $(n-1)\hat{\gamma}(s, t) \sim \text{WP}(n-1, \gamma)$ are independent. In addition,

$$\mathrm{E}\|\Delta\|^2 = \mathrm{tr}(\gamma), \quad \mathrm{Etr}(\hat{\gamma}) = \mathrm{tr}(\gamma). \tag{4.56}$$

Then it is natural to test (4.41) using the following test statistic:

$$F_n = \frac{\|\Delta\|^2}{\mathrm{tr}(\hat{\gamma})} = \frac{n\int_{\mathcal{T}}[\bar{y}(t) - \eta_0(t)]^2 dt}{\mathrm{tr}(\hat{\gamma})}. \tag{4.57}$$

When the variation of $\mathrm{tr}(\hat{\gamma})$ is not taken into account, the above test statistic F_n is equivalent to the L^2-norm-based test statistic T_n defined in (4.50). Therefore, the F-type test is distinguished from the L^2-norm-based test in that the variation of $\mathrm{tr}(\hat{\gamma})$ is taken into account by a way described below. By Theorem 4.2 and under H_0,

$$\begin{aligned} \|\Delta\|^2 &\stackrel{d}{=} \sum_{r=1}^{m} \lambda_r A_r, \; A_r \stackrel{i.i.d.}{\sim} \chi_1^2, \\ \mathrm{tr}(\hat{\gamma}) &\stackrel{d}{=} \sum_{r=1}^{m} \lambda_r E_r, \; E_r \stackrel{i.i.d.}{\sim} \chi_{n-1}^2, \end{aligned}$$

Table 4.2: *Testing (4.39) for the conceptive progesterone data by the F-type test.*

Method	$[a,b]$	F_n	$\hat{\kappa}$	$(n-1)\hat{\kappa}$	P-value
Naive	$[-8,0]$	0.113	1.13	23.78	0.771
	$[0,15]$	101	1.26	26.43	$2.24e-11$
	$[-8,15]$	64.3	1.59	33.51	$3.19e-11$
Bias-reduced	$[-8,0]$	0.113	1.15	24.22	0.776
	$[0,15]$	101	1.30	27.33	$1.09e-11$
	$[-8,15]$	64.3	1.71	35.83	$7.54e-12$

Note: The resolution number of an individual function is $M = 1,000$. The P-values by the F-type test are comparable with those by the L^2-norm-based test as presented in Table 4.1.

where A_r, E_r are all independent. Therefore, under H_0, F_n is an F-type mixture. That is, the null distribution of F_n can be approximated using the methods described in Section 4.4. In particular, by (4.56) and the two-cumulant matched F-approximation approach, we have

$$F_n \sim F_{\hat{\kappa},(n-1)\hat{\kappa}} \text{ approximately,}$$

where $\hat{\kappa}$ is defined in (4.52) or (4.55) with $\hat{\gamma}(s,t)$ given in (4.42) based on the sample (4.40). The null hypothesis in (4.41) is rejected when $F_n > F_{\hat{\kappa},(n-1)\hat{\kappa}}(1-\alpha)$ for a given significance level α.

Remark 4.11 *By Theorem 4.17, when $(n-1)\hat{\kappa}$ is very large, $tr(\hat{\gamma})$ will tend to $tr(\gamma)$ in probability quickly so that the test result by the F-type test will be asymptotically the same as that by the L^2-norm-based test discussed in the previous subsection.*

Example 4.17 *Table 4.2 displays the F-type test for the one-sample problem (4.39) with the conceptive progesterone data. The P-values are about the same as those presented in Table 4.1 for the L^2-norm-based test. This is obviously due to the fact that the approximate degrees of freedom $(n-1)\hat{\kappa}$ of the denominator of F_n are rather large so that $tr(\hat{\gamma})$ converges to $tr(\gamma)$ rather fast. Again, the naive method and the bias-reduced method produced quite similar results.*

4.5.4 Bootstrap Test

When the sample (4.40) is not Gaussian and n is small, the null distributions of the L^2-norm-based test and the F-type test are generally not tractable. In these cases, we may use some nonparametric bootstrap approach to bootstrap the critical values of the L^2-norm-based test and the F-type test described in the previous two subsections, resulting in the L^2-norm-based bootstrap test and the F-type bootstrap test respectively.

Based on the bootstrap sample (4.49) randomly generated from (4.40), we can construct the sample mean and sample covariance functions as $\bar{y}^*(t)$ and

Table 4.3 *Testing (4.39) for the conceptive progesterone data by the L^2-norm-based and F-type bootstrap tests.*

	L^2-norm-based bootstrap test		F-type bootstrap test	
$[a, b]$	T_n	P-value	F_n	P-value
$[-8, 0]$	29.96	0.823	0.113	0.834
$[0, 15]$	40,767	0	101	0
$[-8, 15]$	40,797	0	64.3	0

Note: The number of bootstrap replicates is $N = 5,000$. The resolution number of an individual function is $M = 1,000$. Note that for the one-sample problem (4.39) over the time period $[-8, 0]$, the bootstrapped P-values are slightly larger than those P-values obtained using the two-cumulant matched χ^2-approximation and the two-cumulant matched F-approximation presented in Tables 4.1 and 4.2, respectively.

$\hat{\gamma}^*(s,t)$, computed as in (4.42). The L^2-norm-based bootstrap test computes the bootstrap test statistic $T_n^* = \|\Delta^*\|$ with the bootstrapped pivotal test function $\Delta^*(t) = \sqrt{n}\,[\bar{y}^*(t) - \bar{y}(t)]$. Repeat this process a large number of times so that one can obtain a bootstrap sample of T_n^* that can be used to estimate the $100(1-\alpha)$-percentile of T_n. Similarly, the F-type bootstrap test computes the bootstrap test statistic $F_n^* = \frac{\|\Delta^*\|}{\text{tr}(\hat{\gamma}^*)}$. Repeat this process a large number of times so that one can obtain a bootstrap sample of F_n^* that allows to estimate the $100(1-\alpha)$-percentile of F_n.

Remark 4.12 *The bootstrap tests are easy to understand and implement but they are time consuming. They can be used when the sample size is small.*

Example 4.18 *Table 4.3 displays two bootstrap tests for the one-sample problem (4.39) with the conceptive progesterone data. The statistics T_n and F_n of the L^2-norm-based test and the F-type test were used. The associated bootstrapped P-values are displayed. It is seen that the bootstrap test results are consistent with those from the L^2-norm-based test and the F-type test presented in Tables 4.1 and 4.2, respectively. Notice that for the one-sample problem (4.39) within the time period $[-8, 0]$, the bootstrapped P-values are slightly larger than those P-values obtained using the two-cumulant matched χ^2-approximation and the two-cumulant matched F-approximation.*

4.5.5 Numerical Implementation

In practice, we have to discretize the continuous functions $\Delta(t)$ and $\hat{\gamma}(s,t)$ in the computation of $\|\Delta\|^2, \text{tr}(\hat{\gamma})$ and $\text{tr}(\hat{\gamma}^{\otimes 2})$. Let the resolution number be M, a large number, say, $M = 1,000$, and let $t_1, t_2, \cdots, t_M$ be M resolution time points that are equally spaced in $\mathcal{T}$. Then the functional sample (4.40) is discretized accordingly as

$$\mathbf{y}_i = [y_i(t_1), y_i(t_2), \cdots, y_i(t_M)]^T, \; i = 1, 2, \cdots, n, \tag{4.58}$$

and the one-sample problem (4.41) is discretized as

$$H_0 : \boldsymbol{\eta} = \boldsymbol{\eta}_0, \quad \text{versus} \quad H_1 : \boldsymbol{\eta} \neq \boldsymbol{\eta}_0, \tag{4.59}$$

where $\boldsymbol{\eta} = [\eta(t_1), \cdots, \eta(t_M)]^T$ and $\boldsymbol{\eta}_0 = [\eta_0(t_1), \cdots, \eta_0(t_M)]^T$ denote the vectors consisting of the values of $\eta(t)$ and $\eta_0(t)$ evaluated at the resolution time points, respectively. Let

$$\boldsymbol{\Delta} = \begin{bmatrix} \Delta(t_1) \\ \Delta(t_2) \\ \vdots \\ \Delta(t_M) \end{bmatrix}, \quad \text{and } \hat{\boldsymbol{\Gamma}} = \begin{bmatrix} \hat{\gamma}(t_1, t_1) & \cdots & \hat{\gamma}(t_1, t_M) \\ \hat{\gamma}(t_2, t_1) & \cdots & \hat{\gamma}(t_2, t_M) \\ \vdots & \vdots & \vdots \\ \hat{\gamma}(t_M, t_1) & \cdots & \hat{\gamma}(t_M, t_M) \end{bmatrix}, \tag{4.60}$$

consist of the evaluated values of $\Delta(t)$ and $\hat{\gamma}(s,t)$ at the M resolution time points. Then we have

$$\begin{aligned} T_n &= \|\Delta\|^2 = \textstyle\int_{\mathcal{T}} \Delta^2(t)dt \approx \frac{v(\mathcal{T})}{M} \sum_{i=1}^M \Delta^2(t_i) \\ &= \frac{v(\mathcal{T})}{M} \|\boldsymbol{\Delta}\|^2 = \frac{v(\mathcal{T})}{M} T_n^0, \end{aligned} \tag{4.61}$$

where $v(\mathcal{T})$ denotes the volume of $\mathcal{T}$ and $T_n^0 = \|\boldsymbol{\Delta}\|^2$ denotes the usual squared L^2-norm of $\boldsymbol{\Delta}$. When $\mathcal{T} = [a, b]$, one has $v(\mathcal{T}) = b - a$. Similarly, we have

$$\text{tr}(\hat{\gamma}) = \int_{\mathcal{T}} \hat{\gamma}(t,t)dt \approx \frac{v(\mathcal{T})}{M} \sum_{i=1}^M \hat{\gamma}(t_i, t_i) = \frac{v(\mathcal{T})}{M} \text{tr}(\hat{\boldsymbol{\Gamma}}). \tag{4.62}$$

In addition, we have

$$\text{tr}(\hat{\gamma}^{\otimes 2}) = \int_{\mathcal{T}^2} \hat{\gamma}^2(s,t)dsdt \approx \frac{v(\mathcal{T}^2)}{M^2} \sum_{i=1}^M \sum_{j=1}^M \hat{\gamma}^2(t_i, t_j) = \frac{v(\mathcal{T}^2)}{M^2} \text{tr}(\hat{\boldsymbol{\Gamma}}^2).$$

When $\mathcal{T} = [a, b]$, which is often the case, one has $v(\mathcal{T}^2) = v^2(\mathcal{T}) = (b-a)^2$. In this case, the estimated parameters $\hat{\beta}$ and $\hat{\kappa}$ defined in (4.52) for the Welch-Satterthwaite χ^2-approximation can be approximately expressed as

$$\hat{\beta} \approx \frac{v(\mathcal{T})}{M} \frac{\text{tr}(\hat{\boldsymbol{\Gamma}}^2)}{\text{tr}(\hat{\boldsymbol{\Gamma}})} = \frac{v(\mathcal{T})}{M} \hat{\beta}^0, \quad \text{and } \hat{\kappa} \approx \frac{\text{tr}^2(\hat{\boldsymbol{\Gamma}})}{\text{tr}(\hat{\boldsymbol{\Gamma}}^2)} = \hat{\kappa}^0,$$

where $\hat{\beta}^0$ and $\hat{\kappa}^0$ denote the estimated β and κ when the Welch-Satterthwaite χ^2-approximation is applied to the discretized one-sample problem (4.59) based on the discretized one sample (4.58). It follows that

$$P\left(T_n \geq \hat{\beta}\chi^2_{\hat{\kappa}}\right) \approx P\left(\frac{v(\mathcal{T})}{M} T_n^0 \geq \frac{v(\mathcal{T})}{M} \hat{\beta}^0 \chi^2_{\hat{\kappa}^0}\right) = P\left(T_n^0 \geq \hat{\beta}^0 \chi^2_{\hat{\kappa}^0}\right).$$

We then have the following remark.

Remark 4.13 *When we conduct the L^2-norm-based test, the constant factor $\frac{v(\mathcal{T})}{M}$ in T_n and $\hat{\beta}$ can be omitted at the same time in computation. This will not affect the test result.*

The above remark is also true for the F-type test. In fact, by (4.61) and (4.62), the F-type test statistic (4.57) can be approximately expressed as

$$F_n = \frac{\|\Delta\|^2}{\text{tr}(\hat{\gamma})} \approx \frac{\|\boldsymbol{\Delta}\|^2}{\text{tr}(\hat{\boldsymbol{\Gamma}})} \equiv F_n^0, \tag{4.63}$$

where F_n^0 denotes the test statistic for the F-type test applied to the discretized one-sample problem (4.59) based on the discretized one sample (4.58) so that

$$P(F_n \geq F_{\hat{\kappa},(n-1)\hat{\kappa}}) \approx P(F_n^0 \geq F_{\hat{\kappa}^0,(n-1)\hat{\kappa}^0}).$$

In FDA, the resolution M should be large enough to represent an individual function well. We feel that $M = 1,000$ is often sufficient to this end as indicated by the examples presented in next subsection.

Remark 4.14 *From the above, it is seen that in practice, we have to discretize the functional data so that the various tests described in this section are actually applied to the discretized sample (4.58) by computing $T_n^0, F_n^0, tr(\hat{\boldsymbol{\Gamma}}), tr(\hat{\boldsymbol{\Gamma}}^2), \hat{\beta}^0, \hat{\kappa}^0$, etc. Therefore, when the functional data are very noisy and cannot be reconstructed by the methods proposed in Chapter 3 but they can be observed simultaneously over a common grid of time points so that a sample of vectors like (4.58) can be obtained, then the various tests described in this section can be applied directly to the observed functional data. This is true for all the methodologies investigated in this book.*

Remark 4.15 *When M is large and n is relatively small, we can simplify the computation using the following technique. Note that*

$$\hat{\gamma}(s,t) = (n-1)^{-1} \sum_{i=1}^{n} \hat{v}_i(s)\hat{v}_i(t) = (n-1)^{-1}\hat{\mathbf{v}}(s)^T\hat{\mathbf{v}}(t),$$

where $\hat{\mathbf{v}}(t) = [\hat{v}_1(t), \cdots, \hat{v}_n(t)]^T$ with $\hat{v}_i(t) = y_i(t) - \bar{y}(t), i = 1, 2, \cdots, n$ being the estimated subject-effect functions. It follows that

$$\hat{\boldsymbol{\Gamma}} = (n-1)^{-1}\hat{\mathbf{V}}\hat{\mathbf{V}}^T,$$

where $\hat{\mathbf{V}} = [\hat{\mathbf{v}}(t_1), \hat{\mathbf{v}}(t_2), \cdots, \hat{\mathbf{v}}(t_M)]^T : M \times n$. Thus

$$\begin{aligned} tr(\hat{\boldsymbol{\Gamma}}) &= (n-1)^{-1} tr\left(\hat{\mathbf{V}}\hat{\mathbf{V}}^T\right) = (n-1)^{-1} tr(\mathbf{S}), \\ tr(\hat{\boldsymbol{\Gamma}}^2) &= (n-1)^{-2} tr\left[(\hat{\mathbf{V}}\hat{\mathbf{V}}^T)^2\right] = (n-1)^{-2} tr(\mathbf{S}^2), \end{aligned}$$

where $\mathbf{S} = \hat{\mathbf{V}}^T\hat{\mathbf{V}}$ is an $n \times n$ matrix so that the needed operations for computing $tr(\hat{\boldsymbol{\Gamma}})$ and $tr(\hat{\boldsymbol{\Gamma}}^2)$ are $O(n)$ and $O(n^2)$ instead of $O(M)$ and $O(M^2)$, respectively. This saves a lot of computation.

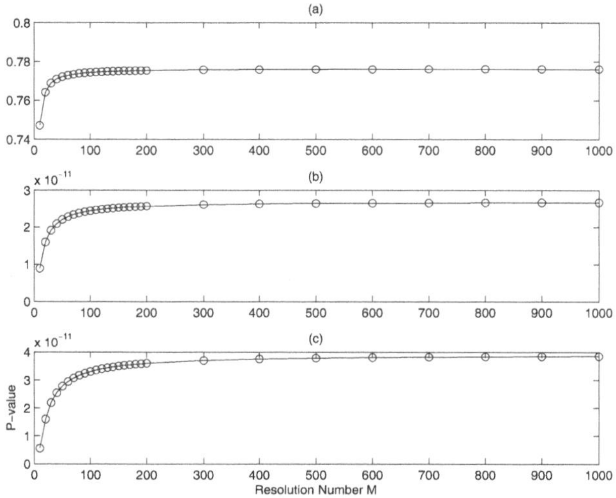

Figure 4.13 *Plots of P-value of the F-type test for testing (4.39) with the conceptive progesterone data against resolution number M for (a) $[a, b] = [-8, 0]$, (b) $[a, b] = 0, 15]$, and (c) $[a, b] = [-8, 15]$. It appears that for these three examples, with increasing M, the P-values are increasing up to their limit values. For small M, the associated P-values are far away from their limit values while for $M = 1,000$, the P-values are very close to their limit values.*

4.5.6 *Effect of Resolution Number*

In this subsection, we investigate the effect of the resolution number M by examples. The basic conclusion is that the resolution number M should not be too small; otherwise, the vector obtained by discretizing an individual function cannot represent the function properly, possibly resulting in misleading results. Figure 4.13 shows such examples where the P-values of the F-type test against $M = 10, 20, 30, \cdots, 180, 190, 200, 300, 400, \cdots, 900, 1,000$ for testing (4.39) with the conceptive progesterone data are depicted. For these three examples, with increasing M, the P-values are increasing up to their limit values. For other cases, the P-values may be decreasing down to their limit values. Note that for small M, the associated P-values are far away from their limit values. According to our experience, we feel that when $M = 1,000$, the P-values will be very close to their true values. That is why we use $M = 1,000$ throughout this book.

4.6 Technical Proofs

In this section, we outline the proofs of some main results described in the previous sections.

Proof of Theorem 4.1 Without loss of generality, we show the case for $k=3$. For other values of k, the proofs are along the same lines. By (4.4) and (4.1), when m is a finite fixed number, we have

$$
\begin{aligned}
\operatorname{tr}(\gamma^{\otimes 3}) &= \int_{\mathcal{T}^3} \gamma(u_3,u_1)\gamma(u_1,u_2)\gamma(u_2,u_3)du_1du_2du_3 \\
&= \int_{\mathcal{T}^3} \left\{\sum_{r_1=1}^{m} \lambda_{r_1}\phi_{r_1}(u_3)\phi_{r_1}(u_1)\right\}\left\{\sum_{r_2=1}^{m} \lambda_{r_2}\phi_{r_2}(u_3)\phi_{r_2}(u_1)\right\} \\
&\quad \times \left\{\sum_{r_3=1}^{m} \lambda_{r_3}\phi_{r_3}(u_3)\phi_{r_3}(u_1)\right\} du_1du_2du_3 \\
&= \sum_{r_1=1}^{m}\sum_{r_2=1}^{m}\sum_{r_3=1}^{m} \lambda_{r_1}\lambda_{r_2}\lambda_{r_3} \\
&\quad \times \int_{\mathcal{T}^3} \phi_{r_1}(u_3)\phi_{r_1}(u_1)\phi_{r_2}(u_3)\phi_{r_2}(u_1)\phi_{r_3}(u_3)\phi_{r_3}(u_1)du_1du_2du_3 \\
&= \sum_{r_1=1}^{m}\sum_{r_2=1}^{m}\sum_{r_3=1}^{m} \lambda_{r_1}\lambda_{r_2}\lambda_{r_3} \int_{\mathcal{T}} \phi_{r_3}(u_1)\phi_{r_1}(u_1)du_1 \\
&\quad \times \int_{\mathcal{T}} \phi_{r_1}(u_2)\phi_{r_2}(u_2)du_2 \int_{\mathcal{T}} \phi_{r_2}(u_3)\phi_{r_3}(u_3)du_3.
\end{aligned}
$$

By the orthonormality of the eigenfunctions over $\mathcal{T}$, the nonzero terms in the above expression are associated with $r_1 = r_2 = r_3 = r$ only. That is, $\operatorname{tr}(\gamma^{\otimes 3}) = \sum_{r=1}^{m} \lambda_r^3$, as desired.

When $m = \infty$, it is more involved and the proof involves the well-known dominated convergence theorem. Set $g_M(u_1,u_2,u_3) = \gamma_M(u_3,u_1)\gamma_M(u_1,u_2)\gamma_M(u_2,u_3)$, where $\gamma_M(s,t) = \sum_{r=1}^{M} \lambda_r\phi_r(s)\phi_r(t)$ for some $M>0$. By the Cauchy-Schwarz inequality, we have

$$
\begin{aligned}
|\gamma_M(s,t)| &\le \left[\sum_{r=1}^{M} \lambda_r\phi_r^2(s)\sum_{r=1}^{M} \lambda_r\phi_r^2(t)\right]^{1/2} \\
&\le \left[\sum_{r=1}^{\infty} \lambda_r\phi_r^2(s)\sum_{r=1}^{\infty} \lambda_r\phi_r^2(t)\right]^{1/2} = \left[h_0(s)h_0(t)\right]^{1/2},
\end{aligned}
$$

where $h_0(s) = \sum_{r=1}^{\infty} \lambda_r\phi_r^2(s)$. It follows that

$$
\begin{aligned}
|g_M(u_1,u_2,u_3)| &\le \left[h_0(u_3)h_0(u_1)h_0(u_1)h_0(u_2)h_0(u_2)h_0(u_3)\right]^{1/2} \\
&= h_0(u_1)h_0(u_2)h_0(u_3) \equiv h(u_1,u_2,u_3).
\end{aligned}
$$

In addition,

$$
\int_{\mathcal{T}^3} h(u_1,u_2,u_3)du_1du_2du_3 = \left[\int_{\mathcal{T}} h_0(u)du\right]^3 = \operatorname{tr}^3(\gamma) < \infty.
$$

Then by Lebesgue's dominated convergence theorem, we have

$$\begin{aligned}
&\mathrm{tr}(\gamma^{\otimes 3}) = \int_{\mathcal{T}^3} \lim_{M\to\infty} g_M(u_1,u_2,u_3)du_1du_2du_3\\
=\ &\lim_{M\to\infty} \int_{\mathcal{T}^3} g_M(u_1,u_2,u_3)du_1du_2du_3\\
=\ &\lim_{M\to\infty} \sum_{r_1,r_2,r_3=1}^{M} \lambda_{r_1}\lambda_{r_2}\lambda_{r_3} \int_{\mathcal{T}} \phi_{r_3}(u_1)\phi_{r_1}(u_1)du_1\\
&\int_{\mathcal{T}} \phi_{r_1}(u_2)\phi_{r_2}(u_2)du_2 \int_{\mathcal{T}} \phi_{r_2}(u_3)\phi_{r_3}(u_3)du_3\\
=\ &\sum_{r=1}^{\infty} \lambda_r^3.
\end{aligned}$$

The theorem is proved.

Proof of Theorem 4.2 Under the given conditions, $y(t)$ has the Karhunen-Loéve expansion (4.2). When m is a finite fixed number, we have

$$\begin{aligned}
\|y\|^2 &= \int_{\mathcal{T}} \eta^2(t)dt + 2\sum_{r=1}^{m} \xi_r \int_{\mathcal{T}} \eta(t)\phi_r(t)dt + \int_{\mathcal{T}} (\sum_{r=1}^{m} \xi_r\phi_r(t))^2 dt\\
&= \|\eta\|^2 + 2\sum_{r=1}^{m} \xi_r\delta_r + \sum_{r=1}^{m} \xi_r^2\\
&= \sum_{r=1}^{m} (\xi_r+\delta_r)^2 + \sum_{r=m+1}^{\infty} \delta_r^2\\
&= \sum_{r=1}^{m} \lambda_r A_r + \sum_{r=m+1}^{\infty} \delta_r^2,
\end{aligned}$$

where $A_r = (\xi_r+\delta_r)^2/\lambda_r \sim \chi_1^2(\lambda_r^{-1}\delta_r^2), r = 1,2,\cdots,m$ are independent as $(\xi_r+\delta_r)/\sqrt{\lambda_r} \sim N(\delta_r/\sqrt{\lambda_r}, 1)$. The expression (4.8) follows. When $m=\infty$, applying Parseval's theorem, we have

$$\|y\|^2 = \sum_{r=1}^{\infty} (\xi_r+\delta_r)^2 = \sum_{r=1}^{\infty} \lambda_r A_r,$$

where again $A_r \sim \chi_1^2(\lambda_r^{-1}\delta_r^2), r = 1,2,\cdots,\infty$. Set $\eta(t) \equiv 0$, the expression (4.9) follows. The theorem is proved.

Proof of Theorem 4.3 The theorem follows immediately from Isserlis' theorem or Wicks theorem for computing high-order moments of the multivariate normal distribution in terms of its covariance matrix.

Proof of Theorem 4.4 By the definition (4.11), for each $i = 1,2,\cdots,k$, we can set $W_i(s,t) \stackrel{d}{=} \sum_{j=1}^{n_i} v_{ij}(s)v_{ij}(t)$, where $v_{ij} \stackrel{i.i.d.}{\sim} \mathrm{GP}(0,\gamma), j = 1,2,\cdots,n_i$. As $W_i(s,t)$ are independent, we have

$$W_1(s,t) + W_2(s,t) + \cdots W_k(s,t) \stackrel{d}{=} \sum_{i=1}^{k}\sum_{j=1}^{n_i} v_{ij}(s)v_{ij}(t),$$

where $v_{ij} \stackrel{i.i.d.}{\sim} \mathrm{GP}(0,\gamma), j = 1,2,\cdots,n_i; i = 1,2,\cdots,k$. By the definition (4.11), we have

$$W_1(s,t) + W_2(s,t) + \cdots W_k(s,t) \sim \mathrm{WP}(n_1 + n_2 + \cdots + n_k, \gamma),$$

as desired. The theorem is proved.

Proof of Theorem 4.5 Let $W_i(s,t) = v_i(s)v_i(t), i = 1,2,\cdots,n$, where $v_i(t), i = 1,2,\cdots,n, \stackrel{i.i.d.}{\sim} \mathrm{GP}(0,\gamma)$. Then $W(s,t) = \sum_{i=1}^n W_i(s,t)$ and $W_i(s,t) \stackrel{i.i.d.}{\sim} \mathrm{WP}(1,\gamma)$. As $\mathrm{E}W_1(s,t) = \gamma(s,t)$, (a) follows. By (4.12), we have

$$\mathrm{tr}(W_i) = \int_{\mathcal{T}} v_i^2(t)dt = \sum_{r=1}^{\infty} \xi_{ir}^2 \stackrel{d}{=} \sum_{r=1}^{m} \lambda_r A_{ir}, \tag{4.64}$$

where $A_{ir} = \xi_{ir}^2/\lambda_r \stackrel{i.i.d.}{\sim} \chi_1^2$ for all i and r. As $\mathrm{tr}(W) = \sum_{i=1}^n \mathrm{tr}(W_i)$, (b) follows from the fact that $A_r = \sum_{i=1}^n A_{ir} \stackrel{i.i.d.}{\sim} \chi_n^2$. The expressions in (c) follow directly from (b). Noticing that $\mathrm{Etr}(W^{\otimes 2}) = \int_{\mathcal{T}^2} \mathrm{E}W^2(s,t)dsdt$ and

$$\begin{aligned}
\mathrm{E}W^2(s,t) &= \mathrm{Var}(W(s,t)) + \mathrm{E}^2(W(s,t)) = n\mathrm{Var}(W_1(s,t)) + n^2\gamma^2(s,t) \\
&= n\mathrm{E}W_1^2(s,t) - n\mathrm{E}^2(W_1(s,t)) + n^2\gamma^2(s,t) \\
&= n\mathrm{E}v_1^2(s)v_1^2(t) + n(n-1)\gamma^2(s,t),
\end{aligned}$$

we have

$$\begin{aligned}
\mathrm{Etr}(W^{\otimes 2}) &= n\mathrm{E}\left(\int_{\mathcal{T}} v_1^2(t)dt\right)^2 + n(n-1)\mathrm{tr}(\gamma^{\otimes 2}) \\
&= n\mathrm{Etr}^2(W_1) + n(n-1)\mathrm{tr}(\gamma^{\otimes 2}) \\
&= n(n+1)\mathrm{tr}(\gamma^{\otimes 2}) + n\mathrm{tr}^2(\gamma),
\end{aligned}$$

as desired, where we use the result $\mathrm{Etr}^2(W_1) = 2\mathrm{tr}(\gamma^{\otimes 2}) + \mathrm{tr}^2(\gamma)$ from Part (c). The proof is completed.

Proof of Theorem 4.6 It follows directly from Theorem 4.5.

Proof of Theorem 4.7 Under the given conditions, set $\mathbf{z}(t) = \mathbf{U}^T\mathbf{y}(t)$, where $\mathbf{U}$ is the orthonormal matrix given in (4.13). Then $\mathbf{z}(t) \sim \mathrm{GP}(\mathbf{U}^T\boldsymbol{\eta}, \gamma\mathbf{I}_n)$. That is, the components $z_1(t),\cdots,z_n(t)$ are independent Gaussian processes with $z_i(t) \sim \mathrm{GP}(\eta_{zi},\gamma)$, where $\eta_{zi}(t)$ is the ith component of $\boldsymbol{\eta}_z(t) = \mathbf{U}^T\boldsymbol{\eta}(t)$. The expression (4.15) follows immediately. When $\boldsymbol{\eta}(t) \equiv 0$, $\mathbf{U}^T\boldsymbol{\eta}(t) \equiv 0$ so that $\mathbf{z}(t) \sim \mathrm{GP}(\mathbf{0}, \gamma\mathbf{I}_n)$. It follows that $W_i(s,t) = z_i(s)z_i(t), i = 1,2,\cdots,n \stackrel{i.i.d.}{\sim} \mathrm{WP}(1,\gamma)$. Then by (4.14), the expression (4.16) follows immediately as desired.

Proof of Theorem 4.8 Under the given conditions, $\mathbf{A}$ has the SVD (4.17).

The theorem follows directly from Theorem 4.7.

Proof of Theorem 4.9 Let $\mathbf{A}_i$ have rank r_i so that it has the Cholesky decomposition $\mathbf{A}_i = \mathbf{S}_i^T\mathbf{S}_i$, where $\mathbf{S}_i : r_i \times n$ is a lower triangular matrix with strictly positive diagonal entries. Set $\mathbf{z}_i(t) = \mathbf{S}_i\mathbf{y}(t), i = 1, 2, \cdots, k$. We first show the "if " part by assuming $\mathbf{A}_i\mathbf{A}_j = \mathbf{0}, 1 \le i < j \le k$. As $\mathbf{B}_i = \mathbf{S}_i\mathbf{S}_i^T, i = 1, 2, \cdots, k$ are positive definite matrices, $\mathbf{A}_i\mathbf{A}_j = \mathbf{S}_i^T\mathbf{S}_i\mathbf{S}_j^T\mathbf{S}_j = \mathbf{0}, 1 \le i < j \le k$ implies that $\mathbf{S}_i\mathbf{S}_j^T = \mathbf{0}$ for $1 \le i < j \le k$. It follows that

$$\begin{aligned} \mathrm{Cov}(\mathbf{z}_i(s), \mathbf{z}_j(t)) &= \mathrm{Cov}(\mathbf{S}_i\mathbf{y}(s), \mathbf{S}_j\mathbf{y}(t)) \\ &= \gamma(s,t)\mathbf{S}_i\mathbf{S}_j^T = \mathbf{0}, 1 \le i < j \le k. \end{aligned} \tag{4.65}$$

Set $\mathbf{z}(t) = [\mathbf{z}_1(t)^T, \mathbf{z}_2(t)^T, \cdots, \mathbf{z}_k(t)^T]^T = \mathbf{S}\mathbf{y}(t)$, where $\mathbf{S} = [\mathbf{S}_1^T, \mathbf{S}_2^T, \cdots, \mathbf{S}_k^T]^T$. Then we have

$$\mathrm{Cov}(\mathbf{z}(s), \mathbf{z}(t)) = \gamma(s,t)\mathrm{diag}(\mathbf{B}_1, \mathbf{B}_2, \cdots, \mathbf{B}_k).$$

As $\mathbf{y}(t) \sim \mathrm{GP}(\boldsymbol{\eta}, \gamma\mathbf{I}_n)$, we have that $\mathbf{z}_i(t), i = 1, 2, \cdots, k$ are also Gaussian processes and are independent of each other. It follows that the quadratic forms $q_i(s,t) = \mathbf{z}_i(s)^T\mathbf{z}_i(t) = \mathbf{y}(s)^T\mathbf{A}_i\mathbf{y}(t), i = 1, 2, \cdots, k$ are independent of each other. We now prove the "only if " part by assuming $q_i(s,t) = \mathbf{z}_i(s)^T\mathbf{z}_i(t) = \mathbf{y}(s)^T\mathbf{A}_i\mathbf{y}(t), i = 1, 2, \cdots, k$ are independent of each other. It follows that $q_i(t,t) = \mathbf{z}_i(t)^T\mathbf{z}_i(t), i = 1, 2, \cdots, k$ are independent of each other. This shows that $z_i(t), i = 1, 2, \cdots, k$ are independent of each other. By (4.65), we then have $\mathbf{S}_i\mathbf{S}_j^T = \mathbf{0}, 1 \le i < j \le k$ as $\gamma(s,t)$ is not always 0. This implies that $\mathbf{A}_i\mathbf{A}_j = \mathbf{S}_i^T\mathbf{S}_i\mathbf{S}_j^T\mathbf{S}_j = \mathbf{0}, 1 \le i < j \le k$. The theorem is proved.

Proof of Theorem 4.10 Set $q(s,t) = \mathbf{y}(s)^T\mathbf{A}\mathbf{y}(t)$. By Theorem 4.8, we have

$$q(t,t) = \|\mathbf{w}(t)\|^2, \ \mathbf{w}(t) = [\mathbf{I}_k, \mathbf{0}]\mathbf{U}^T\mathbf{y}(t) \sim \mathrm{GP}_k(\boldsymbol{\eta}_w, \gamma\mathbf{I}_k),$$

where $\boldsymbol{\eta}_w(t) = [\mathbf{I}_k, \mathbf{0}]\mathbf{U}^T\boldsymbol{\eta}(t)$. Write $\mathbf{w}(t) = [w_1(t), \cdots, w_k(t)]^T$ and $\boldsymbol{\eta}_w(t) = [\eta_{w1}(t), \cdots, \eta_{wk}(t)]^T$. Then $w_i(t), i = 1, 2, \cdots, k$ are independent Gaussian processes. It follows that

$$T = \mathrm{tr}(q) = \int_{\mathcal{T}} \|\mathbf{w}(t)\|^2 dt = \sum_{i=1}^{k} \int_{\mathcal{T}} w_i^2(t)dt.$$

By Theorem 4.2, we have

$$\int_{\mathcal{T}} w_i^2(t)dt = \sum_{r=1}^{m} \lambda_r A_{ir} + \sum_{r=m+1}^{\infty} \delta_{ir}^2,$$

where $A_{ir} \sim \chi_1^2(\lambda_r^{-1}\delta_{ir}^2), r = 1, 2, \cdots, m$, and $\delta_{ir} = \int_{\mathcal{T}} \eta_{wi}(t)\phi_r(t)dt, r = 1, 2, \cdots$, with $\lambda_r, r = 1, 2, \cdots$, and $\phi_r(t), r = 1, 2, \cdots$ being the eigenvalues and eigenfunctions of $\gamma(s,t)$ and m the number of all the positive eigenvalues of $\gamma(s,t)$. As $w_i(t), i=1,2,\cdots,k$ are independent, we have

$$\begin{aligned} T &= \sum_{i=1}^{k}\int_{\mathcal{T}} w_i^2(t)dt \\ &= \sum_{i=1}^{k}\left\{\sum_{r=1}^{m}\lambda_r A_{ir} + \sum_{r=m+1}^{\infty}\delta_{ir}^2\right\} \\ &= \sum_{r=1}^{m}\lambda_r\sum_{i=1}^{k}A_{ir} + \sum_{r=m+1}^{\infty}\sum_{i=1}^{k}\delta_{ir}^2 \\ &= \sum_{r=1}^{m}\lambda_r A_r + \sum_{r=m+1}^{\infty}\delta_r^2, \end{aligned}$$

where $A_r = \sum_{i=1}^k A_{ir} \sim \chi_k^2(\lambda_r^{-1}\delta_r^2)$ and $\delta_r^2 = \sum_{i=1}^k \delta_{ir}^2 = \|\int_{\mathcal{T}} \boldsymbol{\eta}_w(t)\phi_r(t)dt\|^2$. The theorem is proved.

Proof of Theorem 4.11 First notice that $T = \text{tr}(q) = \sum_{i=1}^n a_{ii}z_{ii} + \sum_{i\neq j} a_{ij}z_{ij}$, where $z_{ij} = \int_{\mathcal{T}} y_i(t)y_j(t)dt = \text{tr}(q_{ij})$, where $q_{ij}(s,t) = y_i(s)y_j(t)$. As $y_1(t),\cdots,y_n(t)$ are mutually independent with $y_i(t) \sim \text{GP}(0,\gamma_i)$ and $\text{tr}(\gamma_i) < \infty$, we have $\text{E}z_{ij} = 0$ for $i \neq j$ and $q_{ii}(s,t) \sim \text{WP}(1,\gamma_i)$. By Theorem 4.5, we have $\text{E}(z_{ii}) = \text{tr}(\gamma_i)$ and $\text{Var}(z_{ii}) = 2\text{tr}(\gamma_i^{\otimes 2})$. It follows that $\text{E}(T) = \sum_{i=1}^n a_{ii}\text{tr}(\gamma_i)$. Now $T - \text{E}(T) = \sum_{i=1}^n a_{ii}[z_{ii} - \text{E}(z_{ii})] + \sum_{i\neq j} a_{ij}z_{ij}$. In addition, when $i \neq j$, we have $\text{E}\{[z_{ii} - \text{E}z_{ii}]z_{ij}\} = 0$. Therefore,

$$\text{Var}(T) = \sum_{i=1}^{n} a_{ii}^2\text{Var}(z_{ii}) + \sum_{i\neq j}\sum_{\alpha\neq\beta} a_{ij}a_{\alpha\beta}\text{E}(z_{ij}z_{\alpha\beta}).$$

When there is at least one pair of i,j,α and β are not equal, we have $\text{E}(z_{ij}z_{\alpha\beta}) = 0$. By the fact that $a_{ij} = a_{ji}$, when $\alpha = i, \beta = j$ or when $\alpha = j, \beta = i$, we have

$$\begin{aligned} a_{ij}a_{\alpha\beta}\text{E}\left(z_{ij}z_{\alpha\beta}\right) &= a_{ij}^2\text{E}\int_{\mathcal{T}}\int_{\mathcal{T}} y_i(s)y_i(t)y_j(s)y_j(t)dsdt \\ &= a_{ij}^2\int_{\mathcal{T}}\int_{\mathcal{T}}\gamma_i(s,t)\gamma_j(s,t)dsdt = a_{ij}^2\text{tr}(\gamma_i\otimes\gamma_j). \end{aligned}$$

Therefore, we have

$$\text{Var}(T) = 2\sum_{i=1}^{n} a_{ii}^2\text{tr}(\gamma_i^{\otimes 2}) + 2\sum_{i\neq j}\text{tr}(\gamma_i\otimes\gamma_j) = 2\sum_{i=1}^{n}\sum_{j=1}^{n} a_{ij}^2\text{tr}(\gamma_i\otimes\gamma_j).$$

The theorem is proved.

Proof of Theorem 4.12 See Laha and Rohatgi (1979, p. 474) or van der Vaart and Wellner (1996, p. 50–51).

Proof of Theorem 4.13 First of all, we have

$$\mathrm{E}\|q_1\|^2 = \mathrm{E}\int_{\mathcal{T}}\int_{\mathcal{T}} q_1^2(s,t)dsdt = \mathrm{E}\int_{\mathcal{T}}\int_{\mathcal{T}} y_1^2(s)y_1^2(t)dsdt = \mathrm{E}\|y_1\|^4 < \infty.$$

Then the theorem follows directly from Theorem 4.12 by noticing that $q_i(s,t) = y_i(s)y_i(t), i = 1, 2, \cdots, n$ are i.i.d. with $\mathrm{E}q_1(s,t) = \gamma(s,t)$ and

$$\begin{aligned}\mathrm{Cov}\Big(q_1(s_1,t_1), q_1(s_2,t_2)\Big) &= \mathrm{E}q_1(s_1,t_1)q_1(s_2,t_2) - \mathrm{E}q_1(s_1,t_1)\mathrm{E}q_1(s_2,t_2)\\ &= \mathrm{E}y_1(s_1)y_1(t_1)y_1(s_2)y_1(t_2) - \gamma(s_1,t_1)\gamma(s_2,t_2).\end{aligned}$$

Proof of Theorem 4.14 Under the given conditions, the functional sample (4.40) is Gaussian. The first assertion follows from the fact that $\mathrm{E}\hat{\eta}(t) = \eta(t)$ and $\mathrm{Cov}(\hat{\eta}(s), \hat{\eta}(t)) = \gamma(s,t)/n$. To show the second assertion, let $\mathbf{y}(t) = [y_1(t), y_2(t), \cdots, y_n(t)]^T$. Then $\mathbf{y}(t) \sim \mathrm{GP}_n(\eta\mathbf{1}_n, \gamma\mathbf{I}_n)$. We have

$$(n-1)\hat{\gamma}(s,t) = \mathbf{y}(s)^T(\mathbf{I}_n - \mathbf{J}_n/n)\mathbf{y}(t) = \mathbf{v}(s)^T(\mathbf{I}_n - \mathbf{J}_n/n)\mathbf{v}(t),$$

where $\mathbf{J}_n = \mathbf{1}_n\mathbf{1}_n^T$ is an $n \times n$ matrix of ones and $\mathbf{v}(t) = \mathbf{y}(t) - \eta(t)\mathbf{1}_n \sim \mathrm{GP}_n(\mathbf{0}, \gamma\mathbf{I}_n)$. Notice that $\mathbf{I}_n - \mathbf{J}_n/n$ is an idempotent matrix of rank $n-1$. The second assertion then follows from Theorem 4.8 immediately.

Proof of Theorem 4.15 Under the given conditions, we have $\mathrm{E}\|y_1\|^2 = \|\eta\|^2 + \mathrm{tr}(\gamma) < \infty$ and the functional sample (4.40) is i.i.d. The assertion follows from Theorem 4.12 immediately.

Proof of Theorem 4.16 Let $\mathbf{y}(t) = [y_1(t), \cdots, y_n(t)]^T$. Then $\mathbf{y}(t) \sim \mathrm{SP}_n(\eta\mathbf{1}_n, \gamma\mathbf{I}_n)$. Set $\mathbf{v}(t) = \mathbf{y}(t) - \eta(t)\mathbf{1}_n$. Then $\mathbf{v}(t) \sim \mathrm{SP}_n(\mathbf{0}, \gamma\mathbf{I}_n)$. Set $\mathbf{J}_n = \mathbf{1}_n\mathbf{1}_n^T$. Then

$$\begin{aligned}\hat{\gamma}(s,t) &= (n-1)^{-1}\mathbf{y}(s)^T(\mathbf{I}_n - \mathbf{J}_n/n)\mathbf{y}(t)\\ &= (n-1)^{-1}\mathbf{v}(s)^T(\mathbf{I}_n - \mathbf{J}_n/n)\mathbf{v}(t)\\ &= (n-1)^{-1}\textstyle\sum_{i=1}^n z_i(s,t) - \frac{n}{n-1}\bar{v}(s)\bar{v}(t),\end{aligned}$$

where $\bar{v}(t) = n^{-1}\sum_{i=1}^n v_i(t)$ and $z_i(s,t) = v_i(s)v_i(t), i = 1, 2, \cdots, n$ are i.i.d. with $\mathrm{E}(z_1(s,t)) = \gamma(s,t)$ and

$$\begin{aligned}\varpi[(s_1,t_1),(s_2,t_2)] &= \mathrm{cov}(z_1(s_1,t_1), z_1(s_2,t_2))\\ &= \mathrm{E}\left[v_1(s_1)v_1(t_1)v_1(s_2)v_1(t_2)\right] - \gamma(s_1,t_1)\gamma(s_2,t_2).\end{aligned}$$

By the central limit theorem of i.i.d. stochastic processes, Theorem 4.12, as $n \to \infty$, we have

$$\sqrt{n}\left[n^{-1}\sum_{i=1}^n z_i(s,t) - \gamma(s,t)\right] \xrightarrow{d} \mathrm{GP}(0, \varpi),$$

as by Assumption OS3, we have

$$\mathrm{E}\|z_1\|^2 = \mathrm{E}\int_{\mathcal{T}^2}[v_1(s)v_1(t)]^2 ds dt = \mathrm{E}\left[\int_{\mathcal{T}} v_1^2(t)dt\right]^2 = \mathrm{E}\|v_1\|^4 < \infty.$$

It remains to show that $\bar{v}(t) = o_{UP}(1)$, that is, $\bar{v}(t)$ converges to 0 in probability uniformly over $\mathcal{T}$. This is actually true as $\mathrm{E}\bar{v}(t) \equiv 0$ and $\mathrm{Cov}(\bar{v}(s), \bar{v}(t)) = \gamma(s,t)/n \leq \gamma(t,t)/n \leq \rho/n$. Thus under Assumption OS4, we have $\bar{v}(t) = o_{UP}(1)$. The theorem is then proved.

Proof of Theorem 4.17 Under the given conditions, by Theorem 4.16, as $n \to \infty$, we have $\mathrm{E}[\hat{\gamma}(s,t) - \gamma(s,t)]^2 = \frac{\omega[(s,t),(s,t)]}{n}[1 + o(1)]$. By Assumptions OS4 and OS5, we have

$$|\omega[(s,t),(s,t)]| \leq \mathrm{E}[v_1^2(s)v_1^2(t)] + \gamma^2(s,t) \leq C + \rho, \quad \text{for all } (s,t) \in \mathcal{T}^2.$$

Thus, we have $\hat{\gamma}(s,t) = \gamma(s,t) + O_{UP}(n^{-1/2})$, $(s,t) \in \mathcal{T}^2$, where O_{UP} means "uniformly bounded in probability." It follows that $\hat{\gamma}(s,t) \xrightarrow{p} \gamma(s,t)$ uniformly over $\mathcal{T}^2$. Therefore,

$$\begin{aligned}
\lim_{n\to\infty} \mathrm{tr}(\hat{\gamma}) &= \int_{\mathcal{T}} \lim_{n\to\infty} \hat{\gamma}(t,t)dt = \int_{\mathcal{T}} \gamma(t,t)dt = \mathrm{tr}(\gamma), \\
\lim_{n\to\infty} \mathrm{tr}(\hat{\gamma}^{\otimes 2}) &= \int_{\mathcal{T}}\int_{\mathcal{T}} \lim_{n\to\infty} \hat{\gamma}^2(s,t)dsdt \\
&= \int_{\mathcal{T}}\int_{\mathcal{T}} \gamma^2(s,t)dsdt = \mathrm{tr}(\gamma^{\otimes 2}).
\end{aligned}$$

It follows from (4.52) and (4.54) that as $n \to \infty$, $\hat{\beta} \xrightarrow{p} \beta$ and $\hat{\kappa} \xrightarrow{p} \kappa$. The theorem is proved.

4.7 Concluding Remarks and Bibliographical Notes

It is quite natural to regard functional data as realizations of an underlying stochastic process. Not all stochastic processes can be modeled by Gaussian or Wishart processes. However, for a large functional data set, the distribution of its sample mean function can be well approximated by a Gaussian process. The sample covariance function of a Gaussian process is a Wishart process. The Gaussian process and the Wishart process play roles in functional data analysis similar to the roles played by the multi-normal distribution and the Wishart distribution in multivariate data analysis.

Both the squared L^2-norm of a Gaussian process and the trace of a Wishart process are χ^2-type mixtures. The ratio of any two independent χ^2-type mixtures is an F-type mixture. The two-cumulant matched χ^2-approximation is known as the Welch-Satterthwaite χ^2-approximation. It at least dates back to Welch (1947) and Satterthwaite (1941, 1946), among others. It is also known as Box's χ^2-approximation (Box 1954a,b). The two-cumulant matched χ^2-approximation plays a central role in Welch's approximate degrees of freedom test for the multi-sample Behrens-Fisher problem. The three-cumulant matched χ^2-approximation for χ^2-type mixtures with positive coefficients and central χ^2-variables was studied by Solomon and Stephens (1977) and Buckley and Eagleson (1988). It was extended for general χ^2-type mixtures with possible negative coefficients and noncentral χ^2-variables by Zhang (2005), where he gave a theoretical error bound for the approximation. The distribution of a general χ^2-type mixture with both positive and negative coefficients can be well approximated by the three-cumulant matched χ^2-approximation. However, the two-cumulant matched χ^2-approximation fails to give a good approximation in this case.

The one-sample problem for functional data was studied by Mas (2007) using a penalization approach. The L^2-norm-based test for comparing two nested functional linear models was first proposed by Faraway (1997). He approximated the associated null distribution by a nonparametric bootstrap method. The L^2-norm-based test for a general linear hypothesis testing problem is investigated in Zhang and Chen (2007). The F-type test for comparing two nested functional linear models was proposed and studied by Shen and Faraway (2004). It was extended for a general linear hypothesis testing problem by Zhang (2011a), where some theoretical results were derived. Further results will be presented in later chapters.

4.8 Exercises

1. Assume the SVD (4.1) holds. Derive the two formulas in (4.6).
2. Simulate a large sample from the χ^2-type mixture T_a as defined in (4.23). Compute the associated first four cumulants, skewness, and kurtosis of T_a based on the generated sample. Compare these quantities with their theoretical values computed using (4.28).
3. Show that $\mathrm{tr}^2(\hat{\gamma})$ and $\mathrm{tr}(\hat{\gamma}^{\otimes 2})$ are biased for $\mathrm{tr}^2(\gamma)$ and $\mathrm{tr}(\gamma^{\otimes 2})$, respectively. When $\mathrm{tr}(\gamma) < \infty$, however, they are asymptotically unbiased.
4. Simulate a large sample from the F-type mixture F_a as defined in (4.34). Plot the histogram of the generated sample. Compute the associated first four cumulants, skewness and kurtosis of F_a.
5. For the nonconceptive progesterone data introduced in Section 1.2.1 of Chapter 1, it is of interest to test the following one-sample testing problem:

$$\begin{array}{lll} & H_0 & : \eta(t) \equiv -0.90, t \in [a, b], \\ \text{versus} & H_1 & : \eta(t) \neq -0.90, \text{ for some } t \in [a, b], \end{array} \tag{4.66}$$

where $[a, b] = [-8, 0], [0, 15]$, or $[-8, 15]$. Apply the pointwise t-test, z-test, and bootstrap test to test (4.66).

6. Apply the L^2-norm-based test and the F-type test to test (4.66) for the nonconceptive progesterone data. Comment if the conclusions made here are similar to those obtained from Exercise 5.

Chapter 5

ANOVA for Functional Data

5.1 Introduction

At the end of Chapter 4, we studied the one-sample problem for functional data where we described some basic hypothesis tests for functional data, including the pointwise, L^2-norm-based, F-type, and bootstrap tests. In this chapter, we show how to extend these tests to more complicated designs where two or more functional samples are involved. We start with the two-sample problem for functional data. It is the simplest multi-sample problem. It allows us to understand hypothesis testing problems for functional data with more insight. Studies about one-way ANOVA and two-way ANOVA are then described in Sections 5.3 through 5.4. Technical proofs of the main results are outlined in Section 5.5. Some concluding remarks and bibliographical notes are given in Section 5.6. Section 5.7 is devoted to some exercise problems related to this chapter.

Throughout this chapter, we assume that all the samples involved have a common covariance function. Analysis of variance for heteroscedastic functional data will be discussed in Chapter 9.

5.2 Two-Sample Problem

We use the following example to motivate the two-sample problem for functional data.

Example 5.1 *Figure 5.1 displays the reconstructed individual curves of the progesterone data: (a) nonconceptive and (b) conceptive, obtained by the local linear reconstruction method described in Section 3.2.3 of Chapter 3 with bandwidth* $h^* = 1.40$ *selected by the GCV rule (3.10). The progesterone data were introduced in Section 1.2.1 of Chapter 1. The nonconceptive progesterone curves were from the women who were not pregnant after the ovulation day (Day 0) when they discharged their ova, while the conceptive progesterone curves were from those women who were pregnant. The horizontal axis shows the days before and after the ovulation day. Of interest is to know if there is a significant difference between the mean functions of the nonconceptive and conceptive progesterone curves before or after the ovulation day or over the*

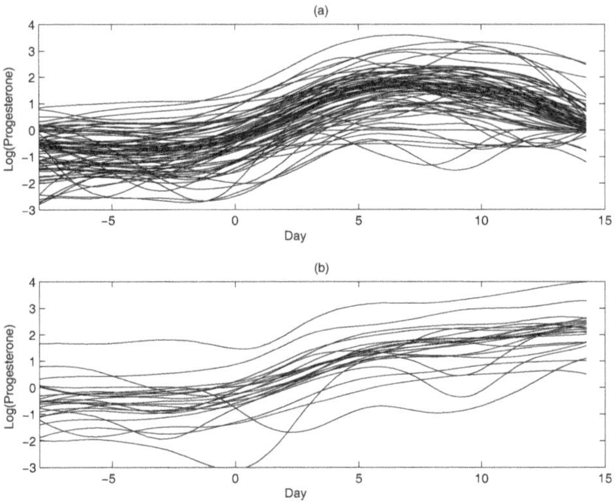

Figure 5.1 *Reconstructed individual curves of the progesterone data: (a) nonconceptive and (b) conceptive, obtained by the local linear reconstruction method described in Section 3.2.3 of Chapter 3 with bandwidth $h^* = 1.40$ selected by the GCV rule (3.10).*

whole experimental period. This knowledge may be used to detect if a woman is pregnant after the ovulation day.

A general two-sample problem for functional data with a common covariance function can be formulated as follows. Suppose we have two functional samples

$$y_{11}(t), \cdots, y_{1n_1}(t) \stackrel{i.i.d.}{\sim} \text{SP}(\eta_1, \gamma), \;\; y_{21}(t), \cdots, y_{2n_2}(t) \stackrel{i.i.d.}{\sim} \text{SP}(\eta_2, \gamma), \tag{5.1}$$

where $\eta_1(t)$ and $\eta_2(t)$ are the unknown mean functions of the two samples, and $\gamma(s,t)$ is their common covariance function, which is usually unknown. We wish to test the following hypotheses:

$$\begin{array}{lll} & H_0 & : \eta_1(t) \equiv \eta_2(t), t \in \mathcal{T}, \\ \text{versus} & H_1 & : \eta_1(t) \neq \eta_2(t), \;\; \text{for some } t \in \mathcal{T}, \end{array} \tag{5.2}$$

where $\mathcal{T}$ is the time period of interest, often a finite interval $[a,b]$ say with $-\infty < a < b < \infty$.

Based on the two functional samples (5.1), the unbiased estimators of the mean functions $\eta_1(t)$, $\eta_2(t)$ and the common covariance functions $\gamma(s,t)$ are given by

$$\begin{array}{rcl} \hat{\eta}_i(t) & = & \bar{y}_i(t) = n_i^{-1} \sum_{j=1}^{n_i} y_{ij}(t), \; i = 1, 2, \\ \hat{\gamma}(s,t) & = & (n-1)^{-1} \sum_{i=1}^{2} \sum_{j=1}^{n_i} \left[y_{ij}(s) - \bar{y}_i(s)\right] \left[y_{ij}(t) - \bar{y}_i(t)\right], \end{array} \tag{5.3}$$

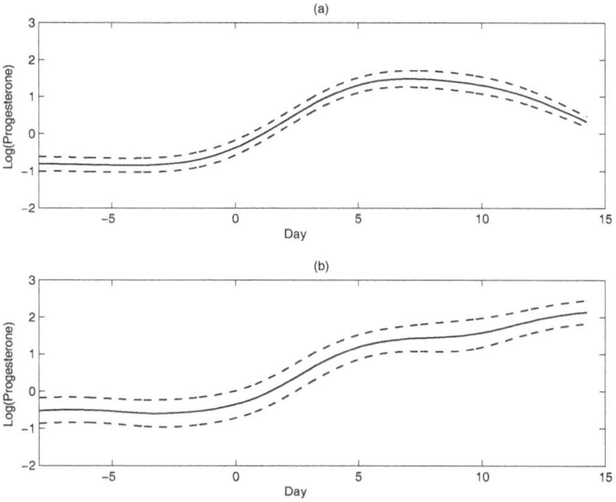

Figure 5.2 *Sample mean functions (solid) of the progesterone data: (a) nonconceptive and (b) conceptive, together with their 95% pointwise confidence bands (dashed). It appears that the two sample mean functions are very similar within eight days before and after the ovulation day (Day 0) but they are very different after the eighth day following the ovulation day.*

which are known as the sample mean and pooled sample covariance functions of the two samples, respectively, where and throughout this section, $n = n_1 + n_2$ denotes the total sample size of the two samples.

Example 5.2 *Figure 5.2 shows the sample mean functions of the progesterone data: (a) nonconceptive and (b) conceptive, together with their* 95% *pointwise confidence bands. It is seen that the two sample mean functions are very similar within eight days before and after the ovulation day but they are very different after the eighth day following the ovulation day. A formal test is needed to verify if these are true statistically.*

Remark 5.1 *In the two samples (5.1), we assume that the two samples have a common covariance function. This equal-covariance function assumption may not be always satisfied. If it is not satisfied, the two-sample problem (5.2) is known as the two-sample Behrens-Fisher problem for functional data (Zhang, Liang, and Xiao 2010). This problem will be discussed in Chapter 9.*

Example 5.3 *For the progesterone data, the equal-covariance function assumption may be approximately satisfied. Figure 5.3 shows the sample covariance functions of the nonconceptive progesterone curves and the conceptive progesterone curves, respectively, in panels (a) and (b). It is seen that the two*

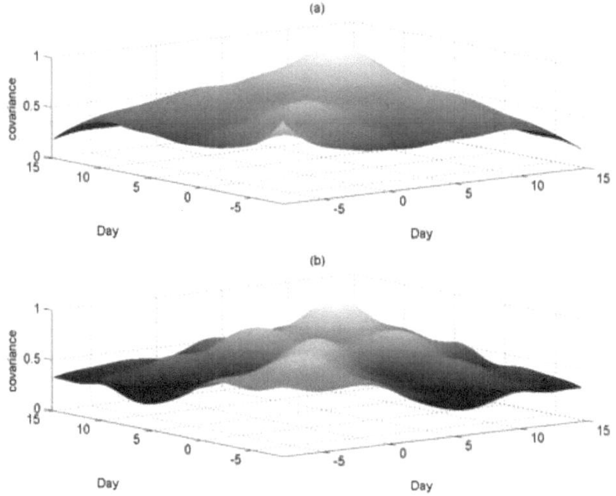

Figure 5.3 *Sample covariance functions of the progesterone data: (a) nonconceptive and (b) conceptive. It appears that the two sample covariance functions are similar in shape.*

Table 5.1 *Traces of the sample covariance functions* $\hat{\gamma}_1(s,t)$ *and* $\hat{\gamma}_2(s,t)$ *and their cross-square functions* $\hat{\gamma}_1^{\otimes 2}(s,t)$ *and* $\hat{\gamma}_2^{\otimes 2}(s,t)$ *calculated with resolution* $M = 1,000$ *over various periods.*

$[a,b]$	$[-8,0]$		$[-8,8]$		$[-8,15]$	
i	1	2	1	2	1	2
$\text{tr}(\hat{\gamma}_i)$	251	264	512	536	728	760
$\text{tr}(\hat{\gamma}_i^{\otimes 2})$	51,076	62,308	191,475	191,688	368,392	361,410

sample covariance functions look quite similar in shape. This similarity may be further verified by some numerical results as displayed in Table 5.1. It is seen that $tr(\hat{\gamma}_1) \approx tr(\hat{\gamma}_2)$ *and* $tr(\hat{\gamma}_1^{\otimes 2}) \approx tr(\hat{\gamma}_2^{\otimes 2})$, *where* $tr(f) = \int_{\mathcal{T}} f(t,t)dt$ *denotes the trace of a bivariate function* $f(s,t)$ *as defined in the previous chapter. Table 5.2 displays the traces of the pooled sample covariance function* $\hat{\gamma}(s,t)$ *(5.3) and its cross-square function* $\hat{\gamma}^{\otimes 2}(s,t)$ *over various periods. By a comparison of Tables 5.1 and 5.2, it is seen that* $tr(\hat{\gamma}), tr(\hat{\gamma}_1)$, *and* $tr(\hat{\gamma}_2)$

Table 5.2 *Traces of the pooled sample covariance function* $\hat{\gamma}(s,t)$ *and its cross-square function* $\hat{\gamma}^{\otimes 2}(s,t)$ *calculated with resolution* $M = 1,000$ *over various periods.*

[a,b]	$[-8,0]$	$[-8,8]$	$[-8,15]$
$\text{tr}(\hat{\gamma})$	255	518	736
$\text{tr}(\hat{\gamma}^{\otimes 2})$	53,577	190,990	365,544

are approximately equal and $tr(\hat{\gamma}^{\otimes 2})$, $tr(\hat{\gamma}_1^{\otimes 2})$, *and* $tr(\hat{\gamma}_2^{\otimes 2})$ *are approximately equal. Nevertheless, a formal test for the equal-covariance function assumption should also be conducted. This formal test will be discussed in Chapter 10.*

5.2.1 *Pivotal Test Function*

To test the two-sample problem (5.2) based on the two samples (5.1), a natural pivotal test function is

$$\Delta(t) = \sqrt{n_1 n_2/n}\left[\bar{y}_1(t) - \bar{y}_2(t)\right], \tag{5.4}$$

which is the scaled mean function difference of the two samples. When the null hypothesis is valid, this quantity will be small and it will be large otherwise. Therefore, it is appropriate to use $\Delta(t)$ as a pivotal test function for the two-sample problem (5.2). Notice that $\Delta(t)$ has its mean and covariance functions as

$$\begin{aligned} \eta_\Delta(t) = \mathrm{E}\Delta(t) = \sqrt{n_1 n_2/n}\left[\eta_1(t) - \eta_2(t)\right], \quad \text{and} \\ \mathrm{Cov}\left[\Delta(s), \Delta(t)\right] = \gamma(s,t). \end{aligned} \tag{5.5}$$

Under the null hypothesis of (5.2), we have $\mathrm{E}\Delta(t) \equiv 0, t \in \mathcal{T}$.

Before we can study any procedures for testing (5.2), we need to study the properties of $\Delta(t)$. Let $\mathcal{L}^2(\mathcal{T})$ denote the set of all integrable functions on $\mathcal{T}$. For easy reference, we list the following assumptions:

Two-Sample Problem Assumptions (TS)

1. The two samples (5.1) are with $\eta_1(t), \eta_2(t) \in \mathcal{L}^2(\mathcal{T})$ and $\mathrm{tr}(\gamma) < \infty$.
2. The two samples (5.1) are Gaussian.
3. As $n \to \infty$, the sample sizes satisfy $n_1/n \to \tau$ such that $\tau \in (0,1)$.
4. The subject-effect functions $v_{ij}(t) = y_{ij}(t) - \eta_i(t), j = 1, 2, \cdots, n_i; i = 1, 2$ are i.i.d..
5. The subject-effect function $v_{11}(t)$ satisfies

$$\mathrm{E}\|v_{11}\|^4 = \mathrm{E}\left[\int_{\mathcal{T}} v_{11}^2(t)dt\right]^2 < \infty.$$

6. The maximum variance $\rho = \max_{t \in \mathcal{T}} \gamma(t,t) < \infty$.
7. The expectation $\mathrm{E}[v_{11}^2(s)v_{11}^2(t)]$ is uniformly bounded. That is, for any $(s,t) \in \mathcal{T}^2$, we have $\mathrm{E}[v_{11}^2(s)v_{11}^2(t)] < C < \infty$, where C is some constant independent of $(s,t) \in \mathcal{T}^2$.

Assumptions TS1 and TS2 are regular while Assumption TS3 requires that the two sample sizes n_1, n_2 tend to ∞ proportionally. Assumptions TS3, TS4, and TS5 are convenient conditions imposed for investigating the asymptotical properties of $\hat{\eta}_1(t), \hat{\eta}_2(t)$, and $\hat{\gamma}(s,t)$ for non-Gaussian functional data. In fact, Assumption TS3 guarantees that as $n \to \infty$, both the sample mean functions

$\bar{y}_1(t)$ and $\bar{y}_2(t)$ will converge to Gaussian processes weakly. Assumptions TS2 and TS5 imply Assumption TS6. Assumption TS6 is also satisfied when $v_{11}(t)$ is uniformly bounded in probability over $\mathcal{T}$, which is a finite interval. First of all, we have the following useful result.

Theorem 5.1 *Under Assumptions TS1 and TS2, we have*

$$\Delta(t) \sim GP(\eta_\Delta, \gamma), \quad \textit{and} \quad (n-2)\hat{\gamma}(s,t) \sim WP(n-2, \gamma). \tag{5.6}$$

Theorem 5.1 shows that under the Gaussian assumption TS2, $\Delta(t)$ is a Gaussian process and $(n-2)\hat{\gamma}(s,t)$ is a Wishart process. It is the key for constructing various tests for (5.2) when the two samples (5.1) are Gaussian. In fact, under H_0 and the Gaussian assumption TS2, by Theorem 5.1, it is easy to see that

$$\sqrt{n_1 n_2/n}\Big[\bar{y}_1(t) - \bar{y}_2(t)\Big] \sim \mathrm{GP}(0, \gamma). \tag{5.7}$$

Based on this, we can easily construct various tests for (5.2). Theorem 5.1 holds for finite functional samples but requires that the Gaussian assumption TS2 be satisfied. In practice, this Gaussian assumption TS2 is not always valid. When Assumptions TS3 and TS4 are satisfied, however, by the central limit theorem for i.i.d. random functions, Theorem 4.12 of Chapter 4, we can show that $\Delta(t)$ is asymptotically a Gaussian process.

Theorem 5.2 *Under Assumptions TS1, TS3, and TS4, as $n \to \infty$, we have*

$$\Delta(t) - \eta_\Delta(t) \xrightarrow{d} GP(0, \gamma),$$

where $\eta_\Delta(t)$ is as defined in (5.5).

The above theorem can be used for constructing various tests for (5.2) when the two samples (5.1) are not Gaussian but with large sample sizes. To study the asymptotic distribution of $\hat{\gamma}(s,t)$, we need Assumptions TS5 and TS6. The following theorem shows that the pooled sample covariance function $\hat{\gamma}(s,t)$ is asymptotically Gaussian and $\sqrt{n}$-consistent. This knowledge will be used in Theorem 5.4 stated in next subsection.

Theorem 5.3 *Under Assumptions TS1 and TS3 through TS6, as $n \to \infty$, we have*

$$\sqrt{n}\,\{\hat{\gamma}(s,t) - \gamma(s,t)\} \xrightarrow{d} GP(0, \varpi),$$

where $\varpi\,\{(s_1,t_1),(s_2,t_2)\} = E\{v_{11}(s_1)v_{11}(t_1)v_{11}(s_2)v_{11}(t_2)\} - \gamma(s_1,t_1)\gamma(s_2,t_2)$.

5.2.2 Methods for Two-Sample Problems

In this subsection, we describe various tests for the two-sample problem (5.2). They are the natural generalizations of those tests for the one-sample problem

(4.41) described in Chapter 4.

Pointwise Tests We here describe pointwise t-, z-, and bootstrap tests for the two-sample problem (5.2) under various conditions. The key idea of a pointwise test is to test the null hypothesis at each time point $t \in \mathcal{T}$. For any fixed $t \in \mathcal{T}$, the sub-problem is

$$H_{0t} : \eta_1(t) = \eta_2(t), \quad \text{versus} \quad H_{1t} : \eta_1(t) \neq \eta_2(t). \tag{5.8}$$

Based on the sample mean functions and the pooled sample covariance function given in (5.3), the pivotal test statistic for (5.8) is

$$z(t) = \frac{[\bar{y}_1(t) - \bar{y}_2(t)]}{\sqrt{(1/n_1 + 1/n_2)\hat{\gamma}(t,t)}} = \frac{\Delta(t)}{\sqrt{\hat{\gamma}(t,t)}}, \tag{5.9}$$

where $\Delta(t)$ is the pivotal test function defined in (5.4).

In many situations, the two samples (5.1) may be approximately Gaussian. That is, Assumption TS2 is approximately satisfied. By Theorem 5.1 and under H_{0t}, we have

$$z(t) \sim t_{n-2}, \quad t \in \mathcal{T}. \tag{5.10}$$

Then, we can conduct the pointwise t-test by rejecting H_{0t} whenever $|z(t)| > t_{n-2}(1-\alpha/2)$ or by reporting the pointwise P-values computed based on the t-distribution (5.10). Alternatively, we can construct the $100(1-\alpha)\%$ pointwise confidence bands for the mean function difference $\eta_1(t) - \eta_2(t)$ as

$$[\bar{y}_1(t) - \bar{y}_2(t)] \pm t_{n-2}(1-\alpha/2)\sqrt{(1/n_1 + 1/n_2)\,\hat{\gamma}(t,t)}, \; t \in \mathcal{T}, \tag{5.11}$$

where $t_{n-2}(1-\alpha/2)$ denotes the $100(1-\alpha/2)$-percentile of the t-distribution with $n-2$ degrees of freedom. The pointwise t-test aims to conduct t-test at each time point $t \in \mathcal{T}$ based on the t-distribution (5.10). The Gaussian assumption is required to be approximately satisfied.

When the Gaussian assumption is not satisfied, for large samples, one may use the pointwise z-test instead. As $n_1, n_2 \to \infty$, by Theorem 5.2, we have

$$z(t) \xrightarrow{d} N(0,1), \text{ for any fixed } t \in \mathcal{T}. \tag{5.12}$$

Thus, one can conduct the pointwise z-test by rejecting H_{0t} whenever $|z(t)| > z_{1-\alpha/2}$ or computing the pointwise P-values using (5.12). Alternatively, one can construct the $100(1-\alpha)\%$ asymptotic pointwise confidence bands for $\eta_1(t) - \eta_2(t)$ obtained from (5.11) by replacing the t-distribution critical value $t_{n-2}(1-\alpha/2)$ by $z_{1-\alpha/2}$, the $100(1-\alpha/2)$-percentile of the standard normal distribution.

When the two samples (5.1) are not Gaussian while both n_1 and n_2 are small, the above pointwise t- and z-tests are not preferred. In this case, one may resort to a pointwise bootstrap test. Let $v_{ij}^*(t), j = 1, 2, \cdots, n_i; i = 1, 2,$ be bootstrapped from the estimated subject-effect functions $\hat{v}_{ij}(t) = y_{ij}(t) - \hat{\eta}_i(t), j = 1, 2, \cdots, n_i; i = 1, 2$. Set

$$y_{ij}^*(t) = \hat{\eta}_i(t) + v_{ij}^*(t), j = 1, 2, \cdots, n_i; i = 1, 2. \tag{5.13}$$

Then we can compute the sample mean functions and the pooled sample covariance function $\bar{y}_i^*(t), i = 1, 2,$ and $\hat{\gamma}^*(s, t)$ as in (5.3) but now based on the two bootstrapped samples (5.13). For the pointwise bootstrap test, we compute

$$z^*(t) = \frac{([\bar{y}_1^*(t) - \bar{y}_2^*(t)] - [\bar{y}_1(t) - \bar{y}_2(t)])}{\sqrt{(\frac{1}{n_1} + \frac{1}{n_2})\hat{\gamma}^*(t, t)}}.$$

Notice that it is important to subtract the original sample mean function difference $\bar{y}_1(t) - \bar{y}_2(t)$ from the bootstrap sample mean function difference $\bar{y}_1^*(t) - \bar{y}_2^*(t)$ in computation of $z^*(t)$ so that the distribution of $z^*(t)$ can mimic the null distribution of $z(t)$ well. Repeat the above bootstrapping process a large number of times, calculate its $100(1 - \alpha/2)$-percentile, and then conduct the pointwise bootstrap test or construct the pointwise bootstrap confidence bands accordingly.

Example 5.4 *Figure 5.4 (a) displays the pivotal test function (solid) with the 5% upper and lower critical lines of the pointwise t-test (dashed), the pointwise z-test (dot-dashed), and the bootstrap test (dotted). It is seen that the critical lines are close to each other, implying that the three pointwise tests largely give similar test results. It is also seen that the pivotal test function runs within the upper and lower critical lines before Day 11, implying that the mean functions (in* log*-scale) of the nonconceptive and conceptive progesterone curves have very small differences before Day 11. However, after Day 11, the pivotal test function moves away from the lower critical lines, implying that the mean function differences of the nonconceptive and conceptive progesterone curves are significant as then.*

Example 5.5 *Alternatively, we can conduct the above pointwise tests using the pointwise P-values. Figure 5.4 (b) displays the pointwise P-values of the pointwise t-test (dashed), the pointwise z-test (dot-dashed), and the pointwise bootstrap test (dotted), together with the 5% significance level line (solid). It is seen that these pointwise P-values are close to each other at each t, indicating that the three pointwise testing procedures produced similar test results. It is seen that these pointwise P-values are smaller than* 5% *only after Day 11, implying that the mean functions of the nonconceptive and conceptive progesterone curves have very small differences before Day 11 and they are significantly different as then.*

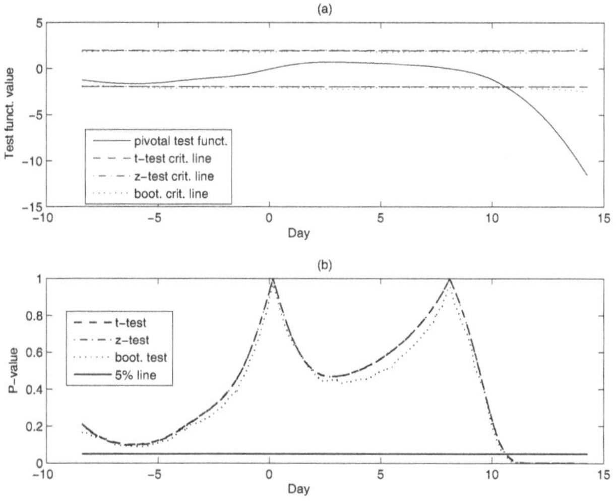

Figure 5.4 *Pointwise t-, z-, and bootstrap tests for the two-sample problem (5.2) for the progesterone data: (a) using the pivotal test function (solid), together with the 5% upper and lower critical lines of the pointwise t-test (dashed), the pointwise z-test (dot-dashed), and the pointwise bootstrap test (dotted); and (b) using the pointwise P-values of the pointwise t-test (dashed), the pointwise z-test (dot-dashed), and the pointwise bootstrap test (dotted), together with the 5% significance level line (solid). Both figures indicate that the mean function differences of the nonconceptive and conceptive progesterone curves are significant after Day 11.*

One can also conduct the pointwise tests by constructing the pointwise confidence intervals (5.11). Similar conclusions can be made. We leave this to the reader as an exercise.

L^2-Norm-Based Test For the two-sample problem (5.2), the L^2-norm-based test uses the squared L^2-norm of the pivotal test function $\Delta(t)$ (5.4) as the test statistic:

$$T_n = \int_{\mathcal{T}} \Delta^2(t)dt = \frac{n_1 n_2}{n} \int_{\mathcal{T}} [\bar{y}_1(t) - \bar{y}_2(t)]^2 dt. \tag{5.14}$$

Under the null hypothesis in (5.2), when the two functional samples (5.1) are Gaussian, by Theorem 5.1, we have $\Delta(t) \sim \mathrm{GP}(0, \gamma)$ and when the two samples (5.1) are large and satisfy Assumptions TS1, TS3, and TS4, then by Theorem 5.2, we have $\Delta(t) \sim \mathrm{GP}(0, \gamma)$ asymptotically. Therefore, under the conditions of Theorem 5.1 or under the conditions of Theorem 5.2, by Theorem 4.2 of Chapter 4 and under the null hypothesis in (5.2), we have or approximately have

$$T_n \stackrel{d}{=} \sum_{r=1}^{m} \lambda_r A_r,\ A_r \stackrel{i.i.d.}{\sim} \chi_1^2,$$

where $\lambda_1, \lambda_2, \cdots, \lambda_m$ are all the positive eigenvalues of the common covariance function $\gamma(s,t)$. It follows that the null distribution of T_n can be approximated using the methods described in Section 4.3 of Chapter 4. In fact, by the Welch-Satterthwaite χ^2-approximation method described there, we have

$$T_n \sim \beta\chi_d^2 \text{ approximately,} \quad \text{where } \beta = \frac{\text{tr}(\gamma^{\otimes 2})}{\text{tr}(\gamma)},\ d = \kappa = \frac{\text{tr}^2(\gamma)}{\text{tr}(\gamma^{\otimes 2})}. \tag{5.15}$$

Again, we emphasize that the parameter β is usually referred to the scale parameter as it is related to the scale of the test statistic T_n of the L^2-norm-based test while the parameter κ is referred to as the degree of freedom adjustment factor (Shen and Faraway 2004), which depends on the common covariance function $\gamma(s,t)$ only. In practice, the parameters β and κ must be estimated based on the two samples (5.1). A natural way to do this is to replace $\text{tr}(\gamma), \text{tr}^2(\gamma)$ and $tr(\gamma^{\otimes 2})$ with their estimators so that

$$\hat{\beta} = \frac{\text{tr}(\hat{\gamma}^{\otimes 2})}{\text{tr}(\hat{\gamma})},\ \hat{d} = \hat{\kappa} = \frac{\text{tr}^2(\hat{\gamma})}{\text{tr}(\hat{\gamma}^{\otimes 2})}, \tag{5.16}$$

where $\hat{\gamma}(s,t)$ is the pooled sample covariance function given in (5.3). In this case, we have

$$T_n \sim \hat{\beta}\chi_{\hat{d}}^2 \text{ approximately.} \tag{5.17}$$

The above method is usually known as the naive method, which often works well for highly correlated data but it is biased as both $\text{tr}^2(\hat{\gamma})$ and $\text{tr}(\hat{\gamma}^{\otimes 2})$ are biased for $\text{tr}^2(\gamma)$ and $\text{tr}(\gamma^{\otimes 2})$. When the functional samples (5.1) are Gaussian, by Theorem 4.6 of Chapter 4, we can obtain the unbiased estimators of $\text{tr}^2(\gamma)$ and $\text{tr}(\gamma^{\otimes 2})$ as

$$\begin{array}{c} \frac{(n-2)(n-1)}{(n-3)n}\left[\text{tr}^2(\hat{\gamma}) - \frac{2\text{tr}(\hat{\gamma}^{\otimes 2})}{n-1}\right], \text{ and} \\ \frac{(n-2)^2}{(n-3)n}\left[\text{tr}(\hat{\gamma}^{\otimes 2}) - \frac{\text{tr}^2(\hat{\gamma})}{n-2}\right], \end{array} \tag{5.18}$$

respectively, as by Theorem 5.1, we have $(n-2)\hat{\gamma}(s,t) \sim \text{WP}(n-2,\gamma)$. Replacing $\text{tr}^2(\gamma)$ and $\text{tr}(\gamma^{\otimes 2})$ in (5.15) by their unbiased estimator defined above results in the so-called bias-reduced method for estimating the parameters β and d. The resulting estimators are

$$\begin{array}{rcl} \hat{\beta} & = & \dfrac{\frac{(n-2)^2}{(n-3)n}\left[\text{tr}(\hat{\gamma}^{\otimes 2}) - \frac{\text{tr}^2(\hat{\gamma})}{n-2}\right]}{\text{tr}(\hat{\gamma})}, \\ \hat{d} & = & \hat{\kappa} = \dfrac{(n-1)\left[\text{tr}^2(\hat{\gamma}) - \frac{2\text{tr}(\hat{\gamma}^{\otimes 2})}{n-1}\right]}{(n-2)\left[\text{tr}(\hat{\gamma}^{\otimes 2}) - \frac{\text{tr}^2(\hat{\gamma})}{n-2}\right]}. \end{array} \tag{5.19}$$

When n_1 and n_2 are large, the estimators $\hat{\beta}$ and $\hat{d}$ obtained by the bias-reduced method are approximately equal to those by the naive method. Furthermore, we can show that with the sample sizes n_1, n_2 growing to infinity, the estimators $\hat{\boldsymbol{\beta}}$ and $\hat{\kappa}$ are consistent in the sense of the following theorem. This result is expected.

Theorem 5.4 *Under Assumptions TS1 and TS3 through TS7, as $n \to \infty$, we have $tr(\hat{\gamma}) \xrightarrow{p} tr(\gamma)$ and $tr(\hat{\gamma}^{\otimes 2}) \xrightarrow{p} tr(\gamma^{\otimes 2})$. Furthermore, as $n \to \infty$, we have*

$$\hat{\beta} \xrightarrow{p} \beta, \quad \hat{\kappa} \xrightarrow{p} \kappa,$$

where $\hat{\beta}$ and $\hat{\kappa}$ are the naive or bias-reduced estimators of β and κ, respectively.

The L^2-norm-based test can then be conducted by rejecting the null hypothesis in (5.2) whenever $T_n > \hat{\beta}\chi^2_{\hat{d}}(1-\alpha)$ for any given significance level α where $\hat{\beta}$ and $\hat{d}$ are given in (5.16) or in (5.19). Alternatively, the two-sample test can be conducted by computing the P-value $P(\chi^2_{\hat{d}} \geq T_n/\hat{\beta})$.

Example 5.6 *Table 5.3 displays the L^2-norm-based test for the two-sample problem (5.2) for the progesterone data over various periods. The functions were discretized with resolution $M = 1{,}000$ as described in Section 4.5.5 of Chapter 4. The quantities $T_n, \hat{\beta}$, and $\hat{\kappa}$ are numerically computed as there but we can ignore the constant term $v(\mathcal{T})/M$ in both the approximate expressions of T_n and $\hat{\beta}$ for ease computation. The values of $\hat{\beta}$ and $\hat{\kappa}$ can be computed using the values of $tr(\hat{\gamma})$ and $tr(\hat{\gamma}^{\otimes 2})$ given in Table 5.2. For example, over the time period $[-8, 0]$, from Table 5.2 we have $tr(\hat{\gamma}) = 255$ and $tr(\hat{\gamma}^{\otimes 2}) = 53{,}577$. By the naive method, we have*

$$\begin{aligned} \hat{\beta} &= \frac{tr(\hat{\gamma}^{\otimes 2})}{tr(\hat{\gamma})} = 53{,}577/255 = 210, \\ \hat{\kappa} &= \frac{tr^2(\hat{\gamma})}{tr(\hat{\gamma}^{\otimes 2})} = 255^2/53{,}577 = 1.214. \end{aligned}$$

Using the formulas (5.18) we can also compute the unbiased estimators of $tr^2(\gamma)$ and $tr(\gamma^{\otimes 2})$. From these, we can compute the bias-reduced estimators of β and κ.

It is seen from Table 5.3 that both the naive and bias-reduced methods produced very similar results and the null hypothesis (5.2) is not significant before Day 8, suggesting that the mean functions (in log*-scale) of the nonconceptive progesterone curves and the conceptive progesterone curves are very likely to be the same before Day 8 while over the whole experimental period from Day −8 to Day 15, they are significantly different from each other. Using the L^2-norm-based test, one can also verify that the two mean functions have no significant differences before Day 10. This is a good exercise left to the reader.*

Table 5.3 *The L^2-norm-based test for the two-sample problem (5.2) for the progesterone data with resolution $M = 1,000$.*

Method	$[a, b]$	T_n	$\hat{\beta}$	$\hat{d} = \hat{\kappa}$	P-value
Naive	$[-8, 0]$	407.4	210.4	1.21	0.208
	$[-8, 8]$	513.2	368.6	1.41	0.347
	$[-8, 15]$	3,751	496.7	1.48	0.012
Bias-reduced	$[-8, 0]$	407.4	205.2	1.22	0.203
	$[-8, 8]$	513.2	358.9	1.42	0.343
	$[-8, 15]$	3,751	483.2	1.50	0.011

Note: The P-values by the naive method are generally comparable with those by the bias-reduced method although the latter are generally smaller than the former.

F-Type Test Notice that the L^2-norm-based test discussed earlier does not take into account the variation of the pooled sample covariance function $\hat{\gamma}(s,t)$ as defined in (5.3) but it works well even when the Gaussian assumption is not satisfied but the sample sizes n_1 and n_2 are sufficiently large. When the two samples (5.1) are Gaussian, we can partially take the variations of $\hat{\gamma}(s,t)$ into account by the so-called F-type test described below.

For the two-sample problem (5.2), recall that the L^2-norm-based test is based on the squared L^2-norm $\|\Delta\|^2$ of the pivotal test function $\Delta(t) = \sqrt{\frac{n_1 n_2}{n}}(\bar{y}_1(t) - \bar{y}_2(t))$. Under the null hypothesis and the Gaussian assumption, Theorem 5.1 states that $\Delta(t) \sim \text{GP}(0, \gamma)$ and $(n-2)\hat{\gamma}(s,t) \sim \text{WP}(n-2, \gamma)$ and they are independent. In addition, we have

$$\text{E}\|\Delta\|^2 = \text{tr}(\gamma), \quad \text{and} \quad \text{Etr}(\hat{\gamma}) = \text{tr}(\gamma). \tag{5.20}$$

Therefore, it is natural to test (5.2) using the following F-type test statistic:

$$F_n = \frac{\|\Delta\|^2}{\text{tr}(\hat{\gamma})} = \frac{\frac{n_1 n_2}{n}\int_{\mathcal{T}}[\bar{y}_1(t) - \bar{y}_2(t)]^2 dt}{\text{tr}(\hat{\gamma})}. \tag{5.21}$$

When the variation of $\text{tr}(\hat{\gamma})$ is not taken into account, the distribution of F_n is essentially the same as that of the L^2-norm-based test statistic T_n defined in (5.14). To take this variation into account, the Gaussian assumption is sufficient. In fact, under the Gaussian assumption and the null hypothesis, by Theorem 4.2, we have

$$\begin{array}{l} \|\Delta\|^2 \stackrel{d}{=} \sum_{r=1}^m \lambda_r A_r, \ A_r \stackrel{i.i.d.}{\sim} \chi_1^2, \\ \text{tr}(\hat{\gamma}) \stackrel{d}{=} \left(\sum_{r=1}^m \lambda_r E_r\right)/(n-2), \ E_r \stackrel{i.i.d.}{\sim} \chi_{n-2}^2, \end{array} \tag{5.22}$$

where A_r, E_r are all independent and $\lambda_1, \lambda_2, \cdots, \lambda_m$ are all the positive eigenvalues of $\gamma(s,t)$. Equivalently, we can write

$$F_n \stackrel{d}{=} \frac{\sum_{r=1}^m \lambda_r A_r}{\sum_{r=1}^m \lambda_r E_r/(n-2)}.$$

It follows that under the Gaussian assumption and the null hypothesis in

Table 5.4 *The F-type test for the two-sample problem (5.2) for the progesterone data with resolution $M = 1,000$.*

Method	$[a, b]$	F_n	$\hat{\kappa}$	$(n-2)\hat{\kappa}$	P-value
Naive	$[-8, 0]$	1.60	1.21	108	0.211
	$[-8, 8]$	0.99	1.41	125	0.349
	$[-8, 15]$	5.10	1.48	132	0.014
Bias-reduced	$[-8, 0]$	1.60	1.22	108	0.211
	$[-8, 8]$	0.99	1.42	126	0.350
	$[-8, 15]$	5.10	1.50	134	0.014

Note: The P-values by the naive method are generally comparable with those by the bias-reduced method.

(5.2), F_n is an F-type mixture as discussed in Section 4.4 in Chapter 4. Therefore, the null distribution of F_n can be approximated using the methods described there. In particular, by (5.20) and the two-cumulant matched F-approximation method described there, we have

$$F_n \sim F_{\kappa,(n-2)\kappa} \text{ approximately,}$$

where $\kappa = \frac{\text{tr}^2(\gamma)}{\text{tr}(\gamma^{\otimes 2})}$ as defined in (5.15) for the L^2-norm-based test. In addition, by the naive method, $\hat{\kappa}$ is given in (5.16) and by the bias-reduced method, it is given in (5.19). The F-type test is then conducted by rejecting the null hypothesis for the two-sample problem (5.2) whenever $F_n > F_{\hat{\kappa},(n-2)\hat{\kappa}}(1-\alpha)$ for any given significance level α. Notice that when $(n-2)\hat{\kappa}$ tends to ∞, the test result by the F-type test will be the same as that by the L^2-norm-based test described earlier.

Example 5.7 *Table 5.4 displays the F-type test for the two-sample problem (5.2) for the progesterone data. The P-values are about the same as those presented in Table 5.3 for the L^2-norm-based test. This is obviously due to the fact that the approximate degrees of freedom $(n-2)\hat{\kappa}$ of the denominator of F_n are rather large. Also, the naive method and the bias-reduced method produced quite similar results.*

Bootstrap Tests When the two samples (5.1) are not Gaussian and when the sample sizes n_1 and n_2 are small, the distributions of the L^2-norm-based test and the F-type test are in general not tractable. In this case, we may bootstrap their critical values, resulting in the L^2-norm-based bootstrap test and the F-type bootstrap test. Based on the two bootstrapped samples (5.13), we can compute the two sample mean functions $\bar{y}_1^*(t), \bar{y}_2^*(t)$, and the pooled sample covariance function $\hat{\gamma}^*(s,t)$ as in (5.3) but based on the bootstrapped two samples (5.13).

For the L^2-norm-based bootstrap test, we compute the bootstrap test statistic $T_n^* = \|\Delta^*\|^2$ with the pivotal test function

$$\Delta^*(t) = \sqrt{\frac{n_1 n_2}{n}} \left[(\bar{y}_1^*(t) - \bar{y}_2^*(t)) - (\bar{y}_1(t) - \bar{y}_2(t))\right].$$

Table 5.5 *The bootstrap tests for the two-sample problem (5.2) for the progesterone data with resolution $M = 1,000$.*

	L^2-norm-based bootstrap test		F-type bootstrap test	
$[a,b]$	T_n	P-value	F_n	P-value
$[-8,0]$	407.4	0.203	1.60	0.214
$[-8,8]$	513.2	0.340	0.99	0.344
$[-8,15]$	3,751	0.014	5.10	0.016

Note: The number of bootstrap replicates is $N = 10,000$. The P-values by the bootstrap method are generally comparable with those by the L^2-norm-based test and by the F-type test, which are presented in Tables 5.3 and 5.4, respectively.

The above bootstrap test statistic is defined in that way as given the original two samples (5.1), the bootstrapped two samples (5.13) have the mean functions $\bar{y}_1(t)$ and $\bar{y}_2(t)$, respectively. Repeat the above bootstrap process a large number of times so that one can obtain a bootstrap sample of T_n^* that can be used to estimate the $100(1-\alpha)$-percentile of $T_n = \|\Delta\|^2$. Similarly, for the F-type bootstrap test, we compute the bootstrap test statistic $F_n^* = \frac{\|\Delta^*\|}{\text{tr}(\hat{\gamma}^*)}$. Repeat this process a large number of times so that one can obtain a bootstrap sample of F_n^* that allows one to estimate the $100(1-\alpha)$-percentile of F_n.

Example 5.8 *Table 5.5 displays the L^2-norm-based and F-type bootstrap tests for the two-sample problem (5.2) for the progesterone data with resolution $M = 1,000$. The number of bootstrap replicates is $N = 10,000$. The associated bootstrapped P-values are displayed. It is seen that the bootstrap test results are consistent with those by the L^2-norm-based test and by the F-type test presented in Tables 5.3 and 5.4, respectively.*

5.3 One-Way ANOVA

In classical linear models, one-way analysis of variance (ANOVA) is a technique used to compare means of three or more samples or groups. The samples or groups are grouped according to a categorical variable, known as a factor. The number of different values taken by the factor is called the level of the factor. In this section, we aim to extend this technique for functional data analysis.

Example 5.9 *We use the Canadian temperature data (Canadian Climate Program 1982), introduced in Section 1.2.4 of Chapter 1, to motivate the one-way ANOVA model. The Canadian temperature data are the daily temperature records of thirty-five Canadian weather stations over a year (365 days), among which, fifteen are in Eastern, another fifteen in Western and the remaining five in Northern Canada. Figure 5.5 presents their reconstructed individual curves, obtained by applying the local linear kernel smoother with the*

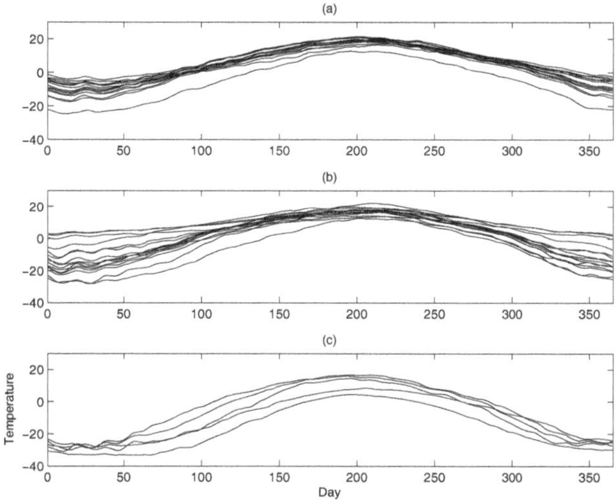

Figure 5.5 *Reconstructed Canadian temperature curves for (a) fifteen Eastern weather stations, (b) fifteen Western weather stations, and (c) five Northern weather stations. The local linear kernel smoother with the well-known Gaussian kernel is applied. The optimal bandwidth* $h^* = 2.79$ *selected by the GCV rule (3.10) is used for all thirty-five temperature curves.*

well-known Gaussian kernel to the individual temperature records of each of the thirty-five weather stations, respectively, but with a common bandwidth $h^* = 2.79$, *selected by the GCV rule (3.10) of Chapter 3. It can be seen that at least at the middle of the year, the temperatures at the Eastern weather stations are comparable with those at the Western weather stations, but they are generally higher than those temperatures at the Northern weather stations. This observation seems reasonable as the Eastern and Western weather stations are located at about the same latitudes while the Northern weather stations are located at higher latitudes. Of interest is to test if the mean temperature curves of the Eastern, Western, and Northern weather stations are the same. Alternatively speaking, we want to test if the mean temperatures of the weather stations over a year are strongly affected by the locations of these weather stations. This motivates a one-way ANOVA problem for functional data, a natural extension of the two-sample problem for functional data discussed in the previous section.*

We can define the one-way ANOVA problem for functional data as follows. Suppose we have k independent samples:

$$y_{i1}(t), \cdots, y_{in_i}(t), \; i = 1, \cdots, k. \tag{5.23}$$

These k samples satisfy

$$y_{ij}(t) = \eta_i(t) + v_{ij}(t),\ v_{ij}(t) \overset{i.i.d.}{\sim} \text{SP}(0, \gamma), \\ j = 1, 2, \cdots, n_i; i = 1, 2, \cdots, k, \tag{5.24}$$

where $\eta_1(t), \eta_2(t), \cdots, \eta_k(t)$ are the unknown group mean functions of the k samples, $v_{ij}(t), j = 1, \cdots, n_i; i = 1, 2, \cdots, k$ are the subject-effect functions, and $\gamma(s, t)$ is the common covariance function. We wish to test the following one-way ANOVA testing problem:

$$H_0 : \eta_1(t) \equiv \eta_2(t) \equiv \cdots \equiv \eta_k(t),\ t \in \mathcal{T}, \tag{5.25}$$

where again $\mathcal{T}$ is some time period of interest, often specified as $[a, b]$ with $-\infty < a < b < \infty$. The above one-way ANOVA problem is also known as the k-sample problem for functional data, extending the two-sample problem for functional data discussed in the previous section.

The one-way ANOVA testing problem (5.25) often aims to check if the effect of a factor or a treatment is statistically significant. This factor or treatment is usually used to group the individual functions into several samples, groups or categories. If the factor or treatment has serious impact on the functional data, the one-way ANOVA problem (5.25) will be statistically significant.

Example 5.10 *In the Canadian temperature data, the factor is "the location of a weather station" that may affect the temperature of the weather station. The number, k, of samples or categories in which a functional data set is grouped is the number of levels of the factor. For the Canadian temperature data, the number of levels is $k = 3$.*

For the one-way ANOVA problem, we are interested in the following three major kinds of tests:

Main-Effect Test Set $\eta_i(t) = \eta(t) + \alpha_i(t), i = 1, 2, \cdots, k$, where $\eta(t)$ is known as the overall mean function of the k samples and $\alpha_i(t)$ is the ith main-effect function for $i = 1, 2, \cdots, k$. Then the model (5.24) can be further written as the following standard one-way ANOVA model for functional data:

$$y_{ij}(t) = \eta(t) + \alpha_i(t) + v_{ij}(t),\ j = 1, 2, \cdots, n_i;\ i = 1, 2, \cdots, k. \tag{5.26}$$

In this formulation, the null hypothesis (5.25) can be equivalently expressed as

$$\alpha_1(t) \equiv \alpha_2(t) \equiv \cdots \equiv \alpha_k(t) \equiv 0, t \in \mathcal{T}, \tag{5.27}$$

that is, to test if the main-effect functions are the same and are equal to 0.

Post Hoc Test When the test (5.27) is accepted, the one-way ANOVA model (5.26) is not significant. When it is rejected, further investigation is often required. For example, one may want to know if any two main-effect functions $\alpha_i(t)$ and $\alpha_j(t)$ are the same, where i and j are any two integers such that $1 \le i < j \le k$. This test can be written as

$$\begin{array}{lll} & H_0 & : \alpha_i(t) \equiv \alpha_j(t), t \in \mathcal{T} \\ \text{versus} & H_1 & : \alpha_i(t) \neq \alpha_j(t), \text{ for some } t \in \mathcal{T}. \end{array} \quad (5.28)$$

The above test is known as a post hoc test. Obviously, it can be equivalently written as

$$\begin{array}{lll} & H_0 & : \eta_i(t) \equiv \eta_j(t), t \in \mathcal{T} \\ \text{versus} & H_1 & : \eta_i(t) \neq \eta_j(t), \text{ for some } t \in \mathcal{T}. \end{array} \quad (5.29)$$

Contrast Test The post hoc tests are special cases of contrast tests. Let $a_1, \cdots, a_k$ be k constants such that they add up to 0, that is, $\sum_{i=1}^k a_i = \mathbf{a}^T \mathbf{1}_k = 0$, where $\mathbf{a} = [a_1, \cdots, a_k]^T$ and $\mathbf{1}_k$ is a column vector of k ones. A contrast is defined as $\sum_{i=1}^k a_i \alpha_i(t) = \mathbf{a}^T \boldsymbol{\alpha}(t)$, a linear combination of the main-effect functions $\alpha_i(t), i = 1, 2, \cdots, k$, where $\boldsymbol{\alpha}(t) = [\alpha_1(t), \cdots, \alpha_k(t)]^T$ consists of all the main-effect functions. A simple contrast is the difference of two main-effect functions, for example, $\alpha_1(t) - \alpha_2(t)$. Another simple example of contrast is $\alpha_1(t) - 3\alpha_2(t) + 2\alpha_5(t)$ when $k \ge 5$. For a given $\mathbf{a} \in \mathcal{R}^k$ such that $\mathbf{a}^T \mathbf{1}_k = 0$, a contrast test is defined as

$$H_0 : \mathbf{a}^T \boldsymbol{\alpha}(t) \equiv 0, \ t \in \mathcal{T} \quad \text{versus} \quad H_1 : \mathbf{a}^T \boldsymbol{\alpha}(t) \neq 0, \text{ for some } t \in \mathcal{T}. \quad (5.30)$$

As $\mathbf{a}^T \mathbf{1}_k = 0$, the above test can be equivalently written as

$$H_0 : \mathbf{a}^T \boldsymbol{\eta}(t) \equiv 0, t \in \mathcal{T} \quad \text{versus} \quad H_1 : \mathbf{a}^T \boldsymbol{\eta}(t) \neq 0, \text{ for some } t \in \mathcal{T}, \quad (5.31)$$

where $\boldsymbol{\eta}(t) = [\eta_1(t), \eta_2(t), \cdots, \eta_k(t)]^T$.

5.3.1 Estimation of Group Mean and Covariance Functions

Based on the k samples (5.23), the group mean functions $\eta_i(t), i = 1, 2, \cdots, k$ and the common covariance function $\gamma(s,t)$ can be unbiasedly estimated as

$$\begin{array}{rcl} \hat{\eta}_i(t) & = & \bar{y}_{i.}(t) = n_i^{-1} \sum_{j=1}^{n_i} y_{ij}(t), \quad i = 1, 2, \cdots, k, \\ \hat{\gamma}(s,t) & = & (n-k)^{-1} \sum_{i=1}^k \sum_{j=1}^{n_i} [y_{ij}(s) - \bar{y}_{i.}(s)][y_{ij}(t) - \bar{y}_{i.}(t)], \end{array} \quad (5.32)$$

where and throughout this section $n = \sum_{i=1}^k n_i$ denotes the total sample size. The estimated covariance function $\hat{\gamma}(s,t)$ is also known as the pooled sample covariance function. Note that $\hat{\eta}_i(t), i = 1, 2, \cdots, k$ are independent and

$$\mathrm{E}\hat{\eta}_i(t) = \eta_i(t), \quad \mathrm{cov}\left[\hat{\eta}_i(s), \hat{\eta}_i(t)\right] = \gamma(s,t)/n_i, i = 1, 2, \cdots, k.$$

Set $\hat{\boldsymbol{\eta}}(t) = [\hat{\eta}_1(t), \hat{\eta}_2(t), \cdots, \hat{\eta}_k(t)]^T$. It is an unbiased estimator of $\boldsymbol{\eta}(t)$. Then we have $\mathrm{E}\hat{\boldsymbol{\eta}}(t) = \boldsymbol{\eta}(t)$ and $\mathrm{Cov}\left[\hat{\boldsymbol{\eta}}(s), \hat{\boldsymbol{\eta}}(t)\right] = \gamma(s,t)\mathbf{D}$, where $\mathbf{D} =$

$\text{diag}(1/n_1, 1/n_2, \cdots, 1/n_k)$ is a diagonal matrix with diagonal entries $1/n_i, i = 1, 2, \cdots, k$. That is, $\hat{\boldsymbol{\eta}}(t) \sim \text{SP}_k(\boldsymbol{\eta}, \gamma\mathbf{D})$, where $\text{SP}_k(\boldsymbol{\eta}, \boldsymbol{\Gamma})$ denotes a k-dimensional stochastic process having the vector of mean functions $\boldsymbol{\eta}(t)$ and the matrix of covariance functions $\boldsymbol{\Gamma}(s,t)$.

To study some procedures for the main-effect, post hoc, and contrast tests, we need to investigate the properties of $\hat{\boldsymbol{\eta}}(t)$ and $\hat{\gamma}(s,t)$ under various conditions. For this purpose, we list the following assumptions:

One-Way ANOVA Assumptions (KS)

1. The k samples (5.23) are with $\eta_1(t), \eta_2(t), \cdots, \eta_k(t) \in \mathcal{L}^2(\mathcal{T})$ and $\text{tr}(\gamma) < \infty$.
2. The k samples (5.23) are Gaussian.
3. As $n \to \infty$, the k sample sizes satisfy $n_i/n \to \tau_i,\ i = 1, 2, \cdots, k$ such that $\tau_1, \tau_2, \cdots, \tau_k \in (0, 1)$.
4. The subject-effect functions $v_{ij}(t) = y_{ij}(t) - \eta_i(t), j = 1, 2, \cdots, n_i; i = 1, 2, \cdots, k$ are i.i.d..
5. The subject-effect function $v_{11}(t)$ satisfies $\text{E}\|v_{11}\|^4 < \infty$.
6. The maximum variance $\rho = \max_{t \in \mathcal{T}} \gamma(t,t) < \infty$.
7. The expectation $\text{E}[v_{11}^2(s)v_{11}^2(t)]$ is uniformly bounded.

The above assumptions are natural extensions of those assumptions for the two-sample problem presented in the previous section. In particular, Assumption KS3 requires that the k sample sizes $n_1, n_2, \cdots, n_k$ tend to ∞ proportionally and Assumptions KS3, KS4, and KS5 are imposed for investigating the asymptotical properties of $\hat{\eta}_i(t) = \bar{y}_{i\cdot}(t), i = 1, 2, \cdots, k$ and $\hat{\gamma}(s,t)$ for non-Gaussian functional data. Assumption KS3 guarantees that as $n \to \infty$, the sample mean functions $\bar{y}_{i\cdot}(t), i = 1, 2, \cdots, k$ will converge to Gaussian processes weakly, and Assumption KS7 is satisfied when $v_{11}(t)$ is uniformly bounded in probability over the finite interval $\mathcal{T}$.

Theorem 5.5 *Under Assumptions KS1 and KS2, we have*

$$\begin{aligned} \mathbf{D}^{-1/2}\left[\hat{\boldsymbol{\eta}}(t) - \boldsymbol{\eta}(t)\right] &\sim GP_k(\mathbf{0}, \gamma\mathbf{I}_k), \quad \textit{and} \\ (n-k)\hat{\gamma}(s,t) &\sim WP(n-k, \gamma). \end{aligned} \tag{5.33}$$

Theorem 5.5 shows that under the Gaussian assumption KS2, $\hat{\boldsymbol{\eta}}(t)$ is a k-dimensional Gaussian process and $(n-k)\hat{\gamma}(s,t)$ is a Wishart process. It is easy to see that Theorem 5.5 is a natural extension of Theorem 5.1 and is the key for constructing various tests for (5.25) when the k samples (5.23) are Gaussian. We would like to emphasize that under the Gaussian assumption KS2, Theorem 5.5 holds even when the sample sizes $n_1, n_2, \cdots, n_k$ are finite.

When the Gaussian assumption KS2 is not valid, we need more assumptions to derive the asymptotic distribution of $\hat{\boldsymbol{\eta}}(t)$, including that the sample sizes $n_1, n_2, \cdots, n_k$ should be proportionally large. That is, Assumption KS3 should be satisfied.

Theorem 5.6 *Under Assumptions KS1, KS3, and KS4, as $n \to \infty$, we have*

$$\mathbf{D}^{-1/2}\left[\hat{\boldsymbol{\eta}}(t) - \boldsymbol{\eta}(t)\right] \xrightarrow{d} GP_k(\mathbf{0}, \gamma \mathbf{I}_k).$$

Assumptions KS3 and KS4 for Theorem 5.6 are sufficient to guarantee that as $n \to \infty$, each of the sample mean functions $\hat{\eta}_i(t) = \bar{y}_{i.}(t), i = 1, 2, \cdots, k$ converges to a Gaussian process weakly. However, when Assumption KS3 is not satisfied, the matrix $n\mathbf{D}$ may not be invertible when n is large and hence the above theorem may not hold. Therefore, Assumption KS3 is also a necessary condition.

With some further assumptions, we can show that the pooled sample covariance function $\hat{\gamma}(s,t)$ is asymptotically Gaussian. This result also shows that $\hat{\gamma}(s,t)$ is root-n consistent.

Theorem 5.7 *Under Assumptions KS1, KS3, KS4, KS5, and KS6, as $n \to \infty$, we have*

$$\sqrt{n}\left\{\hat{\gamma}(s,t) - \gamma(s,t)\right\} \xrightarrow{d} GP(0, \varpi), \tag{5.34}$$

where $\varpi\{(s_1,t_1),(s_2,t_2)\} = E\{v_{11}(s_1)v_{11}(t_1)v_{11}(s_2)v_{11}(t_2)\} - \gamma(s_1,t_1)\gamma(s_2,t_2)$.

For the main-effect, post hoc, or contrast tests, we do not need to identify the main-effect functions $\alpha_i(t), i = 1, 2, \cdots, k$ defined in (5.26). In fact, they are not identifiable unless some constraint is imposed. If we do want to estimate these main-effect functions, the most commonly used constraint is

$$\sum_{i=1}^{k} n_i \alpha_i(t) = 0, \tag{5.35}$$

involving the k sample sizes. Under this constraint, it is easy to show that the unbiased estimators of the main-effect functions are

$$\hat{\alpha}_i(t) = \bar{y}_{i.}(t) - \bar{y}_{..}(t), \; i = 1, 2, \cdots, k, \tag{5.36}$$

where

$$\bar{y}_{..}(t) = n^{-1} \sum_{i=1}^{k} \sum_{j=1}^{n_i} y_{ij}(t) = n^{-1} \sum_{i=1}^{k} n_i \bar{y}_{i.}(t) \tag{5.37}$$

is the usual sample grand mean function. Under the constraint (5.35), $\bar{y}_{..}(t)$ is an unbiased estimator of the grand mean function $\eta(t)$ defined in (5.26).

Let

$$\begin{aligned} \text{SSH}_n(t) &= \textstyle\sum_{i=1}^{k} n_i [\bar{y}_{i.}(t) - \bar{y}_{..}(t)]^2, \text{ and} \\ \text{SSE}_n(t) &= \textstyle\sum_{i=1}^{k} \sum_{j=1}^{n_i} [y_{ij}(t) - \bar{y}_{i.}(t)]^2, \end{aligned} \tag{5.38}$$

denote the pointwise between-subject and within-subject variations, respectively, where $\bar{y}_{i.}(t), i = 1, 2, \cdots, k$ are the group sample mean functions as defined in (5.32) and $\bar{y}_{..}(t)$ is the sample grand mean function as defined in (5.37). Under the constraint (5.35), it is easy to see that

$$\mathrm{SSH}_n(t) = \sum_{i=1}^{k} n_i \hat{\alpha}_i^2(t), \tag{5.39}$$

where $\hat{\alpha}_i(t), i = 1, 2, \cdots, k$ are the estimated main-effect functions as given in (5.36). It is seen that when the null hypothesis (5.25) is valid, $\mathrm{SSH}_n(t)$ should be small and otherwise it is large. From (5.32), we can see that

$$\mathrm{SSE}_n(t) = (n-k)\hat{\gamma}(t,t). \tag{5.40}$$

We have the following results.

Theorem 5.8 *Suppose Assumptions KS1 and KS2 hold. Then under the null hypothesis (5.25), we have*

$$\begin{array}{rcl} \int_{\mathcal{T}} SSH_n(t)dt & \stackrel{d}{=} & \sum_{r=1}^{m} \lambda_r A_r,\ A_r \stackrel{i.i.d.}{\sim} \chi_{k-1}^2, \\ \int_{\mathcal{T}} SSE_n(t)dt & \stackrel{d}{=} & \sum_{r=1}^{m} \lambda_r E_r,\ E_r \stackrel{i.i.d.}{\sim} \chi_{n-k}^2, \end{array}$$

where $A_r, E_r, r = 1, 2, \cdots, m$ are independent of each other, and $\lambda_1, \cdots, \lambda_m$ are all the positive eigenvalues of $\gamma(s,t)$.

Theorem 5.9 *Suppose Assumptions KS1, KS3, and KS4 hold. Then under the null hypothesis (5.25), as $n \to \infty$, we have*

$$\int_{\mathcal{T}} SSH_n(t)dt \stackrel{d}{\to} \sum_{i=1}^{m} \lambda_r A_r,\ A_r \stackrel{i.i.d.}{\sim} \chi_{k-1}^2,$$

where $\lambda_1, \cdots, \lambda_m$ are all the positive eigenvalues of $\gamma(s,t)$.

Theorem 5.8 shows that under the Gaussian assumption KS2 and the null hypothesis, the integrated SSH and SSE are χ^2-type mixtures. Theorem 5.9 shows that when the sample sizes are proportionally tending to infinity and under the null hypothesis, the integrated SSH is asymptotically a χ^2-type mixture even when the Gaussian assumption KS2 is not valid. These two results will be used to construct various tests for the main-effect test of the one-way ANOVA problem as described in the next subsection.

5.3.2 Main-Effect Test

We are now ready to describe various tests for the main-effect testing problem (5.25).

Pointwise Tests We consider the pointwise F-test, the pointwise χ^2-test, and the pointwise bootstrap test. The pointwise F-test for (5.25) was adopted by Ramsay and Silverman (2005, Section 13.2.2., Chapter 13), naturally extending the classical F-test to the context of functional data analysis. The pointwise F-test is conducted for (5.25) at each $t \in \mathcal{T}$ using the following pointwise F statistic:

$$F_n(t) = \frac{\text{SSH}_n(t)/(k-1)}{\text{SSE}_n(t)/(n-k)}. \tag{5.41}$$

From the classical linear model theory, it is easy to see that when the k samples (5.23) are Gaussian, under the null hypothesis (5.25), we have

$$F_n(t) \sim F_{k-1,n-k}, t \in \mathcal{T}. \tag{5.42}$$

The pointwise F-test is then conducted by rejecting (5.25) at each $t \in \mathcal{T}$ whenever $F_n(t) > F_{k-1,n-k}(1-\alpha)$ for any given significance level α or by computing the pointwise P-values at each $t \in \mathcal{T}$ based on the pointwise F-distribution (5.42).

When the k-samples are not Gaussian, for large samples, one may use the pointwise χ^2-test. It is easy to see that as $n_{\min} = \min_{i=1}^k n_i \to \infty$, asymptotically we have

$$F_n(t) \sim \chi^2_{k-1}/(k-1), \quad t \in \mathcal{T}. \tag{5.43}$$

This is because as $n_{\min} \to \infty$, the denominator $\text{SSE}_n(t)/(n-k) = \hat{\gamma}(t,t)$ of $F_n(t)$ tends to $\gamma(t,t)$ almost surely while the numerator $\text{SSH}_n(t)/(k-1)$ tends to $\gamma(t,t)\chi^2_{k-1}/(k-1)$. The pointwise χ^2-test is conducted by rejecting (5.25) at any given t whenever $F_n(t) > \chi^2_{k-1}(1-\alpha)/(k-1)$ or by computing the pointwise P-values of $F_n(t)$ at any given t based on the distribution (5.43).

When the k samples (5.23) are not Gaussian and the sample sizes $n_1, n_2, \cdots, n_k$ are small, the above pointwise F- and χ^2-tests are not preferred. In this case, one may resort to a pointwise bootstrap test that can be briefly described as follows. Let

$$v^*_{ij}(t), j = 1, 2, \cdots, n_i; i = 1, 2, \cdots, k, \tag{5.44}$$

be k bootstrap samples randomly generated from the estimated subject-effect functions $\hat{v}_{ij}(t) = y_{ij}(t) - \hat{\eta}(t_{ij}), j = 1, 2, \cdots, n_{ij}; i = 1, 2, \cdots, k$. We first compute

$$\bar{v}^*_{i.}(t) = n_i^{-1} \sum_{j=1}^{n_i} v^*_{ij}(t), \quad \bar{v}^*_{..}(t) = n^{-1} \sum_{i=1}^{k} \sum_{j=1}^{n_i} v^*_{ij}(t).$$

We then compute the following bootstrapped $\text{SSH}_n(t)$ and $\text{SSE}_n(t)$ as

$$\begin{aligned} \text{SSH}^*_n(t) &= \textstyle\sum_{i=1}^k n_i[\bar{v}^*_{i.}(t) - \bar{v}^*_{..}(t)]^2, \text{ and} \\ \text{SSE}^*_n(t) &= \textstyle\sum_{i=1}^k \sum_{j=1}^{n_i} [v^*_{ij}(t) - \bar{v}^*_{i.}(t)]^2. \end{aligned} \tag{5.45}$$

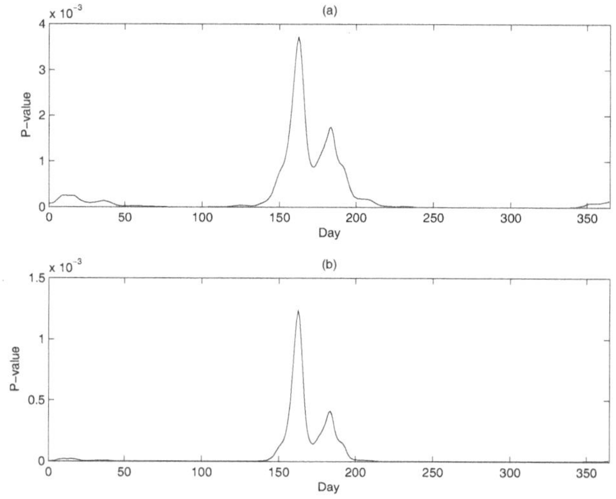

Figure 5.6 *Pointwise tests for the main-effect testing problem (5.25) with the Canadian temperature data with (a) pointwise F-test and (b) pointwise χ^2-test. It appears that the P-values of the pointwise F-test are generally larger than those of the pointwise χ^2-test.*

To mimic the pointwise F-test statistic $F_n(t)$ (5.41), the pointwise bootstrapped test statistic is computed as

$$F_n^*(t) = \frac{\text{SSH}_n^*(t)/(k-1)}{\text{SSE}_n^*(t)/(n-k)}.$$

Repeat the above bootstrapping process a large number of times and calculate the $100(1-\alpha)$-percentile of $F_n^*(t)$, which is used to estimate the $100(1-\alpha)$-percentile of $F_n(t)$. The pointwise bootstrap test is then conducted accordingly.

Example 5.11 *Figure 5.6 displays the pointwise F- and χ^2-tests for the main-effect testing problem (5.25) with the Canadian temperature data. It is seen that both tests suggest to reject the null hypothesis (5.25) at each point with strong evidence. The P-value of the pointwise F-test is less than* 0.004 *while the P-value of the pointwise χ^2-test is less than* 0.0015. *As the sample sizes $n_1 = 15, n_2 = 15, n_3 = 5$ are not large, the P-values of the pointwise F-test are more trustful if the Canadian temperature data are Gaussian.*

L^2-Norm-Based Test The L^2-norm-based test for the two-sample problem (5.2) can now be extended for the main-effect testing problem (5.25). The associated test statistic is defined as the integral of the pointwise between-subject variations:

$$T_n = \int_{\mathcal{T}} \mathrm{SSH}_n(t)dt = \sum_{i=1}^{k} n_i \int_{\mathcal{T}} [\bar{y}_{i.}(t) - \bar{y}_{..}(t)]^2 dt. \tag{5.46}$$

Under the null hypothesis (5.25) and under the conditions of Theorem 5.8 or under the conditions of Theorem 5.9, we have or approximately have

$$T_n = \sum_{r=1}^{m} \lambda_r A_r, \; A_r \overset{i.i.d.}{\sim} \chi^2_{k-1},$$

where $\lambda_r, r = 1, 2, \cdots, m$ are all the positive eigenvalues of $\gamma(s,t)$. It follows that we can approximate the null distribution of T_n by the Welch-Satterthwaite χ^2-approximation method described in Section 4.3 of Chapter 4. By that method, we obtain

$$T_n \sim \beta\chi^2_{(k-1)\kappa} \text{ approximately, where } \beta = \frac{\mathrm{tr}(\gamma^{\otimes 2})}{\mathrm{tr}(\gamma)}, \quad \kappa = \frac{\mathrm{tr}^2(\gamma)}{\mathrm{tr}(\gamma^{\otimes 2})}. \tag{5.47}$$

As mentioned before, the parameter β is usually referred to as the scale parameter as it is related to the scale of the test statistic T_n, while the parameter κ is referred to as the degrees of freedom adjustment factor (Shen and Faraway 2004). In practice, these two parameters must be estimated based on the functional data (5.23). As usual, there are two methods that can be used for this purpose. One is called the naive method and the other is called the bias-reduced method. With the unbiased estimator $\hat{\gamma}(s,t)$ given in (5.32), by the naive method, we have

$$\hat{\beta} = \frac{\mathrm{tr}(\hat{\gamma}^{\otimes 2})}{\mathrm{tr}(\hat{\gamma})}, \quad \hat{\kappa} = \frac{\mathrm{tr}^2(\hat{\gamma})}{\mathrm{tr}(\hat{\gamma}^{\otimes 2})}, \tag{5.48}$$

and by the bias-reduced method, we have

$$\hat{\beta} = \frac{\widehat{\mathrm{tr}(\gamma^{\otimes 2})}}{\mathrm{tr}(\hat{\gamma})}, \quad \hat{\kappa} = \frac{\widehat{\mathrm{tr}^2(\gamma)}}{\widehat{\mathrm{tr}(\gamma^{\otimes 2})}}, \tag{5.49}$$

with

$$\begin{aligned} \widehat{\mathrm{tr}^2(\gamma)} &= \frac{(n-k)(n-k+1)}{(n-k-1)(n-k+2)} \left[\mathrm{tr}^2(\hat{\gamma}) - \frac{2\mathrm{tr}(\hat{\gamma}^{\otimes 2})}{n-k+1}\right], \\ \widehat{\mathrm{tr}(\gamma^{\otimes 2})} &= \frac{(n-k)^2}{(n-k-1)(n-k+2)} \left[\mathrm{tr}(\hat{\gamma}^{\otimes 2}) - \frac{\mathrm{tr}^2(\hat{\gamma})}{n-k}\right]. \end{aligned} \tag{5.50}$$

Notice that the unbiased estimators (5.50) of $\mathrm{tr}^2(\gamma)$ and $\mathrm{tr}(\gamma^{\otimes 2})$ can be obtained when the Gaussian assumption is valid as under the Gaussian assumption KS2, by Theorem 5.5, we have $(n-k)\hat{\gamma}(s,t) \sim \mathrm{WP}(n-k, \gamma)$. Then applying Theorem 4.6 result in (5.50).

Again, we can show that with the sample sizes $n_1, n_2, \cdots, n_k$ growing to infinity proportionally, the estimators $\hat{\boldsymbol{\beta}}$ and $\hat{\kappa}$ are consistent.

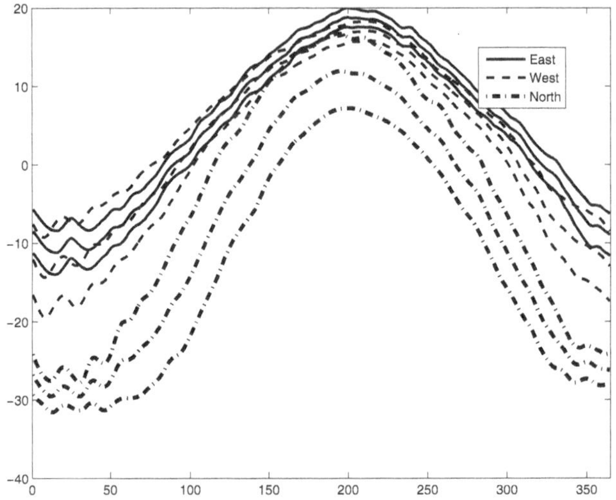

Figure 5.7 *Sample group mean functions of the Canadian temperature data with their 95% pointwise confidence bands for Eastern (solid), Western (dashed), and Northern (dot-dashed) weather stations. It appears that the mean temperature differences over time between the Eastern and the Western weather stations are much smaller than those between the Eastern (or the Western) and the Northern weather stations.*

Theorem 5.10 *Under Assumptions KS1 and KS3 through KS7, as $n \to \infty$, we have $tr(\hat{\gamma}) \xrightarrow{p} tr(\gamma)$ and $tr(\hat{\gamma}^{\otimes 2}) \xrightarrow{p} tr(\gamma^{\otimes 2})$. Furthermore, as $n \to \infty$, we have*

$$\hat{\beta} \xrightarrow{p} \beta, \quad \hat{\kappa} \xrightarrow{p} \kappa,$$

where $\hat{\beta}$ and $\hat{\kappa}$ are the naive or bias-reduced estimators of β and κ.

The L^2-norm-based test for (5.25) is then conducted by rejecting the null hypothesis of (5.25) whenever $T_n > \hat{\beta}\chi^2_{(k-1)\hat{\kappa}}(1-\alpha)$ for any given significance level α. Alternatively, the main-effect test (5.25) can be conducted by computing the P-value $P(\chi^2_{(k-1)\hat{\kappa}} \geq T_n/\hat{\beta})$. It is worthwhile to notice that here there is a need to distinguish between k and κ and for easy presentation, we sometimes use $d = (k-1)\kappa$ and $\hat{d} = (k-1)\hat{\kappa}$ to denote the approximate degrees of freedom of T_n and its estimator.

Example 5.12 *Figure 5.7 superimposes the sample group mean temperature functions of the Eastern (solid), Western (dashed), and Northern (dot-dashed) weather stations, together with their* 95% *pointwise confidence bands. Based on the* 95% *pointwise confidence bands, some informal conclusions can be made. First of all, over the whole year ($[a, b] = [1, 365]$), the differences in the mean temperature functions of the Eastern and the Western weather stations are*

Table 5.6 *Traces of the pooled sample covariance functions* $\hat{\gamma}(s,t)$ *and its cross-square function* $\hat{\gamma}^{\otimes 2}(s,t)$ *of the Canadian temperature data, calculated with resolution* $M = 1,000$ *over various seasons.*

	Spring	Summer	Fall	Winter	Whole year
[a,b]	$[60, 151]$	$[152, 243]$	$[244, 334]$	$[335, 365]\&[1, 59]$	$[1, 365]$
tr($\hat{\gamma}$)	2,481	859	1,342	4,572	9,255
tr($\hat{\gamma}^{\otimes 2}$)	5,081,346	686,818	1,519,203	20,318,974	58,152,749

much less significant than the differences of the mean temperature functions of the Eastern and the Northern weather stations, or of the Western and the Northern weather stations. This is because the 95% *pointwise confidence band of the Eastern weather station temperature curves cover (before Day 151) or stay close (after Day 151) to the mean temperature function of the Western weather stations; however, the* 95% *pointwise confidence bands of the Eastern and Western weather station temperature curves stay far away from the mean temperature function of the Northern weather stations. Second, the significance of the differences of the mean temperature functions of the Eastern and the Western weather stations for different seasons are different. In the Spring (usually defined as the months of March, April and May or* $[a, b] = [60, 151]$*; see Table 5.6), the mean temperature functions are nearly the same, but this is not the case in the Summer (June, July and August or* $[a, b] = [152, 243]$*) or in the Autumn (September, October, and November or* $[a, b] = [244, 334]$*). These conclusions can be made more clear by testing the hypothesis testing problem (5.25) with* $t \in \mathcal{T} = [a, b]$ *using the* L^2*-norm-based test statistic* T_n *(5.46) and with* a, b *properly specified.*

Table 5.6 displays the traces of the pooled sample covariance function $\hat{\gamma}(s,t)$ *(5.32) and its cross-squared function* $\hat{\gamma}^{\otimes 2}(s,t)$ *over various time periods or seasons (Spring, Summer, Fall, Winter, and Whole Year). These quantities allow one to compute the approximate null distributions of the* L^2*-norm-based test using the naive method or the bias-reduced method described in (5.48) and (5.49), respectively. For example, for the Spring season, we have* $tr(\hat{\gamma}) = 2,481$ *and* $tr(\hat{\gamma}^{\otimes 2}) = 5,081,346$*. In addition, we have* $k = 3$ *and* $n = n_1 + n_2 + n_3 =$ 35. *Then by the naive method described in (5.48), we have*

$$\hat{\beta} = 5,081,346/2,481 = 2,048, \quad \hat{\kappa} = 2,481^2/5,081,346 = 1.2114. \qquad (5.51)$$

This implies that $\hat{d} = (k-1)\hat{\kappa} = 2 \times 1.2114 = 2.4228$ *and hence* $T_n \sim 2,048\chi^2_{2.4228}$ *approximately. To use the bias-reduced method for computing* $\hat{\beta}$*,* $\hat{\kappa}$*, and* $\hat{d}$*, we first use (5.50) to compute*

$$\begin{aligned}
\widehat{tr^2(\hat{\gamma})} &= \tfrac{(35-3)(35-3+1)}{(35-3-1)(35-3+2)}[2,481^2 - 2 \times 5,081,346/(35-3+1)] \\
&= 5.8585e6, \\
\widehat{tr(\hat{\gamma}^{\otimes 2})} &= \tfrac{(35-3)^2}{(35-3-1)(35-3+2)}[5,081,346 - 2,481^2/(35-3)] \\
&= 4.7498e6.
\end{aligned}$$

Table 5.7 *The L^2-norm-based test for the one-way ANOVA problem (5.25) for the Canadian temperature data with resolution $M = 1,000$.*

Method	Time period	T_n	$\hat{\beta}$	$\hat{d}$	P-value
Naive	Spring	8.58e4	2.05e3	2.42	1.67e − 9
	Summer	1.87e4	7.99e2	2.15	1.01e − 5
	Fall	7.60e4	1.13e3	2.37	5.44e − 15
	Winter	1.22e5	4.44e3	2.06	1.25e − 6
	Whole year	3.02e5	6.28e3	2.95	1.86e − 10
Bias-reduced	Spring	8.58e4	1.91e3	2.47	4.24e − 10
	Summer	1.87e4	7.50e2	2.17	4.86e − 6
	Fall	7.60e4	1.06e3	2.41	5.55e − 16
	Winter	1.22e5	4.18e3	2.06	5.29e − 7
	Whole year	3.02e5	5.82e3	3.05	3.43e − 11

Note: The P-values by the naive method are generally comparable with those by the bias-reduced method although the former are generally larger than the latter.

Then

$$\hat{\beta} = 4.7498e6/2,481 = 1,914, \quad \hat{\kappa} = 5.8585e6/4.7498e6 = 1.2334, \tag{5.52}$$

so that $\hat{d} = (k-1)\hat{\kappa} = 2 \times 1.2334 = 2.4668$ *and hence by the bias-reduced method,* $T_n \sim 1,914\chi^2_{2.4668}$ *approximately.*

Table 5.7 displays the test results of the L^2-norm-based test for the one-way ANOVA problem (5.25) for the Canadian temperature data with resolution $M = 1,000$ over various seasons. The parameters β, κ, and d are estimated by the naive method and the bias-reduced method. It is seen that all the tests are highly significant, as expected. It is also seen that the test results based on the naive method and the bias-reduced method are consistent although their P-values are not exactly the same.

F-Type Test When the k samples (5.23) are Gaussian, we can conduct an F-type test for the main-effect test (5.25). The F-type test statistic is defined as

$$F_n = \frac{\int_{\mathcal{T}} \mathrm{SSH}_n(t)dt/(k-1)}{\int_{\mathcal{T}} \mathrm{SSE}_n(t)dt/(n-k)}.$$

Under the null hypothesis (5.25) and by Theorems 5.8, we have

$$F_n \stackrel{d}{=} \frac{\sum_{r=1}^m \lambda_r A_r/(k-1)}{\sum_{r=1}^m \lambda_r E_r/(n-k)}, \tag{5.53}$$

where $A_r \stackrel{i.i.d.}{\sim} \chi^2_{k-1}, E_r \stackrel{i.i.d.}{\sim} \chi^2_{n-k}$ and they are all independent; $\lambda_1, \lambda_2, \cdots, \lambda_m$ are all the positive eigenvalues of $\gamma(s,t)$. It follows that the null distribution of F_n can be approximated by the two-cumulant matched F-approximation method described in Section 4.4 of Chapter 4. By that method, we have

$$F_n \sim F_{(k-1)\hat{\kappa},(n-k)\hat{\kappa}} \text{ approximately,} \tag{5.54}$$

Table 5.8 *The F-type test for the one-way ANOVA problem (5.25) for the Canadian temperature data with resolution $M = 1,000$.*

Method	Time period	F_n	$\hat{d}_1$	$\hat{d}_2$	P-value
Naive	Spring	17.30	2.42	38.76	$3.09e-4$
	Summer	10.91	2.15	34.41	$5.12e-3$
	Fall	28.30	2.37	37.99	$1.29e-6$
	Winter	13.31	2.06	32.91	$1.39e-3$
	Whole year	16.33	2.95	47.13	$9.15e-4$
Bias-reduced	Spring	17.30	2.47	39.46	$3.27e-4$
	Summer	10.91	2.17	34.65	$5.21e-3$
	Fall	28.30	2.41	38.61	$1.37e-6$
	Winter	13.31	2.06	30.01	$1.40e-3$
	Whole year	16.33	3.05	48.85	$10.29e-4$

Note: The P-values by the naive method are generally comparable with those by the bias-reduced method.

where by the naive method, $\hat{\kappa}$ is given in (5.48) and by the bias-reduced method, $\hat{\kappa}$ is given in (5.49). For simplicity, we sometimes use $d_1 = (k-1)\kappa, d_2 = (n-k)\kappa$ and $\hat{d}_1 = (k-1)\hat{\kappa}, \hat{d}_2 = (n-k)\hat{\kappa}$ to denote the approximate degrees of freedom of F_n and their estimators.

The null hypothesis for the main-effect test (5.25) is rejected whenever $F_n > F_{\hat{d}_1,\hat{d}_2}(1-\alpha)$ for any given significance level α. Notice that unlike the L^2-norm-based test, the F-type test takes into account the variation of the pooled sample covariance function $\hat{\gamma}(s,t)$ partially so that in terms of size controlling, it is expected that the F-type test will outperform the L^2-norm-based test when the Gaussian assumption is satisfied.

Notice that for the F-type test, we only need to compute F_n and $\hat{\kappa}$ to get the approximate null distribution.

Example 5.13 *Consider the one-way ANOVA for the Canadian temperature data for the Spring season. We already obtained $\hat{\kappa} = 1.2114$ [see (5.51)] by the naive method and $\hat{\kappa} = 1.2334$ [see(5.52)] by the bias-reduced method. Then for the F-type test, by the naive method, we have*

$$\begin{aligned}\hat{d}_1 &= (k-1)\hat{\kappa} = 2 \times 1.2114 = 2.4228,\\ \hat{d}_2 &= (n-k)\hat{\kappa} = (35-3) \times 1.2114 = 38.7648,\end{aligned}$$

and by the bias-reduced method, we have

$$\begin{aligned}\hat{d}_1 &= (k-1)\hat{\kappa} = 2 \times 1.2334 = 2.4668,\\ \hat{d}_2 &= (n-k)\hat{\kappa} = (35-3) \times 1.2334 = 39.4688.\end{aligned}$$

It follows that by the naive method, $F_n \sim F_{2.4228,38.7648}$ approximately, and by the bias-reduced method, $F_n \sim F_{2.4668,39.4688}$ approximately.

Example 5.14 *Table 5.8 displays the test results of the F-type test for the one-way ANOVA of the Canadian temperature data with resolution* $M = 1,000$ *over various seasons. The parameter* κ *is estimated by the naive method and the bias-reduced method. It is seen that all the tests are highly significant, as expected. It is also seen that the test results based on the naive method and the bias-reduced method are comparable. Notice that the P-values of the* F*-type test listed in Table 5.8 are much larger than those of the* L^2*-norm-based test listed in Table 5.7.*

Bootstrap Tests When the k samples (5.23) are not Gaussian, the F-type test described earlier may not be applicable. In this case, some parametric or nonparametric bootstrap methods may be used.

When the sample sizes $n_1, n_2, \cdots, n_k$ are large, one can apply some parametric bootstrap (PB) methods for testing the main-effect test (5.27). From (5.146) in the proof of Theorem 5.9, we can see that under the null hypothesis, as $n \to \infty$, we have

$$T_n = \int_{\mathcal{T}} \text{SSH}_n(t)dt \xrightarrow{d} \sum_{i=1}^{k-1} \int_{\mathcal{T}} w_i^2(t)dt,$$

where $w_i(t), i = 1, 2, \cdots, k-1$ are the $k-1$ components of $\mathbf{w}(t) \sim \text{GP}_{k-1}(\mathbf{0}, \gamma \mathbf{I}_{k-1})$. That is, $w_i(t), i = 1, \cdots, k-1 \overset{i.i.d.}{\sim} \text{GP}(0, \gamma)$ which are known except $\gamma(s,t)$. The unbiased estimator $\hat{\gamma}(s,t)$ of $\gamma(s,t)$ is given in (5.32). In this case, we can adopt the PB method of Cuevas, Febrero, and Fraiman (2004) to obtain an approximate critical value of T_n for any given significance level α. Its key idea is to re-sample the Gaussian processes $w_i(t), i = 1, \cdots, k-1$ from $\text{GP}(0, \hat{\gamma})$ a large number of times so that a large sample of T_n under the null hypothesis can be obtained. The associated PB algorithm can be described as follows:

PB Algorithm for One-Way ANOVA (I)

1. Compute $\hat{\gamma}(s,t)$ using (5.32) based on the k samples (5.23).
2. Re-sample the Gaussian processes $w_i^*(t), i = 1, 2, \cdots, k-1$ from $\text{GP}(0, \hat{\gamma})$.
3. Compute $T_B^* = \sum_{i=1}^{k-1} \int_{\mathcal{T}} [w_i^*(t)]^2 dt$.
4. Repeat Steps 2 and 3 a large number of times to obtain a sequence of T_B^* whose sample percentiles can be used to approximate the percentiles of T_n.

It is obvious that the above PB method is time consuming as it needs to re-sample Gaussian processes a large number of times. By Theorem 5.9, the asymptotic random expression of T_n under the null hypothesis (5.27) is a χ^2-type mixture:

$$T^* = \sum_{r=1}^{m} \lambda_r A_r, \quad A_r \overset{i.i.d.}{\sim} \chi^2_{k-1}.$$

It is seen that the distribution of T^* is known except for the unknown eigenvalues $\lambda_r, r = 1, 2, \cdots, m$ of $\gamma(s,t)$. These unknown eigenvalues can be estimated by the eigenvalues $\hat{\lambda}_r, r = 1, 2, \cdots, \hat{m}$ of $\hat{\gamma}(s,t)$, where $\hat{m}$ is the number of the positive eigenvalues of $\hat{\gamma}(s,t)$. Then the PB method proposed in Zhang and Chen (2007) can be adopted here to generate a large sample of $T_B^* = \sum_{r=1}^{\hat{m}} \hat{\lambda}_r A_r, \quad A_r \overset{i.i.d.}{\sim} \chi^2_{k-1}$. The associated PB algorithm can be described as follows:

PB Algorithm for One-Way ANOVA (II)

1. Compute $\hat{\gamma}(s,t)$ using (5.32) based on the k samples (5.23).
2. Compute the positive eigenvalues $\hat{\lambda}_r, r = 1, 2, \cdots, \hat{m}$ of $\hat{\gamma}(s,t)$.
3. Re-sample $A_r, i = 1, 2, \cdots, k-1$ from χ^2_{k-1}.
4. Compute $T_B^* = \sum_{r=1}^{\hat{m}} \hat{\lambda}_r A_r$.
5. Repeat Steps 3 and 4 a large number of times to obtain a sequence of T_B^* whose sample percentiles can be used to approximate the percentiles of T_n.

This PB method is less intensive than the PB method described earlier as no Gaussian processes need to be re-sampled. It generally works well but is still time consuming as it needs to estimate the eigenvalues of $\hat{\gamma}(s,t)$ and to re-sample $A_r, r = 1, 2, \cdots, \hat{m}$, from χ^2_{k-1} a large number of times.

Alternatively, Cuevas, Febrero, and Fraiman (2004) proposed a PB method for the main-effect test (5.27) that can be described as follows. They proposed to use the L^2-norm-based test statistic T_n (5.46) for testing the one-way ANOVA problem (5.25). But for easy treatment, they used the following test statistic instead:

$$V_n = \sum_{1 \le i < j \le k} n_i \int_{\mathcal{T}} [\bar{y}_{i\cdot}(t) - \bar{y}_{j\cdot}(t)]^2 dt. \tag{5.55}$$

They imposed Assumption KS3, that is, as $n \to \infty$,

$$\frac{n_i}{n} \to \tau_i \in (0,1), i = 1, 2, \cdots, k. \tag{5.56}$$

Under the above condition and under the null hypothesis (5.25), they showed that

$$V_n \xrightarrow{d} \sum_{1 \le i < j \le k} \int_{\mathcal{T}} [w_i(t) - \sqrt{\tau_i/\tau_j} w_j(t)]^2 dt, \tag{5.57}$$

where $w_i(t), i = 1, 2, \cdots, k \overset{i.i.d.}{\sim} \text{GP}(0, \gamma)$. Cuevas, Febrero, and Fraiman (2004) computed the P-value or the empirical critical value of V_n by re-sampling $w_i(t), i = 1, 2, \cdots, k$ from $\text{GP}(0, \hat{\gamma})$ a large number of times, where

$\hat{\gamma}(s,t)$ is the pooled sample covariance function given in (5.32). In summary, their PB algorithm can be described as follows:

PB Algorithm for One-Way ANOVA (III)

1. Compute $\hat{\gamma}(s,t)$ using (5.32) based on the k samples (5.23).
2. Re-sample the Gaussian processes $w_i^*(t), i = 1, 2, \cdots, k$ from $\text{GP}(0, \hat{\gamma})$.
3. Compute $V_B^* = \sum_{1\le i<j\le k} \int_{\mathcal{T}} [w_i^*(t) - \sqrt{\tau_i/\tau_j} w_j^*(t)]^2 dt$, where $\tau_i = n_i/n, i = 1, 2, \cdots, k$.
4. Repeat Steps 2 and 3 a large number of times to obtain a sequence of V_B^* whose sample percentiles can be used to approximate the percentiles of V_n.

The idea of the above PB test can also be applied to the case when the k samples (5.23) actually have different covariance functions. But this PB test is very time consuming, as noticed by Cuevas, Febrero, and Fraiman (2004).

The above PB test can also be applied to T_n (5.46) directly. In fact, for any $t \in \mathcal{T}$, we have

$$\begin{aligned} \text{SSH}_n(t) &= \sum_{i=1}^k n_i[\bar{y}_{i\cdot}(t) - \bar{y}_{\cdot\cdot}(t)]^2 dt \\ &= n^{-1} \sum_{1\le i<j\le k} n_i n_j [\bar{y}_{i\cdot}(t) - \bar{y}_{j\cdot}(t)]^2. \end{aligned} \tag{5.58}$$

It follows that we can write

$$T_n = n^{-1} \sum_{1\le i<j\le k} n_i n_j \int_{\mathcal{T}} [\bar{y}_{i\cdot}(t) - \bar{y}_{j\cdot}(t)]^2 dt. \tag{5.59}$$

Under the condition (5.56) and under the null hypothesis (5.25), by Theorem 5.6, we can show that

$$T_n \xrightarrow{d} \sum_{1\le i<j\le k} \int_{\mathcal{T}} \left[\sqrt{\tau_j} w_i(t) - \sqrt{\tau_i} w_j(t)\right]^2 dt, \tag{5.60}$$

where $w_i(t), i = 1, \cdots, k \overset{i.i.d.}{\sim} \text{GP}(0, \gamma)$. Then the P-value or the empirical critical value of T_n can then be computed similarly using the PB algorithm described above.

Remark 5.2 *The above PB methods for the one-way ANOVA problem are generally not recommended. This is because they require large samples. When this is the case, the L^2-norm-based test works reasonably well for Gaussian and non-Gaussian data while requiring much less computational effort.*

Table 5.9 *The L^2-norm-based and F-type bootstrap tests for the one-way ANOVA problem (5.25) with the Canadian temperature data with resolution $M = 1,000$.*

	L^2-norm-based bootstrap test		F-type bootstrap test	
Time period	T_n	P-value	F_n	P-value
Spring	85,815	0	17.30	$3e-4$
Summer	18,748	0	10.91	$2.93e-2$
Fall	76,007	0	28.30	0
Winter	121,670	0	13.31	$6e-4$
Whole year	302,240	0	16.33	0

Note: The number of bootstrap replicates is $N = 10,000$. The effect of the number of bootstrap replicates $N = 10,000$ on the P-values of the L^2-norm-based and F-type bootstrap tests is noted.

When the k samples (5.23) are not Gaussian and when the sample sizes $n_1, \cdots, n_k$ are small, the L^2-norm-based test, the F-type test, and the above PB methods are not preferred. In this case, a nonparametric bootstrap approach can be used to bootstrap the critical values of T_n or F_n, resulting in an L^2-norm-based bootstrap test and an F-type bootstrap test respectively.

Let $v_{ij}^*(t), j = 1, 2, \cdots, n_i; i = 1, \cdots, k$, be k bootstrap samples randomly generated from the estimated subject-effect functions $\hat{v}_{ij}(t) = y_{ij}(t) - \hat{\eta}_i(t), j = 1, 2, \cdots, n_{ij}; i = 1, 2, \cdots, k$. Set

$$y_{ij}^*(t) = \hat{\eta}_i(t) + v_{ij}^*(t), j = 1, 2, \cdots, n_{ij}; i = 1, 2, \cdots, k. \tag{5.61}$$

Then we can compute the k sample group mean functions $\bar{y}_{1.}^*(t), \cdots, \bar{y}_{k.}^*(t)$, the sample grand mean function $\bar{y}_{..}^*(t)$, and the pooled sample covariance function $\hat{\gamma}^*(s,t)$ as in (5.32) but based on the k bootstrap samples (5.61). Then we can compute

$$\begin{array}{rcl} \text{SSH}_n^*(t) & = & \sum_{i=1}^k n_i \left\{[\bar{y}_{i.}^*(t) - \bar{y}_{..}^*(t)] - [\bar{y}_{i.}(t) - \bar{y}_{..}(t)]\right\}^2, \\ \text{SSE}_n^*(t) & = & (n-k)\hat{\gamma}^*(t,t). \end{array}$$

For the L^2-norm-based bootstrap test or the F-type bootstrap test, we compute

$$T_n^* = \int_{\mathcal{T}} \text{SSH}_n^*(t)dt, \text{ or } F_n^* = \frac{\int_{\mathcal{T}} \text{SSH}_n^*(t)dt/(k-1)}{\int_{\mathcal{T}} \text{SSE}_n^*(t)dt/(n-k)}.$$

Repeat this process a large number of times to obtain a bootstrap sample of T_n^* or F_n^* that can be used to estimate the $100(1-\alpha)$-percentile of T_n or F_n. The L^2-norm-based bootstrap test or the F-type bootstrap test can then be conducted accordingly.

Example 5.15 *Table 5.9 displays the test results of the L^2-norm-based and F-type bootstrap tests for the one-way ANOVA problem (5.25) for the Canadian temperature data with resolution $M = 1,000$ over various seasons. To save time, the number of bootstrap replicates is only $10,000$. Recall that the*

P-values of the L^2-norm-based test as listed in Table 5.7 are much smaller than those of the F-type test as listed in Table 5.8. Interestingly, from Table 5.9, it is seen that the P-values of the L^2-norm-based bootstrap test are all 0, which are much smaller than those P-values of the F-type bootstrap test. It seems that the structures of the test statistics of the L^2-norm-based and F-type bootstrap tests may affect the test results. Further study is warranted.

5.3.3 Tests of Linear Hypotheses

In the previous subsection we presented some methods for the main-effect test (5.27). In this subsection, we study how to test the post hoc test (5.28) and the contrast test (5.30) in a unified framework. That is, given the k samples (5.23), we want to test the following general linear hypothesis testing (GLHT) problem:

$$H_0: \mathbf{C}\boldsymbol{\eta}(t) \equiv \mathbf{c}(t),\ t \in \mathcal{T}, \quad \text{versus} \quad H_1: \mathbf{C}\boldsymbol{\eta}(t) \neq \mathbf{c}(t), t \in \mathcal{T}, \tag{5.62}$$

where $\mathbf{C}: q \times k$ is a known coefficient matrix with $\text{rank}(\mathbf{C}) = q$, and $\mathbf{c}(t): q \times 1$ is a known constant function, often specified as $\mathbf{0}$. In fact, the post hoc test (5.29) and the contrast test (5.31) can be written in the form of the GLHT problem (5.62) if we set $\mathbf{c}(t) \equiv \mathbf{0}$, and set $\mathbf{C} = \mathbf{e}_{i,k} - \mathbf{e}_{j,k}$ and $\mathbf{C} = \mathbf{a}$, respectively, where again $\mathbf{e}_{r,k}$ denotes a k-dimensional unit vector whose rth entry is 1 and others 0. In addition, the main-effect test (5.25) can also be written into the form of the GLHT problem (5.62) by setting $\mathbf{c}(t) \equiv \mathbf{0}$ and $\mathbf{C} = [\mathbf{I}_{k-1}, -\mathbf{1}_{k-1}]$.

Remark 5.3 *The matrix $\mathbf{C}$ is often a contrast matrix with rows summing up to 0, that is, each row being a contrast. It is well known that for a hypothesis testing problem that can be written in the form of the GLHT problem (5.62), the associated contrast matrix $\mathbf{C}$ is not unique. For example, $\tilde{\mathbf{C}} = \Big(-\mathbf{1}_{k-1}, \mathbf{I}_{k-1}\Big)$ is also a contrast matrix for the main-effect testing problem (5.25). It is known from Kshirsagar (1972, Chapter 5, Section 4) that for any two contrast matrices $\tilde{\mathbf{C}}$ and $\mathbf{C}$ for the same null hypothesis, there is a nonsingular matrix $\mathbf{P}$ such that*

$$\tilde{\mathbf{C}} = \mathbf{PC}. \tag{5.63}$$

On the other hand, it is easy to show that for any nonsingular matrix $\mathbf{P}$, $\mathbf{PC}$ is also a contrast matrix as long as $\mathbf{C}$ is a contrast matrix. For easy reference, we call (5.63) a contrast transformation.

Notice that we have $\text{E}\left[\mathbf{C}\hat{\boldsymbol{\eta}}(t) - \mathbf{c}(t)\right] = \mathbf{C}\boldsymbol{\eta}(t) - \mathbf{c}(t)$ and

$$\text{Cov}\Big[\mathbf{C}\hat{\boldsymbol{\eta}}(s) - \mathbf{c}(s), \mathbf{C}\hat{\boldsymbol{\eta}}(t) - \mathbf{c}(t)\Big] = \gamma(s,t)\mathbf{CDC}^T,$$

where $\mathbf{D} = \text{diag}(\frac{1}{n_1}, \frac{1}{n_2}, \cdots, \frac{1}{n_k})$ as defined earlier. As $\mathbf{CDC}^T$ is a square matrix of full rank, we then arrive at the following pivotal test function:

$$\mathbf{z}(t) = \left(\mathbf{CDC}^T\right)^{-1/2} \left[\mathbf{C}\hat{\boldsymbol{\eta}}(t) - \mathbf{c}(t)\right]. \tag{5.64}$$

It is easy to see that

$$\mathbf{z}(t) \sim \text{SP}_q(\boldsymbol{\eta}_z, \gamma \mathbf{I}_q), \tag{5.65}$$

where

$$\boldsymbol{\eta}_z(t) = (\mathbf{CDC}^T)^{-1/2}\left[\mathbf{C}\boldsymbol{\eta}(t) - \mathbf{c}(t)\right]. \tag{5.66}$$

Under the null hypothesis in (5.62), $\boldsymbol{\eta}_z(t) \equiv 0, t \in \mathcal{T}$. The squared L^2-norm $\|\mathbf{z}(t)\|^2$ of $\mathbf{z}(t)$ at $t \in \mathcal{T}$ can then be used as the pointwise sum of squares due to hypothesis:

$$\text{SSH}_n(t) = \left[\mathbf{C}\hat{\boldsymbol{\eta}}(t) - \mathbf{c}(t)\right]^T \left(\mathbf{CDC}^T\right)^{-1} \left[\mathbf{C}\hat{\boldsymbol{\eta}}(t) - \mathbf{c}(t)\right], \tag{5.67}$$

which, together with $\text{SSE}_n(t) = (n-k)\hat{\gamma}(t,t)$, the pointwise sum of squares due to errors, will be used to define various tests for the GLHT problem (5.62).

Remark 5.4 *The pivotal test function $\mathbf{z}(t)$ (5.64) is constructed in a way such that each component of $\mathbf{z}(t)$ is a process with the same covariance function $\gamma(s,t)$ as indicated in (5.65). In addition, it is easy to check that the squared L^2-norm $SSH_n(t)$ (5.67) of $\mathbf{z}(t)$ is invariant when $\mathbf{C}$ and $\mathbf{c}(t)$ are replaced, respectively, with*

$$\tilde{\mathbf{C}} = \mathbf{PC}, \quad \text{and} \quad \tilde{\mathbf{c}}(t) = \mathbf{Pc}(t), \tag{5.68}$$

where $\mathbf{P}$ is a full-rank matrix of size $q \times q$ such as the one defined in (5.63). This property is important for the GLHT problem (5.62) as for a hypothesis testing problem that can be written in the form of (5.62), the associated $\mathbf{C}$ and $\mathbf{c}(t)$ are not uniquely defined as indicated by Remark 5.3.

Theorem 5.11 *Under Assumptions KS1 and KS2 and the null hypothesis in (5.62), we have*

$$\int_{\mathcal{T}} SSH_n(t)dt \stackrel{d}{=} \sum_{r=1}^m \lambda_r A_r, \quad \int_{\mathcal{T}} SSE_n(t)dt \stackrel{d}{=} \sum_{r=1}^m \lambda_r E_r,$$

where $A_r, r = 1, 2, \cdots, m \stackrel{i.i.d.}{\sim} \chi_q^2$ and $E_r, r = 1, 2, \cdots, m \stackrel{i.i.d.}{\sim} \chi_{n-k}^2$ are independent, and $\lambda_1, \cdots, \lambda_m$ are all the positive eigenvalues of $\gamma(s,t)$.

In the above theorem, it is shown that under the Gaussian assumption and the null hypothesis, both the integrated SSH and SSE are χ^2-type mixtures. Notice that the random expression of $\int_{\mathcal{T}} \text{SSE}_n(t)dt$ was given in Theorem 5.8. The purpose for including it in the above theorem is to emphasize the fact that $A_r, r = 1, 2, \cdots, m$ and $E_r, r = 1, 2, \cdots, m$ are independent. This fact is needed for conveniently introducing the F-type test for the GLHT problem (5.62). When the Gaussian assumption KS2 is not satisfied, as expected, we can show that under some mild conditions, the integrated SSH is asymptotically a χ^2-type mixture, as shown by the following theorem.

Theorem 5.12 *Under Assumptions KS1, KS3, KS4, and the null hypothesis in (5.62), as $n \to \infty$, we have*

$$\int_{\mathcal{T}} SSH_n(t)dt \xrightarrow{d} \sum_{r=1}^{m} \lambda_r A_r,$$

where $A_r, r = 1, 2, \cdots, m \overset{i.i.d.}{\sim} \chi_q^2$ and $\lambda_1, \cdots, \lambda_m$ are all the positive eigenvalues of $\gamma(s, t)$.

We are now ready to describe various tests for the GLHT problem (5.62). These tests are essentially the same as those for the main-effect testing problem presented earlier.

Pointwise Tests We describe a pointwise F-test and a pointwise χ^2-test here. The test statistic of the pointwise F-test is defined as

$$F_n(t) = \frac{\text{SSH}_n(t)/q}{\text{SSE}_n(t)/(n-k)}. \tag{5.69}$$

When the k samples (5.23) are Gaussian, under the null hypothesis in (5.62), we have

$$F_n(t) \sim F_{q,n-k}, t \in \mathcal{T}.$$

The pointwise F-test can be conducted accordingly. When the Gaussian assumption is not valid, for large samples, one may use the pointwise χ^2-test. For large samples, that is, under Assumptions KS3 and KS4, it is standard to show that

$$F_n(t) \xrightarrow{d} \chi_q^2/q, \quad t \in \mathcal{T}.$$

The pointwise χ^2-test can be conducted accordingly.

When the k samples (5.23) are not Gaussian and $n_1, \cdots, n_k$ are small, the above pointwise F and χ^2-tests are not preferred. In this case, one may resort to some bootstrap approaches as described at the end of this subsection.

L^2-Norm-Based Test For the GLHT problem (5.62), the L^2-norm-based test uses the following test statistic

$$T_n = \int_{\mathcal{T}} \text{SSH}_n(t)dt.$$

Under the null hypothesis in (5.62) and under the conditions of Theorem 5.11 or under the conditions of Theorem 5.12, we have or approximately have

$$T_n \overset{d}{=} \sum_{r=1}^{m} \lambda_r A_r, \ A_r \overset{i.i.d.}{\sim} \chi_q^2,$$

where $\lambda_1, \lambda_2, \cdots, \lambda_m$ are all the positive eigenvalues of $\gamma(s,t)$. Then the null distribution of T_n can be approximated by the Welch-Satterthwaite χ^2-approximation method using the methods described in Section 4.3 of Chapter 4. In fact, by this method, we have

$$T_n \sim \hat{\beta}\chi^2_{q\hat{\kappa}} \text{ approximately}$$

where by the naive method, $\hat{\beta}$ and $\hat{\kappa}$ are given in (5.48) and by the bias-reduced method, they are given in (5.49). The L^2-norm-based test can then be conducted accordingly.

F-Type Test As for the main-effect testing problem (5.25), for Gaussian data, we can also conduct an F-type test for the GLHT problem (5.62) using the following F-type test statistic:

$$F_n = \frac{\int_{\mathcal{T}} \mathrm{SSH}_n(t)dt/q}{\int_{\mathcal{T}} \mathrm{SSE}_n(t)dt/(n-k)}.$$

By Theorem 5.11 and under the null hypothesis in (5.62), we have

$$F_n \stackrel{d}{=} \frac{\sum_{r=1}^m \lambda_r A_r/q}{\sum_{r=1}^m \lambda_r E_r/(n-k)},$$

where $A_r, r = 1, \cdots, m \stackrel{i.i.d.}{\sim} \chi^2_q$ and $E_r, r = 1, \cdots, m \stackrel{i.i.d.}{\sim} \chi^2_{n-k}$ are independent, and $\lambda_1, \lambda_2, \cdots, \lambda_m$ are all the positive eigenvalues of $\gamma(s,t)$. That is, under the null hypothesis, F_n is an F-type mixture. It follows that the null distribution of F_n can be approximated by the two-cumulant matched F-approximation method described in Section 4.4 of Chapter 4. In fact, by that method, we have

$$F_n \sim F_{q\hat{\kappa},(n-k)\hat{\kappa}} \text{ approximately},$$

where by the naive method, $\hat{\kappa}$ is given in (5.48) and by the bias-reduced method, $\hat{\kappa}$ is given in (5.49). The F-type test can then be conducted accordingly.

Bootstrap Tests When the k samples (5.23) are not Gaussian and $n_1, \cdots, n_k$ are small, then some bootstrap tests may be preferred. Here we describe a pointwise bootstrap test, an L^2-norm-based bootstrap test, and an F-type bootstrap test.

Let $v^*_{ij}(t), j = 1, 2, \cdots, n_i; i = 1, 2, \cdots, k$, be k bootstrap samples randomly generated from the estimated subject-effect functions $\hat{v}_{ij}(t) = y_{ij}(t) - \hat{\eta}_i(t), j = 1, 2, \cdots, n_i; i = 1, 2, \cdots, k$. Set $y^*_{ij}(t) = \hat{\eta}_i(t) + v^*_{ij}(t), j = 1, 2, \cdots, n_i; i = 1, 2, \cdots, k$. Then we can compute the sample mean functions, and the pooled sample covariance function as $\hat{\eta}^*_i(t) = \bar{y}^*_{i.}(t), i = 1, 2, \cdots, k$ and $\hat{\gamma}^*(s,t)$ based on the above k bootstrap samples. We then compute

$$\mathrm{SSH}^*_n(t) = [\hat{\boldsymbol{\eta}}^*(t) - \hat{\boldsymbol{\eta}}(t)]^T \mathbf{C}^T \left(\mathbf{CDC}^T\right)^{-1} \mathbf{C}\,[\hat{\boldsymbol{\eta}}^*(t) - \hat{\boldsymbol{\eta}}(t)],$$
$$\mathrm{SSE}^*_n(t) = (n-k)\hat{\gamma}^*(t,t),$$

where $\hat{\boldsymbol{\eta}}^*(t) = [\hat{\eta}_1^*(t), \cdots, \hat{\eta}_k^*(t)]^T$.

For the pointwise bootstrap test, the L^2-norm-based bootstrap test, or the F-type bootstrap test, we compute

$$F_n^*(t) = \frac{\text{SSH}_n^*(t)/q}{\text{SSE}_n^*(t)/(n-k)}, \; T_n^* = \int_{\mathcal{T}} \text{SSH}_n^*(t)dt, \quad \text{or}$$
$$F_n^* = \frac{\int_{\mathcal{T}} \text{SSH}_n^*(t)dt/q}{\int_{\mathcal{T}} \text{SSE}_n^*(t)dt/(n-k)}.$$

Repeat the above bootstrapping process a large number of times, calculate the $100(1-\alpha)$-percentile of $F_n^*(t), T_n^*$ or F_n^*, and then conduct the pointwise bootstrap test, the L^2-norm-based bootstrap test, or the F-type bootstrap test accordingly.

5.4 Two-Way ANOVA

In the previous section, we studied one-way ANOVA for functional data, involving one categorical variable or factor. In many situations, functional data may involve two or more factors (categorical variables), each with several levels (categories). In this section, we study two-way ANOVA for functional data. A two-way ANOVA model for functional data involves two factors. We use the following example to motivate a two-way ANOVA model for functional data.

Example 5.16 *We use the left-cingulum data introduced in Section 1.2.6 of Chapter 1 as a motivating and illustrating example. The data set was collected for thirty-nine children from 9 to 19 years old over arc length from* -60 *to* 60, *aiming to study if the Radial Diffusibility (RD) in the left-cingulum is affected by age and family of a child. Figure 1.10 of Chapter 1 displays the thirty-nine left-cingulum curves over the arc length. An outlying left-cingulum curve is spotted in the figure and is removed for the data analysis conducted in this section. In the left-cingulum data, the response variable is "RD" while the covariates include "GHR" and "AGE," where GHR stands for Genetic High Risk and AGE is the age of a child in the study. The GHR variable is a categorical variable, taking two values. When GHR = 1, it means that the child is from a family with at least 1 direct relative with schizophrenia disease and when GHR = 0, the child is from a normal family. The AGE variable is a continuous variable. In this section, the AGE variable is transformed into a categorical variable with two levels: children with* $AGE < 15$ *and children with* $AGE \geq 15$ *so that the resulting left-cingulum data can be modeled by an unbalanced two-way ANOVA model:*

$$\begin{array}{c} y_{ijk}(t) = \eta_0(t) + \alpha_i(t) + \beta_j(t) + \theta_{ij}(t) + v_{ijk}(t), \; t \in [-60, 60], \\ k = 1, 2, \cdots, n_{ij}, \; i = 1, 2, \; j = 1, 2, \end{array} \tag{5.70}$$

where $\eta_0(t)$ *denotes the overall mean function;* $\alpha_i(t), i = 1, 2$ *denotes the main-effect functions of GHR;* $\beta_j(t), j = 1, 2$, *the main-effect functions of AGE;*

$\theta_{ij}(t), i = 1,2, j = 1,2$, the interaction-effect functions between GHR and AGE; and $[n_{11}, n_{12}, n_{21}, n_{22}] = [15, 10, 7, 6]$. We want to know if the main-effects of GHR, AGE, and their interaction-effects are statistically significant. That is, is the variable RD affected by the categorical variables GHR and AGE? Equivalently, we want to test the following null hypotheses:

$$\begin{array}{rcl} H_A & : & \alpha_1(t) \equiv \alpha_2(t) \equiv 0, \ t \in [-60, 60], \\ H_B & : & \beta_1(t) \equiv \beta_2(t) \equiv 0, \ t \in [-60, 60], \\ H_{AB} & : & \theta_{11}(t) \equiv \theta_{12}(t) \equiv \theta_{21}(t) \equiv \theta_{22}(t) \equiv 0, \ t \in [-60, 60]. \end{array} \tag{5.71}$$

In general, a two-way ANOVA model can be defined as follows. Let a two-way experiment have two factors A and B, with a and b levels, respectively. There are a total of ab factorial combinations or cells. Suppose at the (i,j)th cell that we have a random functional sample:

$$y_{ijk}(t), t \in \mathcal{T}, k = 1, 2, \cdots, n_{ij}, \tag{5.72}$$

satisfying the following cell model:

$$y_{ijk}(t) = \eta_{ij}(t) + v_{ijk}(t), t \in \mathcal{T}, \ k = 1, \cdots, n_{ij}, \tag{5.73}$$

where $\mathcal{T}$ is the support of t, and $\eta_{ij}(t)$ is the cell mean function of the random sample at the (i,j)th cell. All these ab samples are assumed to be independent of each other and all the subject-effect functions

$$v_{ijk}(t), k = 1, 2, \cdots, n_{ij}; i = 1, 2, \cdots, a; j = 1, 2, \cdots, b \overset{i.i.d.}{\sim} \text{SP}(0, \gamma), \tag{5.74}$$

where $\gamma(s,t)$ is the common covariance function for all the samples.

For a two-way ANOVA model, the cell mean functions $\eta_{ij}(t)$ will decompose into the form

$$\begin{array}{c} \eta_{ij}(t) = \eta_0(t) + \alpha_i(t) + \beta_j(t) + \theta_{ij}(t), \ t \in \mathcal{T}, \\ i = 1, 2, \cdots, a; j = 1, 2, \cdots, b, \end{array} \tag{5.75}$$

where $\eta_0(t)$ is the grand mean function; $\alpha_i(t)$ and $\beta_j(t)$ are the ith and jth main-effect functions of factors A and B, respectively; and $\theta_{ij}(t)$ is the (i,j)th interaction effect function between factors A and B so that the model (5.73) can be further written as the following two-way ANOVA model:

$$\begin{array}{rcl} y_{ijk}(t) & = & \eta_0(t) + \alpha_i(t) + \beta_j(t) + \theta_{ij}(t) + v_{ijk}(t), \ t \in \mathcal{T}, \\ & & k = 1, 2, \cdots, n_{ij}; i = 1, 2, \cdots, a; j = 1, 2, \cdots, b, \end{array} \tag{5.76}$$

where the subject-effect functions $v_{ijk}(t), k = 1, 2, \cdots, n_{ij}; i = 1, 2, \cdots, a; j = 1, 2, \cdots, b$ satisfy (5.74). For this model, we are interested in the following null hypotheses:

$$\begin{array}{rcl} H_{0A} & : & \alpha_i(t) \equiv 0, \ i = 1, 2, \cdots, a, \ t \in \mathcal{T}, \\ H_{0AB} & : & \alpha_i(t) \equiv 0, i = 1, 2, \cdots, a, \ t \in \mathcal{T}, \\ & & \beta_j(t) \equiv 0, \ j = 1, 2, \cdots, b, \ t \in \mathcal{T}, \\ H_{0I} & : & \theta_{ij}(t) \equiv 0, \ i = 1, 2, \cdots a; j = 1, 2, \cdots, b; \ t \in \mathcal{T}. \end{array} \tag{5.77}$$

The first null hypothesis aims to test if the main-effects of Factor A are significant. A test of the main-effects of Factor B can be similarly handled. The second null hypothesis aims to test if the main-effects of the two factors are simultaneously 0. The latter one aims to test if there is an interaction-effect between the two factors.

5.4.1 Estimation of Cell Mean and Covariance Functions

For two-way ANOVA, the cell mean functions $\eta_{ij}(t), i = 1, \cdots, a; j = 1, \cdots, b$ are estimable, that is, they can be well defined provided $n_{ij} > 1, i = 1, \cdots, a; j = 1, \cdots, b$ (see Section 5.4.5 for the case when all $n_{ij} = 1$). Their unbiased estimators are given by the following sample cell mean functions of the ab cell samples (5.73):

$$\hat{\eta}_{ij}(t) = \bar{y}_{ij\cdot}(t) = n_{ij}^{-1} \sum_{k=1}^{n_{ij}} y_{ijk}(t), \quad i = 1, 2, \cdots, a; j = 1, 2, \cdots, b. \tag{5.78}$$

Based on this, we can also estimate the common covariance function unbiasedly by the following pooled sample covariance function

$$\hat{\gamma}(s,t) = (n - ab)^{-1} \sum_{i=1}^{a} \sum_{j=1}^{b} \sum_{k=1}^{n_{ij}} [y_{ijk}(s) - \bar{y}_{ij\cdot}(s)][y_{ijk}(t) - \bar{y}_{ij\cdot}(t)], \tag{5.79}$$

where and throughout, $n = \sum_{i=1}^{a} \sum_{j=1}^{b} n_{ij}$.

Example 5.17 *Figure 5.8 displays the sample cell mean functions of the left-cingulum data with their 95% pointwise confidence bands. A comparison of the sample cell mean functions in the left panels and those in the right panels may hint that the main-effect of AGE may be quite significant. Similarly, a comparison of the sample cell mean functions in the upper panels and those in the lower panels may suggest that the main-effect of GHR may be less significant.*

Figure 5.9 presents the sample covariance function of the left-cingulum data. It appears that the left-cingulum data are less correlated because the sample variances $\hat{\gamma}(t,t), t \in \mathcal{T}$ are much larger than the sample covariances $\hat{\gamma}(s,t), s \neq t$. For further discussion, with resolution $M = 1,000$, we list the traces of the sample covariance function and its squared function as follows:

$$tr(\hat{\gamma}) = 274.72, \quad tr(\hat{\gamma}^{\otimes 2}) = 15,305. \tag{5.80}$$

Put all the cell mean functions into a vector as

$$\boldsymbol{\eta}(t) = [\eta_{11}(t), \cdots, \eta_{1b}(t), \cdots, \eta_{a1}(t), \cdots, \eta_{ab}(t)]^T.$$

Then its unbiased estimator is

$$\hat{\boldsymbol{\eta}}(t) = [\hat{\eta}_{11}(t), \cdots, \hat{\eta}_{1b}(t), \cdots, \hat{\eta}_{a1}(t), \cdots, \hat{\eta}_{ab}(t)]^T.$$

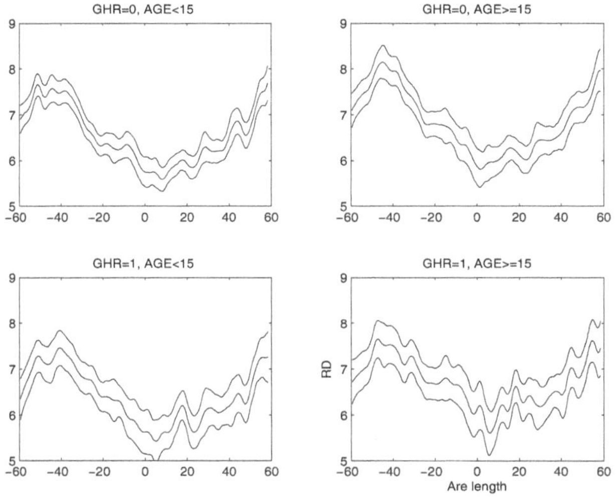

Figure 5.8 *Sample cell mean functions of the left-cingulum data with their* 95% *pointwise confidence bands. A comparison of the sample cell mean functions in the left panels and those in the right panels may hint that the main-effect of AGE may be quite significant.*

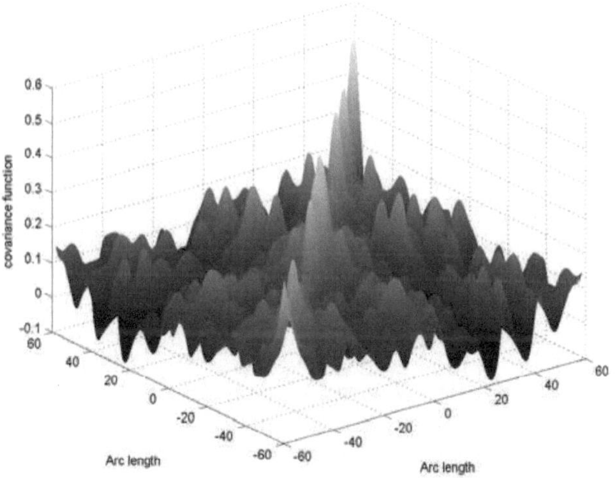

Figure 5.9 *Pooled sample covariance function of the left-cingulum data. It appears that the left-cingulum data are less correlated because the pooled sample variances* $\hat{\gamma}(t,t), t \in \mathcal{T}$ *are much larger than the pooled sample covariances* $\hat{\gamma}(s,t), s \neq t$.

As all the ab cell samples are independent and have the common covariance function $\gamma(s,t)$, we have

$$\begin{array}{c}\hat{\boldsymbol{\eta}}(t) \sim \mathrm{SP}(\boldsymbol{\eta}, \gamma\mathbf{D}), \quad \text{where} \\ \mathbf{D} = \mathrm{diag}(1/n_{11}, \cdots, 1/n_{1b}, \cdots, 1/n_{a1}, \cdots, 1/n_{ab}).\end{array} \tag{5.81}$$

To study various procedures for testing the main-effect and interaction-effect tests as listed in (5.77), we need to investigate the properties of $\hat{\boldsymbol{\eta}}(t)$ and $\hat{\gamma}(s,t)$ under various conditions. To this end, we list the following assumptions for two-way ANOVA:

Two-Way ANOVA Assumptions (TW)

1. The ab samples (5.72) are with $\eta_{ij}(t) \in \mathcal{L}^2(\mathcal{T}), i = 1, \cdots, a; j = 1, \cdots, b$, and $\mathrm{tr}(\gamma) < \infty$.
2. The ab samples (5.72) are Gaussian.
3. As $n \to \infty$, the ab sample sizes satisfy $n_{ij}/n \to \tau_{ij}$, $i = 1, \cdots, a; j = 1, \cdots, b$ such that $\tau_{ij} \in (0,1)$ $i = 1, \cdots, a; j = 1, \cdots, b$.
4. The subject-effect functions $v_{ijk}(t), k = 1, 2, \cdots, n_{ij}; i = 1, 2, \cdots, a; j = 1, \cdots, b$ are i.i.d..
5. The subject-effect function, $v_{111}(t)$ satisfies $\mathrm{E}\|v_{111}\|^4 < \infty$.
6. The maximum variance $\rho = \max_{t \in \mathcal{T}} \gamma(t,t) < \infty$.
7. The expectation $\mathrm{E}[v_{111}^2(s)v_{111}^2(t)]$ is uniformly bounded.

The above assumptions are natural generalizations of those assumptions listed for the one-way ANOVA problem presented in the previous section. Based on these assumptions, we can dig out some useful properties of $\hat{\boldsymbol{\eta}}(t)$ and $\hat{\gamma}(s,t)$.

Theorem 5.13 *Under Assumptions TW1 and TW2, we have*

$$\begin{array}{c}\mathbf{D}^{-1/2}\left[\hat{\boldsymbol{\eta}}(t) - \boldsymbol{\eta}(t)\right] \sim GP_{ab}(\mathbf{0}, \gamma\mathbf{I}_{ab}), \quad \textit{and} \\ (n-ab)\hat{\gamma}(s,t) \sim WP(n-ab, \gamma).\end{array}$$

It is easy to see that Theorem 5.13 is a natural extension of Theorem 5.5. It says that when the Gaussian assumption TW2 is valid, $\hat{\boldsymbol{\eta}}(t)$ is an (ab)-dimensional Gaussian process while $(n-ab)\hat{\gamma}(s,t)$ is a Wishart process. They are the keys for constructing various tests for the various testing problems given in (5.77) when the ab samples (5.72) are Gaussian.

As before, for non-Gaussian functional data, more assumptions are needed. The following two theorems show that under some proper conditions, both $\hat{\boldsymbol{\eta}}(t)$ and $\hat{\gamma}(s,t)$ are asymptotically Gaussian.

Theorem 5.14 *Under Assumptions TW1, TW3, and TW4, as $n \to \infty$, we have*

$$\mathbf{D}^{-1/2}\left[\hat{\boldsymbol{\eta}}(t) - \boldsymbol{\eta}(t)\right] \xrightarrow{d} GP_{ab}(\mathbf{0}, \gamma\mathbf{I}_{ab}).$$

Theorem 5.15 *Under Assumptions TW1 and TW3 through TW6, as $n \to \infty$, we have*

$$\sqrt{n}\,\{\hat{\gamma}(s,t) - \gamma(s,t)\} \xrightarrow{d} GP(0, \varpi),$$

where $\varpi\{(s_1,t_1),(s_2,t_2)\} = E\{v_{11}(s_1)v_{11}(t_1)v_{11}(s_2)v_{11}(t_2)\} - \gamma(s_1,t_1)\gamma(s_2,t_2)$.

In Theorem 5.14, Assumption TW3 is needed to guarantee that as $n \to \infty$, each of the sample cell mean functions $\hat{\eta}_{ij}(t) = \bar{y}_{ij\cdot}(t), i = 1,\cdots,a; j = 1,\cdots,b$ converges to a Gaussian process weakly. While in Theorem 5.15, Assumptions TW5 and TW6 are imposed so that $\hat{\gamma}(s,t)$ is asymptotically a bivariate Gaussian process.

5.4.2 Main and Interaction Effect Functions

Notice that the two-way ANOVA model (5.76) is not identifiable as the functional parameters $\eta_0(t), \alpha_i(t), \beta_j(t)$, and $\theta_{ij}(t)$ are not uniquely defined. To overcome this difficulty, we have to impose some side constraints. Given a sequence of positive weights $w_{ij}, i = 1,2,\cdots,a; j = 1,2,\cdots,b$, the following side constraints may be imposed:

$$\begin{array}{l} \sum_{i=1}^a w_{i\cdot}\alpha_i(t) = 0, \;\; \sum_{j=1}^b w_{\cdot j}\beta_j(t) = 0, \\ \sum_{i=1}^a w_{ij}\theta_{ij}(t) = 0, \;\; j = 1,2,\cdots,b-1, \\ \sum_{j=1}^b w_{ij}\theta_{ij}(t) = 0, \;\; i = 1,2,\cdots,a-1, \\ \sum_{i=1}^a \sum_{j=1}^b w_{ij}\theta_{ij}(t) = 0, \end{array} \quad (5.82)$$

where $w_{i\cdot} = \sum_{j=1}^b w_{ij}$ and $w_{\cdot j} = \sum_{i=1}^a w_{ij}$.

When the weights can be written as

$$w_{ij} = g_i h_j, i = 1,\cdots,a; j = 1,\cdots,b, \quad (5.83)$$

such that $g_i > 0$, $\sum_{i=1}^a g_i = 1$ and $h_j > 0$, $\sum_{j=1}^b h_j = 1$, we can easily identify the parameters $\eta_0(t), \alpha_i(t), \beta_j(t)$, and $\theta_{ij}(t)$ as

$$\begin{array}{ll} \eta_0(t) = \sum_{i=1}^a \sum_{j=1}^b g_i h_j \eta_{ij}(t), & \alpha_i(t) = \sum_{j=1}^b h_j \eta_{ij}(t) - \eta_0(t), \\ \beta_j(t) = \sum_{i=1}^a g_i \eta_{ij}(t) - \eta_0(t), & \theta_{ij}(t) = \eta_{ij}(t) - \alpha_i(t) - \beta_j(t) - \eta_0(t). \end{array}$$

Let $\mathbf{g} = [g_1,\cdots,g_a]^T$ and $\mathbf{h} = [h_1,\cdots,h_b]^T$. Denote again a p-dimensional unit vector whose rth component is 1 and others are 0 as $\mathbf{e}_{r,p}$. Then we further have

$$\begin{array}{ll} \eta_0(t) = [\mathbf{g}^T \otimes \mathbf{h}^T]\boldsymbol{\eta}(t), & \alpha_i(t) = [(\mathbf{e}_{i,a} - \mathbf{g})^T \otimes \mathbf{h}^T]\boldsymbol{\eta}(t), \\ \beta_j(t) = [\mathbf{g}^T \otimes (\mathbf{e}_{j,b} - \mathbf{h})^T]\boldsymbol{\eta}(t), & \theta_{ij}(t) = [(\mathbf{e}_{i,a} - \mathbf{g})^T \otimes (\mathbf{e}_{j,b} - \mathbf{h})^T]\boldsymbol{\eta}(t), \\ & i = 1,2,\cdots,a; j = 1,2,\cdots,b, \end{array}$$

where $\otimes$ denotes the usual Kronecker product operation. Set

$$\begin{array}{rcl} \boldsymbol{\alpha}(t) & = & [\alpha_1(t),\cdots,\alpha_a(t)]^T, \;\; \boldsymbol{\beta}(t) = [\beta_1(t),\cdots,\beta_b(t)]^T, \\ \boldsymbol{\theta}(t) & = & [\theta_{11}(t),\cdots,\theta_{1b}(t),\cdots,\theta_{a1}(t),\cdots,\theta_{ab}(t)]^T, \end{array}$$

and set

$$\begin{aligned}\mathbf{A}_a &= (\mathbf{I}_a - \mathbf{1}_a\mathbf{g}^T) \otimes \mathbf{h}^T,\\ \mathbf{A}_b &= \mathbf{g}^T \otimes (\mathbf{I}_b - \mathbf{1}_b\mathbf{h}^T),\\ \mathbf{A}_{ab} &= (\mathbf{I}_a - \mathbf{1}_a\mathbf{g}^T) \otimes (\mathbf{I}_b - \mathbf{1}_b\mathbf{h}^T).\end{aligned} \tag{5.84}$$

Notice that the matrices $\mathbf{A}_a, \mathbf{A}_b$, and $\mathbf{A}_{ab}$ are not full-rank matrices, having ranks $(a-1), (b-1)$, and $(a-1)(b-1)$, respectively. Then in matrix notation, we can further write

$$\boldsymbol{\alpha}(t) = \mathbf{A}_a\boldsymbol{\eta}(t), \quad \boldsymbol{\beta}(t) = \mathbf{A}_b\boldsymbol{\eta}(t), \quad \boldsymbol{\theta}(t) = \mathbf{A}_{ab}\boldsymbol{\eta}(t). \tag{5.85}$$

Based on the estimators of the cell mean functions given in (5.78), the estimated effect functions are then given by

$$\hat{\boldsymbol{\alpha}}(t) = \mathbf{A}_a\hat{\boldsymbol{\eta}}(t), \quad \hat{\boldsymbol{\beta}}(t) = \mathbf{A}_b\hat{\boldsymbol{\eta}}(t), \quad \hat{\boldsymbol{\theta}}(t) = \mathbf{A}_{ab}\hat{\boldsymbol{\eta}}(t). \tag{5.86}$$

It is worthwhile to mention the following remarks.

Remark 5.5 *The estimated effect functions $\hat{\boldsymbol{\alpha}}(t), \hat{\boldsymbol{\beta}}(t)$, and $\hat{\boldsymbol{\theta}}(t)$ depend on the weights $w_{ij}, i = 1, \cdots, a; j = 1, \cdots, b$. For different weights, their values may be different.*

Remark 5.6 *The estimators of the effect functions in (5.86) and the estimator of the grand mean function $\hat{\eta}_0(t) = [\mathbf{g} \otimes \mathbf{h}]^T\hat{\boldsymbol{\eta}}(t)$ can also be obtained by minimizing the following weighted least squares criterion:*

$$\sum_{i=1}^{a}\sum_{j=1}^{b}\sum_{k=1}^{n_{ij}} \frac{w_{ij}}{n_{ij}} \int_{\mathcal{T}} [y_{ijk}(t) - \eta_0(t) - \alpha_i(t) - \beta_j(t) - \theta_{ij}(t)]^2\, dt,$$

subject to the side conditions (5.82) and that the weights have the structure (5.83).

Remark 5.7 *There are a few methods that can be used to specify the weights $w_{ij}, i = 1, 2, \cdots, a; j = 1, 2, \cdots, b$. For classical two-way ANOVA, see some examples by Fujikoshi (1993). In this section we use the following two simple methods: the equal-weight method and the size-adapted-weight method. Both the methods specify the weights as $w_{ij} = g_i h_j, i = 1, 2, \cdots, a; j = 1, 2, \cdots, b$ with the equal-weight method specifying $\mathbf{g}$ and $\mathbf{h}$ with*

$$g_i = 1/a, h_j = 1/b, i = 1, 2, \cdots, a; j = 1, 2, \cdots, b, \tag{5.87}$$

while the size-adapted-weight method specifies $\mathbf{g}$ and $\mathbf{h}$ with

$$g_i = \sum_{j=1}^{b} n_{ij}/n, i = 1, 2, \cdots, a; \quad h_j = \sum_{i=1}^{a} n_{ij}/n, j = 1, 2, \cdots, b. \tag{5.88}$$

When the two-way ANOVA design is balanced, that is, when all the cell sizes $n_{ij}, i = 1, 2, \cdots, a; j = 1, 2, \cdots, b$ are the same, the size-adapted-weight method reduces to the equal-weight method.

5.4.3 Tests of Linear Hypotheses

Notice that under the side constraints (5.82), the three null hypotheses in (5.77) can be equivalently written as

$$\begin{array}{ll} H_{0A}: & \mathbf{H}_a\boldsymbol{\alpha}(t) \equiv 0, t \in \mathcal{T}, \\ H_{0AB}: & \mathbf{H}_a\boldsymbol{\alpha}(t) \equiv 0,\ \mathbf{H}_b\boldsymbol{\beta}(t) \equiv 0, t \in \mathcal{T}, \\ H_{0I}: & \mathbf{H}_i\boldsymbol{\theta}(t) \equiv 0, t \in \mathcal{T}, \end{array} \tag{5.89}$$

where

$$\mathbf{H}_a = \Big(\mathbf{I}_{a-1}, -\mathbf{1}_{a-1}\Big), \mathbf{H}_b = \Big(\mathbf{I}_{b-1}, -\mathbf{1}_{b-1}\Big), \quad \text{and } \mathbf{H}_i = \mathbf{H}_a \otimes \mathbf{H}_b.$$

Notice that the null hypothesis H_{0AB} can be further written as

$$H_{0AB}: \mathbf{H}_{ab}[\boldsymbol{\alpha}(t)^T, \boldsymbol{\beta}(t)^T]^T \equiv 0, t \in \mathcal{T}, \text{ where } \mathbf{H}_{ab} = \mathrm{diag}(\mathbf{H}_a, \mathbf{H}_b).$$

The coefficient matrices $\mathbf{H}_a, \mathbf{H}_{ab}$, and $\mathbf{H}_i$ are contrast matrices with each row summing up to 0. They are full-rank matrices, having ranks $(a-1), (a-1)+(b-1)$, and $(a-1)(b-1)$, respectively. By (5.85), it is seen that each of the testing problems associated with the three null hypotheses given in (5.89) can be equivalently expressed in the form of the GLHT problem as defined in (5.91) with $\mathbf{C}$, respectively, being

$$\mathbf{C}_a = \mathbf{H}_a\mathbf{A}_a,\ \ \mathbf{C}_{ab} = \mathbf{H}_{ab}[\mathbf{A}_a^T, \mathbf{A}_b^T]^T,\ \ \mathbf{C}_i = \mathbf{H}_i\mathbf{A}_{ab}. \tag{5.90}$$

For the two-way ANOVA model (5.76), the GLHT problem can be expressed as:

$$\begin{array}{rrl} & H_0 & : \ \mathbf{C}\boldsymbol{\eta}(t) \equiv \mathbf{c}(t),\ t \in \mathcal{T}, \\ \text{versus} & H_1 & : \ \mathbf{C}\boldsymbol{\eta}(t) \neq \mathbf{c}(t), \ \text{ for some } t \in \mathcal{T}, \end{array} \tag{5.91}$$

where $\mathbf{C}: q \times (ab)$ is a known coefficient matrix of full rank with $\mathrm{rank}(\mathbf{C}) = q$, and $\mathbf{c}(t): q \times 1$ is a vector of known constant functions that is often specified as $\mathbf{0}$.

Remark 5.8 *The matrix* $\mathbf{C}$ *in (5.91) can often be written as* $\mathbf{C}_0\mathbf{Q}$*, where* $\mathbf{C}_0$ *is a contrast matrix and* $\mathbf{Q}$ *is some proper real matrix. For example, the matrices* $\mathbf{C}_a, \mathbf{C}_{ab}$*, and* $\mathbf{C}_i$ *defined in (5.90) are in such a form. As mentioned in Remark 5.3, by Kshirsagar (1972, Chapter 5, Section 4), there is a non-singular matrix* $\mathbf{P}$ *such that* $\tilde{\mathbf{C}}_0 = \mathbf{P}\mathbf{C}_0$*. It follows that we have*

$$\tilde{\mathbf{C}} = \tilde{\mathbf{C}}_0\mathbf{Q} = \mathbf{P}\mathbf{C}_0\mathbf{Q} = \mathbf{P}\mathbf{C}. \tag{5.92}$$

As before, to construct some proper test statistic for the GLHT problem (5.91), we first need to find a proper pivotal test function. To this end, notice that we have $\mathrm{E}\left[\mathbf{C}\hat{\boldsymbol{\eta}}(t) - \mathbf{c}(t)\right] = \mathbf{C}\boldsymbol{\eta}(t) - \mathbf{c}(t)$ and

$$\mathrm{Cov}\left[\mathbf{C}\hat{\boldsymbol{\eta}}(s) - \mathbf{c}(s), \mathbf{C}\hat{\boldsymbol{\eta}}(t) - \mathbf{c}(t)\right] = \gamma(s,t)\mathbf{C}\mathbf{D}\mathbf{C}^T,$$

where $\mathbf{D}$ is the diagonal matrix defined in (5.81). As $\mathbf{CDC}^T$ is a square matrix of full rank, we then arrive at the following pivotal test function:

$$\mathbf{z}(t) = \left(\mathbf{CDC}^T\right)^{-1/2}\left[\mathbf{C}\hat{\boldsymbol{\eta}}(t) - \mathbf{c}(t)\right]. \tag{5.93}$$

It is seen that we have

$$\mathbf{z}(t) \sim \text{SP}_q(\boldsymbol{\eta}_z, \gamma\mathbf{I}_q), \tag{5.94}$$

where

$$\boldsymbol{\eta}_z(t) = (\mathbf{CDC}^T)^{-1/2}\left[\mathbf{C}\boldsymbol{\eta}(t) - \mathbf{c}(t)\right]. \tag{5.95}$$

Under the null hypothesis in (5.91), $\boldsymbol{\eta}_z(t) \equiv 0, t \in \mathcal{T}$. The squared L^2-norm $\|\mathbf{z}(t)\|^2$ of $\mathbf{z}(t)$ can then be used as the pointwise sum of squared errors due to hypothesis (SSH):

$$\text{SSH}_n(t) = \left[\mathbf{C}\hat{\boldsymbol{\eta}}(t) - \mathbf{c}(t)\right]^T \left(\mathbf{CDC}^T\right)^{-1}\left[\mathbf{C}\hat{\boldsymbol{\eta}}(t) - \mathbf{c}(t)\right], \tag{5.96}$$

which, together with the pointwise sum of squared errors

$$\text{SSE}_n(t) = \sum_{i=1}^{a}\sum_{j=1}^{b}\sum_{k=1}^{n_{ij}}\left[y_{ijk}(t) - \bar{y}_{ij.}(t)\right]^2 = (n-ab)\hat{\gamma}(t,t), \tag{5.97}$$

will be used to define various tests for the GLHT problem (5.91). Notice that the above relationship is useful as it allows us to compute $\int_{\mathcal{T}}\text{SSE}_n(t)dt$ by the trace of the pooled sample covariance function. In fact, we have

$$\int_{\mathcal{T}}\text{SSE}_n(t)dt = (n-ab)\text{tr}(\hat{\gamma}). \tag{5.98}$$

Remark 5.9 *The pivotal test function $\mathbf{z}(t)$ (5.93) is constructed in a way such that each component of $\mathbf{z}(t)$ is a process with the same covariance function $\gamma(s,t)$ as indicated in (5.94). In addition, it is easy to check that the squared L^2-norm $SSH_n(t)$ (5.96) of $\mathbf{z}(t)$ is invariant when $\mathbf{C}$ and $\mathbf{c}(t)$ are replaced, respectively, with*

$$\tilde{\mathbf{C}} = \mathbf{PC} \quad \textit{and} \quad \tilde{\mathbf{c}}(t) = \mathbf{Pc}(t), \tag{5.99}$$

where $\mathbf{P}$ is a full-rank matrix of size $q \times q$ such as the one defined in (5.92). This property is important for the GLHT problem (5.91) as for a hypothesis testing problem that can be written in the form of the GLHT problem (5.91), the associated $\mathbf{C}$ and $\mathbf{c}(t)$ are not uniquely defined, as indicated by Remark 5.8.

Example 5.18 *For the left-cingulum data, we have $n = 39-1 = 38$ after the outlying left-cingulum curve is removed and $a = 2, b = 2$. By (5.80), we have $tr(\hat{\gamma}) = 265.48$. Then using formula (5.98), we have*

$$\int_{\mathcal{T}} SSE_n(t)dt = (38-4)\times 274.72 = 9{,}340.5. \tag{5.100}$$

Remark 5.10 *The GLHT problem (5.91) for the two-way ANOVA model (5.76) is rather similar to the GLHT problem (5.62) for the one-way ANOVA model (5.26). Therefore, the key ideas of the pointwise tests, the L^2-norm-based test, the F-type test, and the bootstrap test for the GLHT problem (5.91) are almost the same as those of their counterparts for the GLHT problem (5.62). Those readers who have read the GLHT problem for the one-way ANOVA model in the previous section can skip what follows in this subsection.*

Let $\lambda_1, \lambda_2, \cdots, \lambda_m$ be all the positive eigenvalues of $\gamma(s,t)$. As expected, it is easy to show that under the Gaussian assumption TW2 and the null hypothesis, both the integrated SSH and SSE are χ^2-type mixtures.

Theorem 5.16 *Under Assumptions TW1 and TW2 and the null hypothesis in (5.91), we have*

$$\int_{\mathcal{T}} SSH_n(t)dt \stackrel{d}{=} \sum_{r=1}^m \lambda_r A_r, \quad \int_{\mathcal{T}} SSE_n(t)dt \stackrel{d}{=} \sum_{r=1}^m \lambda_r E_r,$$

where $A_r, r = 1, \cdots, m \stackrel{i.i.d.}{\sim} \chi_q^2$ are independent of $E_r, r = 1, 2, \cdots, m \stackrel{i.i.d.}{\sim} \chi_{n-ab}^2$.

When the Gaussian assumption is not satisfied, the above theorem is of course not valid. However, as expected, when the cell sizes are sufficiently large, then by the central limit theorem for i.i.d. stochastic processes, Theorem 4.12 of Chapter 4, we can easily show that the integrated SSH is asymptotically a χ^2-type mixture.

Theorem 5.17 *Under Assumptions TW1, TW3, and TW4 and the null hypothesis in (5.91), as $n \to \infty$, we have*

$$\int_{\mathcal{T}} SSH_n(t)dt \stackrel{d}{\to} \sum_{r=1}^m \lambda_r A_r,$$

where $A_r, r = 1, \cdots, m, \stackrel{i.i.d.}{\sim} \chi_q^2$.

Based on the above two theorems, we are now ready to describe various tests for the GLHT problem (5.91).

Pointwise Tests We state the pointwise F-test and the pointwise χ^2-test. The test statistic of the pointwise F-test is defined as

$$F_n(t) = \frac{\text{SSH}_n(t)/q}{\text{SSE}_n(t)/(n-ab)}. \tag{5.101}$$

When the ab samples (5.72) are Gaussian, under the null hypothesis in (5.91), we have $F_n(t) \sim F_{q,n-ab}, t \in \mathcal{T}$. The pointwise F-test can be conducted accordingly.

When the Gaussian assumption is not valid, for large samples, one can use the pointwise χ^2-test. For large samples, that is, as $\min_{i,j} n_{ij} \to \infty$, we have $F_n(t) \xrightarrow{d} \chi_q^2/q, \;\; t \in \mathcal{T}$. The pointwise χ^2-test can be conducted accordingly.

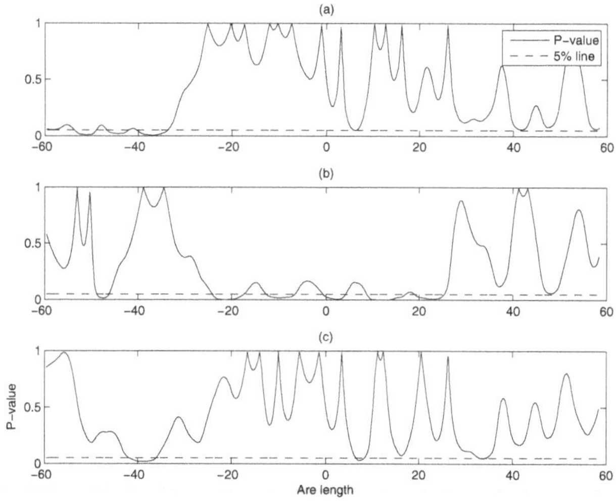

Figure 5.10 *Two-way ANOVA for the left-cingulum data by the pointwise F-test for (a) GHR, (b) AGE, and (c) GHR×AGE. The pointwise P-values were computed based on the null distribution $F_{1,34}$. In each panel, the 5% significance level line (dashed) is shown.*

Example 5.19 *Figure 5.10 displays the two-way ANOVA for the left-cingulum data by the pointwise F-test. The pointwise P-values for testing the main-effects of GHR and AGE and their interaction-effects are shown in the upper, middle, and lower panels, respectively. These pointwise P-values were computed based on the null distribution $F_{1,34}$. It is seen that*

1. *The main-effect of GHR is significant only over three small subintervals at the left end of the arc length support.*
2. *The main-effect of AGE is significant over a number medium-size subintervals on the arc length support.*
3. *The interaction-effect between GHR and AGE is significant only over three very small subintervals on the arc length support.*

L^2-Norm-Based Test For the GLHT problem (5.91), the L^2-norm-based test uses the test statistic

$$T_n = \int_{\mathcal{T}} \text{SSH}_n(t)dt. \tag{5.102}$$

Under the null hypothesis in (5.91) and under the conditions of Theorem 5.16 or the conditions of Theorem 5.17, we have or approximately have

$$T_n \stackrel{d}{=} \sum_{r=1}^{m} \lambda_r A_r, \ A_r \stackrel{i.i.d.}{\sim} \chi_q^2,$$

where $\lambda_1, r = 1, 2, \cdots, m$ are all the positive eigenvalues of $\gamma(s,t)$. Then the null distribution of T_n can be approximated by the Welch-Satterthwaite χ^2-approximation method described in Section 4.3 of Chapter 4. In fact, by that method, we have

$$T_n \sim \beta\chi_{q\kappa}^2 \text{ approximately, where } \beta = \frac{\text{tr}(\gamma^{\otimes 2})}{\text{tr}(\gamma)}, \quad \kappa = \frac{\text{tr}^2(\gamma)}{\text{tr}(\gamma^{\otimes 2})}. \tag{5.103}$$

As mentioned in the previous sections, the parameters β and κ can be estimated by a naive method and a bias-reduced method based on the functional data (5.72). With the pooled sample covariance function $\hat{\gamma}(s,t)$ given in (5.79), by the naive method,

$$\hat{\beta} = \frac{\text{tr}(\hat{\gamma}^{\otimes 2})}{\text{tr}(\hat{\gamma})}, \quad \hat{\kappa} = \frac{\text{tr}^2(\hat{\gamma})}{\text{tr}(\hat{\gamma}^{\otimes 2})}, \tag{5.104}$$

and by the bias-reduced method,

$$\hat{\beta} = \frac{\widehat{\text{tr}(\gamma^{\otimes 2})}}{\text{tr}(\hat{\gamma})}, \quad \hat{\kappa} = \frac{\widehat{\text{tr}^2(\gamma)}}{\widehat{\text{tr}(\gamma^{\otimes 2})}}. \tag{5.105}$$

with

$$\begin{aligned} \widehat{\text{tr}^2(\gamma)} &= \frac{(n-ab)(n-ab+1)}{(n-ab-1)(n-ab+2)}\left[\text{tr}^2(\hat{\gamma}) - \frac{2\text{tr}(\hat{\gamma}^{\otimes 2})}{n-ab+1}\right], \\ \widehat{\text{tr}(\gamma^{\otimes 2})} &= \frac{(n-ab)^2}{(n-ab-1)(n-ab+2)}\left[\text{tr}(\hat{\gamma}^{\otimes 2}) - \frac{\text{tr}^2(\hat{\gamma})}{n-ab}\right]. \end{aligned} \tag{5.106}$$

The above unbiased estimators of $\text{tr}^2(\gamma)$ and $\text{tr}(\gamma^{\otimes 2})$ are obtained by applying Theorem 5.13 and Theorem 4.6 of Chapter 4. The consistency of $\hat{\beta}$ and $\hat{\kappa}$ is given in the following theorem.

Theorem 5.18 *Under Assumptions TW1 and TW3 through TW7, as $n \to \infty$, we have $tr(\hat{\gamma}) \xrightarrow{p} tr(\gamma)$ and $tr(\hat{\gamma}^{\otimes 2}) \xrightarrow{p} tr(\gamma^{\otimes 2})$. Furthermore, as $n \to \infty$, we have*

$$\hat{\beta} \xrightarrow{p} \beta, \quad \hat{\kappa} \xrightarrow{p} \kappa,$$

where $\hat{\beta}$ and $\hat{\kappa}$ are the naive or bias-reduced estimators of β and κ.

The L^2-norm-based test is conducted by rejecting the null hypothesis in (5.91) whenever $T_n > \hat{\beta}\chi_{q\hat{\kappa}}^2(1-\alpha)$ for a given significance level α.

Table 5.10 *Two-way ANOVA for the left-cingulum data by the L^2-norm-based test with resolution $M = 1,000$.*

Weighting method	Effect	T_n	P-value
Equal-weight	GHR	7.5906	0.1747
	AGE	15.5160	0.0079
	GHR×AGE	2.5733	0.7575
Size-adapted-weight	GHR	7.4279	0.1849
	AGE	16.326	0.0057
	GHR×AGE	2.5733	0.7575

Note: The null distributions were approximated by the naive method with $\hat{\beta} = 55.71$ and $\hat{d} = \hat{\kappa} = 4.931$. The P-values with the equal-weight method are generally comparable with those with the size-adapted-weight method and the P-value of the interaction-effect is independent of the weighting method.

Example 5.20 *For the left-cingulum data, by (5.80), we have $tr(\hat{\gamma}) = 274.72$ and $tr(\hat{\gamma}^{\otimes 2}) = 15,305$. Then by the naive method using (5.104), we have*

$$\begin{aligned} \hat{\beta} &= \frac{tr(\hat{\gamma}^{\otimes 2})}{tr(\hat{\gamma})} = \frac{15305}{274.72} = 55.71, \\ \hat{d} = \hat{\kappa} &= \frac{tr^2(\hat{\gamma})}{tr(\hat{\gamma}^{\otimes 2})} = \frac{274.72^2}{15305} = 4.931. \end{aligned} \tag{5.107}$$

Table 5.10 presents the two-way ANOVA results for the left-cingulum data by the L^2-norm-based test with resolution $M = 1,000$. The equal-weight and size-adapted-weight methods were considered. It is seen that under both the weighting methods, the L^2-norm-based test results suggest that the main-effect of GHR and the interaction-effect between GHR and AGE are not significant, while the main-effect of AGE is highly significant. These conclusions are consistent with those observed from the pointwise F-test presented in Figure 5.10. In addition, these results can be explained biologically. It is known that the radial diffusibility grows with brain development.

F-Type Test As before, under the Gaussian assumption, the F-type test for the GLHT problem (5.91) uses the following test statistic:

$$F_n = \frac{\int_{\mathcal{T}} \text{SSH}_n(t)dt/q}{\int_{\mathcal{T}} \text{SSE}_n(t)dt/(n-ab)}. \tag{5.108}$$

Under the null hypothesis in (5.91) and by Theorem 5.16, we have

$$F_n \stackrel{d}{=} \frac{\sum_{r=1}^m \lambda_r A_r/q}{\sum_{r=1}^m \lambda_r E_r/(n-ab)},$$

where $A_r \stackrel{i.i.d.}{\sim} \chi_q^2, E_r \stackrel{i.i.d.}{\sim} \chi_{n-ab}^2$ and they are all independent, and $\lambda_1, \lambda_2, \cdots, \lambda_m$ are all the positive eigenvalues of $\gamma(s,t)$. It follows that the null distribution of F_n can be approximated by the two-cumulant matched

Table 5.11 *Two-way ANOVA for the left-cingulum data by the F-type test with resolution $M = 1,000$.*

Weighting method	Effect	F_n	P-value
Equal-weight	GHR	1.5393	0.1811
	AGE	3.1464	0.0099
	GHR×AGE	0.5219	0.7571
Size-adapted-weight	GHR	1.5063	0.1912
	AGE	3.3107	0.0073
	GHR×AGE	0.5219	0.7571

Note: The null distributions were approximated by the naive method with $\hat{d}_1 = \hat{\kappa} = 4.9312$ and $\hat{d}_2 = (n - ab)\hat{\kappa} = 167.66$. The P-values by the F-type test are generally comparable with those by the L^2-norm-based test as presented in Table 5.10.

F-approximation method described in Section 4.4 of Chapter 4. In fact, by that method, we have

$$F_n \sim F_{q\hat{\kappa},(n-ab)\hat{\kappa}} \text{ approximately,}$$

where by the naive method, $\hat{\kappa}$ is given in (5.104) and by the bias-reduced method, $\hat{\kappa}$ is given in (5.105). We reject the null hypothesis in (5.91) whenever $F_n > F_{q\hat{\kappa},(n-ab)\hat{\kappa}}(1 - \alpha)$ for any given significance level α. Notice that when $(n - ab)\hat{\kappa}$ are large, the conclusion by the F-type test will be about the same as that by the L^2-norm-based test discussed earlier.

Example 5.21 *For the left-cingulum data, we have $n = 38, a = b = 2$. By (5.107), we have $\hat{\kappa} = 4.931$ by the naive method so that $\hat{d}_1 = \hat{\kappa} = 4.931$ and $\hat{d}_2 = (n-ab)\hat{\kappa} = 34\times 4.931 = 167.66$. Table 5.11 presents the two-way ANOVA results for the left-cingulum data by the F-type test with resolution $M = 1,000$. The equal-weight and size-adapted-weight methods were considered. It is seen that under both weighting methods, the F-type test results suggest that the main-effect of GHR and the interaction-effect between GHR and AGE are not significant while the main-effect of AGE is highly significant. These conclusions are consistent with those obtained by the L^2-norm-based test presented in Table 5.10.*

Bootstrap Tests When the ab samples (5.72) are not Gaussian, the two-cumulant matched F-approximation method may not be applicable for approximating the null distribution of the F-type test. When the ab samples (5.72) are not Gaussian and the sample sizes $n_{ij}, i = 1, 2, \cdots, a; j = 1, 2, \cdots, b$ are too small, the Welch-Satterthwaite χ^2-approximation method may not be applicable for approximating the null distribution of the L^2-norm-based test. In these cases, we can use some nonparametric bootstrap approaches to bootstrap the critical values of the pointwise test, the L^2-norm-based test, and the F-type test for the GLHT problem (5.91), resulting in the so-called the pointwise bootstrap test, the L^2-norm-based bootstrap test, and the F-type bootstrap test, respectively.

Let $v_{ijk}^*(t), k = 1, \cdots, n_{ij}; i = 1, \cdots, a; j = 1, \cdots, b$ be ab bootstrap samples randomly generated from the estimated subject-effect functions:

$$\hat{v}_{ijk}(t) = y_{ijk}(t) - \hat{\eta}_{ij}(t),\ k = 1, \cdots, n_{ij};\ i = 1, \cdots, a; j = 1, \cdots, b.$$

Set $y_{ijk}^*(t) = \hat{\eta}_{ij}(t) + v_{ijk}^*(t), k = 1, \cdots, n_{ij}; i = 1, \cdots, a, j = 1, \cdots, b$. We can then compute the sample cell mean functions and the pooled sample covariance function as $\hat{\eta}_{ij}^*(t) = \bar{y}_{ij.}^*(t), i = 1, 2, \cdots, a; j = 1, 2, \cdots, b$ and $\hat{\gamma}^*(s,t)$ based on the ab bootstrap samples. We then compute

$$\begin{aligned}\mathrm{SSH}_n^*(t) &= \left[\hat{\boldsymbol{\eta}}^*(t) - \hat{\boldsymbol{\eta}}(t)\right]^T \mathbf{C}^T \left(\mathbf{CDC}^T\right)^{-1} \mathbf{C}\left[\hat{\boldsymbol{\eta}}^*(t) - \hat{\boldsymbol{\eta}}(t)\right],\\ \mathrm{SSE}_n^*(t) &= (n - ab)\hat{\gamma}^*(t,t),\end{aligned}$$

where $\hat{\boldsymbol{\eta}}^*(t) = [\hat{\eta}_{11}^*(t), \cdots, \hat{\eta}_{1b}^*(t), \cdots, \hat{\eta}_{a1}^*(t), \cdots, \hat{\eta}_{ab}^*(t)]^T$. Then for the pointwise bootstrap test, the L^2-norm-based bootstrap test, or the F-type bootstrap test, we compute, respectively,

$$\begin{aligned}F_n^*(t) &= \frac{\mathrm{SSH}_n^*(t)/q}{\mathrm{SSE}_n^*(t)/(n-ab)},\ T_n^* = \textstyle\int_{\mathcal{T}} \mathrm{SSH}_n^*(t)dt,\ \text{or}\\ F_n^* &= \frac{\int_{\mathcal{T}} \mathrm{SSH}_n^*(t)dt/q}{\int_{\mathcal{T}} \mathrm{SSE}_n^*(t)dt/(n-ab)}.\end{aligned}$$

Repeat the above bootstrapping process a large number of times, calculate the $100(1-\alpha)$-percentile for $F_n^*(t), T_n^*$, or F_n^*, and then conduct the pointwise bootstrap test, the L^2-norm-based bootstrap test, or the F-type bootstrap test accordingly.

5.4.4 Balanced Two-Way ANOVA with Interaction

In some situations, the two-way ANOVA model (5.76) is balanced in the sense that the cell sizes $n_{ij} \equiv c$ for all $i = 1, \cdots, a; j = 1, \cdots, b$. In this subsection, keep in mind that we focus on the case when $n_{ij} \equiv c > 1$ so that $n = abc$, and study how to test the null hypotheses listed in (5.77). The methodologies discussed in the previous subsections are applicable for such balanced two-way ANOVA models. In this subsection, we aim to simplify those methodologies for such balanced two-way ANOVA models.

Notice that for balanced two-way ANOVA models, both the equal-weight method and the size-adapted-weight method lead to the same side constraints:

$$\begin{aligned}&\textstyle\sum_{i=1}^a \alpha_i(t) \equiv 0,\ \sum_{j=1}^b \beta_j(t) \equiv 0,\\ &\textstyle\sum_{j=1}^b \theta_{ij}(t) \equiv 0,\ i = 1, \cdots, a-1,\\ &\textstyle\sum_{i=1}^a \theta_{ij}(t) \equiv 0,\ j = 1, \cdots, b-1,\\ &\textstyle\sum_{i=1}^a \sum_{j=1}^b \theta_{ij}(t) \equiv 0.\end{aligned} \tag{5.109}$$

Under the above side constraints, the usual least squares estimators of the functional parameters $\eta_0(t), \alpha_i(t), \beta_j(t)$, and $\theta_{ij}(t)$ are given by

$$\begin{array}{rcl}\hat{\eta}_0(t) &=& \bar{y}_{...}(t),\\ \hat{\alpha}_i(t) &=& \bar{y}_{i..}(t)-\bar{y}_{...}(t),\ i=1,\cdots,a,\\ \hat{\beta}_j(t) &=& \bar{y}_{.j.}(t)-\bar{y}_{...}(t),\ j=1,\cdots,b,\\ \hat{\theta}_{ij}(t) &=& \bar{y}_{ij.}(t)-\bar{y}_{i..}(t)-\bar{y}_{.j.}(t)+\bar{y}_{...}(t),\\ && i=1,\cdots,a,\ j=1,\cdots,b,\end{array}$$

where

$$\begin{array}{rcl}\bar{y}_{ij.}(t) &=& \sum_{k=1}^c y_{ijk}(t)/c,\ i=1,\cdots,a; j=1,\cdots,b,\\ \bar{y}_{i..}(t) &=& \sum_{j=1}^b\sum_{k=1}^c y_{ijk}(t)/(bc),\ i=1,\cdots,a,\\ \bar{y}_{.j.}(t) &=& \sum_{i=1}^a\sum_{k=1}^c y_{ijk}(t)/(ac),\ j=1,\cdots,b,\\ \bar{y}_{...}(t) &=& \sum_{i=1}^a\sum_{j=1}^b\sum_{k=1}^c y_{ijk}(t)/(abc).\end{array}\tag{5.110}$$

Define $\bar{v}_{ij.}(t), \bar{v}_{i..}(t), \bar{v}_{.j.}(t)$, and $\bar{v}_{...}(t)$ similarly. Then we can further express $\hat{\eta}_0(t), \hat{\alpha}_i(t), \hat{\beta}_j(t)$, and $\hat{\theta}_{ij}(t)$ as

$$\begin{array}{rcl}\hat{\eta}_0(t) &=& \eta_0(t)+\bar{v}_{...}(t),\\ \hat{\alpha}_i(t) &=& \alpha_i(t)+\bar{v}_{i..}(t)-\bar{v}_{...}(t),\ i=1,\cdots,a,\\ \hat{\beta}_j(t) &=& \beta_j(t)+\bar{v}_{.j.}(t)-\bar{v}_{...}(t),\ j=1,\cdots,b,\\ \hat{\theta}_{ij}(t) &=& \theta_{ij}(t)+\bar{v}_{ij.}(t)-\bar{v}_{i..}(t)-\bar{v}_{.j.}(t)+\bar{v}_{...}(t),\\ && i=1,\cdots,a,\ j=1,\cdots,b.\end{array}\tag{5.111}$$

In addition, we have

$$\begin{array}{rcl}\hat{v}_{ijk}(t) &=& y_{ijk}(t)-\bar{y}_{ij.}(t)=v_{ijk}(t)-\bar{v}_{ij.}(t),\\ && i=1,\cdots,a; j=1,\cdots,b; k=1,\cdots,c.\end{array}\tag{5.112}$$

For any fixed $t \in \mathcal{T}$, let $\text{SST}_n(t)$ denote the pointwise total sum of squares; let $\text{SSA}_n(t), \text{SSB}_n(t)$, $\text{SSI}_n(t)$ denote the pointwise sum of squares due to factor A, factor B, and their interaction, respectively; and let $\text{SSE}_n(t)$ denote the pointwise sum of squares due to errors. Then, by the classical balanced two-way ANOVA, we have the following decomposition:

$$\text{SST}_n(t) = \text{SSA}_n(t) + \text{SSB}_n(t) + \text{SSI}_n(t) + \text{SSE}_n(t),$$

where

$$\begin{array}{rcl}\text{SST}_n(t) &=& \sum_{i=1}^a\sum_{j=1}^b\sum_{k=1}^c \left[y_{ijk}(t)-\bar{y}_{...}(t)\right]^2,\\ \text{SSA}_n(t) &=& bc\sum_{i=1}^a \left[\bar{y}_{i..}(t)-\bar{y}_{...}(t)\right]^2 = bc\sum_{i=1}^a \hat{\alpha}_i^2(t),\\ \text{SSB}_n(t) &=& ac\sum_{j=1}^b \left[\bar{y}_{.j.}(t)-\bar{y}_{...}(t)\right]^2 = ac\sum_{j=1}^b \hat{\beta}_j^2(t),\\ \text{SSI}_n(t) &=& c\sum_{i=1}^a\sum_{j=1}^b \left[\bar{y}_{ij.}(t)-\bar{y}_{i..}(t)-\bar{y}_{.j.}(t)+\bar{y}_{...}(t)\right]^2\\ &=& c\sum_{i=1}^a\sum_{j=1}^b \hat{\theta}_{ij}^2(t),\\ \text{SSE}_n(t) &=& \sum_{i=1}^a\sum_{j=1}^b\sum_{k=1}^c \left[y_{ijk}(t)-\bar{y}_{ij.}(t)\right]^2.\end{array}\tag{5.113}$$

It is seen that $\text{SSA}_n(t), \text{SSA}_n(t)+\text{SSB}_n(t)$, and $\text{SSI}_n(t)$ are natural pivotal test functions for testing the null hypotheses H_{0A}, H_{0AB}, and H_{0I}, respectively. By (5.111), we further have

$$\begin{aligned}
\mathrm{SSA}_n(t) &= bc\sum_{i=1}^a [\alpha_i(t)+\bar{v}_{i..}(t)-\bar{v}_{...}(t)]^2,\\
\mathrm{SSB}_n(t) &= ac\sum_{j=1}^b [\beta_j(t)+\bar{v}_{.j.}(t)-\bar{v}_{...}(t)]^2,\\
\mathrm{SSI}_n(t) &= c\sum_{i=1}^a\sum_{j=1}^b [\theta_{ij}(t)+\bar{v}_{ij.}(t)\\
&\quad -\bar{v}_{i..}(t)-\bar{v}_{.j.}(t)+\bar{v}_{...}(t)]^2,\\
\mathrm{SSE}_n(t) &= \sum_{i=1}^a\sum_{j=1}^b\sum_{k=1}^c [v_{ijk}(t)-\bar{v}_{ij.}(t)]^2.
\end{aligned} \tag{5.114}$$

Let $\mathbf{v}_{ij}(t) = [v_{ij1}(t), v_{ij2}(t), \cdots, v_{ijc}(t)]^T$ denote the vector of the subject-effect functions at the (i,j)th cell, and let

$$\mathbf{v}(t) = [\mathbf{v}_{11}(t)^T, \cdots, \mathbf{v}_{1b}(t)^T, \cdots, \mathbf{v}_{a1}(t)^T, \cdots, \mathbf{v}_{ab}(t)^T]^T.$$

Then we have

$$\mathrm{SSE}_n(t) = \mathbf{v}(t)^T\left[\mathbf{I}_a\otimes\mathbf{I}_b\otimes(\mathbf{I}_c-\frac{\mathbf{J}_c}{c})\right]\mathbf{v}(t), \tag{5.115}$$

where $\mathbf{I}_r$ denotes the identity matrix of size $r\times r$ and $\mathbf{J}_r$ denotes the $r\times r$ matrix of ones. In addition, we have the following results. Under H_{0A}, we have

$$\mathrm{SSA}_n(t) = \mathbf{v}(t)^T\left[(\mathbf{I}_a-\frac{\mathbf{J}_a}{a})\otimes\frac{\mathbf{J}_b}{b}\otimes\frac{\mathbf{J}_c}{c}\right]\mathbf{v}(t). \tag{5.116}$$

Under H_{0AB}, we have

$$\begin{aligned}
\mathrm{SSA}_n(t) &= \mathbf{v}(t)^T\left[(\mathbf{I}_a-\tfrac{\mathbf{J}_a}{a})\otimes\tfrac{\mathbf{J}_b}{b}\otimes\tfrac{\mathbf{J}_c}{c}\right]\mathbf{v}(t),\\
\mathrm{SSB}_n(t) &= \mathbf{v}(t)^T\left[\tfrac{\mathbf{J}_a}{a}\otimes(\mathbf{I}_b-\tfrac{\mathbf{J}_b}{b})\otimes\tfrac{\mathbf{J}_c}{c}\right]\mathbf{v}(t).
\end{aligned} \tag{5.117}$$

Under H_{0I}, we have

$$\mathrm{SSI}_n(t) = \mathbf{v}(t)^T\left[(\mathbf{I}_a-\frac{\mathbf{J}_a}{a})\otimes(\mathbf{I}_b-\frac{\mathbf{J}_b}{b})\otimes\frac{\mathbf{J}_c}{c}\right]\mathbf{v}(t). \tag{5.118}$$

The matrices

$$\begin{array}{ll}
\mathbf{I}_a\otimes\mathbf{I}_b\otimes(\mathbf{I}_c-\tfrac{\mathbf{J}_c}{c}), & (\mathbf{I}_a-\tfrac{\mathbf{J}_a}{a})\otimes\tfrac{\mathbf{J}_b}{b}\otimes\tfrac{\mathbf{J}_c}{c},\\
\tfrac{\mathbf{J}_a}{a}\otimes(\mathbf{I}_b-\tfrac{\mathbf{J}_b}{b})\otimes\tfrac{\mathbf{J}_c}{c}, & (\mathbf{I}_a-\tfrac{\mathbf{J}_a}{a})\otimes(\mathbf{I}_b-\tfrac{\mathbf{J}_b}{b})\otimes\tfrac{\mathbf{J}_c}{c},
\end{array}$$

are idempotent with ranks $ab(c-1), (a-1), (b-1)$, and $(a-1)(b-1)$, respectively.

Set

$$\mathbf{z}(t) = \sqrt{c}[\bar{v}_{11.}(t), \cdots, \bar{v}_{1b.}(t), \cdots, \bar{v}_{a1.}(t), \cdots, \bar{v}_{ab.}(t)]^T. \tag{5.119}$$

Then we further have the following results:

$$\begin{array}{l}
\text{Under } H_{0A} \text{ or } H_{0AB},\ \mathrm{SSA}_n(t) = \mathbf{z}(t)^T\left[(\mathbf{I}_a-\tfrac{\mathbf{J}_a}{a})\otimes\tfrac{\mathbf{J}_b}{b}\right]\mathbf{z}(t),\\
\text{Under } H_{0AB},\ \mathrm{SSB}_n(t) = \mathbf{z}(t)^T\left[\tfrac{\mathbf{J}_a}{a}\otimes(\mathbf{I}_b-\tfrac{\mathbf{J}_b}{b})\right]\mathbf{z}(t),\\
\text{Under } H_{0I},\ \mathrm{SSI}_n(t) = \mathbf{z}(t)^T\left[(\mathbf{I}_a-\tfrac{\mathbf{J}_a}{a})\otimes(\mathbf{I}_b-\tfrac{\mathbf{J}_b}{b})\right]\mathbf{z}(t).
\end{array} \tag{5.120}$$

The matrices

$$(\mathbf{I}_a - \frac{\mathbf{J}_a}{a}),\ (\mathbf{I}_b - \frac{\mathbf{J}_b}{b}),\ \text{ and, } (\mathbf{I}_a - \frac{\mathbf{J}_a}{a}) \otimes (\mathbf{I}_b - \frac{\mathbf{J}_b}{b}) \tag{5.121}$$

are idempotent with ranks $(a-1), (b-1)$, and $(a-1)(b-1)$, respectively.

As usual, we first derive the random expressions of the integrals of $\mathrm{SSA}_n(t), \mathrm{SSB}_n(t), \mathrm{SSI}_n(t)$, and $\mathrm{SSE}_n(t)$ over $\mathcal{T}$ when the samples are Gaussian and non-Gaussian. The following two theorems show that these random integrals are χ^2-type mixtures when the Gaussian assumption is valid or asymptotically so when the Gaussian assumption is not valid.

Theorem 5.19 *Under Assumptions TW1 and TW2 with* $n_{ij} \equiv c > 1$, *we have*

$$\begin{array}{llll} \text{Under } H_{0A}, & \int_{\mathcal{T}} SSA_n(t)dt \stackrel{d}{=} \sum_{r=1}^m \lambda_r A_r, & A_r \stackrel{i.i.d.}{\sim} \chi^2_{(a-1)}, \\ \text{Under } H_{0AB}, & \int_{\mathcal{T}} SSB_n(t)dt \stackrel{d}{=} \sum_{r=1}^m \lambda_r B_r, & B_r \stackrel{i.i.d.}{\sim} \chi^2_{(b-1)}, \\ \text{Under } H_{0I}, & \int_{\mathcal{T}} SSI_n(t)dt \stackrel{d}{=} \sum_{r=1}^m \lambda_r C_r, & C_r \stackrel{i.i.d.}{\sim} \chi^2_{(a-1)(b-1)}, \\ & \int_{\mathcal{T}} SSE_n(t)dt \stackrel{d}{=} \sum_{r=1}^m \lambda_r E_r, & E_r \stackrel{i.i.d.}{\sim} \chi^2_{ab(c-1)}, \end{array}$$

where $A_r, B_r, C_r, E_r, r = 1, 2, \cdots, m$ *are independent of each other, and* $\lambda_1, \cdots, \lambda_m$ *are all the positive eigenvalues of* $\gamma(s,t)$.

Theorem 5.20 *Under Assumptions TW1 and TW4 with* $n_{ij} \equiv c > 1$, *as* $c \to \infty$, *we have*

$$\begin{array}{llll} \text{Under } H_{0A}, & \int_{\mathcal{T}} SSA_n(t)dt \stackrel{d}{\to} \sum_{r=1}^m \lambda_r A_r, & A_r \stackrel{i.i.d.}{\sim} \chi^2_{(a-1)}, \\ \text{Under } H_{0AB}, & \int_{\mathcal{T}} SSB_n(t)dt \stackrel{d}{\to} \sum_{r=1}^m \lambda_r B_r, & B_r \stackrel{i.i.d.}{\sim} \chi^2_{(b-1)}, \\ \text{Under } H_{0I}, & \int_{\mathcal{T}} SSI_n(t)dt \stackrel{d}{\to} \sum_{r=1}^m \lambda_r C_r, & C_r \stackrel{i.i.d.}{\sim} \chi^2_{(a-1)(b-1)}, \end{array}$$

where $A_r, B_r, C_r, r = 1, 2, \cdots, m$ *are independent of each other, and* $\lambda_1, \cdots, \lambda_m$ *are all the positive eigenvalues of* $\gamma(s,t)$.

In the above two theorems, attention should be paid to the degrees of freedom in the χ^2-type mixtures. Based on the above two theorems, we are now ready to describe various tests for the null hypotheses listed in (5.77).

Pointwise Tests We describe the pointwise F-test and the pointwise χ^2-test. The pointwise F-test for H_{0A}, H_{0AB}, and H_{0I}, respectively, uses the following test statistics:

$$\begin{aligned} F_n^A(t) &= \frac{\mathrm{SSA}_n(t)/(a-1)}{\mathrm{SSE}_n(t)/[ab(c-1)]}, \\ F_n^{AB}(t) &= \frac{[\mathrm{SSA}_n(t)+\mathrm{SSB}_n(t)]/(a+b-2)}{\mathrm{SSE}_n(t)/[ab(c-1)]}, \\ F_n^I(t) &= \frac{\mathrm{SSI}_n(t)/[(a-1)(b-1)]}{\mathrm{SSE}_n(t)/[ab(c-1)]}. \end{aligned} \tag{5.122}$$

When the ab samples (5.72) with $n_{ij} \equiv c > 1$ are Gaussian, we have

$$\begin{array}{lll} \text{Under } H_{0A}, & F_n^A(t) \sim F_{(a-1),ab(c-1)}, & t \in \mathcal{T}, \\ \text{Under } H_{0AB}, & F_n^{AB}(t) \sim F_{(a+b-2),ab(c-1)}, & t \in \mathcal{T}, \\ \text{Under } H_{0I}, & F_n^I(t) \sim F_{(a-1)(b-1),ab(c-1)}, & t \in \mathcal{T}. \end{array}$$

The pointwise F-test can be conducted using the above distributions.

When the Gaussian assumption is not valid, for large samples, one can use the pointwise χ^2-test. In fact, for large samples, that is, as $c \to \infty$, we have

$$\begin{array}{lll} \text{Under } H_{0A}, & F_n^A(t) \xrightarrow{d} \chi^2_{a-1}/(a-1), & t \in \mathcal{T}, \\ \text{Under } H_{0AB}, & F_n^{AB}(t) \xrightarrow{d} \chi^2_{a+b-2}/(a+b-2), & t \in \mathcal{T}, \\ \text{Under } H_{0I}, & F_n^I(t) \xrightarrow{d} \chi^2_{(a-1)(b-1)}/[(a-1)(b-1)], & t \in \mathcal{T}. \end{array}$$

The pointwise χ^2-test can be conducted using the above asymptotical χ^2-distributions.

L^2-Norm-Based Test The L^2-norm-based test for H_{0A}, H_{0AB}, and H_{0I}, respectively, uses the following test statistics:

$$\begin{array}{lll} T_n^A & = & \int_{\mathcal{T}} \mathrm{SSA}_n(t)dt, \\ T_n^{AB} & = & \int_{\mathcal{T}} [\mathrm{SSA}_n(t) + \mathrm{SSB}_n(t)]\, dt, \\ T_n^I & = & \int_{\mathcal{T}} \mathrm{SSI}_n(t)dt. \end{array} \tag{5.123}$$

By Theorems 5.19 and 5.20, the null distributions of T_n^A, T_n^{AB}, and T_n^I can be approximated by the Welch-Satterthwaite χ^2-approximation method described in Section 4.3 of Chapter 4. In fact, we have

$$\begin{array}{lll} T_n^A & \sim \hat{\beta}\chi^2_{(a-1)\hat{\kappa}} & \text{approximately}, \\ T_n^{AB} & \sim \hat{\beta}\chi^2_{(a+b-2)\hat{\kappa}} & \text{approximately}, \\ T_n^I & \sim \hat{\beta}\chi^2_{(a-1)(b-1)\hat{\kappa}} & \text{approximately}, \end{array}$$

where by the naive method, $\hat{\beta}$ and $\hat{\kappa}$ are given in (5.104) and by the bias-reduced method, $\hat{\beta}$ and $\hat{\kappa}$ are given in (5.105) but with $n_{ij} = c > 1$ for all $i = 1, 2, \cdots, a; j = 1, 2, \cdots, b$. The null hypotheses H_{0A}, H_{0AB}, and H_{0I} can then be tested based on the above approximate distributions.

F-Type Test Suppose the ab samples (5.72) are Gaussian with $n_{ij} \equiv c > 1$. Then, we have

$$\mathbf{v}(t) \sim \mathrm{GP}_{abc}(\mathbf{0}, \gamma \mathbf{I}_{abc}).$$

Then the F-type tests for H_{0A}, H_{0AB}, and H_{0I} can, respectively, use the following test statistics:

$$
\begin{aligned}
F_n^A &= \frac{\int_{\mathcal{T}} \mathrm{SSA}_n(t)dt/(a-1)}{\int_{\mathcal{T}} \mathrm{SSE}_n(t)dt/[ab(c-1)]}, \\
F_n^{AB} &= \frac{\int_{\mathcal{T}} [\mathrm{SSA}_n(t)+\mathrm{SSB}_n(t)]dt/(a+b-2)}{\int_{\mathcal{T}} \mathrm{SSE}_n(t)dt/[ab(c-1)]}, \\
F_n^I &= \frac{\int_{\mathcal{T}} \mathrm{SSI}_n(t)dt/[(a-1)(b-1)]}{\int_{\mathcal{T}} \mathrm{SSE}_n(t)dt/[ab(c-1)]}.
\end{aligned} \tag{5.124}
$$

By Theorem 5.19, we have

$$
\begin{aligned}
&\text{Under } H_{0A}, && F_n^A \stackrel{d}{=} \frac{\sum_{r=1}^m \lambda_r A_r/(a-1)}{\sum_{r=1}^m \lambda_r E_r/[ab(c-1)]} \\
&\text{Under } H_{0AB}, && F_n^{AB} \stackrel{d}{=} \frac{\sum_{r=1}^m \lambda_r [A_r+B_r]/(a+b-2)}{\sum_{r=1}^m \lambda_r E_r/[ab(c-1)]}, \\
&\text{Under } H_{0I}, && F_n^I \stackrel{d}{=} \frac{\sum_{r=1}^m \lambda_r C_r/[(a-1)(b-1)]}{\sum_{r=1}^m \lambda_r E_r/[ab(c-1)]},
\end{aligned}
$$

where $A_r \stackrel{i.i.d.}{\sim} \chi^2_{a-1}, B_r \stackrel{i.i.d.}{\sim} \chi^2_{b-1}, C_r \stackrel{i.i.d.}{\sim} \chi^2_{(a-1)(b-1)}$, and $E_r \stackrel{i.i.d.}{\sim} \chi^2_{ab(c-1)}$ are independent of each other; and $\lambda_1, \lambda_2, \cdots, \lambda_m$ are all the positive eigenvalues of $\gamma(s,t)$. It follows that the null distributions of F_n^A, F_n^{AB}, and F_n^I can be approximated by the two-cumulant matched F-approximation method described in Section 4.4 of Chapter 4. We then have

$$
\begin{aligned}
&\text{Under } H_{0A}, && F_n^A \sim F_{(a-1)\hat{\kappa},[ab(c-1)]\hat{\kappa}} && \text{approximately}, \\
&\text{Under } H_{0AB}, && F_n^{AB} \sim F_{(a+b-2)\hat{\kappa},[ab(c-1)]\hat{\kappa}} && \text{approximately}, \\
&\text{Under } H_{0I}, && F_n^I \sim F_{[(a-1)(b-1)]\hat{\kappa},[ab(c-1)]\hat{\kappa}} && \text{approximately},
\end{aligned}
$$

where by the naive method, $\hat{\kappa}$ is given in (5.104) and by the bias-reduced method, $\hat{\kappa}$ is given in (5.105) but with $n_{ij} = c > 1$ for all $i = 1, \cdots, a; j = 1, \cdots, b$. The F-type tests for H_{0A}, H_{0AB}, and H_{0I} can then be conducted accordingly.

Bootstrap Tests When the ab samples (5.72) with $n_{ij} \equiv c$ are not Gaussian and c is small, nonparametric bootstrap approaches can be used to conduct the pointwise tests, the L^2-norm-based tests, and the F-type tests for H_{0A}, H_{0AB} and H_{0I}, resulting in the pointwise bootstrap test, the L^2-norm-based bootstrap test, and the F-type bootstrap test, respectively.

Let $v_{ijk}^*(t), k = 1, 2, \cdots, c; i = 1, 2, \cdots, a; j = 1, 2, \cdots, b$ be the ab bootstrap samples randomly generated from $\hat{v}_{ijk}(t) = y_{ijk}(t) - \bar{y}_{ij.}(t), k = 1, \cdots, c; i = 1, \cdots, a; j = 1, \cdots, b$, respectively. Set $\mathbf{v}_{ij}^*(t) = [v_{ij1}^*(t), \cdots, v_{ijc}^*(t)]^T, i = 1, \cdots, a; j = 1, \cdots, b$ and set $\mathbf{v}^*(t) = [\mathbf{v}_{11}^*(t)^T, \cdots, \mathbf{v}_{1b}^*(t)^T, \cdots, \mathbf{v}_{a1}^*(t)^T, \cdots, \mathbf{v}_{ab}^*(t)^T]^T$. Then we can compute

$$
\mathrm{SSA}_n^*(t) = \mathbf{v}^*(t)^T \left[(\mathbf{I}_a - \frac{\mathbf{J}_a}{a}) \otimes \frac{\mathbf{J}_b}{b} \otimes \frac{\mathbf{J}_c}{c} \right] \mathbf{v}^*(t),
$$

$$\begin{aligned}
\text{SSB}_n^*(t) &= \mathbf{v}^*(t)^T \left[\frac{\mathbf{J}_a}{a} \otimes (\mathbf{I}_b - \frac{\mathbf{J}_b}{b}) \otimes \frac{\mathbf{J}_c}{c}\right] \mathbf{v}^*(t),\\
\text{SSI}_n^*(t) &= \mathbf{v}^*(t)^T \left[(\mathbf{I}_a - \frac{\mathbf{J}_a}{a}) \otimes (\mathbf{I}_b - \frac{\mathbf{J}_b}{b}) \otimes \frac{\mathbf{J}_c}{c}\right] \mathbf{v}^*(t),\\
\text{SSE}_n^*(t) &= \mathbf{v}^*(t)^T \left[\mathbf{I}_a \otimes \mathbf{I}_b \otimes (\mathbf{I}_c - \frac{\mathbf{J}_c}{c})\right] \mathbf{v}^*(t).
\end{aligned}$$

For the pointwise bootstrap test for H_{0A}, H_{0AB}, and H_{0I}, one respectively computes the following test statistics:

$$\begin{aligned}
F_n^{*A}(t) &= \frac{\text{SSA}_n^*(t)/(a-1)}{\text{SSE}_n^*(t)/[ab(c-1)]},\\
F_n^{*AB}(t) &= \frac{[\text{SSA}_n^*(t)+\text{SSB}_n^*(t)]/(a+b-2)}{\text{SSE}_n^*(t)/[ab(c-1)]},\\
F_n^{*I}(t) &= \frac{\text{SSI}_n^*(t)/[(a-1)(b-1)]}{\text{SSE}_n^*(t)/[ab(c-1)]}.
\end{aligned}$$

Repeat this process a large number of times so that one can obtain bootstrap samples of $F_n^{*A}(t), F_n^{*AB}(t)$, and $F_n^{*I}(t)$ that can be used to estimate the $100(1-\alpha)$-percentiles of $F_n^A(t), F_n^{AB}(t)$, and $F_n^I(t)$, respectively.

For the L^2-norm-based bootstrap test for H_{0A}, H_{0AB}, and H_{0I}, one respectively computes the following test statistics:

$$\begin{aligned}
T_n^{*A} &= \int_{\mathcal{T}} \text{SSA}_n^*(t)dt,\\
T_n^{*AB} &= \int_{\mathcal{T}} [\text{SSA}_n^*(t) + \text{SSB}_n^*(t)]\, dt,\\
T_n^{*I} &= \int_{\mathcal{T}} \text{SSI}_n^*(t)dt.
\end{aligned}$$

Repeat this process a large number of times so that one can obtain bootstrap samples of T_n^{*A}, T_n^{*AB}, and T_n^{*I} that can be used to estimate the $100(1-\alpha)$-percentiles of T_n^A, T_n^{AB} and T_n^I, respectively.

Similarly, for the F-type bootstrap test for H_{0A}, H_{0AB}, and H_{0I}, one respectively computes the following test statistics:

$$\begin{aligned}
F_n^{*A} &= \frac{\int_{\mathcal{T}} \text{SSA}_n^*(t)dt/(a-1)}{\int_{\mathcal{T}} \text{SSE}_n^*(t)dt/[ab(c-1)]},\\
F_n^{*AB} &= \frac{\int_{\mathcal{T}}[\text{SSA}_n^*(t)+\text{SSB}_n^*(t)]dt/(a+b-2)}{\int_{\mathcal{T}} \text{SSE}_n^*(t)dt/[ab(c-1)]},\\
F_n^{*I} &= \frac{\int_{\mathcal{T}} \text{SSI}_n^*(t)dt/[(a-1)(b-1)]}{\int_{\mathcal{T}} \text{SSE}_n^*(t)dt/[ab(c-1)]}.
\end{aligned}$$

Repeat this process a large number of times so that one can obtain bootstrap samples of F_n^{*A}, F_n^{*AB}, and F_n^* that can be used to estimate the $100(1-\alpha)$-percentiles of F_n^A, F_n^{AB}, and F_n^I, respectively.

5.4.5 Balanced Two-Way ANOVA without Interaction

In some situations, the two-way ANOVA model (5.76) is balanced with $n_{ij} \equiv 1$ for all $i = 1, \cdots, a; j = 1, \cdots, b$. That is, there is only one observed function per cell. In this case, no interaction between factors A and B can be modeled, resulting in the so-called balanced two-way ANOVA model without interaction:

$$\begin{aligned} y_{ij}(t) &= \eta_0(t) + \alpha_i(t) + \beta_j(t) + v_{ij}(t), \\ v_{ij}(t) &\overset{i.i.d.}{\sim} \text{SP}(0, \gamma), i = 1, \cdots, a; j = 1, \cdots, b. \end{aligned} \quad (5.125)$$

In this case, the interaction-effect between factors A and B is assumed to be ignorable. For this model, interesting are the following two null hypotheses:

$$\begin{array}{lll} H_{0A} & : & \alpha_1(t) \equiv \alpha_2(t) \equiv \cdots \equiv \alpha_a(t) \equiv 0,\ t \in \mathcal{T}, \\ H_{0AB} & : & \alpha_1(t) \equiv \alpha_2(t) \equiv \cdots \equiv \alpha_a(t) \equiv 0, \\ & & \beta_1(t) \equiv \beta_2(t) \equiv \cdots \equiv \beta_b(t) \equiv 0,\ t \in \mathcal{T}. \end{array} \quad (5.126)$$

To identify the functional parameters in (5.125), the side constraints are imposed as

$$\sum_{i=1}^{a} \alpha_i(t) \equiv 0,\ \sum_{j=1}^{b} \beta_j(t) \equiv 0,\ t \in \mathcal{T}. \quad (5.127)$$

Under the above side constraints, the usual least squares estimators of the functional parameters $\eta_0(t), \alpha_i(t)$, and $\beta_j(t)$ are given by

$$\begin{array}{lll} \hat{\eta}_0(t) & = & \bar{y}_{..}(t), \\ \hat{\alpha}_i(t) & = & \bar{y}_{i.}(t) - \bar{y}_{..}(t),\ i = 1, \cdots, a, \\ \hat{\beta}_j(t) & = & \bar{y}_{.j}(t) - \bar{y}_{..}(t),\ j = 1, \cdots, b, \end{array}$$

where

$$\begin{array}{lll} \bar{y}_{i.}(t) & = & \sum_{j=1}^{b} y_{ij}(t)/b,\ i = 1, \cdots, a, \\ \bar{y}_{.j}(t) & = & \sum_{i=1}^{a} y_{ij}(t)/a,\ j = 1, \cdots, b, \\ \bar{y}_{..}(t) & = & \sum_{i=1}^{a} \sum_{j=1}^{b} y_{ij}(t)/(ab). \end{array} \quad (5.128)$$

Define $\bar{v}_{i.}(t), \bar{v}_{.j}(t)$ and $\bar{v}_{..}(t)$ similarly. Then we can further express $\hat{\eta}_0(t), \hat{\alpha}_i(t)$, and $\hat{\beta}_j(t)$ as

$$\begin{array}{lll} \hat{\eta}_0(t) & = & \eta_0(t) + \bar{v}_{..}(t), \\ \hat{\alpha}_i(t) & = & \alpha_i(t) + \bar{v}_{i.}(t) - \bar{v}_{..}(t),\ i = 1, \cdots, a, \\ \hat{\beta}_j(t) & = & \beta_j(t) + \bar{v}_{.j}(t) - \bar{v}_{..}(t),\ j = 1, \cdots, b. \end{array} \quad (5.129)$$

In addition, we have

$$\begin{array}{lll} \hat{v}_{ij}(t) & = & y_{ij}(t) - \hat{\eta}_0(t) - \hat{\alpha}_i(t) - \hat{\beta}_j(t) \\ & = & v_{ij}(t) - v_{i.}(t) - v_{.j}(t) + v_{..}(t), \\ & & i = 1, \cdots, a, j = 1, \cdots, b. \end{array} \quad (5.130)$$

It follows that

$$\hat{\mathbf{v}}(t) = \left[(\mathbf{I}_a - \frac{\mathbf{J}_a}{a}) \otimes (\mathbf{I}_b - \frac{\mathbf{J}_b}{b}) \right] \mathbf{v}(t), \quad (5.131)$$

where

$$\begin{aligned} \mathbf{v}(t) &= [v_{11}(t), \cdots, v_{1b}(t), \cdots, v_{a1}(t), \cdots, v_{ab}(t)]^T, \\ \hat{\mathbf{v}}(t) &= [\hat{v}_{11}(t), \cdots, \hat{v}_{1b}(t), \cdots, \hat{v}_{a1}(t), \cdots, \hat{v}_{ab}(t)]^T. \end{aligned}$$

Therefore, under (5.125), the common covariance function $\gamma(s,t)$ can be unbiasedly estimated as

$$\begin{aligned}\hat{\gamma}(s,t) &= \sum_{i=1}^{a}\sum_{j=1}^{b}\hat{v}_{ij}(s)\hat{v}_{ij}(t)/[(a-1)(b-1)] \\ &= \mathbf{v}(s)^T\left[(\mathbf{I}_a-\tfrac{\mathbf{J}_a}{a})\otimes(\mathbf{I}_b-\tfrac{\mathbf{J}_b}{b})\right]\mathbf{v}(t)/[(a-1)(b-1)].\end{aligned} \tag{5.132}$$

For any fixed $t\in\mathcal{T}$, let $\mathrm{SST}_n(t)$ denote the pointwise total sum of squares, $\mathrm{SSA}_n(t)$ and $\mathrm{SSB}_n(t)$ denote the pointwise sum of squares due to factor A and factor B respectively, and $\mathrm{SSE}_n(t)$ denotes the pointwise sum of squares due to errors. Then, by the classical balanced two-way ANOVA without interaction, we have the following decomposition:

$$\mathrm{SST}_n(t)=\mathrm{SSA}_n(t)+\mathrm{SSB}_n(t)+\mathrm{SSE}_n(t),$$

where

$$\begin{aligned}\mathrm{SST}_n(t) &= \sum_{i=1}^{a}\sum_{j=1}^{b}\left[y_{ij}(t)-\bar{y}_{..}(t)\right]^2, \\ \mathrm{SSA}_n(t) &= b\sum_{i=1}^{a}\left[\bar{y}_{i.}(t)-\bar{y}_{..}(t)\right]^2=b\sum_{i=1}^{a}\hat{\alpha}_i^2(t), \\ \mathrm{SSB}_n(t) &= a\sum_{j=1}^{b}\left[\bar{y}_{.j}(t)-\bar{y}_{..}(t)\right]^2=a\sum_{j=1}^{b}\hat{\beta}_j^2(t), \\ \mathrm{SSE}_n(t) &= \sum_{i=1}^{a}\sum_{j=1}^{b}\left[y_{ij}(t)-\bar{y}_{i.}(t)-\bar{y}_{.j}(t)+\bar{y}_{..}(t)\right]^2.\end{aligned} \tag{5.133}$$

It is seen that $\mathrm{SSA}_n(t)$ and $\mathrm{SSA}_n(t)+\mathrm{SSB}_n(t)$ are natural pivotal test functions for testing the null hypotheses H_{0A} and H_{0AB} respectively. By (5.129), we further have

$$\begin{aligned}\mathrm{SSA}_n(t) &= b\sum_{i=1}^{a}\left[\alpha_i(t)+\bar{v}_{i.}(t)-\bar{v}_{..}(t)\right]^2, \\ \mathrm{SSB}_n(t) &= a\sum_{j=1}^{b}\left[\beta_j(t)+\bar{v}_{.j}(t)-\bar{v}_{..}(t)\right]^2, \\ \mathrm{SSE}_n(t) &= \sum_{i=1}^{a}\sum_{j=1}^{b}\left[v_{ij}(t)-\bar{v}_{i.}(t)-\bar{v}_{.j}(t)-\bar{v}_{..}(t)\right]^2.\end{aligned} \tag{5.134}$$

Then we have

$$\mathrm{SSE}_n(t)=\mathbf{v}(t)^T\left[(\mathbf{I}_a-\frac{\mathbf{J}_a}{a})\otimes(\mathbf{I}_b-\frac{\mathbf{J}_b}{b})\right]\mathbf{v}(t). \tag{5.135}$$

In addition, we have the following results:

$$\begin{aligned}&\text{Under } H_{0A},\ \ \mathrm{SSA}_n(t) &= \mathbf{v}(t)^T\left[(\mathbf{I}_a-\tfrac{\mathbf{J}_a}{a})\otimes\tfrac{\mathbf{J}_b}{b}\right]\mathbf{v}(t), \\ &\text{Under } H_{0B},\ \ \mathrm{SSB}_n(t) &= \mathbf{v}(t)^T\left[\tfrac{\mathbf{J}_a}{a}\otimes(\mathbf{I}_b-\tfrac{\mathbf{J}_b}{b})\right]\mathbf{v}(t).\end{aligned} \tag{5.136}$$

The matrices $(\mathbf{I}_a-\frac{\mathbf{J}_a}{a})\otimes\frac{\mathbf{J}_b}{b}$, $\frac{\mathbf{J}_a}{a}\otimes(\mathbf{I}_b-\frac{\mathbf{J}_b}{b})$ and $(\mathbf{I}_a-\frac{\mathbf{J}_a}{a})\otimes(\mathbf{I}_b-\frac{\mathbf{J}_b}{b})$ are idempotent with ranks $(a-1),(b-1)$ and $(a-1)(b-1)$ respectively. To describe various tests for the null hypotheses H_{0A} and H_{0AB}, we may impose the following Gaussian assumption:

$$\mathbf{v}(t)\sim\mathrm{GP}_{ab}(\mathbf{0},\gamma\mathbf{I}_{ab}). \tag{5.137}$$

Under this Gaussian assumption, for any $t \in \mathcal{T}$, we have

$$\begin{array}{rrcl} \text{Under } H_{0A} & \text{SSA}_n(t) & \sim & \gamma(t,t)\chi^2_{a-1}, \\ \text{Under } H_{0AB} & \text{SSB}_n(t) & \sim & \gamma(t,t)\chi^2_{b-1}, \\ & \text{SSE}_n(t) & \sim & \gamma(t,t)\chi^2_{(a-1)(b-1)}. \end{array} \quad (5.138)$$

Furthermore, $\text{SSA}_n(t), \text{SSB}_n(t)$ and $\text{SSE}_n(t)$ are independent of each other. In addition, we have the following theorem showing that the integrals of $\text{SSA}_n(t), \text{SSB}_n(t)$ and $\text{SSE}_n(t)$ are χ^2-type mixtures.

Theorem 5.21 *Under the Gaussian assumption (5.137), we have*

$$\begin{array}{lll} \text{Under } H_{0A}, & \int_{\mathcal{T}} SSA_n(t)dt \stackrel{d}{=} \sum_{r=1}^m \lambda_r A_r, & A_r \stackrel{i.i.d.}{\sim} \chi^2_{a-1}, \\ \text{Under } H_{0AB}, & \int_{\mathcal{T}} SSB_n(t)dt \stackrel{d}{=} \sum_{r=1}^m \lambda_r B_r, & B_r \stackrel{i.i.d.}{\sim} \chi^2_{b-1}, \\ & \int_{\mathcal{T}} SSE_n(t)dt \stackrel{d}{=} \sum_{r=1}^m \lambda_r E_r, & E_r \stackrel{i.i.d.}{\sim} \chi^2_{(a-1)(b-1)}, \end{array}$$

where $A_r, B_r, E_r, r = 1, 2, \cdots, m$ *are independent of each other, and* $\lambda_1, \cdots, \lambda_m$ *are all the positive eigenvalues of* $\gamma(s,t)$.

Based on the above theorem, we are now ready to describe various tests for the two hypothesis testing problems listed in (5.126).

Pointwise Tests The pointwise F-test for H_{0A} and H_{0AB} respectively uses the following test statistics:

$$\begin{aligned} F_n^A(t) &= \frac{\text{SSA}_n(t)/(a-1)}{\text{SSE}_n(t)/[(a-1)(b-1)]}, \\ F_n^{AB}(t) &= \frac{[\text{SSA}_n(t)+\text{SSB}_n(t)]/(a+b-2)}{\text{SSE}_n(t)/[(a-1)(b-1)]}. \end{aligned} \quad (5.139)$$

Under the Gaussian assumption (5.137) and by (5.138), we have

$$\begin{array}{lll} \text{Under } H_{0A}, & F_n^A(t) \sim F_{a-1,(a-1)(b-1)}, & t \in \mathcal{T}, \\ \text{Under } H_{0AB}, & F_n^{AB}(t) \sim F_{a+b-2,(a-1)(b-1)}, & t \in \mathcal{T}. \end{array}$$

The pointwise F-test can be conducted using the above distributions.

L^2-Norm-Based Test The L^2-norm-based test for H_{0A} and H_{0AB} respectively uses the following test statistics:

$$T_n^A = \int_{\mathcal{T}} \text{SSA}_n(t)dt, \quad T_n^{AB} = \int_{\mathcal{T}} [\text{SSA}_n(t) + \text{SSB}_n(t)]\, dt. \quad (5.140)$$

By Theorem 5.21, we have

$$T_n^A \stackrel{d}{=} \sum_{r=1}^m \lambda_r A_r, \quad T_n^{AB} \stackrel{d}{=} \sum_{r=1}^m \lambda_r (A_r + B_r).$$

It follows that under the Gaussian assumption (5.137), the null distributions of T_n^A and T_n^{AB} can be approximated by the Welch-Satterthwaite χ^2-approximation method described in Section 4.3 of Chapter 4:

$$\begin{array}{lcll} T_n^A & \sim & \hat{\beta}\chi^2_{(a-1)\hat{\kappa}} & \text{approximately,} \\ T_n^{AB} & \sim & \hat{\beta}\chi^2_{(a+b-2)\hat{\kappa}} & \text{approximately,} \end{array} \tag{5.141}$$

where by the naive method,

$$\hat{\beta} = \frac{\text{tr}(\hat{\gamma}^{\otimes 2})}{\text{tr}(\hat{\gamma})}, \quad \hat{\kappa} = \frac{\text{tr}^2(\hat{\gamma})}{\text{tr}(\hat{\gamma}^{\otimes 2})}, \tag{5.142}$$

and by the bias-reduced method,

$$\hat{\beta} = \frac{\widehat{\text{tr}(\gamma^{\otimes 2})}}{\text{tr}(\hat{\gamma})}, \quad \hat{\kappa} = \frac{\widehat{\text{tr}^2(\gamma)}}{\widehat{\text{tr}(\gamma^{\otimes 2})}}, \tag{5.143}$$

with $m = (a-1)(b-1)$,

$$\begin{array}{lcl} \widehat{\text{tr}^2(\gamma)} & = & \frac{m(m+1)}{(m-1)(m+2)}\left[\text{tr}^2(\hat{\gamma}) - \frac{2\text{tr}(\hat{\gamma}^{\otimes 2})}{m+1}\right], \\ \widehat{\text{tr}(\gamma^{\otimes 2})} & = & \frac{m^2}{(m-1)(m+2)}\left[\text{tr}(\hat{\gamma}^{\otimes 2}) - \frac{\text{tr}^2(\hat{\gamma})}{m}\right]. \end{array} \tag{5.144}$$

The above unbiased estimators of $\text{tr}^2(\gamma)$ and $\text{tr}(\gamma^{\otimes 2})$ are obtained by Theorem 4.6 of Chapter 4 and using the fact $m\hat{\gamma}(s,t) \sim WP(m,\gamma)$. In all the above, the pooled sample covariance function $\hat{\gamma}(s,t)$ is given in (5.132).

F-Type Test Under the Gaussian assumption (5.137), the F-type test for H_{0A} and H_{0AB} respectively uses the following test statistics:

$$\begin{array}{lcl} F_n^A & = & \frac{\int_{\mathcal{T}} \text{SSA}_n(t)dt/(a-1)}{\int_{\mathcal{T}} \text{SSE}_n(t)dt/[(a-1)(b-1)]}, \\ F_n^{AB} & = & \frac{\int_{\mathcal{T}}[\text{SSA}_n(t)+\text{SSB}_n(t)]dt/(a+b-2)}{\int_{\mathcal{T}} \text{SSE}_n(t)dt/[(a-1)(b-1)]}. \end{array}$$

As $\mathbf{v}(t) \sim \text{GP}_{ab}(\mathbf{0}, \gamma\mathbf{I}_{ab})$, by (5.136) and (5.138), we have that $\text{SSA}_n(t), \text{SSB}_n(t)$, and $\text{SSE}_n(t)$ are independent of each other. In addition, for fixed t, their degrees of freedom are $(a-1), (b-1)$, and $(a-1)(b-1)$, respectively. Then by Theorem 5.21, we have

$$\begin{array}{ll} \text{Under } H_{0A}, & F_n^A \stackrel{d}{=} \frac{\sum_{r=1}^m \lambda_r A_r/(a-1)}{\sum_{r=1}^m \lambda_r E_r/[(a-1)(b-1)]}, \\ \text{Under } H_{0AB}, & F_n^{AB} \stackrel{d}{=} \frac{\sum_{r=1}^m \lambda_r [A_r+B_r]/(a+b-2)}{\sum_{r=1}^m \lambda_r E_r/[(a-1)(b-1)]}, \end{array}$$

where $A_r \stackrel{i.i.d.}{\sim} \chi^2_{a-1}, B_r \stackrel{i.i.d.}{\sim} \chi^2_{b-1}$ are independent of $E_r \stackrel{i.i.d.}{\sim} \chi^2_{(a-1)(b-1)}$, and

$\lambda_1, \lambda_2, \cdots, \lambda_m$ are all the positive eigenvalues of $\gamma(s,t)$. It follows that the null distribution of F_n^A and F_n^{AB} can be approximated by the two-cumulant matched F-approximation method using the methods described in Section 4.4 of Chapter 4. In fact, we have

$$\begin{array}{lll} \text{Under } H_{0A}, & F_n^A \sim F_{(a-1)\hat{\kappa},[(a-1)(b-1)]\hat{\kappa}} & \text{approximately}, \\ \text{Under } H_{0AB}, & F_n^{AB} \sim F_{(a+b-2)\hat{\kappa},[(a-1)(b-1)]\hat{\kappa}} & \text{approximately}, \end{array}$$

where by the naive method, $\hat{\kappa}$ is given in (5.142) and by the bias-reduced method, $\hat{\kappa}$ is given in (5.143). The F-type tests for H_{0A} and H_{0AB} can then be conducted accordingly.

Bootstrap Tests When the Gaussian assumption (5.137) is not valid, nonparametric bootstrap approaches can be used to conduct the pointwise test, the L^2-norm-based test, and the F-type test for H_{0A} and H_{0AB}, resulting in the so-called the pointwise bootstrap test, the L^2-norm-based bootstrap test, and the F-type bootstrap test, respectively.

Let $v_{ij}^*(t), i = 1, 2, \cdots, a; j = 1, 2, \cdots, b$ be the bootstrap samples randomly generated from $\hat{v}_{ij}(t)$, $i = 1, \cdots, a, j = 1, \cdots, b$ as defined in (5.130). Set $\mathbf{v}^*(t) = [v_{11}^*(t), \cdots, v_{1b}^*(t), \cdots, v_{a1}^*(t), \cdots, v_{ab}^*(t)]^T$. Then we can compute

$$\begin{aligned} \text{SSA}_n^*(t) &= \mathbf{v}^*(t)^T \left[(\mathbf{I}_a - \frac{\mathbf{J}_a}{a}) \otimes \frac{\mathbf{J}_b}{b} \right] \mathbf{v}^*(t), \\ \text{SSB}_n^*(t) &= \mathbf{v}^*(t)^T \left[\frac{\mathbf{J}_a}{a} \otimes (\mathbf{I}_b - \frac{\mathbf{J}_b}{b}) \right] \mathbf{v}^*(t), \\ \text{SSE}_n^*(t) &= \mathbf{v}^*(t)^T \left[(\mathbf{I}_a - \frac{\mathbf{J}_a}{a}) \otimes (\mathbf{I}_b - \frac{\mathbf{J}_b}{b}) \right] \mathbf{v}^*(t). \end{aligned}$$

For the pointwise bootstrap tests for H_{0A} and H_{0AB}, one respectively computes the following test statistics:

$$\begin{aligned} F_n^{*A}(t) &= \frac{\text{SSA}_n^*(t)/(a-1)}{\text{SSE}_n^*(t)/[(a-1)(b-1)]}, \\ F_n^{*AB}(t) &= \frac{[\text{SSA}_n^*(t)+\text{SSB}_n^*(t)]/(a+b-2)}{\text{SSE}_n^*(t)/[(a-1)(b-1)]}. \end{aligned}$$

Repeat this process a large number of times so that one can obtain bootstrap samples of $F_n^{*A}(t)$ and $F_n^{*AB}(t)$ that can be used to estimate the $100(1-\alpha)$-percentiles of $F_n^A(t)$ and $F_n^{AB}(t)$, respectively.

For the L^2-norm-based bootstrap test for H_{0A} and H_{0AB}, one respectively computes the following test statistics:

$$T_n^{*A} = \int_{\mathcal{T}} \text{SSA}_n^*(t)dt, \ \ T_n^{*AB} = \int_{\mathcal{T}} [\text{SSA}_n^*(t) + \text{SSB}_n^*(t)]\, dt.$$

Repeat this process a large number of times so that one can obtain bootstrap samples of T_n^{*A} and T_n^{*AB} that can be used to estimate the $100(1-\alpha)$-percentiles of T_n^A and T_n^{AB}, respectively.

Similarly, for the F-type bootstrap test for H_{0A} and H_{0AB}, one respectively computes the following test statistics:

$$F_n^{*A} = \frac{\int_{\mathcal{T}} \text{SSA}_n^*(t)dt/(a-1)}{\int_{\mathcal{T}} \text{SSE}_n^*(t)dt/[(a-1)(b-1)]},$$

$$F_n^{*AB} = \frac{\int_{\mathcal{T}} [\text{SSA}_n^*(t)+\text{SSB}_n^*(t)]dt/(a+b-2)}{\int_{\mathcal{T}} \text{SSE}_n^*(t)dt/[(a-1)(b-1)]}.$$

Repeat this process a large number of times so that one can obtain bootstrap samples of F_n^{*A} and F_n^{*AB} that can be used to estimate the $100(1-\alpha)$-percentiles of F_n^A and F_n^{AB}, respectively.

5.5 Technical Proofs

In this section, we outline the proofs of the main results described in this chapter.

Proof of Theorem 5.1 Under the Gaussian assumption, the first assertion is obvious. To show the second assertion, notice that $(n-2)\hat{\gamma}(s,t) = (n_1-1)\hat{\gamma}_1(s,t) + (n_2-1)\hat{\gamma}_2(s,t)$, where $\hat{\gamma}_1(s,t)$ and $\hat{\gamma}_2(s,t)$ are the sample covariance functions of the two functional samples (5.1), respectively. By Theorem 4.14 in Chapter 4, we have $(n_1-1)\hat{\gamma}_1(s,t) \sim \text{WP}(n_1-1,\gamma)$, $(n_2-1)\hat{\gamma}_2(s,t) \sim \text{WP}(n_2-1,\gamma)$ and they are independent. By Theorem 4.4, we have $(n-2)\hat{\gamma}(s,t) \sim \text{WP}(n-2,\gamma)$. The theorem is proved.

Proof of Theorem 5.2 Notice that

$$\Delta(t) - \eta_\Delta(t) = \sqrt{n_2/n}\sqrt{n_1}\left[\hat{\eta}_1(t) - \eta_1(t)\right] + \sqrt{n_1/n}\sqrt{n_2}\left[\hat{\eta}_2(t) - \eta_2(t)\right].$$

Under Assumptions TS1, TS3, and TS4, as $n \to \infty$, we have $n_1/n \to \tau, n_2/n \to 1-\tau$ and by Theorem 4.15 in Chapter 4, we have

$$\sqrt{n_1}\left[\hat{\eta}_1(t) - \eta_1(t)\right] \xrightarrow{d} \text{GP}(0,\gamma), \ \sqrt{n_2}\left[\hat{\eta}_2(t) - \eta_2(t)\right] \xrightarrow{d} \text{GP}(0,\gamma).$$

The theorem is then proved.

Proof of Theorem 5.3 Notice that

$$\sqrt{n}\left[\hat{\gamma}(s,t) - \gamma(s,t)\right] = a_n\sqrt{n_1}\left[\hat{\gamma}_1(s,t) - \gamma(s,t)\right] + b_n\sqrt{n_2}\left[\hat{\gamma}_2(s,t) - \gamma(s,t)\right],$$

where $\hat{\gamma}_1(s,t)$ and $\hat{\gamma}_2(s,t)$ are the sample covariance functions of the two functional samples (5.1), respectively, and $a_n = [\sqrt{n}(n_1-1)]/[\sqrt{n_1}(n-2)]$ and $b_n = [\sqrt{n}(n_2-1)]/[\sqrt{n_2}(n-2)]$. As $n \to \infty$, we have $a_n \to \sqrt{\tau}$ and $b_n \to \sqrt{1-\tau}$. In addition, under the given conditions, by Theorem 4.16 in Chapter 4, we have

$$\sqrt{n_1}[\hat{\gamma}_1(s,t) - \gamma(s,t)] \xrightarrow{d} \text{GP}(0,\varpi), \ \ \sqrt{n_2}[\hat{\gamma}_2(s,t) - \gamma(s,t)] \xrightarrow{d} \text{GP}(0,\varpi),$$

where $\varpi\{(s_1,t_1),(s_2,t_2)\} = \mathrm{E}v_{11}(s_1)v_{11}(t_1)v_{11}(s_2)v_{11}(t_2) - \gamma(s_1,t_1)\gamma(s_2,t_2)$. The theorem is then proved.

Proof of Theorem 5.4 Under the given conditions, by Theorem 5.3, as $n\to\infty$, we have $\mathrm{E}[\hat{\gamma}(s,t)-\gamma(s,t)]^2 = \frac{\omega[(s,t),(s,t)]}{n}[1+o(1)]$. By Assumptions TS6 and TS7, we have

$$|\omega[(s,t),(s,t)]| \le \mathrm{E}[v_{11}^2(s)v_{11}^2(t)] + \gamma^2(s,t) \le C+\rho, \quad \text{for all } (s,t)\in\mathcal{T}^2.$$

It follows that, as $n\to\infty$, we have $\hat{\gamma}(s,t)=\gamma(s,t)+O_{UP}(n^{-1/2})$, $(s,t)\in\mathcal{T}^2$, where O_{UP} means "uniformly bounded in probability." Thus, as $n\to\infty$, we have $\hat{\gamma}(s,t)\xrightarrow{p}\gamma(s,t)$ uniformly over $\mathcal{T}^2$. Therefore, as $n\to\infty$, we have

$$\begin{aligned}
\lim_{n\to\infty}\mathrm{tr}(\hat{\gamma}) &= \int_{\mathcal{T}}\lim_{n\to\infty}\hat{\gamma}(t,t)dt = \int_{\mathcal{T}}\gamma(t,t)dt = \mathrm{tr}(\gamma),\\
\lim_{n\to\infty}\mathrm{tr}(\hat{\gamma}^{\otimes 2}) &= \int_{\mathcal{T}}\int_{\mathcal{T}}\lim_{n\to\infty}\hat{\gamma}^2(s,t)dsdt\\
&= \int_{\mathcal{T}}\int_{\mathcal{T}}\gamma^2(s,t)dsdt = \mathrm{tr}(\gamma^{\otimes 2}),
\end{aligned}$$

in probability. It follows from (5.16) and (5.19) that as $n\to\infty$, $\hat{\beta}\xrightarrow{p}\beta$ and $\hat{\kappa}\xrightarrow{p}\kappa$. The theorem is proved.

Proof of Theorem 5.5 Notice that $\mathbf{D}^{-1/2}[\hat{\boldsymbol{\eta}}(t)-\boldsymbol{\eta}(t)] = [z_1(t),z_2(t),\cdots,z_k(t)]^T$, where $z_i(t)=\sqrt{n_i}[\hat{\eta}_i(t)-\eta_i(t)], i=1,2,\cdots,k$ are independent and by Theorem 4.14, we have $z_i(t)\sim \mathrm{GP}(0,\gamma), i=1,2,\cdots,k$. It follows that $\mathbf{D}^{-1/2}[\hat{\boldsymbol{\eta}}(t)-\boldsymbol{\eta}(t)]\sim \mathrm{GP}_k(\mathbf{0},\gamma\mathbf{I}_k)$ as desired. To show the second assertion, notice that $(n-k)\hat{\gamma}(s,t)=\sum_{i=1}^k(n_i-1)\hat{\gamma}_i(s,t)$, where $\hat{\gamma}_i(s,t), i=1,2,\cdots,k$ are the sample covariance functions of the k functional samples (5.23). By Theorem 4.14, we have $(n_i-1)\hat{\gamma}_i(s,t)\sim \mathrm{WP}(n_i-1,\gamma), i=1,2,\cdots,k$ and they are independent. Then by Theorem 4.4, we have $(n-k)\hat{\gamma}(s,t)\sim\mathrm{WP}(n-k,\gamma)$. The theorem is proved.

Proof of Theorem 5.6 Notice that $\mathbf{D}^{-1/2}[\hat{\boldsymbol{\eta}}(t)-\boldsymbol{\eta}(t)] = [z_1(t),z_2(t),\cdots,z_k(t)]^T$, where $z_i(t)=\sqrt{n_i}[\hat{\eta}_i(t)-\eta_i(t)], i=1,2,\cdots,k$ are independent. Under the given conditions and by Theorem 4.15, as $n\to\infty$, we have $z_i(t)\xrightarrow{d}\mathrm{GP}(0,\gamma), i=1,2,\cdots,k$. It follows that $\mathbf{D}^{-1/2}[\hat{\boldsymbol{\eta}}(t)-\boldsymbol{\eta}(t)]\xrightarrow{d}\mathrm{GP}_k(\mathbf{0},\gamma\mathbf{I}_k)$ as desired. The theorem is then proved.

Proof of Theorem 5.7 Notice that

$$\sqrt{n}\left[\hat{\gamma}(s,t)-\gamma(s,t)\right] = \sum_{i=1}^k a_i\sqrt{n_i}\left[\hat{\gamma}_i(s,t)-\gamma(s,t)\right],$$

where $\hat{\gamma}_i(s,t), i=1,2,\cdots,k$ are the sample covariance functions of the k

functional samples (5.23) respectively, and $a_i = [\sqrt{n}(n_i-1)]/[\sqrt{n_i}(n-k)], i = 1, 2, \cdots, k$. As $n \to \infty$, we have $a_i \to \sqrt{\tau_i}$ and $\sum_{i=1}^k a_i^2 \to 1$. In addition, under the given conditions, by Theorem 4.16, we have

$$\sqrt{n_i}[\hat{\gamma}_i(s,t) - \gamma(s,t)] \xrightarrow{d} \text{GP}(0, \varpi),\ i = 1, 2, \cdots, k,$$

where $\varpi((s_1, t_1), (s_2, t_2)) = \text{E}v_{11}(s_1)v_{11}(t_1)v_{11}(s_2)v_{11}(t_2) - \gamma(s_1, t_1)\gamma(s_2, t_2)$. The theorem is then proved.

Proof of Theorem 5.8 Under the one-way ANOVA model (5.26) and the null hypothesis (5.25), we can further express

$$\text{SSH}_n(t) = \sum_{i=1}^k n_i[\bar{v}_{i\cdot}(t) - \bar{v}_{\cdot\cdot}(t)]^2 = \mathbf{z}_n(t)^T(\mathbf{I}_k - \mathbf{b}_n\mathbf{b}_n^T/n)\mathbf{z}_n(t), \tag{5.145}$$

where $\bar{v}_{i\cdot}(t) = n_i^{-1}\sum_{j=1}^{n_i} v_{ij}(t)$ and $\bar{v}_{\cdot\cdot}(t) = n^{-1}\sum_{i=1}^k n_i\bar{v}_{i\cdot}(t)$, $\mathbf{b}_n = [n_1^{1/2}, n_2^{1/2}, \cdots, n_k^{1/2}]^T$, and $\mathbf{z}_n(t) = [n_1^{1/2}\bar{v}_{1\cdot}(t), \cdots, n_k^{1/2}\bar{v}_{k\cdot}(t)]^T$. On the one hand, under Assumption KS2, the k samples (5.23) are Gaussian, we have $\mathbf{z}_n(t) \sim \text{GP}_k(\mathbf{0}, \gamma\mathbf{I}_k)$. On the other hand, it is easy to verify that $\mathbf{I}_k - \mathbf{b}_n\mathbf{b}_n^T/n$ is an idempotent matrix of rank $k-1$. The first assertion of the theorem follows immediately from Theorem 4.10 of Chapter 4.

To show the second assertion of the theorem, notice that by (5.40), we have $\int_{\mathcal{T}} \text{SSE}_n(t)dt = (n-k)\text{tr}(\hat{\gamma})$ and by Theorem 5.5, we have $(n-k)\hat{\gamma}(s,t) \sim \text{WP}(n-k, \gamma)$. By Theorem 4.5(b), we have $\int_{\mathcal{T}} \text{SSE}_n(t)dt \stackrel{d}{=} \sum_{r=1}^m \lambda_r E_r$, $E_r \stackrel{i.i.d.}{\sim} \chi_{n-k}^2$. Notice that $\hat{\eta}_i(t), i = 1, 2, \cdots, k$ and $\hat{\gamma}(s,t)$ are independent. So are $A_r, r = 1, 2, \cdots, m$ and $E_r, r = 1, 2, \cdots, m$. The theorem is proved.

Proof of Theorem 5.9 By (5.145), we have $\text{SSH}_n(t) = \mathbf{z}_n(t)^T(\mathbf{I}_k - \mathbf{b}_n\mathbf{b}_n^T/n)\mathbf{z}_n(t)$, where $\mathbf{z}_n(t)$ and $\mathbf{b}_n$ are as defined in the proof of Theorem 5.8. It is easy to see that as $n \to \infty$, we have

$$\mathbf{I}_k - \mathbf{b}_n\mathbf{b}_n^T/n \to \mathbf{I}_k - \mathbf{b}\mathbf{b}^T,$$

where $\mathbf{b} = \lim_{n\to\infty} \mathbf{b}_n/\sqrt{n} = [\tau_1^{1/2}, \tau_2^{1/2}, \cdots, \tau_k^{1/2}]^T$. It is obvious that $\mathbf{I}_k - \mathbf{b}\mathbf{b}^T$ is an idempotent matrix of rank $k-1$ and has the singular value decomposition

$$\mathbf{I}_k - \mathbf{b}\mathbf{b}^T = \mathbf{U}\text{diag}(\mathbf{I}_{k-1}, 0)\mathbf{U}^T,$$

where $\mathbf{U}$ is an orthonormal matrix. In addition, under the given conditions, by Theorem 5.6, and under the null hypothesis (5.25), as $n \to \infty$, we have $\mathbf{z}_n(t) \xrightarrow{d} \text{GP}_k(\mathbf{0}, \gamma\mathbf{I}_k)$. It follows that we have

$$\text{SSH}_n(t) \xrightarrow{d} \mathbf{z}(t)^T(\mathbf{I}_k - \mathbf{b}\mathbf{b}^T)\mathbf{z}(t) \stackrel{d}{=} \mathbf{w}(t)^T\mathbf{w}(t), \tag{5.146}$$

where $\mathbf{w}(t) \sim \text{GP}_{k-1}(\mathbf{0}, \gamma\mathbf{I}_{k-1})$, consisting of the first $(k-1)$ component of $\mathbf{U}^T\mathbf{z}(t) \sim \text{GP}_k(\mathbf{0}, \gamma\mathbf{I}_k)$. The theorem then follows immediately from Theorem 4.10 of Chapter 4.

Proof of Theorem 5.10 The proof is along the same lines as those for the proof of Theorem 5.4 except now we need to apply Theorem 5.7 to show that $\hat{\gamma}(s,t)$ converges to $\gamma(s,t)$ in probability uniformly over $\mathcal{T}^2$ and to use (5.48) and (5.49) to show that $\hat{\beta} \xrightarrow{p} \beta$ and $\hat{\kappa} \xrightarrow{p} \kappa$. The theorem is proved.

Proof of Theorem 5.11 By (5.64) and (5.67), we have $\int_{\mathcal{T}} \text{SSH}_n(t)dt = \int_{\mathcal{T}} \|\mathbf{z}(t)\|^2 dt$. When the k samples (5.23) are Gaussian and under the null hypothesis in (5.62), we have $\mathbf{z}(t) \sim \text{GP}_q(\mathbf{0}, \gamma\mathbf{I}_q)$. The first assertion of the theorem then follows from Theorem 4.10 of Chapter 4. The second assertion of the theorem is proved in Theorem 5.8. The independence between $A_r, r = 1, 2, \cdots, m$ and $E_r, r = 1, 2, \cdots, m$ is due to the fact that $\boldsymbol{\eta}(t)$ and $\hat{\gamma}(s,t)$ are independent. The theorem is proved.

Proof of Theorem 5.12 Under the given conditions, by Theorem 5.6 and under the null hypothesis in (5.62), as $n \to \infty$, we have $\mathbf{z}(t) \xrightarrow{d} \text{GP}_q(\mathbf{0}, \gamma\mathbf{I}_q)$, where $\mathbf{z}(t)$ is given in (5.64). The theorem then follows immediately from (5.64) and Theorem 4.10 of Chapter 4.

Proof of Theorem 5.13 Notice that

$$\mathbf{D}^{-1/2}[\hat{\boldsymbol{\eta}}(t) - \boldsymbol{\eta}(t)] = [z_{11}(t), \cdots, z_{1b}(t), \cdots, z_{a1}(t), \cdots, z_{ab}(t)]^T,$$

where $z_{ij}(t) = \sqrt{n_{ij}}[\hat{\eta}_{ij}(t) - \eta_{ij}(t)], i = 1, 2, \cdots, a, j = 1, 2, \cdots, b$ are independent and by Theorem 4.14, we have $z_{ij}(t) \sim \text{GP}(0, \gamma), i = 1, 2, \cdots, a, j = 1, 2, \cdots, b$. It follows that $\mathbf{D}^{-1/2}[\hat{\boldsymbol{\eta}}(t) - \boldsymbol{\eta}(t)] \sim \text{GP}_{ab}(\mathbf{0}, \gamma\mathbf{I}_{ab})$, as desired. To show the second assertion, notice that $(n - ab)\hat{\gamma}(s,t) = \sum_{i=1}^{a}\sum_{j=1}^{b}(n_{ij} - 1)\hat{\gamma}_{ij}(s,t)$, where $\hat{\gamma}_{ij}(s,t), i = 1, 2, \cdots, a, j = 1, 2, \cdots, b$ are the sample covariance functions of the ab functional samples (5.72). By Theorem 4.14 we have $(n_{ij} - 1)\hat{\gamma}_{ij}(s,t) \sim \text{WP}(n_{ij} - 1, \gamma), i = 1, 2, \cdots, a, j = 1, 2, \cdots, b$ and they are independent. Then by Theorem 4.4, we have $(n - ab)\hat{\gamma}(s,t) \sim \text{WP}(n - ab, \gamma)$. The theorem is proved.

Proof of Theorem 5.14 Notice that

$$\mathbf{D}^{-1/2}[\hat{\boldsymbol{\eta}}(t) - \boldsymbol{\eta}(t)] = [z_{11}(t), \cdots, z_{1a}(t), \cdots, z_{a1}(t), \cdots, z_{ab}(t)]^T,$$

where $z_{ij}(t) = \sqrt{n_{ij}}[\hat{\eta}_{ij}(t) - \eta_{ij}(t)], i = 1, 2, \cdots, a; j = 1, 2, \cdots, b$ are independent. Under the given conditions and by Theorem 4.15, as $n \to \infty$, we have $z_{ij}(t) \xrightarrow{d} \text{GP}(0, \gamma), i = 1, 2, \cdots, a, j = 1, 2, \cdots, b$. It follows that $\mathbf{D}^{-1/2}[\hat{\boldsymbol{\eta}}(t) - \boldsymbol{\eta}(t)] \xrightarrow{d} \text{GP}_{ab}(\mathbf{0}, \gamma\mathbf{I}_{ab})$, as desired. The theorem is then proved.

Proof of Theorem 5.15 Notice that

$$\sqrt{n}\left[\hat{\gamma}(s,t)-\gamma(s,t)\right]=\sum_{i=1}^{a}\sum_{j=1}^{b}h_{ij}\sqrt{n_{ij}}\left[\hat{\gamma}_{ij}(s,t)-\gamma(s,t)\right],$$

where $\hat{\gamma}_{ij}(s,t), i=1,2,\cdots,a, j=1,2,\cdots,b$ are the sample covariance functions of the ab functional samples (5.72), respectively, and $h_{ij}=[\sqrt{n}(n_{ij}-1)]/[\sqrt{n_{ij}}(n-ab)], i=1,2,\cdots,a, j=1,2,\cdots,b$. As $n\to\infty$, we have, $h_{ij}\to\sqrt{\tau_{ij}}$ and $\sum_{i=1}^{a}\sum_{j=1}^{b}h_{ij}^2\to 1$. In addition, under the given conditions, by Theorem 4.16, we have

$$\sqrt{n_{ij}}[\hat{\gamma}_{ij}(s,t)-\gamma(s,t)]\xrightarrow{d}\text{GP}(0,\varpi),\ i=1,2,\cdots,a, j=1,2,\cdots,b,$$

where $\varpi((s_1,t_1),(s_2,t_2))=\text{E}v_{111}(s_1)v_{111}(t_1)v_{111}(s_2)v_{111}(t_2)-\gamma(s_1,t_1)\gamma(s_2,t_2)$. The theorem is then proved.

Proof of Theorem 5.16 By (5.93) and (5.96), we have $\int_{\mathcal{T}}\text{SSH}_n(t)dt=\int_{\mathcal{T}}\|\mathbf{z}(t)\|^2dt$. When the ab samples (5.72) are Gaussian, under the null hypothesis in (5.91), we have $\mathbf{z}(t)\sim\text{GP}_q(\mathbf{0},\gamma\mathbf{I}_q)$. The assertions of the theorem follow immediately from Theorem 4.10 in Chapter 4.

Proof of Theorem 5.17 Under the given conditions, as $n\to\infty$, by Theorem 5.14, we have $\mathbf{z}(t)\xrightarrow{d}\text{GP}_q(\mathbf{0},\gamma\mathbf{I}_q)$, where $\mathbf{z}(t)$ is given in (5.93). This allows the assertions of the theorem to follow immediately from (5.93) and Theorem 4.10 in Chapter 4.

Proof of Theorem 5.18 The proof is along the same lines as those for the proof of Theorem 5.4 except now we need to apply Theorem 5.15 to show that $\hat{\gamma}(s,t)$ converges to $\gamma(s,t)$ in probability uniformly over $\mathcal{T}^2$ and to use (5.104) and (5.105) to show that $\hat{\beta}\xrightarrow{p}\beta$ and $\hat{\kappa}\xrightarrow{p}\kappa$. The theorem is proved.

Proof of Theorem 5.19 When the ab samples (5.72) are Gaussian with $n_{ij}\equiv c>1$, we have $\mathbf{z}(t)\sim\text{GP}_{ab}(\mathbf{0},\gamma\mathbf{I}_{ab})$, where $\mathbf{z}(t)$ is given in (5.119). The assertions of the theorem follow immediately from the expressions in (5.115) and (5.120) and from Theorems 4.9 and 4.10 in Chapter 4.

Proof of Theorem 5.20 Under the given conditions, as $c\to\infty$, by Theorem 5.14, we have $\mathbf{z}(t)\xrightarrow{d}\text{GP}_{ab}(\mathbf{0},\gamma\mathbf{I}_{ab})$ where $\mathbf{z}(t)$ is given in (5.119). This allows the assertions of the theorem follow immediately from the expressions in (5.120) and from Theorems 4.9 and 4.10 in Chapter 4.

Proof of Theorem 5.21 As $\mathbf{v}(t)\sim\text{GP}_{ab}(\mathbf{0},\gamma\mathbf{I}_{ab})$, the assertions of the theorem follow immediately from the expressions in (5.135) and (5.136) and from Theorems 4.9 and 4.10 of Chapter 4.

5.6 Concluding Remarks and Bibliographical Notes

In this chapter we reviewed various ANOVA models for functional data. The pointwise, L^2-norm-based, F-type, and bootstrap tests were described for the two-sample problem, one-way ANOVA, and two-way ANOVA, respectively. The pointwise tests are essentially the classical tests for various ANOVA models for scalar variables. For simplicity, we adopted the Welch-Satterthwaite χ^2-approximation for approximating the null distribution of the L^2-norm-based test and the two-cumulant matched F-approximation for approximating the null distribution of the F-type test. Other approaches described in Sections 4.3 and 4.4 in Chapter 4 for approximating the distribution of a χ^2-type mixture or an F-type mixture can also be applied here. The bootstrap tests are recommended only when the functional data are not Gaussian and the sample sizes are small as they are generally time consuming.

The pointwise F-test was first adopted in Ramsay and Silverman (2005, Section 13.2.2, Chapter 13) for functional linear regression models with functional responses. The L^2-norm-based test was first studied by Faraway (1997), who proposed a bootstrap approach to approximate the null distribution. Zhang and Chen (2007) extended the L^2-norm-based test for a general linear hypothesis testing problem in the framework of functional linear models with noisy observed functional data. Some further work on the L^2-norm-based test can be found in Zhang, Peng, and Zhang (2010), and Zhang, Liang, and Xiao (2010), among others. The F-type test was first studied by Shen and Faraway (2004). The theoretical properties of the F-type test for functional linear models with functional responses are given in Zhang (2011a), who showed that the F-type test is root-n consistent. The bootstrap tests are investigated in Faraway (1997), Cuevas, Febrero, and Fraiman (2004), Zhang, Peng, and Zhang (2010), and Zhang and Sun (2010), among others. For curve data from stationary Gaussian processes, Fan and Lin (1998) developed an adaptive Neyman test.

Studies on the two-sample problems for functional data include James and Sood (2006), Hall and Van Keilegom (2007), and Zhang, Peng, and Zhang (2010), among others. A two-sample test is developed in Berkes et al. (2009) for detecting changes in the mean function of functional data. One-way ANOVA for functional data has been investigated in Cuevas, Febrero, and Fraiman (2004) by a parametric bootstrap method and in Zhang and Liang (2013) by a globalizing pointwise F-test. Munk et al. (2008) discussed a one- and multi-sample problem for functional data with application to projective shape analysis. Some important work about two-way ANOVA for functional data can be found in Abramovich et al. (2004), Abramovich and Angelini (2006), Antoniadis and Sapatinas (2007), Yang and Nie (2008), and Cuesta-Albertos and Febrero-Bande (2010), among others. Most recently, Crainiceanu, Staicu, Ray, and Punjabi (2012) proposed bootstrap-based inference methods on the mean difference between two correlated functional processes, and Gromenko and Kokoszka (2012) developed a significance test for evaluating the equality of the mean functions in two samples of spatially indexed functional data.

5.7 Exercises

1. For the hypothesis testing problem (5.8), conduct the pointwise t, z or bootstrap tests for the progesterone data by constructing the pointwise confidence intervals (5.11). Comment if the conclusions obtained here are similar to those obtained from the pointwise t, z or bootstrap tests using the P-values presented in Figure 5.4 (b).
2. For the progesterone data, test (5.2) over $[a, b] = [-8, 10]$ using the L^2-norm-based test presented in Section 5.2.2.
3. Let $x_{ij}, j = 1, 2, \cdots, n_i; i = 1, 2, \cdots, k$ be real numbers. Define $\bar{x}_{i.} = n_i^{-1}\sum_{j=1}^{n_i} x_{ij}$ and $\bar{x}_{..} = n^{-1}\sum_{i=1}^k \sum_{j=1}^{n_i} x_{ij}$, where $n = \sum_{i=1}^k n_i$.
   (a) Show that $\bar{x}_{..} = n^{-1}\sum_{i=1}^k n_i\bar{x}_{i.} = n^{-1}\mathbf{b}_n^T\mathbf{z}$, where $\mathbf{b}_n = [n_1^{1/2}, n_2^{1/2}, \cdots, n_k^{1/2}]^T$ and $\mathbf{z} = [n_1^{1/2}\bar{x}_{1.}, n_2^{1/2}\bar{x}_{2.}, \cdots, n_k^{1/2}\bar{x}_{k.}]^T$.
   (b) Show that $\sum_{i=1}^k n_i(\bar{x}_{i.} - \bar{x}_{..})^2 = \mathbf{z}^T(\mathbf{I}_k - \mathbf{b}_n\mathbf{b}_n^T/n)\mathbf{z}$ and $\mathbf{I}_k - \mathbf{b}_n\mathbf{b}_n^T/n$ is an idempotent matrix of rank $(k-1)$.
   (c) Show that $\sum_{i=1}^k n_i(\bar{x}_{i.} - \bar{x}_{..})^2 = n^{-1}\sum_{1\le i<j\le k} n_i n_j(\bar{x}_{i.} - \bar{x}_{j.})^2$.
4. For the left-cingulum data, based on the information presented in Tables 5.10 and 5.11, approximate the null distributions of the L^2-norm-based test and the F-type test using the bias-reduced method and then construct Tables 5.10 and 5.11 accordingly. Compare the new results with those presented in Tables 5.10 and 5.11.

Chapter 6

Linear Models with Functional Responses

6.1 Introduction

In the previous chapter, we reviewed various ANOVA models for functional data. We described the pointwise, L^2-norm-based, F-type, and bootstrap tests for various ANOVA models. These ANOVA models can be regarded as special functional linear models (FLMs) with functional responses and one or several categorical covariates. In this chapter, we show how these statistical tests can be extended to linear models with functional responses and some continuous covariates. Theses covariates can be time independent or dependent. When the covariates are time independent, that is, the covariates do not change over time, the associated FLMs are less involved. We deal with this case in Section 6.2 for Gaussian and non-Gaussian data. In many situations, however, the covariates can also depend on time so that the study of the associated FLMs is more involved. This case is dealt with in Section 6.3 for Gaussian data only. Technical proofs of the main results are given in Section 6.4. Some concluding remarks and bibliographical notes are given in Section 6.5. Exercises for this chapter are given in Section 6.6.

6.2 Linear Models with Time-Independent Covariates

Example 6.1 *To illustrate FLMs, Faraway (1997), Shen and Faraway (2004), and Zhang (2011a) used the ergonomics data introduced in Section 1.2.8 of Chapter 1. The data were collected for a study of motion of an automobile driver to reach some fixed targets in a test car. The angles formed at the right elbow between the upper and lower arms of the driver were measured on an equally spaced grid of points over a period of time, rescaled to the unit interval* $[0, 1]$ *for convenience. The number of the design time points varies from one angle curve to another. Within the test car, there are twenty locations where the motion of the automobile driver wanted to reach. At each location, three replicates were taken. Totally, there were sixty angle response curves recorded over time.*

Figure 6.1 displays six of the reconstructed right elbow angle curves ob-

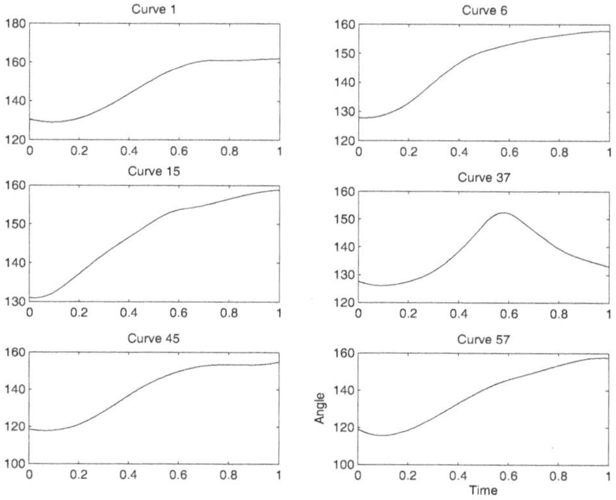

Figure 6.1 *Six selected angle curves from the ergonomics data. Note that Curve 37 is different from others in shape.*

tained using the regression spline method introduced in Chapter 3. These angle curves were fitted using a quadratic regression spline with seven equally spaced knots. The number of knots was selected by the GCV rule (3.10) as described in Section 3.2.2 of Chapter 3. For illustration purposes, these angle curves were evaluated at a grid of $M = 1,000$ *equally spaced time points over* $[0, 1]$. *From this figure, it is seen that Curve 37 has quite a different shape from the remaining curves. It is suspected to be an unusual curve; see also Figure 6.2 where all the ergonomics curves are displayed. For this reason, Curve 37 is removed from the ergonomics data used in this chapter. For techniques of detecting unusual functions, the reader is referred to Chapter 8.*

The angle curves over time are the response functions associated with the coordinates of the twenty locations of the targets in the test car. We use (c_x, c_y, c_z) *to denote the coordinates of the target locations where,* x *represents the left-to-right direction,* y *represents the close-to-far direction, and* z *the down-to-up direction. To model the association between the angle curves and the coordinates of the target locations, Faraway (1997) and Shen and Faraway (2004) suggested the following quadratic FLM:*

$$\begin{aligned} y(t) &= \beta_0(t) + c_x\beta_1(t) + c_y\beta_2(t) + c_z\beta_3(t) + c_x^2\beta_4(t) + c_y^2\beta_5(t) \\ &\quad + c_z^2\beta_6(t) + c_xc_y\beta_7(t) + c_xc_z\beta_8(t) + c_yc_z\beta_9(t) + v(t), \end{aligned} \tag{6.1}$$

where $y(t)$ *and* $v(t)$ *denote the response function (angle curve) and the subject-effect function, which is not explained by the predictor (coordinates of the target locations). Some questions arise naturally. Is the overall regression significant? Is a coefficient function significant? How to select a proper model?*

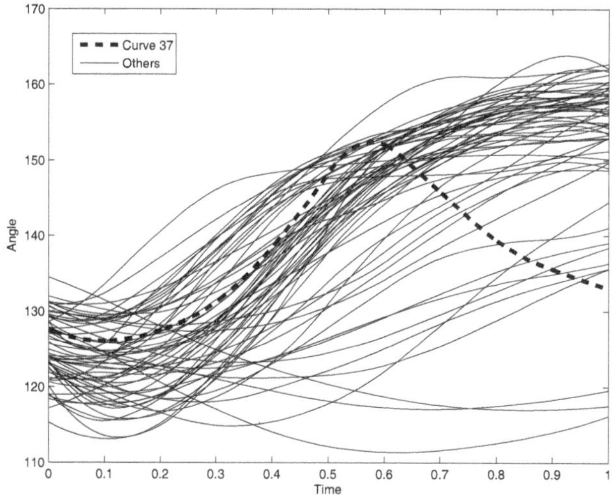

Figure 6.2 *The smoothed right elbow angle curves of the ergonomics data. Curve 37 is highlighted as a wide dashed curve as it has a different shape from the remaining curves.*

How to detect the unusual curves? In this chapter, we focus on the first three problems. We discuss the last problem in Chapter 8.

Let $\mathbf{x}_i = [1, x_{i1}, \cdots, x_{ip}]^T, i = 1, 2, \cdots, n$ be the time independent $(p+1)$-dimensional covariates (predictors) and $y_i(t), i = 1, 2, \cdots, n$ be the response functions. A general functional data set for an FLM can be expressed as

$$[\mathbf{x}_i^T, y_i(t)], \ \ i = 1, 2, \cdots, n. \tag{6.2}$$

Then a general FLM with functional responses and time-independent covariates can be defined as

$$y_i(t) = \mathbf{x}_i^T \boldsymbol{\beta}(t) + v_i(t), \ v_i(t) \overset{i.i.d.}{\sim} \text{SP}(0, \gamma), t \in \mathcal{T}, \ i = 1, 2, \cdots, n, \tag{6.3}$$

where $\boldsymbol{\beta}(t) = [\beta_0(t), \beta_1(t), \cdots, \beta_p(t)]^T$ is the $(p+1)$-dimensional vector of coefficient functions; $v_i(t), i = 1, 2, \cdots, n$, are the subject-effect functions; $\mathcal{T} = [a, b], -\infty < a < b < \infty$ is the support of the design time points; and as before, $\text{SP}(\eta, \gamma)$ denotes a stochastic process with mean function $\eta(t)$ and covariance function $\gamma(s, t)$. Notice that the coefficient function $\beta_0(t)$ is the intercept function of (6.3).

Remark 6.1 *Compared with the FLM (6.3), the FLM (3.39) in Section 3.4 of Chapter 3 has an extra term $\epsilon_i(t)$ that models the measurement error process. This term can be ignored when the functional data used are the reconstructed functions obtained by the smoothing methods described in Chapter 3 and when* *the required conditions as stated in Chapter 3 are approximately satisfied. Otherwise, the term $\epsilon_i(t)$ in (3.39) can be regarded as absorbed into the term $v_i(t)$ in the FLM (6.3). The methodologies described in this book continue to work in this case as we do not assume any particular structure for the covariance function $\gamma(s,t)$ in the statistical inferences discussed in this book.*

In matrix form, (6.3) can be written as

$$\mathbf{y}(t) = \mathbf{X}\boldsymbol{\beta}(t) + \mathbf{v}(t),\ t \in \mathcal{T},\ \mathbf{v}(t) \sim \text{SP}_n(\mathbf{0}, \gamma\mathbf{I}_n), \tag{6.4}$$

where $\mathbf{y}(t) = [y_1(t), \cdots, y_n(t)]^T$, $\mathbf{X} = [\mathbf{x}_1, \cdots, \mathbf{x}_n]^T$, and $\mathbf{v}(t) = [v_1(t), \cdots, v_n(t)]^T$. Notice that $\mathbf{y}(t), \mathbf{X}$ and $\mathbf{v}(t)$ are the vector of response functions, the design matrix, and the vector of subject-effect functions, respectively, and $\text{SP}_n(\boldsymbol{\eta}, \boldsymbol{\Gamma})$ denotes an n-dimensional stochastic process with vector of mean functions $\boldsymbol{\eta}(t)$ and matrix of covariance functions $\boldsymbol{\Gamma}(s,t)$.

6.2.1 Coefficient Function Estimation

Throughout this section, we assume that $\mathbf{X}$ has full rank. The FLMs with a non-full-rank design matrix will be discussed in Chapter 7. Then the pointwise least squares estimator of $\boldsymbol{\beta}(t)$ under (6.4) is

$$\hat{\boldsymbol{\beta}}(t) = [\mathbf{X}^T\mathbf{X}]^{-1}\mathbf{X}^T\mathbf{y}(t),\ t \in \mathcal{T}. \tag{6.5}$$

Set $\mathbf{P}_\mathbf{X} = \mathbf{X}[\mathbf{X}^T\mathbf{X}]^{-1}\mathbf{X}^T$. Then it is well known that $\mathbf{P}_\mathbf{X}$ is a projection matrix so that $\mathbf{P}_\mathbf{X}\mathbf{X} = \mathbf{X}$ and $\mathbf{X}^T\mathbf{P}_\mathbf{X} = \mathbf{X}^T$, and $\mathbf{P}_\mathbf{X}$ is an idempotent matrix of rank $(p+1)$, having $p+1$ eigenvalues being 1 and others being 0.

Notice that for any $t \in \mathcal{T}$, the cross-product

$$[\mathbf{y}(t) - \mathbf{X}\hat{\boldsymbol{\beta}}(t)]^T[\mathbf{X}\hat{\boldsymbol{\beta}}(t) - \mathbf{X}\boldsymbol{\beta}(t)] = 0. \tag{6.6}$$

This implies that the pointwise least squares estimator $\hat{\boldsymbol{\beta}}(t)$ minimizes globally the following integrated squared error:

$$Q(\boldsymbol{\beta}) = \int_\mathcal{T} \|\mathbf{y}(t) - \mathbf{X}\boldsymbol{\beta}(t)\|^2 dt, \tag{6.7}$$

where again $\|\cdot\|$ denotes the usual L^2-norm. In fact, by (6.6), we have

$$\begin{aligned}
Q(\boldsymbol{\beta}) &= \int_\mathcal{T} \|[\mathbf{y}(t) - \mathbf{X}\hat{\boldsymbol{\beta}}(t)] + [\mathbf{X}\hat{\boldsymbol{\beta}}(t) - \mathbf{X}\boldsymbol{\beta}(t)]\|^2 dt \\
&= \int_\mathcal{T} \|\mathbf{y}(t) - \mathbf{X}\hat{\boldsymbol{\beta}}(t)\|^2 dt + \int_\mathcal{T} \|\mathbf{X}\hat{\boldsymbol{\beta}}(t) - \mathbf{X}\boldsymbol{\beta}(t)\|^2 dt \\
&= Q(\hat{\boldsymbol{\beta}}) + \int_\mathcal{T} [\hat{\boldsymbol{\beta}}(t) - \boldsymbol{\beta}(t)]^T[\mathbf{X}^T\mathbf{X}][\hat{\boldsymbol{\beta}}(t) - \boldsymbol{\beta}(t)] dt \\
&\geq Q(\hat{\boldsymbol{\beta}}),
\end{aligned}$$

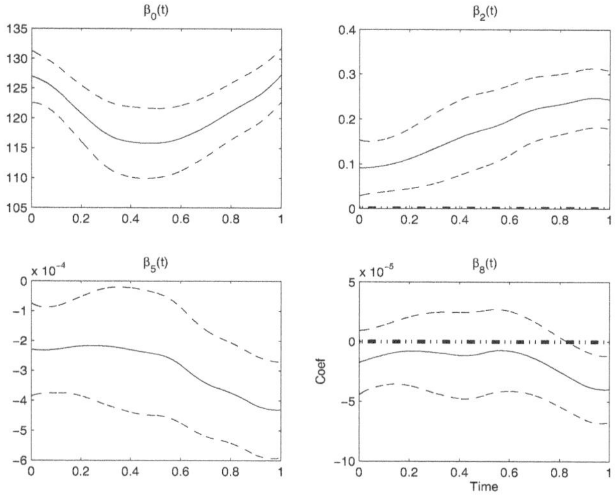

Figure 6.3 *Four fitted coefficient functions (solid) of the ergonomics data with 95% pointwise confidence bands (dashed). The coefficient $\beta_8(t)$ may be less significant as its 95% pointwise confidence bands cover the zero horizontal line (dot-dashed) most of time, and the other three coefficient functions are likely to be highly significant.*

with the equality holding if and only if when $\boldsymbol{\beta}(t) \equiv \hat{\boldsymbol{\beta}}(t), t \in \mathcal{T}$ due to the fact that $\mathbf{X}^T\mathbf{X}$ is a positive definite matrix.

The vector of fitted response functions can be expressed as

$$\hat{\mathbf{y}}(t) = [\hat{y}_1(t), \hat{y}_2(t), \cdots, \hat{y}_n(t)]^T = \mathbf{X}\hat{\boldsymbol{\beta}}(t) = \mathbf{P}_{\mathbf{X}}\mathbf{y}(t), \tag{6.8}$$

and the vector of fitted subject-effect functions is

$$\hat{\mathbf{v}}(t) = \mathbf{y}(t) - \hat{\mathbf{y}}(t) = (\mathbf{I}_n - \mathbf{P}_{\mathbf{X}})\mathbf{y}(t) = (\mathbf{I}_n - \mathbf{P}_{\mathbf{X}})\mathbf{v}(t). \tag{6.9}$$

We have $\mathrm{E}(\hat{\mathbf{v}}(t)) = 0$ and $\mathrm{Cov}(\hat{\mathbf{v}}(s), \hat{\mathbf{v}}(t)) = \gamma(s,t)(\mathbf{I}_n - \mathbf{P}_{\mathbf{X}})$. This implies that $\mathrm{E}\hat{\mathbf{v}}(s)^T\hat{\mathbf{v}}(t) = \mathrm{tr}\left[\mathrm{E}(\hat{\mathbf{v}}(s)\hat{\mathbf{v}}(t)^T)\right] = [n-p-1]\gamma(s,t)$. Therefore, the covariance function $\gamma(s,t)$ can be unbiasedly estimated by

$$\hat{\gamma}(s,t) = [n-p-1]^{-1}\sum_{i=1}^{n}[y_i(s) - \hat{y}_i(s)]\,[y_i(t) - \hat{y}_i(t)]\,. \tag{6.10}$$

Example 6.2 *For the ergonomics data, the FLM (6.1) was fitted using (6.5). To see what the fitted coefficient functions look like, in Figure 6.3 we display four estimated coefficient functions (solid) with their* 95% *pointwise confidence bands (dashed). The pointwise confidence bands were computed based on Theorems 6.1 and 6.2 of the next subsection and using the estimated covariance*

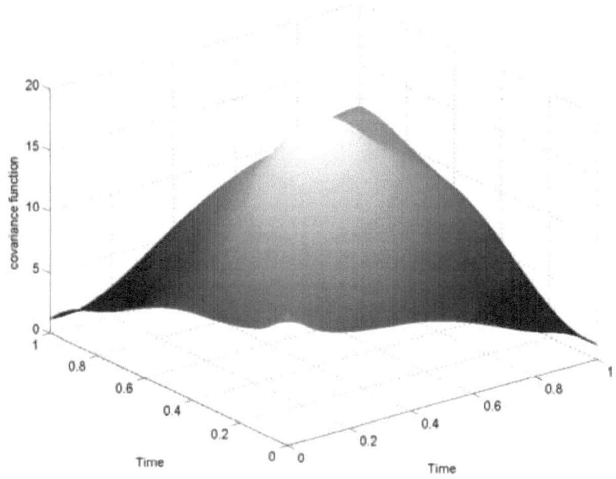

Figure 6.4 *The estimated covariance function of the ergonomics data. It appears that the large variations appear in the middle when the driver tried to reach his target.*

function (6.10). It is seen that the coefficients $\beta_0(t), \beta_2(t)$, *and* $\beta_5(t)$ *are highly significant, while the coefficient function* $\beta_8(t)$ *is less significant. This indicates that the close-to-far direction plays an important role in the angle curves of the ergonomics data.*

The estimated covariance function of the ergonomics data is displayed in Figure 6.4. It is seen that the big variations appear in the middle when the driver tried to reach his target. For further inference, based on resolution $M = 1,000$, *we list the traces of the estimated covariance function and its cross-square function as follows:*

$$tr(\hat{\gamma}) = 14.352, \quad \textit{and} \quad tr(\hat{\gamma}^{\otimes 2}) = 117.57. \tag{6.11}$$

These two quantities will be useful later for statistical inferences for the ergonomics data based on the quadratic FLM (6.1).

6.2.2 Properties of the Estimators

In order to describe various statistical inference procedures for testing the FLM (6.3), we first need to investigate the properties of $\hat{\boldsymbol{\beta}}(t)$ and $\hat{\gamma}(s,t)$ under various conditions. For this purpose, we list the following assumptions:

Linear Model Assumptions (LMA)

1. The coefficient functions are L^2-integrable and the covariance function $\gamma(s,t)$ has finite trace. That is, $\beta_0(t), \beta_2(t), \cdots, \beta_p(t) \in \mathcal{L}^2(\mathcal{T})$ and $\mathrm{tr}(\gamma) < \infty$.

2. The subject-effect functions $v_1(t), \cdots, v_n(t)$ are i.i.d. Gaussian.
3. The covariates $\mathbf{x}_1, \cdots, \mathbf{x}_n$ are i.i.d. so that as $n \to \infty$, $n^{-1}\mathbf{X}^T\mathbf{X} \to \boldsymbol{\Omega} = \mathrm{E}(\mathbf{x}_1\mathbf{x}_1^T)$ almost surely, where $\boldsymbol{\Omega}$ is invertible.
4. The subject-effect functions $v_1(t), \cdots, v_n(t)$ are i.i.d..
5. The subject-effect function $v_1(t)$ satisfies $\mathrm{E}\|v_1\|^4 < \infty$.
6. The maximum variance $\rho = \max_{t\in\mathcal{T}} \gamma(t,t) < \infty$.
7. The expectation $\mathrm{E}[v_1^2(s)v_1^2(t)]$ is uniformly bounded. That is, $\mathrm{E}[v_1^2(s)v_1^2(t)] \le C < \infty$ for all $(s,t) \in \mathcal{T}^2$, where C is some real constant.

The above assumptions are listed for studying the properties of $\hat{\boldsymbol{\beta}}(t)$ and $\hat{\gamma}(s,t)$ for Gaussian and non-Gaussian functional data. First of all, under the Gaussian assumption LMA2, it is easy to show that given the design matrix $\mathbf{X}$, $\hat{\boldsymbol{\beta}}(t)$ is a $(p+1)$-dimensional Gaussian process and $(n-p-1)\hat{\gamma}(s,t)$ is a Wishart process.

Theorem 6.1 *Under Assumptions LMA1 and LMA2, we have*

$$\hat{\boldsymbol{\beta}}(t)|\mathbf{X} \sim GP_{p+1}\left(\boldsymbol{\beta}, \boldsymbol{\Gamma}_{\hat{\boldsymbol{\beta}}}\right), \quad \text{and} \quad (n-p-1)\hat{\gamma}(s,t) \sim WP(n-p-1, \gamma),$$

where $\boldsymbol{\Gamma}_{\hat{\boldsymbol{\beta}}}(s,t) = \gamma(s,t)(\mathbf{X}^T\mathbf{X})^{-1}$.

When the Gaussian assumption is not satisfied, the above theorem is no longer valid. However, when the sample size n is large enough and under some further assumptions, we can show that $\hat{\boldsymbol{\beta}}(t)$ is asymptotically a $(p+1)$-dimensional Gaussian process and $\hat{\gamma}(s,t)$ is asymptotically a bivariate Gaussian process.

Theorem 6.2 *Under Assumptions LMA1, LMA3, and LMA4, as $n \to \infty$, we have*

$$\sqrt{n}\left[\hat{\boldsymbol{\beta}}(t) - \boldsymbol{\beta}(t)\right] \xrightarrow{d} GP_{p+1}(0, \gamma\boldsymbol{\Omega}^{-1}).$$

Theorem 6.3 *Under Assumptions LMA1 and LMA3 through LMA6, as $n \to \infty$, we have*

$$\sqrt{n}\left\{\hat{\gamma}(s,t) - \gamma(s,t)\right\} \xrightarrow{d} GP(0, \varpi), \tag{6.12}$$

where

$$\varpi\left[(s_1,t_1),(s_2,t_2)\right] = E\{v_1(s_1)v_1(t_1)v_1(s_2)v_1(t_2)\} - \gamma(s_1,t_1)\gamma(s_2,t_2).$$

The above three theorems are useful for constructing various tests about the coefficient functions $\beta_0(t), \beta_1(t), \cdots, \beta_p(t)$ and the covariance function $\gamma(s,t)$.

6.2.3 Multiple Correlation Coefficient

In classical linear models, the multiple correlation coefficient is used to assess the quality of the prediction of the response variable by the predictors. It can be interpreted as the proportion of the variation of the response variable explained by the predictors. In this subsection, we extend this concept to the context of FLMs. We define the pointwise multiple correlation coefficient of an FLM as the multiple correlation coefficient of the FLM at any given time point. It can be used to assess the quality of the prediction of the response functions by the predictors at any given time point and to be interpreted as the proportion of the variations of the response functions explained by the predictors at any design time point.

Notice that for any fixed $t \in \mathcal{T}$, the FLM (6.4) is a classical linear model with the response vector $\mathbf{y}(t)$, the design matrix $\mathbf{X}$, and the random error term $\mathbf{v}(t)$. Let $\mathrm{SST}_n(t), \mathrm{SSR}_n(t)$, and $\mathrm{SSE}_n(t)$ denote the pointwise total variations, the pointwise variations explained by the FLM, and the pointwise variations unexplained by the FLM, respectively. Then we have

$$\begin{array}{rcl} \mathrm{SST}_n(t) & = & \sum_{i=1}^n [y_i(t) - \bar{y}(t)]^2 = \mathbf{y}(t)^T(\mathbf{I}_n - \mathbf{J}_n/n)\mathbf{y}(t), \\ \mathrm{SSR}_n(t) & = & \sum_{i=1}^n [\hat{y}_i(t) - \bar{y}(t)]^2 = \mathbf{y}(t)^T(\mathbf{P}_\mathbf{X} - \mathbf{J}_n/n)\mathbf{y}(t), \\ \mathrm{SSE}_n(t) & = & \sum_{i=1}^n [y_i(t) - \hat{y}_i(t)]^2 = \mathbf{y}(t)^T(\mathbf{I}_n - \mathbf{P}_\mathbf{X})\mathbf{y}(t), \end{array} \tag{6.13}$$

where $\mathbf{J}_n = \mathbf{1}_n\mathbf{1}_n^T$ is an $n \times n$ matrix of ones. Then the pointwise multiple correlation coefficient is defined as

$$R_n^2(t) = \frac{\mathrm{SSR}_n(t)}{\mathrm{SST}_n(t)}, \ t \in \mathcal{T}. \tag{6.14}$$

Notice that the degrees of freedom of $\mathrm{SST}_n(t), \mathrm{SSR}_n(t)$, and $\mathrm{SSE}_n(t)$ are $n-1, p$ and $n-p-1$, respectively, so that the pointwise adjusted multiple correlation coefficient can be expressed as

$$\begin{array}{rcl} R_{n,adj}^2(t) & = & 1 - \frac{\mathrm{SSE}_n(t)/(n-p-1)}{\mathrm{SST}_n(t)/(n-1)} \\ & = & 1 - \frac{n-1}{n-p-1}[1 - R_n^2(t)], t \in \mathcal{T}. \end{array} \tag{6.15}$$

To define a global measure for assessing the quality of the prediction of the response function, we notice that the following integrated squares,

$$\int_{\mathcal{T}} \mathrm{SST}_n(t)dt, \ \int_{\mathcal{T}} \mathrm{SSR}_n(t)dt, \ \int_{\mathcal{T}} \mathrm{SSE}_n(t)dt,$$

can be used to measure the total variations, the variations explained by the FLM and the variations not explained by the FLM. Based on this viewpoint, we define the global multiple correlation coefficient as

$$R_n^2 = \frac{\int_{\mathcal{T}} \mathrm{SSR}_n(t)dt}{\int_{\mathcal{T}} \mathrm{SST}_n(t)dt}. \tag{6.16}$$

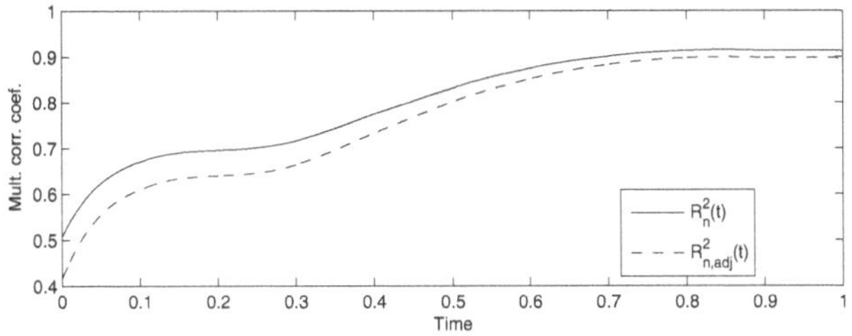

Figure 6.5 *Plots of $R_n^2(t)$ (solid) and $R_{n,adj}^2(t)$ (dashed) of the quadratic FLM (6.1) for the ergonomics data. It seems that a large proportion of the pointwise variations are explained by the quadratic FLM (6.1).*

It is easy to see that $R_n^2 \in [0,1]$ and it is a weighted average of the pointwise multiple correlation coefficients $R_n(t), t \in \mathcal{T}$:

$$R_n^2 = \int_{\mathcal{T}} w_n(t) R_n(t) dt,$$

where $w_n(t) = \text{SST}_n(t)/\int_{\mathcal{T}} \text{SST}_n(s)ds, t \in \mathcal{T}$ is a nonnegative weight function. To take the degrees of freedom into account, we can define the adjusted multiple correlation coefficient as

$$R_{n,adj}^2 = 1 - \frac{n-1}{n-p-1}(1 - R_n^2). \tag{6.17}$$

Example 6.3 *Figure 6.5 displays the plots of $R_n^2(t)$ (solid) and $R_{n,adj}^2(t)$ (dashed) of the quadratic FLM (6.1) for the ergonomics data. It is seen that $R_n^2(t) \in [0.5, 0.91]$ while $R_{n,adj}^2(t) \in [0.40, 0.90]$, showing that a large proportion of the pointwise variations can be explained by the quadratic FLM (6.1). With resolution $M = 1{,}000$, we obtained*

$$\int_{\mathcal{T}} SSR_n(t)dt = 4{,}056.2, \quad \text{and} \quad \int_{\mathcal{T}} SST_n(t)dt = 4{,}759.46. \tag{6.18}$$

By (6.16) and (6.17), we then have

$$R_n^2 = 85.22\%, \quad \text{and} \quad R_{n,adj}^2 = 82.51\%. \tag{6.19}$$

Thus, the FLM (6.1) explains about 85.22% variations of the angle curves of the ergonomics data.

6.2.4 *Comparing Two Nested FLMs*

The concept of nested FLMs is a natural generalization from classical nested linear models. Two FLMs are nested if both contain the same predictors, with one having at least one additional predictor. For example, the following two FLMs are nested:

$$\begin{array}{rcl} \text{Model 1: } \mathbf{y}(t) & = & \mathbf{X}_1\boldsymbol{\beta}_1(t) + \mathbf{v}(t), \\ \text{Model 2: } \mathbf{y}(t) & = & \mathbf{X}\boldsymbol{\beta}(t) + \mathbf{v}(t), \end{array} \tag{6.20}$$

where $\boldsymbol{\beta}(t) = [\boldsymbol{\beta}_1(t)^T, \boldsymbol{\beta}_2(t)^T]^T$, $\mathbf{X} = [\mathbf{X}_1, \mathbf{X}_2] : n \times (1+p)$ with $\mathbf{X}_1 : n \times (1+p-q), \mathbf{X}_2 : n \times q$ so that $\mathbf{X}\boldsymbol{\beta}(t) = \mathbf{X}_1\boldsymbol{\beta}_1(t) + \mathbf{X}_2\boldsymbol{\beta}_2(t)$. The first model in (6.20) usually contains the intercept function and is nested within the second model. Alternatively, the first model is called the reduced FLM and the second model is called the full FLM.

A question arises naturally. How can we know if the more complex full FLM contributes additional information about the association between the response function $\mathbf{y}(t)$ and the predictors in the design matrix $\mathbf{X}$? Answering this question is equivalent to testing the following two nested FLM comparison problem:

$$\begin{array}{rl} & H_0 \ : \mathbf{y}(t) = \mathbf{X}_1\boldsymbol{\beta}_1(t) + \mathbf{v}(t), t \in \mathcal{T}, \\ \text{versus} & H_1 \ : \mathbf{y}(t) = \mathbf{X}\boldsymbol{\beta}(t) + \mathbf{v}(t), t \in \mathcal{T}, \end{array} \tag{6.21}$$

or equivalently, we can write:

$$H_0 : \boldsymbol{\beta}_2(t) \equiv 0, t \in \mathcal{T}, \quad \text{versus} \quad H_1 : \ \boldsymbol{\beta}_2(t) \neq 0, \ \text{ for some } t \in \mathcal{T}.$$

Faraway (1997) and Shen and Faraway (2004) investigated the above problem by proposing a bootstrap test and an F-type test, respectively. For further description, let

$$\begin{array}{rcl} \text{SSE}_{n0}(t) & = & \mathbf{y}(t)^T(\mathbf{I}_n - \mathbf{P}_{\mathbf{X}_1})\mathbf{y}(t), \\ \text{SSE}_n(t) & = & \mathbf{y}(t)^T(\mathbf{I}_n - \mathbf{P}_{\mathbf{X}})\mathbf{y}(t), \\ \text{SSH}_n(t) & = & \text{SSE}_{n0}(t) - \text{SSE}_n(t) = \mathbf{y}(t)^T(\mathbf{P}_{\mathbf{X}} - \mathbf{P}_{\mathbf{X}_1})\mathbf{y}(t), \end{array} \tag{6.22}$$

denote the pointwise SSE (sum of squares due to errors) under the reduced FLM, the full FLM, and the pointwise SSH (sum of squares due to hypothesis), that is, the extra variations in $\mathbf{y}(t)$ explained by the extra predictors in the full FLM. It is easy to show that $\text{SSH}_n(t)$ is always nonnegative due to the fact that $\mathbf{P}_{\mathbf{X}} - \mathbf{P}_{\mathbf{X}_1}$ is an idempotent matrix of rank q. Under the Gaussian assumption LMA2, by Theorem 6.1, it is easy to show that the integrated SSH and SSE are χ^2-type mixtures.

Theorem 6.4 *Assume that Assumptions LMA1 and LMA2 and the null hypothesis in (6.21) hold. Given* $\mathbf{X}$*, we have*

$$\begin{array}{rll} \int_{\mathcal{T}} SSH_n(t)dt & \stackrel{d}{=} \sum_{r=1}^m \lambda_r A_r, & A_r \stackrel{i.i.d.}{\sim} \chi_q^2, \\ \int_{\mathcal{T}} SSE_n(t)dt & \stackrel{d}{=} \sum_{r=1}^m \lambda_r B_r, & B_r \stackrel{i.i.d.}{\sim} \chi_{n-p-1}^2, \end{array}$$

where $A_r, B_r, r = 1, 2, \cdots, m$ *are independent of each other, and* $\lambda_1, \cdots, \lambda_m$ *are all the positive eigenvalues of* $\gamma(s,t)$.

When the Gaussian assumption LMA2 is not satisfied, the above theorem is no longer valid. However, under some further assumptions, we can show that the integrated SSH is asymptotically a χ^2-type mixture.

Theorem 6.5 *Assume that Assumptions LMA1, LMA3, and LMA4 and the null hypothesis (6.21) hold. Given* $\mathbf{X}$*, as* $n \to \infty$*, we have*

$$\int_{\mathcal{T}} SSH_n(t)dt \xrightarrow{d} \sum_{r=1}^{m} \lambda_r A_r,\ A_r \overset{i.i.d.}{\sim} \chi_q^2,$$

where $\lambda_1, \cdots, \lambda_m$ *are all the positive eigenvalues of* $\gamma(s,t)$.

Based on the above two theorems, we are now ready to state various tests for the two nested FLM comparison problem (6.21).

Pointwise Tests We shall describe a pointwise F-test and a pointwise χ^2-test. We first discuss the pointwise F-test. For any fixed $t \in \mathcal{T}$, the two nested FLM comparison problem (6.21) reduces to the classical two nested model comparison problem so that they can be tested using the classical F-test at each time point $t \in \mathcal{T}$. The pointwise F-test uses the following test statistic:

$$F_n(t) = \frac{[\text{SSE}_{n0}(t) - \text{SSE}_n(t)]/q}{\text{SSE}_n(t)/(n-p-1)} = \frac{\text{SSH}_n(t)/q}{\text{SSE}_n(t)/(n-p-1)}, t \in \mathcal{T}. \tag{6.23}$$

It is well known that under the Gaussian assumption LMA2, for any fixed $t \in \mathcal{T}$, $F_n(t) \sim F_{q,n-p-1}$ so that it is easy to conduct the pointwise F-test at each $t \in \mathcal{T}$. In practice, one conducts the pointwise F-test for the two nested FLM comparison problem (6.21) at a grid of preselected time points. If the pointwise F-test leads to rejecting H_0, then at least one of the additional predictors in the full FLM contributes information about the response function.

When the Gaussian assumption is not valid, as $n \to \infty$, it is standard to show that $F_n(t) \xrightarrow{d} \chi_q^2/q$ so that the pointwise χ^2-test can be conducted accordingly.

L^2-Norm-Based Test As usual, the L^2-norm-based test for (6.21) uses the following test statistic

$$T_n = \int_{\mathcal{T}} \text{SSH}_n(t)dt = \int_{\mathcal{T}} \mathbf{y}(t)^T(\mathbf{P_X} - \mathbf{P_{X_1}})\mathbf{y}(t)dt. \tag{6.24}$$

Let $\mathbf{w}(t) = (\mathbf{P_X} - \mathbf{P_{X_1}})\mathbf{y}(t)$. Then we can write $T_n = \int_{\mathcal{T}} \|\mathbf{w}(t)\|^2 dt$, where $\|\mathbf{w}(t)\|$ denotes the usual L^2-norm of $\mathbf{w}(t)$. It is clear that under the null hypothesis in (6.21), T_n will be small; otherwise, it will be large. Therefore, it is natural to use T_n as a reasonable test statistic for testing (6.21).

By Theorems 6.4 and 6.5, we can approximate the null distribution of T_n

using the methods described in Section 4.3 of Chapter 4. The simplest method is the Welch-Satterthwaite χ^2-approximation method, by which we obtain

$$\begin{array}{c} T_n \sim \beta\chi_d^2 \text{ approximately, where} \\ \beta = \frac{\text{tr}(\gamma^{\otimes 2})}{\text{tr}(\gamma)}, \ \kappa = \frac{\text{tr}^2(\gamma)}{\text{tr}(\gamma^{\otimes 2})}, \ d = q\kappa. \end{array} \tag{6.25}$$

In practice, the parameters β, κ, and d must be estimated based on the functional data (6.2). As usual, we here give two methods, a naive method and a bias-reduced method. The key idea of the naive method is to replace $\gamma(s,t)$ with its unbiased estimator $\hat{\gamma}(s,t)$ given in (6.10) so that

$$\hat{\beta} = \frac{\text{tr}(\hat{\gamma}^{\otimes 2})}{\text{tr}(\hat{\gamma})}, \ \hat{\kappa} = \frac{\text{tr}^2(\hat{\gamma})}{\text{tr}(\hat{\gamma}^{\otimes 2})}, \ \hat{d} = q\hat{\kappa}, \tag{6.26}$$

and hence

$$T_n \sim \hat{\beta}\chi_{\hat{d}}^2 \text{ approximately.} \tag{6.27}$$

The naive method often works well but both $\text{tr}^2(\hat{\gamma})$ and $\text{tr}(\hat{\gamma}^{\otimes 2})$ are biased for $\text{tr}^2(\gamma)$ and $\text{tr}(\gamma^{\otimes 2})$. To overcome this problem, we can use the bias-reduced method. When the response functions $y_i(t), i = 1, 2, \cdots, n$ are Gaussian, by Theorem 4.6, we can obtain the unbiased estimators of $\text{tr}^2(\gamma)$ and $\text{tr}(\gamma^{\otimes 2})$ as

$$\begin{array}{rcl} \widehat{\text{tr}^2(\gamma)} & = & \frac{(n-p-1)(n-p)}{(n-p-2)(n-p+1)}\left[\text{tr}^2(\hat{\gamma}) - \frac{2\text{tr}(\hat{\gamma}^{\otimes 2})}{n-p}\right], \\ \widehat{\text{tr}(\gamma^{\otimes 2})} & = & \frac{(n-p-1)^2}{(n-p-2)(n-p+1)}\left[\text{tr}(\hat{\gamma}^{\otimes 2}) - \frac{\text{tr}^2(\hat{\gamma})}{n-p-1}\right], \end{array} \tag{6.28}$$

respectively, as by Theorem 6.1, we have $(n-p-1)\hat{\gamma}(s,t) \sim \text{WP}(n-p-1, \gamma)$. Replacing $\text{tr}^2(\gamma)$ and $\text{tr}(\gamma^{\otimes 2})$ in (6.25) by their unbiased estimators defined above results in the so-called bias-reduced estimators of β, κ, and d:

$$\hat{\beta} = \frac{\widehat{\text{tr}(\gamma^{\otimes 2})}}{\text{tr}(\hat{\gamma})}, \ \hat{\kappa} = \frac{\widehat{\text{tr}^2(\gamma)}}{\widehat{\text{tr}(\gamma^{\otimes 2})}}, \ \hat{d} = q\hat{\kappa}. \tag{6.29}$$

The null hypothesis of (6.21) is rejected whenever $T_n > \hat{\beta}\chi_{\hat{d}}^2(1-\alpha)$ for any given significance level α, where $\hat{\beta}$ and $\hat{d}$ are obtained by either the naive method or the bias-reduced method. Alternatively, the L^2-norm-based test for (6.21) can be conducted by computing the P-value $P(\chi_{\hat{d}}^2 \geq \hat{T}_n/\hat{\beta})$ where $\hat{T}_n$ is the value of T_n computed based on the functional data (6.2).

Notice that with increasing the sample size n, the estimators $\hat{\beta}$ and $\hat{\kappa}$ are consistent, as shown in the following theorem.

Theorem 6.6 *Under Assumptions LMA1 and LMA3 through LMA7, as* $n \to \infty$, *we have* $tr(\hat{\gamma}) \xrightarrow{p} tr(\gamma)$ *and* $tr(\hat{\gamma}^{\otimes 2}) \xrightarrow{p} tr(\gamma^{\otimes 2})$. *Furthermore, as* $n \to \infty$, *we have*

$$\hat{\beta} \xrightarrow{p} \beta, \ \hat{\kappa} \xrightarrow{p} \kappa,$$

where $\hat{\beta}$ *and* $\hat{\kappa}$ *are the naive or bias-reduced estimators of* β *and* κ.

Example 6.4 *For the ergonomics data, it is easy to compute the naive estimators of* β *and* κ *using the values of* $tr(\hat{\gamma})$ *and* $tr(\hat{\gamma}^{\otimes 2})$ *given in (6.18). In fact, we have*

$$\hat{\beta} = 117.57/14.352 = 8.192, \quad \text{and} \quad \hat{\kappa} = 14.352^2/117.57 = 1.752. \tag{6.30}$$

Therefore, for the two nested FLM comparison problem (6.21) for the ergonomics data, by the naive method, we always have the following null distribution:

$$T_n \sim 8.192\chi^2_{1.752q} \text{ approximately}, \tag{6.31}$$

where q *is the number of extra covariates in the full FLM. For the bias-reduced method, notice that* $n = 60 - 1 = 59$ *after an outlying curve is removed from the ergonomics data and* $p = 9$. *By (6.28), we first obtain*

$$\begin{array}{rcl} \widehat{tr^2(\gamma)} & = & \frac{49\times 50}{48\times 51}(14.352^2 - \frac{2\times 117.57}{50}) = 201.44, \\ \widehat{tr(\gamma^{\otimes 2})} & = & \frac{49^2}{48\times 51}(117.57 - 14.352^2/49) = 111.19. \end{array} \tag{6.32}$$

Then the bias-reduced estimators of β *and* κ *are*

$$\hat{\beta} = 111.19/14.352 = 7.7474, \quad \text{and} \quad \hat{\kappa} = 201.44/111.19 = 1.8117. \tag{6.33}$$

Therefore, for the two nested FLM comparison problem (6.21) for the ergonomics data, by the bias-reduced method, we always have the following null distribution:

$$T_n \sim 7.7474\chi^2_{1.8117q} \text{ approximately}. \tag{6.34}$$

F-Type Test When the response functions $y_i(t), i = 1, 2, \cdots, n$ are Gaussian, we can conduct an F-type test for (6.21). As usual, the F-type test statistic is defined as

$$F_n = \frac{\int_{\mathcal{T}} \text{SSH}_n(t)dt/q}{\int_{\mathcal{T}} \text{SSE}_n(t)dt/(n-p-1)}. \tag{6.35}$$

Under the null hypothesis in (6.21), by Theorem 6.4, we have

$$F_n \stackrel{d}{=} \frac{\sum_{r=1}^m \lambda_r A_r/q}{\sum_{r=1}^m, \lambda_r B_r/(n-p-1)}, \quad A_r \stackrel{i.i.d.}{\sim} \chi^2_q, \; B_r \stackrel{i.i.d.}{\sim} \chi^2_{n-p-1},$$

where $A_r, B_r, r = 1, 2, \cdots, m$ are independent and $\lambda_1, \lambda_2, \cdots, \lambda_m$ are all the positive eigenvalues of $\gamma(s,t)$. It follows that the null distribution of F_n can be estimated using the methods described in Section 4.4 of Chapter 4. In fact, by the two-cumulant matched F-approximation method described there, we have

$$F_n \sim F_{q\kappa,(n-p-1)\kappa} \text{ approximately},$$

where $\kappa = \frac{\text{tr}^2(\gamma)}{\text{tr}(\gamma^{\otimes 2})}$ as defined in (6.25) for the L^2-norm-based test. Again, the naive method and the bias-reduced method described earlier for the L^2-norm-based test can be used to obtain the estimator of κ, namely $\hat{\kappa}$. In fact, the naive estimator of κ is given in (6.26) and its bias-reduced estimator is given in (6.29). Therefore, we have

$$F_n \sim F_{q\hat{\kappa},(n-p-1)\hat{\kappa}} \text{ approximately.} \tag{6.36}$$

The null hypothesis in (6.21) is rejected whenever $F_n > F_{q\hat{\kappa},(n-p-1)\hat{\kappa}}(1-\alpha)$ for any given significance level α. Notice that when $n-p-1$ is very large, the F-type test will tend to the L^2-norm-based test.

One advantage of the F-type test is that it takes into account the variation of the pooled sample covariance function $\hat{\gamma}(s,t)$ partially so that in terms of size controlling, the F-type test should outperform the L^2-norm-based test when the Gaussian assumption LMA2 is satisfied.

Example 6.5 *For the ergonomics data, the naive estimator of κ is $\hat{\kappa} = 1.752$ as given in (6.30). Then by the F-type test and under the null hypothesis, we always have*

$$F_n \sim F_{1.752q,1.752(n-p-1)} \text{ approximately,} \tag{6.37}$$

where $n = 59$ and $p = 9$ under the quadratic FLM (6.1). The bias-reduced estimator of κ is $\hat{\kappa} = 1.8117$ as given in (6.30). Then by the F-type test and under the null hypothesis, we always have

$$F_n \sim F_{1.8117q,1.8117(n-p-1)} \text{ approximately.} \tag{6.38}$$

Bootstrap Tests When the functional data are not Gaussian, it is not recommended to use the two-cumulant matched F-approximation method for the null distribution approximation of the F-type test; and when the sample size n is small and the Gaussian assumption is not valid, it is also not recommended to use the Welch-Satterthwaite χ^2-approximation method for the null distribution approximation of the L^2-norm-based test. In these cases, we can use nonparametric bootstrap tests. We briefly describe them as follows. Let $v_i^*(t), i = 1, 2, \cdots, n$ be the bootstrap sample randomly generated from the estimated subject-effect functions $\hat{v}_i(t), i = 1, 2, \cdots, n$ as defined in (6.9). Set $\mathbf{v}^*(t) = [v_1^*(t), \cdots, v_n^*(t)]^T$. Then we can compute $\text{SSH}_n^*(t) = \mathbf{v}^*(t)^T(\mathbf{P}_{\mathbf{X}} - \mathbf{P}_{\mathbf{X}_1})\mathbf{v}^*(t)$ and $\text{SSE}_n^*(t) = \mathbf{v}^*(t)^T(\mathbf{I}_n - \mathbf{P}_{\mathbf{X}})\mathbf{v}^*(t)$. The pointwise bootstrap test statistic $F_n^*(t)$, the L^2-norm-based bootstrap test statistic T_n^*, or the F-type bootstrap test statistic F_n^* can then be respectively calculated as

$$F_n^*(t) = \frac{\text{SSH}_n^*(t)/q}{\text{SSE}_n^*(t)/(n-p-1)},\ T_n^* = \int_{\mathcal{T}} \text{SSH}_n^*(t)dt, \text{ or}$$
$$F_n^* = \frac{\int_{\mathcal{T}} \text{SSH}_n^*(t)dt/q}{\int_{\mathcal{T}} \text{SSE}_n^*(t)dt/(n-p-1)}.$$

Repeat the above bootstrapping process a large number of times, calculate its $100(1-\alpha)$-percentile, and then conduct the pointwise F-test, the L^2-norm-based test, or the F-type test accordingly. The above nonparametric tests are simple to implement but they are generally time consuming.

6.2.5 *Significance of All the Non-Intercept Coefficient Functions*

After fitting the FLM (6.3), one often wants to know if the FLM (6.3) is significant. This is equivalent to testing if all the non-intercept coefficient functions are 0:

$$\begin{array}{lll} & H_0 & : \beta_1(t) \equiv \beta_2(t) \equiv \cdots \equiv \beta_p(t) \equiv 0, \ t \in \mathcal{T}, \\ \text{versus} & H_1 & : H_0 \text{ is false.} \end{array} \qquad (6.39)$$

This test is usually referred to as the overall test for the FLM (6.3). It can be written in the form of the two nested FLM comparison problem (6.21) as

$$\begin{array}{lll} & H_0 & : \mathbf{y}(t) = \mathbf{1}_n\beta_0(t) + \mathbf{v}(t), t \in \mathcal{T}, \\ \text{versus} & H_1 & : \mathbf{y}(t) = \mathbf{X}\boldsymbol{\beta}(t) + \mathbf{v}(t), \ t \in \mathcal{T}. \end{array}$$

That is, the reduced FLM is the intercept model with $\mathbf{X}_1 = \mathbf{1}_n$. Therefore, when applying the various tests described in the previous subsection for the overall testing problem (6.39), we should keep in mind that $q = (p+1) - 1 = p$ and

$$\text{SSH}_n(t) = \mathbf{y}(t)^T(\mathbf{P_X} - \mathbf{J}_n/n)\mathbf{y}(t) = \sum_{i=1}^n [\hat{y}_i(t) - \bar{y}(t)]^2 = \text{SSR}_n(t),$$

where $\mathbf{J}_n = \mathbf{1}_n\mathbf{1}_n^T$ is the $n \times n$ matrix of ones, and $\text{SSR}_n(t)$ denotes the pointwise SSR (sum of squares due to regression). Then the various tests for (6.39) can be conducted as follows:

1. For the pointwise F-test, one uses $F_n(t) = \text{SSR}_n(t)/\text{SSE}_n(t) \sim F_{p,n-p-1}, t \in \mathcal{T}$. For the pointwise χ^2-test, one uses $F_n(t) \sim \chi_p^2/p, t \in \mathcal{T}$.
2. For the L^2-norm-based test, one uses $T_n = \int_{\mathcal{T}} \text{SSR}_n(t)dt \sim \hat{\beta}\chi^2_{p\hat{\kappa}}$ approximately, where $\hat{\beta}$ and $\hat{\kappa}$ are given in (6.26) or in (6.29).
3. For the F-type test, one uses $F_n = \frac{\int_{\mathcal{T}} \text{SSR}_n(t)dt/p}{\int_{\mathcal{T}} \text{SSE}_n(t)dt/(n-p-1)} \sim F_{p\hat{\kappa},(n-p-1)\hat{\kappa}}$ approximately, where $\hat{\kappa}$ are given in (6.26) or in (6.29).

Example 6.6 *For the ergonomics data, consider conducting the L^2-norm-based test for the overall testing problem (6.39) based on the quadratic FLM (6.1). By (6.18), we have $\hat{T}_n = 4,056.2$. By the naive method and by (6.31), under the null hypothesis, we have $T_n \sim 8.192\chi^2_{15.768}$ approximately. Based on this approximate null distribution, the associated 95% critical value is $8.192\chi^2_{15.768}(.95) = 212.95$. Therefore, the quadratic FLM (6.1) is highly significant for the ergonomics data. Similarly, we can conduct the F-type test for the overall testing problem (6.39) based on the quadratic FLM (6.1). The conclusion is exactly the same.*

Notice that the overall testing problem (6.39) is equivalent to testing if the underlying pointwise multiple correlation coefficient function $R^2(t)$ is zero:

$$H_0 : R^2(t) \equiv 0, t \in \mathcal{T}, \quad \text{versus} \quad H_1 : R^2(t) \neq 0, \text{ for some } t \in \mathcal{T}, \tag{6.40}$$

or test if the underlying global multiple correlation coefficient R^2 is zero:

$$H_0 : R^2 = 0, \quad \text{versus} \quad H_1 : R^2 \neq 0.$$

The above problem can be tested using the global multiple correlation coefficient

$$R_n^2 = \frac{\int_{\mathcal{T}} \text{SSR}_n(t)dt}{\int_{\mathcal{T}} \text{SST}_n(t)dt} = \frac{F_n}{F_n + (n-p-1)/p}.$$

Let $\hat{R}_n^2$ be the value of R_n^2 based on the functional data (6.2). Then

$$\begin{aligned} P\left(R_n^2 \geq \hat{R}_n^2\right) &= P\left(\tfrac{F_n}{F_n+(n-p-1)/p} \geq \hat{R}_n^2\right) \\ &= P\left(F_n \geq \frac{\hat{R}_n^2/p}{(1-\hat{R}_n^2)/(n-p-1)}\right). \end{aligned}$$

Therefore, the problem (6.40) can be tested using the approximated null distribution of the F-type test.

6.2.6 *Significance of a Single Coefficient Function*

When fitting the FLM (6.3), it is often of interest to know which covariates are crucial to be included. This is equivalent to testing if a particular coefficient function, for example, $\beta_j(t)$ for some $j = 0, 1, \cdots, p$, is significant:

$$H_0 : \beta_j(t) \equiv 0, t \in \mathcal{T}, \quad \text{versus} \quad H_1 : \beta_j(t) \neq 0, \text{ for some } t \in \mathcal{T}. \tag{6.41}$$

To test the above problem, one can apply the various tests for the two nested FLM comparison problem (6.21) by fitting the full FLM and the reduced FLM, including all the covariates except the covariate associated with the coefficient function $\beta_j(t)$. It is very tedious to repeat this process for each of the coefficient functions. Fortunately, this can be avoided by a simple trick as described below.

Let $\mathbf{X}^{(-j)}$ be the design matrix obtained from the original design matrix $\mathbf{X}$ by excluding the column associated with $\beta_j(t)$. Then according to the classical linear regression model theory, we have

$$\text{SSH}_n(t) = \mathbf{y}(t)^T(\mathbf{P}_{\mathbf{X}} - \mathbf{P}_{\mathbf{X}^{(-j)}})\mathbf{y}(t) = \frac{\hat{\beta}_j^2(t)}{(\mathbf{X}^T\mathbf{X})_{jj}^{-1}}, \tag{6.42}$$

where $\hat{\beta}_j(t)$ is the estimator of $\beta_j(t)$ based on the full FLM and $(\mathbf{X}^T\mathbf{X})_{jj}^{-1}$ denotes the jth diagonal entry of $(\mathbf{X}^T\mathbf{X})^{-1}$. Recall that $\text{SSE}_n(t)$ is the pointwise SSE of the full FLM. The various tests for (6.41) can be conducted as follows:

1. For the pointwise F-test, one uses

$$F_{nj}(t) = \frac{\hat{\beta}_j^2(t)/(\mathbf{X}^T\mathbf{X})_{jj}^{-1}}{\text{SSE}_n(t)/(n-p-1)} \sim F_{1,n-p-1}, t \in \mathcal{T}.$$

For the pointwise χ^2-test, one uses $F_{nj}(t) \sim \chi_1^2, t \in \mathcal{T}$, when the sample size n is large.

2. For the L^2-norm-based test, one uses

$$T_{nj} = \frac{\|\hat{\beta}_j\|^2}{(\mathbf{X}^T\mathbf{X})_{jj}^{-1}} \sim \hat{\beta}\chi^2_{\hat{\kappa}} \text{ approximately,} \tag{6.43}$$

where $\|\hat{\beta}_j\|^2 = \int_{\mathcal{T}} \hat{\beta}_j^2(t)dt$ is the usual squared L^2-norm of the estimated coefficient function $\hat{\beta}_j(t)$, and the estimated parameters $\hat{\beta}$ and $\hat{\kappa}$ are given in (6.26) or in (6.29), obtained by the naive method or the bias-reduced method.

3. For the F-type test, one uses

$$\begin{aligned} F_{nj} &= \frac{\|\hat{\beta}_j\|^2/(\mathbf{X}^T\mathbf{X})_{jj}^{-1}}{\int_{\mathcal{T}} \text{SSE}_n(t)dt/(n-p-1)} \\ &\sim F_{\hat{\kappa},(n-p-1)\hat{\kappa}} \text{ approximately,} \end{aligned} \tag{6.44}$$

where $\hat{\kappa}$ are given in (6.26) or in (6.29).

Example 6.7 *Table 6.1 presents the individual coefficient tests of the quadratic FLM (6.1) for the ergonomics data with resolution* $M = 1,000$.

Table 6.1 *Individual coefficient tests of the quadratic FLM (6.1) for the ergonomics data with resolution* $M = 1,000$.

	L^2-norm-based test		F-type test	
	$8.192\chi^2_{1.752}(.95) = 45.07$		$F_{1.752,85.85}(.95) = 3.25$	
Coefficient function	T_n	P-value	F_n	P-value
$\beta_0(t)$	31,257	0	2,177.80	0
$\beta_1(t)$	181.34	$1.05e-5$	12.64	$3.67e-5$
$\beta_2(t)$	356.78	$2.17e-10$	24.86	$1.47e-8$
$\beta_3(t)$	106.77	$1.06e-3$	7.44	$1.69e-3$
$\beta_4(t)$	104.78	$1.20e-3$	7.30	$1.88e-3$
$\beta_5(t)$	152.57	$6.22e-5$	10.63	$1.54e-4$
$\beta_6(t)$	48.43	$4.04e-2$	3.37	$4.49e-2$
$\beta_7(t)$	116.05	$5.95e-4$	8.09	$1.03e-3$
$\beta_8(t)$	20.85	$2.35e-1$	1.45	$2.39e-1$
$\beta_9(t)$	70.02	$1.04e-2$	4.88	$1.29e-2$

Note: The null distributions of the L^2*-norm-based test and the* F*-type test were obtained using the naive method. The estimated parameters* $\hat{\beta} = 8.192$ *and* $\hat{\kappa} = 1.752$ *are given in (6.30).*

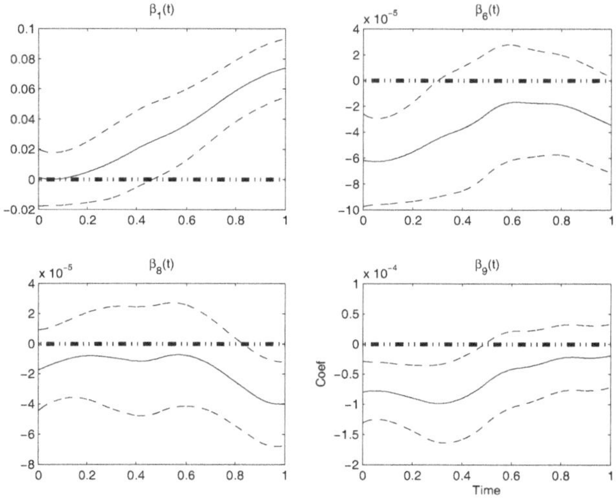

Figure 6.6 *Four fitted coefficient functions (solid) of the ergonomics data with 95% pointwise confidence bands (dashed). The coefficient $\beta_8(t)$ may be less significant as its 95% pointwise confidence bands cover the zero horizontal line (wide dot-dashed) most of time, and the other three coefficient functions are likely to be highly significant.*

The null distributions of the L^2-norm-based test and the F-type test were obtained using the naive method. The estimated parameters $\hat{\beta} = 8.192$ and $\hat{\kappa} = 1.752$ are given in (6.30). It is seen that the coefficient function $\beta_8(t)$ is not significant, the coefficient function $\beta_6(t)$ is at the margin of the 5% level of significance, and all other coefficient functions are highly significant at the 5% level of significance. These results are in agreement with Figure 6.6 where the estimated coefficient functions $\hat{\beta}_1(t), \hat{\beta}_6(t), \hat{\beta}_8(t)$, and $\beta_9(t)$ (solid) are presented with their 95% confidence bounds (dashed). For example, the 95% pointwise confidence bands of $\beta_8(t)$ contain the zero horizontal line (wide dot-dashed line) most of the time, showing that $\beta_8(t)$ may be not significant.

6.2.7 Tests of Linear Hypotheses

In the previous three subsections, we described the two nested model comparison problem and its two special cases. This two nested FLM comparison problem can be further extended to the following GLHT problem:

$$\begin{array}{rll} & H_0 & : \mathbf{C}\boldsymbol{\beta}(t) \equiv \mathbf{c}(t),\ t \in \mathcal{T}, \\ \text{versus} & H_1 & : \mathbf{C}\boldsymbol{\beta}(t) \neq \mathbf{c}(t), \ \text{ for some } t \in \mathcal{T}, \end{array} \tag{6.45}$$

where $\mathbf{C}$ is a given $q \times (p+1)$ full rank matrix, and $\mathbf{c}(t) = [c_1(t), \cdots, c_q(t)]^T$ a vector of given functions. Zhang and Chen (2007) studied the above GLHT problem by proposing an L^2-norm-based test and Zhang (2011a) studied an F-type test.

First of all, we have $\mathrm{E}\left[\mathbf{C}\hat{\boldsymbol{\beta}}(t) - \mathbf{c}(t)\right] = \mathbf{C}\boldsymbol{\beta}(t) - \mathbf{c}(t)$ and

$$\mathrm{Cov}\left[\mathbf{C}\hat{\boldsymbol{\beta}}(s) - \mathbf{c}(s), \mathbf{C}\hat{\boldsymbol{\beta}}(t) - \mathbf{c}(t)\right] = \gamma(s,t)\mathbf{C}(\mathbf{X}^T\mathbf{X})^{-1}\mathbf{C}^T.$$

As $\mathbf{C}(\mathbf{X}^T\mathbf{X})^{-1}\mathbf{C}^T$ is a full rank matrix of size $q \times q$, the pivotal test function for the GLHT problem (6.45) can be specified as:

$$\mathbf{z}(t) = \left[\mathbf{C}(\mathbf{X}^T\mathbf{X})^{-1}\mathbf{C}^T\right]^{-1/2}\left[\mathbf{C}\hat{\boldsymbol{\beta}}(t) - \mathbf{c}(t)\right], \tag{6.46}$$

with the mean function vector and the covariance function matrix as

$$\begin{aligned} \boldsymbol{\eta}_z(t) &= \left[\mathbf{C}(\mathbf{X}^T\mathbf{X})^{-1}\mathbf{C}^T\right]^{-1/2}\left[\mathbf{C}\boldsymbol{\beta}(t) - \mathbf{c}(t)\right], \\ \boldsymbol{\Gamma}_z(s,t) &= \gamma(s,t)\mathbf{I}_q. \end{aligned} \tag{6.47}$$

Theorem 6.7 *Under Assumptions LMA1 and LMA2, we have*

$$\mathbf{z}(t) - \boldsymbol{\eta}_z(t) \sim GP_q(\mathbf{0}, \gamma\mathbf{I}_q). \tag{6.48}$$

Theorem 6.7 indicates that under the Gaussian assumption LMA2, the components of $\mathbf{z}(t)$ are independent Gaussian processes and have the common covariance function $\gamma(s,t)$. This is also asymptotically true even when the response functions $y_i(t), i = 1, 2, \cdots, n$ are not Gaussian with the sample size n diverging.

Theorem 6.8 *Assume that Assumptions LMA1, LMA3, and LMA4 hold. Given $\mathbf{X}$, as $n \to \infty$, we have*

$$\mathbf{z}(t) - \boldsymbol{\eta}_z(t) \xrightarrow{d} GP_q(0, \gamma\mathbf{I}_q). \tag{6.49}$$

To describe various tests for the GLHT problem (6.45), set

$$\begin{aligned} \mathrm{SSH}_n(t) &= \|\mathbf{z}(t)\|^2 \\ &= \left[\mathbf{C}\hat{\boldsymbol{\beta}}(t) - \mathbf{c}(t)\right]^T\left[\mathbf{C}(\mathbf{X}^T\mathbf{X})^{-1}\mathbf{C}^T\right]^{-1}\left[\mathbf{C}\hat{\boldsymbol{\beta}}(t) - \mathbf{c}(t)\right]. \end{aligned} \tag{6.50}$$

Remark 6.2 *The pivotal test function $\mathbf{z}(t)$ (6.46) is constructed in a way such that each component of $\mathbf{z}(t)$ is a process with the same covariance $\gamma(s,t)$ as indicated in (6.47). In addition, it is easy to check that the squared L^2-norm $SSH_n(t)$ (6.50) of $\mathbf{z}(t)$ is invariant when $\mathbf{C}$ and $\mathbf{c}(t)$ are replaced, respectively, with*

$$\tilde{\mathbf{C}} = \mathbf{P}\mathbf{C}, \quad \tilde{\mathbf{c}}(t) = \mathbf{P}\mathbf{c}(t), \tag{6.51}$$

where $\mathbf{P}$ is a full-rank matrix of size $q \times q$. This property is important for the GLHT problem (6.45) as for a hypothesis testing problem that can be written in the form of (6.45), the associated $\mathbf{C}$ and $\mathbf{c}(t)$ are not uniquely defined.

It is easy to show that under the Gaussian assumption LMA2, the integrated SSH is a χ^2-type mixture.

Theorem 6.9 *Under Assumptions LMA1 and LMA2 and the null hypothesis in (6.45), we have*

$$\int_{\mathcal{T}} SSH_n(t)dt \stackrel{d}{=} \sum_{r=1}^{m} \lambda_r A_r,\ A_r \stackrel{i.i.d.}{\sim} \chi_q^2,$$

where $\lambda_1, \cdots, \lambda_m$ are all the positive eigenvalues of $\gamma(s,t)$.

Under some further assumptions, we can show that the integrated SSH is asymptotically a χ^2-type mixture even when the Gaussian assumption LMA2 is not satisfied.

Theorem 6.10 *Assume that Assumptions LMA1, LMA3, and LMA4 hold. Given* $\mathbf{X}$ *and under the null hypothesis in (6.45), as $n \to \infty$, we have*

$$\int_{\mathcal{T}} SSH_n(t)dt \stackrel{d}{\to} \sum_{r=1}^{m} \lambda_r A_r,\ A_r \stackrel{i.i.d.}{\sim} \chi_q^2,$$

where $\lambda_1, \cdots, \lambda_m$ are all the positive eigenvalues of $\gamma(s,t)$.

Based on the above two theorems, we are now ready to present various tests for the GLHT problem (6.45) that are along the same lines as those given in Section 6.2.4 for the two nested FLM comparison problem.

Pointwise Tests We describe a pointwise F-test and a pointwise χ^2-test for the GLHT problem (6.45). The pointwise F-test uses the test statistic $F_n(t) = \frac{\text{SSH}_n(t)}{\text{SSE}_n(t)}, t \in \mathcal{T}$, where $\text{SSE}_n(t)$ is the pointwise sum of squared errors defined in (6.22). When the response functions $y_i(t), i = 1, 2, \cdots, n$ are Gaussian, under the null hypothesis in (6.45), we have

$$F_n(t) \sim F_{q,n-p-1}, t \in \mathcal{T}.$$

The pointwise F-test is then conducted accordingly.

When the response functions $y_i(t), i = 1, 2, \cdots, n$ are non-Gaussian, under some mild conditions, as $n \to \infty$, asymptotically we have

$$F_n(t) \stackrel{d}{\to} \chi_q^2/q,\ \ t \in \mathcal{T}.$$

The pointwise χ^2-test is then conducted accordingly.

L^2-Norm-Based Test The L^2-norm-based test for (6.45) uses the test statistic

$$T_n = \int_{\mathcal{T}} \text{SSH}_n(t)dt.$$

By Theorems 6.9 and 6.10, we can approximate the null distribution of T_n by the Welch-Satterthwaite χ^2-approximation method and obtain

$$T_n \sim \hat{\beta}\chi^2_{q\hat{\kappa}} \text{ approximately,} \tag{6.52}$$

where $\hat{\beta}$ and $\hat{\kappa}$ are given in (6.26) or in (6.29). The L^2-norm-based test for (6.45) can then be conducted accordingly.

Remark 6.3 *For the ergonomics data, the approximate null distributions of T_n by the naive method and the bias-reduced method are already given in (6.31) and (6.34) respectively, having nothing to do with the constant matrix $\mathbf{C}$ and the constant function $\mathbf{c}(t)$.*

F-Type Test When the response functions $y_i(t), i = 1, 2, \cdots, n$ are Gaussian, we can conduct the F-type test for (6.45). The F-type test statistic is defined as

$$F_n = \frac{\int_{\mathcal{T}} \mathrm{SSH}_n(t)dt/q}{\int_{\mathcal{T}} \mathrm{SSE}_n(t)dt/(n-p-1)}.$$

Under the null hypothesis (6.45) and by Theorems 6.1, 6.9, and 4.5, we have

$$F_n \stackrel{d}{=} \frac{\sum_{r=1}^m \lambda_r A_r/q}{\sum_{r=1}^m \lambda_r B_r/(n-p-1)},$$

where $A_r \stackrel{i.i.d.}{\sim} \chi^2_q, B_r \stackrel{i.i.d.}{\sim} \chi^2_{n-p-1}$ and they are all independent, and $\lambda_1, \lambda_2, \cdots, \lambda_m$ are all the positive eigenvalues of $\gamma(s,t)$. It follows that the null distribution of F_n can be approximated by the two-cumulant matched F-approximation method described in Section 4.4 of Chapter 4. We then have

$$F_n \sim F_{q\hat{\kappa},(n-p-1)\hat{\kappa}} \text{ approximately,}$$

where $\hat{\kappa}$ is given in (6.26) or in (6.29). The F-type test for (6.45) is then conducted accordingly.

Remark 6.4 *Again, for the ergonomics data, the approximate null distributions of F_n by the naive method and the bias-reduced method were already given in (6.37) and (6.38), respectively.*

Bootstrap Tests As for the two-model comparison problem, when the functional data are not Gaussian and the sample size n is small, we will use nonparametric bootstrap tests. This is because when the Gaussian assumption is not valid, it is not recommended to use the two-cumulant matched F-approximation method for the null distribution approximation of the F-type test; and when the sample size n is small and the Gaussian assumption is not valid, it is also not recommended to use the Welch-Satterthwaite χ^2-approximation method for the null distribution approximation of the L^2-norm-based test. We briefly describe the nonparametric bootstrap tests as follows. Let $v_i^*(t), i = 1, 2, \cdots, n$, be the bootstrap sample randomly generated from the estimated subject-effect functions $\hat{v}_i(t), i = 1, 2, \cdots, n$ as defined in (6.9). Set $\mathbf{v}^*(t) = [v_1^*(t), \cdots, v_n^*(t)]^T$. Compute $\mathbf{y}^*(t) = \mathbf{X}\hat{\boldsymbol{\beta}}(t) + \mathbf{v}^*(t)$ and $\hat{\boldsymbol{\beta}}^*(t) = (\mathbf{X}^T\mathbf{X})^{-1}\mathbf{X}^T\mathbf{y}^*(t)$. Then we compute $\text{SSH}_n^*(t) = \left[\hat{\boldsymbol{\beta}}^*(t) - \hat{\boldsymbol{\beta}}(t)\right]^T \mathbf{C}^T[\mathbf{C}(\mathbf{X}^T\mathbf{X})^{-1}\mathbf{C}^T]^{-1}\mathbf{C}\left[\hat{\boldsymbol{\beta}}^*(t) - \hat{\boldsymbol{\beta}}(t)\right]$ and $\text{SSE}_n^*(t) = \mathbf{y}^*(t)^T(\mathbf{I}_n - \mathbf{P}_\mathbf{X})\mathbf{y}^*(t)$. The pointwise bootstrap test statistic $F_n^*(t)$, the L^2-norm-based bootstrap test statistic, or the F-type bootstrap test statistic can then be computed as

$$F_n^*(t) = \frac{\text{SSH}_n^*(t)/q}{\text{SSE}_n^*(t)/(n-p-1)}, \quad T_n^* = \int_{\mathcal{T}} \text{SSH}_n^*(t)dt, \quad \text{or}$$
$$F_n^* = \frac{\int_{\mathcal{T}} \text{SSH}_n^*(t)dt/q}{\int_{\mathcal{T}} \text{SSE}_n^*(t)dt/(n-p-1)}.$$

Repeat the above bootstrapping process a large number of times, calculate its $100(1-\alpha)$-percentile, and then conduct the pointwise bootstrap F-test, the L^2-norm-based bootstrap test, or the F-type bootstrap test accordingly.

6.2.8 *Variable Selection*

From Table 6.1, we can see that the coefficient function $\beta_8(t)$ is not significant. Should we remove it from the quadratic FLM (6.1)? Or generally speaking, should we delete those insignificant coefficient functions in an FLM? Unfortunately, there is no simple answer. Some believe that, for a FLM, it makes more sense to keep all or none of a set of terms of the same order, while others believe that one should delete those insignificant coefficient functions and just retain the significant coefficient functions to yield a parsimonious and efficient model in order to improve the prediction performance of the predictors and to obtain a better understanding of the underlying process that generates the functional data. Although this problem warrants further study, we show how to do variable selection in FLMs in this subsection.

Various variable selection methods, for example, forward selection, backward elimination, and stepwise procedures in classical linear models can be naturally adopted for FLMs by replacing the classical F-test with the L^2-norm-based test or the F-type test described in the previous subsections and, in particular, in Section 6.2.6 for covariate selection. As in classical linear models, the simplest model is the intercept FLM, which does not include any covariates at all, and the most complicated model is the full FLM, which includes all the covariates under consideration. The key idea of these procedures is to select those "significant" covariates into the FLM. As in classical linear models, the significance of a covariate is usually with respect to the significance level prespecified for covariate selection. Let α_{in} and α_{out} specify the significance levels for adding a covariate to or deleting a covariate from the current FLM. The three procedures can then be briefly described as follows:

- **Forward Selection** The forward selection procedure starts with the intercept FLM. Test the addition of each covariate using one of the methods described in Section 6.2.6 by either the L^2-norm-based test (6.43) or the F-type test (6.44), add the covariate that improves the FLM the most and is "significant" with respect to α_{in}. Repeat this process until no further covariate candidates can be added.
- **Backward Elimination** The backward elimination procedure starts with the full FLM, test the deletion of each covariate using one of the methods described in Section 6.2.6 by either the L^2-norm-based test (6.43) or the F-type test (6.44), delete the covariate that improves the FLM the most by being deleted and is not "significant" with respect to α_{out}. Repeat this process until no further covariates can be deleted.
- **Stepwise** The stepwise procedure is a combination of the above two procedures with α_{in} and α_{out}. It starts with adding those covariates that are significant with respect to α_{in} until no further covariates can be added. It then moves to delete those covariates that are insignificant with respect to α_{out} until no further covariates can be deleted. This process is then repeated until no further covariates can be added or deleted.

Remark 6.5 *As in classical linear models, in order to get an FLM with better prediction power using the forward selection procedure, we can take* $\alpha_{in} = 15\%$ *or* 25%*, which is usually much larger than* 5%*, the usual significance level specified for hypothesis testing. We can treat* α_{out} *similarly for the backward elimination procedure. For the stepwise method, as in variable selection in classical linear models, it is often helpful to require that* $\alpha_{in} \geq \alpha_{out}$ *so that a larger model is selected before the backward elimination procedure starts.*

Example 6.8 *Consider the variable selection for the ergonomics data by the forward selection method using* $\alpha_{in} = 25\%$*. The bias-reduced* F*-type test is used and the resolution number is* $M = 1,000$*. Table 6.2 shows the P-values of the selected coefficient functions in each step. The second column shows the P-values of the coefficient functions added in Step 1, showing that the term* b_i *(the coefficient function* $\beta_2(t)$*) in the quadratic FLM (6.1) is introduced. The P-values in Columns 3 through 9 for Steps 2 through 8 show the introduction of the coefficient functions* $\beta_5(t), \beta_7(t), \beta_1(t), \beta_4(t), \beta_9(t), \beta_6(t)$ *one by one until no further coefficient functions with their P-values less than* $\alpha_{in} = 25\%$ *can be introduced. The last two rows of Table 6.2 show the values of* R_n^2 *and* $R_{n,adj}^2$ *for each step, allowing to see how these values are varying during the introduction of the coefficient functions. Notice that the coefficient function* $\beta_8(t)$ *was never introduced in the first eight steps.*

Example 6.9 *Consider now the variable selection for the ergonomics data by the backward elimination method using* $\alpha_{out} = 25\%$*. Again, the bias-reduced*

Table 6.2: *Functional variable selection for the ergonomics data.*

	P-values for step							
Coef. func.	1	2	3	4	5	6	7	8
$\beta_0(t)$	0.000	0.000	0.000	0.000	0.000	0.000	0.000	0.000
$\beta_1(t)$					0.001	0.000	0.000	0.000
$\beta_2(t)$	0.000	0.000	0.000	0.000	0.000	0.000	0.000	0.000
$\beta_3(t)$				0.001	0.000	0.000	0.004	0.004
$\beta_4(t)$						0.025	0.017	0.006
$\beta_5(t)$		0.000	0.000	0.000	0.000	0.000	0.000	0.001
$\beta_6(t)$								0.086
$\beta_7(t)$			0.000	0.000	0.001	0.007	0.004	0.004
$\beta_8(t)$								
$\beta_9(t)$							0.162	0.023
R_n^2	0.605	0.694	0.759	0.791	0.819	0.833	0.839	0.848
$R_{n,adj}^2$	0.598	0.683	0.746	0.776	0.802	0.813	0.817	0.824

Note: The forward selection method with $\alpha_{in} = 25\%$ and the bias-reduced F-type test are used. The resolution number for the ergonomics data is $M = 1,000$. Note that the coefficient function $\beta_8(t)$ is never introduced into any of the FLMs before the procedure is terminated.

Table 6.3: *Functional variable selection for the ergonomics data.*

	P-values for step	
Coef. func.	1	2
$\beta_0(t)$	0.000	0.000
$\beta_1(t)$	0.000	0.000
$\beta_2(t)$	0.000	0.000
$\beta_3(t)$	0.002	0.004
$\beta_4(t)$	0.003	0.006
$\beta_5(t)$	0.000	0.001
$\beta_6(t)$	0.057	0.086
$\beta_7(t)$	0.001	0.004
$\beta_8(t)$	0.285	
$\beta_9(t)$	0.017	0.023
R_n^2	0.852	0.848
$R_{n,adj}^2$	0.825	0.824

Note: The backward elimination method with $\alpha_{in} = 25\%$ and the bias-reduced F-type test are used. The resolution number for the ergonomics data is $M = 1,000$. Note that the coefficient function $\beta_8(t)$ is first removed from the full FLM and the procedure is then terminated.

F-type test is used and the resolution number is $M = 1,000$. Table 6.3 shows the P-values of the remaining coefficient functions in each step. The second column shows the P-values of the coefficient functions in Step 1, showing that the coefficient function $\beta_8(t)$ in the quadratic FLM (6.1) should be deleted as its P-value is larger than $\alpha_{out} = 25\%$. The third column shows the P-values for Step 2, showing that there are no coefficient functions with their P-values larger than α_{out} and hence the backward elimination process stops at this step. The final model is the same as one obtained in Example 6.8 by the forward selection method although, in general, the final model by the backward elimination method may be different from the one obtained by the forward selection method. Notice that the coefficient function $\beta_8(t)$ was the first coefficient function deleted.

6.3 Linear Models with Time-Dependent Covariates

In many situations, the covariates are time dependent. That is, they are varying over time. In this case, let $\mathbf{x}_i(t) = [1, x_{i1}(t), \cdots, x_{ip}(t)]^T, i = 1, 2, \cdots, n$ be the covariate functions and $y_i(t), i = 1, 2, \cdots, n$ be the response functions. We can then denote the associated functional data as follows:

$$[\mathbf{x}_i(t)^T, y_i(t)], i = 1, 2, \cdots, n. \tag{6.53}$$

To model the above functional data, we can use the following FLM with time-dependent covariates:

$$y_i(t) = \mathbf{x}_i(t)^T \boldsymbol{\beta}(t) + v_i(t),\ v_i(t) \overset{i.i.d.}{\sim} \text{SP}(0, \gamma), t \in \mathcal{T}, i = 1, 2, \cdots, n, \tag{6.54}$$

where all the variables, except the time-dependent covariates $\mathbf{x}_i(t), i = 1, 2, \cdots, n$, are the same as those defined in the FLM with time-independent covariates (6.3). In matrix form, (6.54) can be rewritten as

$$\mathbf{y}(t) = \mathbf{X}(t)\boldsymbol{\beta}(t) + \mathbf{v}(t),\ t \in \mathcal{T}, \tag{6.55}$$

where $\mathbf{X}(t) = [\mathbf{x}_1(t), \cdots, \mathbf{x}_n(t)]^T$ and other variables are the same as those defined in (6.4).

6.3.1 Estimation of the Coefficient Functions

Throughout this section, we assume that $\mathbf{X}(t)$ has full rank for each $t \in \mathcal{T}$ of interest. Then the pointwise least squares estimator of $\boldsymbol{\beta}(t)$ under (6.55) is

$$\hat{\boldsymbol{\beta}}(t) = [\mathbf{X}(t)^T\mathbf{X}(t)]^{-1}\mathbf{X}(t)^T\mathbf{y}(t),\ t \in \mathcal{T}. \tag{6.56}$$

Set $\mathbf{P}_{\mathbf{X}(t)} = \mathbf{X}(t)[\mathbf{X}(t)^T\mathbf{X}(t)]^{-1}\mathbf{X}(t)^T$. Then it is well known that $\mathbf{P}_{\mathbf{X}(t)}$ is an idempotent matrix of rank $(p+1)$. For any $t \in \mathcal{T}$, the cross-product

$$[\mathbf{y}(t) - \mathbf{X}(t)\hat{\boldsymbol{\beta}}(t)]^T[\mathbf{X}(t)\hat{\boldsymbol{\beta}}(t) - \mathbf{X}(t)\boldsymbol{\beta}(t)] = 0,$$

which shows that the pointwise least squares estimator $\hat{\boldsymbol{\beta}}(t)$ minimizes the following integrated squared error:

$$Q(\boldsymbol{\beta}) = \int_{\mathcal{T}} \|\mathbf{y}(t) - \mathbf{X}(t)\boldsymbol{\beta}(t)\|^2 dt$$

globally.

The vector of fitted response functions can be expressed as

$$\hat{\mathbf{y}}(t) = [\hat{y}_1(t), \hat{y}_2(t), \cdots, \hat{y}_n(t)]^T = \mathbf{X}(t)\hat{\boldsymbol{\beta}}(t) = \mathbf{P}_{\mathbf{X}(t)}\mathbf{y}(t), \tag{6.57}$$

and the vector of fitted subject-effect functions is

$$\hat{\mathbf{v}}(t) = \mathbf{y}(t) - \hat{\mathbf{y}}(t) = (\mathbf{I}_n - \mathbf{P}_{\mathbf{X}(t)})\mathbf{y}(t) = (\mathbf{I}_n - \mathbf{P}_{\mathbf{X}(t)})\mathbf{v}(t). \tag{6.58}$$

Then we have

$$\mathrm{E}(\hat{\mathbf{v}}(t)) = 0 \text{ and } \mathrm{Cov}(\hat{\mathbf{v}}(s), \hat{\mathbf{v}}(t)) = \gamma(s,t)\left(\mathbf{I}_n - \mathbf{P}_{\mathbf{X}(s)})(\mathbf{I}_n - \mathbf{P}_{\mathbf{X}(t)}\right).$$

Thus, $\mathrm{E}[\hat{\mathbf{v}}(s)^T\hat{\mathbf{v}}(t)] = \gamma(s,t)\left\{n - 2(p+1) + \mathrm{tr}[\mathbf{P}_{\mathbf{X}(s)}\mathbf{P}_{\mathbf{X}(t)}]\right\}$. An unbiased estimator of the covariance function $\gamma(s,t)$ is then given by

$$\hat{\gamma}(s,t) = \frac{\sum_{i=1}^n [y_i(s) - \hat{y}_i(s)]\,[y_i(t) - \hat{y}_i(t)]}{n - 2(p+1) + \mathrm{tr}[\mathbf{P}_{\mathbf{X}(s)}\mathbf{P}_{\mathbf{X}(t)}]}. \tag{6.59}$$

When $X(t) \equiv \mathbf{X}$, that is, the covariates are time independent, (6.59) reduces to (6.10) as expected.

6.3.2 Compare Two Nested FLMs

Consider comparing the following two nested FLMs:

$$\begin{array}{rll} & H_0 & : \mathbf{y}(t) = \mathbf{X}_1(t)\boldsymbol{\beta}_1(t) + \mathbf{v}(t), t \in \mathcal{T} \\ \text{versus} & H_1 & : \mathbf{y}(t) = \mathbf{X}(t)\boldsymbol{\beta}(t) + \mathbf{v}(t), t \in \mathcal{T}, \end{array} \tag{6.60}$$

where $\mathbf{X}(t)\boldsymbol{\beta}(t) = \mathbf{X}_1(t)\boldsymbol{\beta}_1(t) + \mathbf{X}_2(t)\boldsymbol{\beta}_2(t)$ and $\mathbf{X}_2(t)$ is a full rank matrix of size $n \times q$. For this purpose, set

$$\begin{array}{rcl} \mathrm{SSE}_{n0}(t) & = & \mathbf{y}(t)^T(\mathbf{I}_n - \mathbf{P}_{\mathbf{X}_1(t)})\mathbf{y}(t), \\ \mathrm{SSE}_n(t) & = & \mathbf{y}(t)^T(\mathbf{I}_n - \mathbf{P}_{\mathbf{X}(t)})\mathbf{y}(t), \\ \mathrm{SSH}_n(t) & = & \mathrm{SSE}_{n0}(t) - \mathrm{SSE}_n(t) = \mathbf{y}(t)^T(\mathbf{P}_{\mathbf{X}(t)} - \mathbf{P}_{\mathbf{X}_1(t)})\mathbf{y}(t), \end{array} \tag{6.61}$$

which denote the pointwise SSE under the reduced FLM, the full FLM, and the extra variations in $\mathbf{y}(t)$ explained by the extra predictors in the full FLM.

Pointwise Tests We shall consider a pointwise F-test and a pointwise χ^2-test. The pointwise F-test uses the following test statistic:

$$F_n(t) = \frac{\mathrm{SSH}_n(t)/q}{\mathrm{SSE}_n(t)/(n-p-1)},\ t \in \mathcal{T}.$$

Under the Gaussian assumption, that is, the response functions $y_i(t), i = 1, 2, \cdots, n$ are Gaussian, for any fixed $t \in \mathcal{T}$, $F_n(t) \sim F_{q,n-p-1}$ so that it is easy to conduct the pointwise F-test at each $t \in \mathcal{T}$. When the Gaussian assumption is not valid, as $n \to \infty$, it is standard to show that for any $t \in \mathcal{T}$, we have $F_n(t) \xrightarrow{d} \chi_q^2/q$ so that we can conduct the pointwise χ^2-test accordingly.

L^2-Norm-Based Test The test statistic of the L^2-norm-based test for (6.60) is usually constructed as

$$\begin{aligned} T_n^0 &= \int_{\mathcal{T}} \mathrm{SSH}_n(t)dt \\ &= \int_{\mathcal{T}} \mathbf{y}(t)^T(\mathbf{P}_{\mathbf{X}(t)} - \mathbf{P}_{\mathbf{X}_1(t)})\mathbf{y}(t)dt = \int_{\mathcal{T}} \mathbf{y}(t)^T\mathbf{H}(t)\mathbf{y}(t)dt, \end{aligned}$$

where $H(t) = \mathbf{P}_{\mathbf{X}(t)} - \mathbf{P}_{\mathbf{X}_1(t)}$. Unfortunately, for the FLM with time-dependent covariates (6.54), it is rather challenging to derive the null distribution of T_n^0. To overcome this difficulty, let us evaluate the covariate and response functions (6.53) at the following equally spaced design time points:

$$t_j, j = 1, 2, \cdots, M. \tag{6.62}$$

When M is large, we have

$$T_n^0 \approx \frac{v(\mathcal{T})}{M} \sum_{j=1}^M \mathbf{y}(t_j)^T\mathbf{H}(t_j)\mathbf{y}(t_j) = \frac{v(\mathcal{T})}{M} T_n,$$

where $v(\mathcal{T})$ denotes the volume of $\mathcal{T}$, and

$$T_n = \sum_{j=1}^M \mathbf{y}(t_j)^T\mathbf{H}(t_j)\mathbf{y}(t_j). \tag{6.63}$$

Set

$$\mathbf{y} = \begin{pmatrix} \mathbf{y}(t_1) \\ \mathbf{y}(t_2) \\ \cdots \\ \mathbf{y}(t_M) \end{pmatrix}, \quad \mathbf{H} = \begin{pmatrix} \mathbf{H}(t_1) & \mathbf{0} & \cdots & \mathbf{0} \\ \mathbf{0} & \mathbf{H}(t_2) & \cdots & \mathbf{0} \\ \vdots & \vdots & \vdots & \vdots \\ \mathbf{0} & \mathbf{0} & \cdots & \mathbf{H}(t_M) \end{pmatrix}. \tag{6.64}$$

That is, $\mathbf{y}$ is an $(nM) \times 1$ vector, obtained by stacking $\mathbf{y}(t_j), j = 1, 2, \cdots, M$ one by one and $\mathbf{H}$ is an $(nM) \times (nM)$ block diagonal matrix with block matrices $\mathbf{H}(t_j), j = 1, 2, \cdots, M$ being its block diagonal entries. Then we can further express T_n as a quadratic form of $\mathbf{y}$:

$$T_n = \mathbf{y}^T\mathbf{H}\mathbf{y}. \tag{6.65}$$

It is standard to show that $\mathrm{Cov}(\mathbf{y}) = \Gamma_M \otimes \mathbf{I}_n$. That is,

$$\mathrm{Cov}(\mathbf{y}) = \begin{pmatrix} \gamma(t_1, t_1)\mathbf{I}_n & \gamma(t_1, t_2)\mathbf{I}_n & \cdots & \gamma(t_1, t_M)\mathbf{I}_n \\ \gamma(t_2, t_1)\mathbf{I}_n & \gamma(t_2, t_2)\mathbf{I}_n & \cdots & \gamma(t_2, t_M)\mathbf{I}_n \\ \vdots & \vdots & \vdots & \vdots \\ \gamma(t_M, t_1)\mathbf{I}_n & \gamma(t_M, t_2)\mathbf{I}_n & \cdots & \gamma(t_M, t_M)\mathbf{I}_n \end{pmatrix}, \tag{6.66}$$

where $\mathbf{\Gamma}_M$ is an $M \times M$ matrix whose (i,j)th entry is $\gamma(t_i, t_j)$ for $i, j = 1, 2, \cdots, M$. It is standard to show that $T_n^0 = \lim_{M\to\infty} \frac{v(\mathcal{T})}{M} T_n$. For large M, it is then natural to use T_n as the test statistic of the L^2-norm-based test for (6.60).

Remark 6.6 *In practice, we may not be able to reconstruct $\mathbf{x}_i(t), i = 1, \cdots, n$ and $\mathbf{y}_i(t), i = 1, \cdots, n$ from the observed functional data. In this case, if all the covariate and response functions (6.53) are observed at the same grid of the distinct design time points (6.62) (may not be equally spaced), we still can use T_n for (6.60) but the relationship between T_n and T_n^0 is not clear.*

In order to describe various statistical inference procedures for the FLM with time-dependent covariates (6.54), we need impose the following strong assumptions:

Linear Model Assumptions (LMB)

1. The coefficient functions are L^2-integrable and the covariance function $\gamma(s,t)$ has finite trace. That is, $\beta_0(t), \beta_2(t), \cdots, \beta_p(t) \in \mathcal{L}^2(\mathcal{T})$ and $\text{tr}(\gamma) < \infty$.
2. The subject-effect functions $v_1(t), \cdots, v_n(t)$ are i.i.d. Gaussian.

Remark 6.7 *It is clear that we impose stronger assumptions for the FLMs with time-dependent covariates than those imposed for the FLMs with time-independent covariates studied in the previous section. This is because the FLMs with time-dependent covariates are much more complicated than the FLMs with time-independent covariates.*

Under the Gaussian assumption and the null hypothesis, we can show that T_n is a χ^2-type mixture whose distribution can be well approximated by the Welch-Satterthwaite χ^2-approximation method described in Section 4.3 of Chapter 4.

Theorem 6.11 *Assume that Assumptions LMB1 and LMB2 and the null hypothesis in (6.60) hold. Given $\mathbf{H}$, we have*

$$T_n \stackrel{d}{=} \sum_{r=1}^{nM} \lambda_r A_r, \quad A_r \stackrel{i.i.d.}{\sim} \chi_1^2,$$

where $\lambda_1, \lambda_2, \cdots, \lambda_{nM}$ are the eigenvalues of $\mathbf{W} = (\mathbf{\Gamma}_M^{1/2} \otimes \mathbf{I}_n)\mathbf{H}(\mathbf{\Gamma}_M^{1/2} \otimes \mathbf{I}_n)$.

Based on the above theorem, under the null hypothesis in (6.60) and by the Welch-Satterthwaite χ^2-approximation method described in Section 4.3 of Chapter 4, we have

$$T_n \sim \beta \chi_d^2 \text{ approximately}, \tag{6.67}$$

where

$$\beta = \frac{\sum_{r=1}^{nM} \lambda_r^2}{\sum_{r=1}^{nM} \lambda_r}, \quad d = \frac{(\sum_{r=1}^{nM} \lambda_r)^2}{\sum_{r=1}^{nM} \lambda_r^2}. \tag{6.68}$$

It is very time consuming to directly compute β and d using the expressions (6.68) with the eigenvalues $\lambda_1, \cdots, \lambda_{nM}$ of $\mathbf{W}$ due to the big size of $\mathbf{W}$. For example, when $n = 30$ and $M = 1,000$, which are not uncommon in FDA, $\mathbf{W}$ has 900 million entries. In this case, it needs a lot of memory and time to find the eigenvalues of $\mathbf{W}$.

Fortunately, it is standard to show that $\sum_{r=1}^{nM} \lambda_r = \mathrm{tr}(\mathbf{W})$ and $\sum_{r=1}^{nM} \lambda_r^2 = \mathrm{tr}(\mathbf{W}^2)$ so that we can avoid finding the eigenvalues of $\mathbf{W}$ by computing

$$\beta = \frac{\mathrm{tr}(\mathbf{W}^2)}{\mathrm{tr}(\mathbf{W})}, \quad d = \frac{\mathrm{tr}^2(\mathbf{W})}{\mathrm{tr}(\mathbf{W}^2)}.$$

In addition, applying the property $\mathrm{tr}(\mathbf{AB}) = \mathrm{tr}(\mathbf{BA})$, we have

$$\mathrm{tr}(\mathbf{W}) = \mathrm{tr}\left(\mathbf{H}(\mathbf{\Gamma}_M \otimes \mathbf{I}_n)\right), \quad \mathrm{tr}(\mathbf{W}^2) = \mathrm{tr}\left([\mathbf{H}(\mathbf{\Gamma}_M \otimes \mathbf{I}_n)]^2\right).$$

In this case, we can further express β and d as

$$\beta = \frac{\mathrm{tr}([\mathbf{H}(\mathbf{\Gamma}_M \otimes \mathbf{I}_n)]^2)}{\mathrm{tr}(\mathbf{H}(\mathbf{\Gamma}_M \otimes \mathbf{I}_n))}, \quad d = \frac{\mathrm{tr}^2(\mathbf{H}(\mathbf{\Gamma}_M \otimes \mathbf{I}_n))}{\mathrm{tr}([\mathbf{H}(\mathbf{\Gamma}_M \otimes \mathbf{I}_n)]^2)}. \tag{6.69}$$

By (6.64), (6.66), and some simple algebra, we can express $\mathbf{H}(\mathbf{\Gamma}_M \otimes \mathbf{I}_n)$ as

$$\begin{pmatrix} \mathbf{H}(t_1)\gamma(t_1,t_1) & \mathbf{H}(t_1)\gamma(t_1,t_2) & \cdots & \mathbf{H}(t_1)\gamma(t_1,t_M) \\ \mathbf{H}(t_2)\gamma(t_2,t_1) & \mathbf{H}(t_2)\gamma(t_2,t_2) & \cdots & \mathbf{H}(t_2)\gamma(t_2,t_M) \\ \vdots & \vdots & \vdots & \vdots \\ \mathbf{H}(t_M)\gamma(t_M,t_1) & \mathbf{H}(t_M)\gamma(t_M,t_2) & \cdots & \mathbf{H}(t_M)\gamma(t_M,t_M) \end{pmatrix}.$$

Noticing that $\mathrm{tr}[H(t_j)] = q$ and $\gamma(t_i,t_j) = \gamma(t_j,t_i)$, we have

$$\begin{aligned} \mathrm{tr}(\mathbf{H}(\mathbf{\Gamma}_M \otimes \mathbf{I}_n)) &= \textstyle\sum_{j=1}^M \mathrm{tr}\left[\mathbf{H}(t_j)\gamma(t_j,t_j)\right] = q\sum_{j=1}^M \gamma(t_j,t_j), \\ \mathrm{tr}([\mathbf{H}(\mathbf{\Gamma}_M \otimes \mathbf{I}_n)]^2) &= \textstyle\sum_{i=1}^M \sum_{j=1}^M \mathrm{tr}\left[\mathbf{H}(t_i)\gamma(t_i,t_j)\mathbf{H}(t_j)\gamma(t_j,t_i)\right] \\ &= \textstyle\sum_{i=1}^M \sum_{j=1}^M \mathrm{tr}\left[\mathbf{H}(t_i)\mathbf{H}(t_j)\right]\gamma^2(t_i,t_j). \end{aligned} \tag{6.70}$$

By the above expressions, the naive estimators of β and d are given by

$$\begin{aligned} \hat{\beta} &= \frac{\sum_{i=1}^M \sum_{j=1}^M \mathrm{tr}[\mathbf{H}(t_i)\mathbf{H}(t_j)]\hat{\gamma}^2(t_i,t_j)}{q\sum_{j=1}^M \hat{\gamma}(t_j,t_j)}, \\ \hat{d} &= \frac{[q\sum_{j=1}^M \hat{\gamma}(t_j,t_j)]^2}{\sum_{i=1}^M \sum_{j=1}^M \mathrm{tr}[\mathbf{H}(t_i)\mathbf{H}(t_j)]\hat{\gamma}^2(t_i,t_j)}, \end{aligned} \tag{6.71}$$

where $\hat{\gamma}(s,t)$ is given in (6.59). We then have

$$T_n \sim \hat{\beta}\chi^2_{\hat{d}} \text{ approximately.} \tag{6.72}$$

The L^2-norm-based test for (6.60) can then be conducted accordingly.

F-Type Test As usual, the F-type test for (6.60) uses the following test statistic:

$$F_n^0 = \frac{\int_{\mathcal{T}} \mathrm{SSH}_n(t)dt/q}{\int_{\mathcal{T}} \mathrm{SSE}_n(t)dt/(n-p-1)} = \frac{\int_{\mathcal{T}} \mathbf{y}(t)^T\mathbf{H}(t)\mathbf{y}(t)dt/q}{\int_{\mathcal{T}} \mathbf{y}(t)^T\mathbf{G}(t)\mathbf{y}(t)dt/(n-p-1)},$$

where $\mathbf{G}(t) = \mathbf{I}_n - \mathbf{P}_{\mathbf{x}(t)}$ and $\mathbf{H}(t)$ and $\mathbf{y}(t)$ are the same as those defined earlier for the L^2-norm-based test. Again, it is difficult to find the null distribution of F_n^0. Similar to the treatment of the L^2-norm-based test described earlier, we can evaluate the covariate and response functions (6.53) at the equally spaced design time points (6.62). When M is large, we have

$$F_n^0 \approx \frac{\sum_{j=1}^M \mathbf{y}(t_j)^T\mathbf{H}(t_j)\mathbf{y}(t_j)/q}{\sum_{j=1}^M \mathbf{y}(t_j)^T\mathbf{G}(t_j)\mathbf{y}(t_j)/(n-p-1)} = F_n,$$

where F_n can be further written as

$$F_n = \frac{\mathbf{y}^T\mathbf{H}\mathbf{y}/q}{\mathbf{y}^T\mathbf{G}\mathbf{y}/(n-p-1)}, \tag{6.73}$$

with $\mathbf{y}, \mathbf{H}$ defined earlier for the L^2-norm-based test and

$$\mathbf{G} = \mathrm{diag}(\mathbf{G}(t_1), \mathbf{G}(t_2), \cdots, \mathbf{G}(t_M)).$$

A new challenge is that the distribution of F_n cannot be approximated using the two-cumulant matched F-approximation method described in Section 4.4 of Chapter 4 as the numerator and the denominator of F_n are not independent even under the Gaussian assumption. This is because $\mathrm{Cov}(\mathbf{Hy}, \mathbf{Gy}) = \mathbf{H}\mathrm{Cov}(\mathbf{y})\mathbf{G} = \mathbf{H}(\mathbf{\Gamma}_M \otimes \mathbf{I}_n)\mathbf{G} \neq 0$, which can be verified by writing $\mathbf{H}(\mathbf{\Gamma}_M \otimes \mathbf{I}_n)\mathbf{G}$ as

$$\begin{pmatrix} \mathbf{H}(t_1)\mathbf{G}(t_1)\gamma(t_1,t_1) & \cdots & \mathbf{H}(t_1)\mathbf{G}(t_M)\gamma(t_1,t_M) \\ \mathbf{H}(t_2)\mathbf{G}(t_1)\gamma(t_2,t_1) & \cdots & \mathbf{H}(t_2)\mathbf{G}(t_M)\gamma(t_2,t_M) \\ \vdots & \vdots & \vdots \\ \mathbf{H}(t_M)\mathbf{G}(t_1)\gamma(t_M,t_1) & \cdots & \mathbf{H}(t_M)\mathbf{G}(t_M)\gamma(t_M,t_M) \end{pmatrix}.$$

To overcome this difficulty, let $\hat{F}_n$ denote the value of F_n computed based on the functional data (6.53) and we write

$$P(F_n \geq \hat{F}_n) = P\left(\frac{\mathbf{y}^T\mathbf{H}\mathbf{y}/q}{\mathbf{y}^T\mathbf{G}\mathbf{y}/(n-p-1)} \geq \hat{F}_n\right) = P(S_n \geq 0), \tag{6.74}$$

where $S_n = \mathbf{y}^T(\mathbf{H} - \rho\mathbf{G})\mathbf{y}$ and $\rho = \frac{q\hat{F}_n}{n-p-1}$. We have the following theorem about the null distribution of S_n.

Theorem 6.12 *Assume that Assumptions LMB1 and LMB2 and the null hypothesis in (6.60) hold. Given $\hat{F}_n, \mathbf{H}$, and $\mathbf{G}$, we have*

$$S_n \stackrel{d}{=} \sum_{r=1}^{nM} \lambda_r A_r, \quad A_r \stackrel{i.i.d.}{\sim} \chi_1^2,$$

where $\lambda_1, \lambda_2, \cdots, \lambda_{nM}$ are the eigenvalues of $\mathbf{W} = (\mathbf{\Gamma}_M^{1/2} \otimes \mathbf{I}_n)(\mathbf{H} - \rho\mathbf{G})(\mathbf{\Gamma}_M^{1/2} \otimes \mathbf{I}_n)$.

Theorem 6.12 is due to Xu, Shen, Yang, and Shoptaw (2011). Notice that not all the eigenvalues of $\mathbf{W}$ are positive or negative. According to Section 4.3 of Chapter 4, it is not proper to approximate the null distribution of S_n by the Welch-Satterwaithe χ^2-approximation approach; rather, we should use the three-cumulant matched χ^2-approximation method (Zhang 2005) described there to approximate the null distribution of S_n. That is,

$$S_n \sim \beta\chi_d^2 + \beta_0 \text{ approximately,}$$

where the parameters β, d and β_0 are obtained as

$$\beta = \frac{\sum_{r=1}^{nM} \lambda_r^3}{\sum_{r=1}^{nM} \lambda_r^2}, \quad d = \frac{[\sum_{r=1}^{nM} \lambda_r^2]^3}{[\sum_{r=1}^{nM} \lambda_r^3]^2}, \quad \beta_0 = \sum_{r=1}^{nM} \lambda_r - \frac{[\sum_{r=1}^{nM} \lambda_r^2]^2}{\sum_{r=1}^{nM} \lambda_r^3},$$

by matching the first three cumulants of S_n and $R \sim \beta\chi_d^2 + \beta_0$; see Section 4.3 in Chapter 4 for more details.

As $\sum_{r=1}^{nM} \lambda_r^l = \text{tr}(\mathbf{W}^l) \equiv \kappa_l, l = 1, 2, 3$, we can re-express the parameters β, d, and β_0 as

$$\begin{aligned} \beta &= \frac{\text{tr}(\mathbf{W}^3)}{\text{tr}(\mathbf{W}^2)} = \frac{\kappa_3}{\kappa_2}, \quad d = \frac{\text{tr}^3(\mathbf{W}^2)}{\text{tr}^2(\mathbf{W}^3)} = \frac{\kappa_2^3}{\kappa_3^2}, \\ \beta_0 &= \text{tr}(\mathbf{W}) - \frac{\text{tr}^2(\mathbf{W}^2)}{\text{tr}(\mathbf{W}^3)} = \kappa_1 - \frac{\kappa_2^2}{\kappa_3}. \end{aligned}$$

By some simple algebra, we have $\kappa_l = \text{tr}([(\mathbf{H} - \rho\mathbf{G})(\mathbf{\Gamma}_M \otimes \mathbf{I}_p)]^l), l = 1, 2, 3$. Set $\mathbf{D}(t) = \mathbf{H}(t) - \rho\mathbf{G}(t)$. Then the naive estimators of $\kappa_l, l = 1, 2, 3$ are

$$\begin{aligned} \hat{\kappa}_1 &= \sum_{i=1}^M \text{tr}\left[\mathbf{D}(t_i)\right] \hat{\gamma}(t_i, t_i) = q(1 - \hat{F}_n) \sum_{i=1}^M \hat{\gamma}(t_i, t_i), \\ \hat{\kappa}_2 &= \sum_{i=1}^M \sum_{j=1}^M \text{tr}\left[\mathbf{D}(t_i)\mathbf{D}(t_j)\right] \hat{\gamma}^2(t_i, t_j), \\ \hat{\kappa}_3 &= \sum_{i=1}^M \sum_{j=1}^M \sum_{k=1}^M \text{tr}\left[\mathbf{D}(t_i)\mathbf{D}(t_j)\mathbf{D}(t_k)\right] \hat{\gamma}(t_i, t_j)\hat{\gamma}(t_j, t_k)\hat{\gamma}(t_k, t_i), \end{aligned}$$

where $\hat{\gamma}(s, t)$ is given in (6.59). It follows that $S_n \sim \hat{\beta}\chi_{\hat{d}}^2 + \hat{\beta}_0$ approximately, where

$$\hat{\beta} = \frac{\hat{\kappa}_3}{\hat{\kappa}_2}, \quad \hat{d} = \frac{\hat{\kappa}_2^3}{\hat{\kappa}_3^2}, \quad \hat{\beta}_0 = \hat{\kappa}_1 - \frac{\hat{\kappa}_2^2}{\hat{\kappa}_3}. \tag{6.75}$$

Therefore, we can compute the P-value of the F-type test as $P(F_n \geq \hat{F}_n) =$

$P(S_n \geq 0) \approx P(\hat{\beta}\chi^2_{\hat{d}} + \hat{\beta}_0 \geq 0)$, where

$$P\left(\hat{\beta}\chi^2_{\hat{d}} + \hat{\beta}_0 \geq 0\right) = \begin{cases} P\left(\chi^2_{\hat{d}} \geq -\hat{\beta}_0/\hat{\beta}\right), & \text{when } \hat{\beta} > 0, \\ P\left(\chi^2_{\hat{d}} \leq -\hat{\beta}_0/\hat{\beta}\right), & \text{when } \hat{\beta} < 0. \end{cases}$$

The key idea of the above F-type test is the same as that of the quasi F-test proposed by Xu et al. (2011). But the estimated covariance function used in Xu et al. (2011) is not unbiased and they did not discuss how to compute $\hat{\beta}, \hat{d}$, and $\hat{\beta}_0$ when the resolution M is large.

Bootstrap Tests When the functional data are not Gaussian, it is very challenging to derive the null distributions of T_n and F_n. In this case, we can use nonparametric bootstrap tests. Let $v_i^*(t), i = 1, 2, \cdots, n$, be the bootstrap sample randomly generated from the estimated subject-effect functions $\hat{v}_i(t), i = 1, 2, \cdots, n$ as defined in (6.58). Set $\mathbf{v}^*(t) = [v_1^*(t), \cdots, v_n^*(t)]^T$. Then we can compute

$$\text{SSH}_n^*(t) = \mathbf{v}^*(t)^T(\mathbf{P}_{\mathbf{X}(t)} - \mathbf{P}_{\mathbf{X}_1(t)})\mathbf{v}^*(t), \quad \text{SSE}_n^*(t) = \mathbf{v}^*(t)^T(\mathbf{I}_n - \mathbf{P}_{\mathbf{X}(t)})\mathbf{v}^*(t).$$

The pointwise bootstrap test statistic $F_n^*(t)$, the L^2-norm-based bootstrap test statistic T_n^*, or the F-type bootstrap test statistic can be respectively calculated as

$$F_n^*(t) = \frac{\text{SSH}_n^*(t)/q}{\text{SSE}_n^*(t)/(n-p-1)}, \quad T_n^* = \int_{\mathcal{T}} \text{SSH}_n^*(t)dt,$$
$$F_n^* = \frac{\int_{\mathcal{T}} \text{SSH}_n^*(t)dt/q}{\int_{\mathcal{T}} \text{SSE}_n^*(t)dt/(n-p-1)}.$$

Repeat the above bootstrapping process a large number of times, calculate its $100(1-\alpha)$-percentile, and then conduct the pointwise test, the L^2-norm-based test, or the F-type test accordingly.

6.3.3 *Tests of Linear Hypotheses*

Based on the FLM with time-dependent covariates (6.55), the GLHT problem is still the one defined in (6.45). We still have $\text{E}[\mathbf{C}\hat{\boldsymbol{\beta}}(t) - \mathbf{c}(t)] = \mathbf{C}\boldsymbol{\beta}(t) - \mathbf{c}(t)$ but

$$\text{Cov}\left[\mathbf{C}\hat{\boldsymbol{\beta}}(s) - \mathbf{c}(s), \mathbf{C}\hat{\boldsymbol{\beta}}(t) - \mathbf{c}(t)\right] = \gamma(s,t)\mathbf{S}_0(s,t),$$

where $\mathbf{S}_0(s,t) = \mathbf{C}[\mathbf{X}(s)^T\mathbf{X}(s)]^{-1}\mathbf{X}(s)^T\mathbf{X}(t)[\mathbf{X}(t)^T\mathbf{X}(t)]^{-1}\mathbf{C}^T$ is also time dependent. Set

$$\mathbf{S}(t) = \mathbf{S}_0(t,t) = \mathbf{C}\left[\mathbf{X}(t)^T\mathbf{X}(t)\right]^{-1}\mathbf{C}^T. \tag{6.76}$$

As $\mathbf{S}(t)$ is a full rank matrix of size $q \times q$, the pivotal test function for the GLHT problem (6.45) can be specified as

$$\mathbf{z}(t) = \mathbf{S}(t)^{-1/2}\left[\mathbf{C}\hat{\boldsymbol{\beta}}(t) - \mathbf{c}(t)\right]. \tag{6.77}$$

Simple calculation gives the vector of mean functions and the matrix of covariance functions of $\mathbf{z}(t)$ as

$$\begin{aligned} \boldsymbol{\eta}_z(t) &= \mathbf{S}(t)^{-1/2}\left[\mathbf{C}\boldsymbol{\beta}(t) - \mathbf{c}(t)\right], \\ \boldsymbol{\Gamma}_z(s,t) &= \gamma(s,t)\mathbf{S}(s)^{-1/2}\mathbf{S}_0(s,t)\mathbf{S}(t)^{-1/2}. \end{aligned}$$

Notice that for defining various tests for the GLHT problem (6.45), the pointwise sum of squared errors, $\mathrm{SSE}_n(t)$, is still the one defined in (6.61) while the pointwise sum of squared errors due to hypothesis, $\mathrm{SSH}_n(t)$, can be defined as the usual squared L^2-norm of the pivotal test function $\mathbf{z}(t)$ in (6.77):

$$\mathrm{SSH}_n(t) = \|\mathbf{z}(t)\|^2 = \left[\mathbf{C}\hat{\boldsymbol{\beta}}(t) - \mathbf{c}(t)\right]^T \mathbf{S}(t)^{-1}\left[\mathbf{C}\hat{\boldsymbol{\beta}}(t) - \mathbf{c}(t)\right]. \tag{6.78}$$

Remark 6.8 *The pivotal test function $\mathbf{z}(t)$ (6.77) is constructed in a way such that the squared L^2-norm $SSH_n(t)$ (6.78) of $\mathbf{z}(t)$ is invariant when $\mathbf{C}$ and $\mathbf{c}(t)$ are replaced respectively with $\tilde{\mathbf{C}}$ and $\tilde{\mathbf{c}}(t)$ as defined in (6.51). This property is important for the GLHT problem (6.45) under the FLM with time-dependent covariates (6.54).*

We are now ready to present various tests for the GLHT problem (6.45) under the FLMs with time-dependent covariates.

Pointwise Tests Again we can consider a pointwise F-test and a pointwise χ^2-test for the GLHT problem (6.45). The pointwise F-test is defined as $F_n(t) = \frac{\mathrm{SSH}_n(t)}{\mathrm{SSE}_n(t)}$. When the functional data are Gaussian, under the null hypothesis in (6.45), we have $F_n(t) \sim F_{q,n-p-1}, t \in \mathcal{T}$. The pointwise F-test is conducted accordingly. When the functional data are not Gaussian, under some mild conditions, as $n \to \infty$, asymptotically we have $F_n(t) \xrightarrow{d} \chi_q^2/q,\ \ t \in \mathcal{T}$. Then we can conduct the pointwise χ^2-test accordingly.

L^2-Norm-Based Test As before, the L^2-norm-based test for (6.45) uses the test statistic

$$T_n^0 = \int_{\mathcal{T}} \mathrm{SSH}_n(t)dt.$$

Again, it is rather challenging to derive the null distribution of T_n^0. To overcome this difficulty, we evaluate the covariate and response functions (6.53) at the equally spaced design time points (6.62) so that when M is large, we have

$$T_n^0 \approx \frac{v(\mathcal{T})}{M}\sum_{j=1}^M \mathrm{SSH}_n(t_j) = \frac{v(\mathcal{T})}{M}T_n,$$

where again $v(\mathcal{T})$ denotes the volume of $\mathcal{T}$, and

$$T_n = \sum_{j=1}^{M} \mathrm{SSH}_n(t_j) = \sum_{j=1}^{M} [\mathbf{C}\hat{\boldsymbol{\beta}}(t_j) - \mathbf{c}(t_j)]^T \mathbf{S}(t_j)^{-1} [\mathbf{C}\hat{\boldsymbol{\beta}}(t_j) - \mathbf{c}(t_j)]. \quad (6.79)$$

Under the null hypothesis in (6.45), we have $\mathbf{C}\boldsymbol{\beta}(t) = \mathbf{c}(t)$ so that

$$\mathbf{C}\hat{\boldsymbol{\beta}}(t) - \mathbf{c}(t) = \mathbf{C}\left[\hat{\boldsymbol{\beta}}(t) - \boldsymbol{\beta}(t)\right] = \mathbf{C}\left[\mathbf{X}(t)^T\mathbf{X}(t)\right]^{-1}\mathbf{X}(t)^T\mathbf{v}(t).$$

It follows that under the null hypothesis in (6.45), we have

$$T_n = \sum_{j=1}^{M} \mathbf{v}(t_j)^T \mathbf{H}(t_j)\mathbf{v}(t_j) = \mathbf{v}^T\mathbf{H}\mathbf{v}, \quad (6.80)$$

where $\mathbf{H}(t) = \mathbf{X}(t)[\mathbf{X}(t)^T\mathbf{X}(t)]^{-1}\mathbf{C}^T\mathbf{S}(t)^{-1}\mathbf{C}[\mathbf{X}(t)^T\mathbf{X}(t)]^{-1}\mathbf{X}(t)^T$, $\mathbf{v} = [\mathbf{v}(t_1)^T, \cdots, \mathbf{v}(t_M)^T]^T$ and $\mathbf{H} = \mathrm{diag}[\mathbf{H}(t_1), \mathbf{H}(t_2), \cdots, \mathbf{H}(t_M)]$. Under the Gaussian assumption, it is easy to see that

$$\mathbf{v} \sim N_{nM}(\mathbf{0}, \Gamma_M \otimes \mathbf{I}_n), \quad (6.81)$$

where again $\mathbf{\Gamma}_M$ is an $M \times M$ matrix whose (i,j)th entry is $\gamma(t_i, t_j)$ for $i, j = 1, 2, \cdots, M$. As $M \to \infty$, it is easy to see that $\frac{v(\mathcal{T})}{M} T_n \to T_n^0$. Therefore, for large M, we can use T_n as the test statistic of the L^2-norm-based test for (6.45). Notice that Remark 6.6 is still applicable for the L^2-norm-based test for the GLHT problem (6.45) and Theorem 6.11 is applicable for the current test statistic T_n with the matrix $\mathbf{H}$ defined above. Therefore, the null distribution of T_n can be approximated using the method described in the previous subsection for comparing two nested models using the L^2-norm-based test so that the L^2-norm-based test for the GLHT problem (6.45) can be conducted accordingly.

F-Type Test As before, the F-type test for (6.45) uses the test statistic

$$F_n^0 = \frac{\int_{\mathcal{T}} \mathrm{SSH}_n(t)dt/q}{\int_{\mathcal{T}} \mathrm{SSE}_n(t)dt/(n-p-1)}.$$

Again, it is difficult to find the null distribution of F_n^0. Similar to the treatment of the L^2-norm-based test described earlier, we can evaluate the covariate and response functions (6.53) at the equally spaced design time points (6.62). When M is large, we have

$$F_n^0 \approx \frac{\sum_{j=1}^{M} \mathrm{SSH}_n(t_j)/q}{\sum_{j=1}^{M} \mathrm{SSE}_n(t_j)/(n-p-1)} \equiv F_n.$$

Again, the distribution of F_n cannot be approximated using the two-cumulant matched F-approximation method described in Section 4.4 of Chapter 4 as the numerator and the denominator of F_n are not independent. To overcome this difficulty, let $\hat{F}_n$ denote the value of F_n computed based on the functional data (6.53) and we write

$$P(F_n \geq \hat{F}_n) = P\left(\frac{\sum_{j=1}^M \text{SSH}_n(t_j)/q}{\sum_{j=1}^M \text{SSE}_n(t_j)/(n-p-1)} \geq \hat{F}_n \right) = P(S_n \geq 0),$$

where $S_n = \sum_{j=1}^M \text{SSH}_n(t_j) - \rho \sum_{j=1}^M \text{SSE}_n(t_j)$ and $\rho = \frac{q\hat{F}_n}{n-p-1}$. Under the null hypothesis in (6.45) and by (6.79) and (6.80), we have $\sum_{j=1}^M \text{SSH}_n(t_j) = \mathbf{v}^T\mathbf{H}\mathbf{v}$, where $\mathbf{v}, \mathbf{H}$ are defined in (6.80). From the development of the F-type test for the two nested model comparison (6.60), we already have $\sum_{j=1}^M \text{SSE}_n(t_j) = \mathbf{v}^T\mathbf{G}\mathbf{v}$ where $\mathbf{G} = \text{diag}(\mathbf{G}(t_1), \cdots, \mathbf{G}(t_M))$ and $\mathbf{G}(t) = \mathbf{I}_n - \mathbf{P}_{\mathbf{X}(t)}$, as defined before. It follows that

$$S_n = \mathbf{v}^T(\mathbf{H} - \rho\mathbf{G})\mathbf{v},$$

where $\mathbf{v}$ has the normal distribution (6.81). Theorem 6.12 holds for S_n defined here for the GLHT problem (6.45) based on the FLM with time-dependent covariates (6.54) with the matrix $\mathbf{H} - \rho\mathbf{G}$ defined here. The method described there for approximating the null distribution of the F-type test for the two nested FLM comparison (6.60) can also be used to approximate the null distribution of the S_n defined here. The F-type test is then conducted accordingly.

Bootstrap Tests When the functional data are not Gaussian, then nonparametric bootstrap tests should be conducted. Let $v_i^*(t), i = 1, 2, \cdots, n$, be the bootstrap sample randomly generated from the estimated subject-effect functions $\hat{v}_i(t), i = 1, 2, \cdots, n$ as defined in (6.58). Set $\mathbf{y}^*(t) = \mathbf{X}(t)\hat{\boldsymbol{\beta}}(t) + \mathbf{v}^*(t)$, where $\mathbf{v}^*(t) = [v_1^*(t), \cdots, v_n^*(t)]^T$. Then we can compute $\hat{\boldsymbol{\beta}}^*(t) = [\mathbf{X}(t)^T\mathbf{X}(t)]^{-1}\mathbf{X}(t)^T\mathbf{y}^*(t)$ and $\text{SSH}_n^*(t) = \left[\hat{\boldsymbol{\beta}}^*(t) - \hat{\boldsymbol{\beta}}(t)\right]^T \mathbf{C}^T\mathbf{S}(t)^{-1}\mathbf{C}\left[\hat{\boldsymbol{\beta}}^*(t) - \hat{\boldsymbol{\beta}}(t)\right]$ and $\text{SSE}_n^*(t) = \mathbf{y}^*(t)^T(\mathbf{I}_n - \mathbf{P}_{\mathbf{X}(t)})\mathbf{y}^*(t)$, where $\mathbf{S}(t)$ is the one as defined in (6.76). The bootstrap test statistic $F_n^*(t)$, the L^2-norm-based bootstrap test statistic, or the F-type bootstrap test statistic can then be computed as

$$F_n^*(t) = \frac{\text{SSH}_n^*(t)/q}{\text{SSE}_n^*(t)/(n-p-1)}, \quad T_n^* = \int_{\mathcal{T}} \text{SSH}_n^*(t)dt, \text{ or}$$
$$F_n^* = \frac{\int_{\mathcal{T}} \text{SSH}_n^*(t)dt/q}{\int_{\mathcal{T}} \text{SSE}_n^*(t)dt/(n-p-1)}.$$

Repeat the above bootstrapping process a large number of times, calculate its $100(1-\alpha)$-percentile, and then conduct the pointwise test, the L^2-norm-based test, or the F-type test accordingly.

6.4 Technical Proofs

In this section, we outline the proofs of the main results described in this chapter.

Lemma 6.1 *If* $\mathbf{y} \sim N_p(\mathbf{0}, \mathbf{\Sigma})$ *and* $\mathbf{A} : p \times p$ *is a symmetric matrix, then*

$$\mathbf{y}^T\mathbf{A}\mathbf{y} \stackrel{d}{=} \sum_{r=1}^{p} \lambda_r A_r, \; A_r \sim \chi_1^2,$$

where $\lambda_1, r = 1, 2, \cdots, p$ *are the eigenvalues of* $\mathbf{W} = \mathbf{\Sigma}^{1/2}\mathbf{A}\mathbf{\Sigma}^{1/2}$.

Proof of Lemma 6.1 The proof of the lemma can be found in Box (1954a,b). It does no harm to present a simple proof here. Under the assumptions, we have $\mathbf{y} \stackrel{d}{=} \mathbf{\Sigma}^{1/2}\mathbf{z}$, where $\mathbf{z} \sim N_p(\mathbf{0}, \mathbf{I}_p)$. $\mathbf{W} = \mathbf{\Sigma}^{1/2}\mathbf{A}\mathbf{\Sigma}^{1/2}$ is a nonnegative matrix and it has the singular value decomposition $\mathbf{W} = \mathbf{U}\text{diag}(\lambda_1, \lambda_2, \cdots, \lambda_p)\mathbf{U}^T$, where $\lambda_1, \cdots, \lambda_p$ are the eigenvalues of $\mathbf{W}$, and the columns of $\mathbf{U}$ are the orthonormal eigenvectors of $\mathbf{W}$. It follows that $\mathbf{U}^T\mathbf{z} \stackrel{d}{=} \mathbf{z} \sim N_p(\mathbf{0}, \mathbf{I}_p)$ and hence

$$\begin{aligned} \mathbf{y}^T\mathbf{A}\mathbf{y} &= \mathbf{z}^T\mathbf{U}\text{diag}(\lambda_1, \cdots, \lambda_p)\mathbf{U}^T\mathbf{z} \\ &\stackrel{d}{=} \mathbf{z}^T\text{diag}(\lambda_1, \cdots, \lambda_p)\mathbf{z} \\ &\stackrel{d}{=} \textstyle\sum_{r=1}^{p} \lambda_r A_r, \end{aligned}$$

where $A_r = z_r^2 \sim \chi_1^2, r = 1, 2, \cdots, p$ with z_r being the rth entry of $\mathbf{z}$. The lemma is proved.

Proof of Theorem 6.1 Notice that under the Gaussian assumption LMA2, we have $\mathbf{v}(t) \sim \text{GP}_n(0, \gamma\mathbf{I}_n)$. It follows that $\hat{\boldsymbol{\beta}}(t) = \boldsymbol{\beta}(t) + (\mathbf{X}^T\mathbf{X})^{-1}\mathbf{X}^T\mathbf{v}(t) \sim \text{GP}_{p+1}(\boldsymbol{\beta}, \mathbf{\Gamma}_{\hat{\boldsymbol{\beta}}})$, where

$$\mathbf{\Gamma}_{\hat{\boldsymbol{\beta}}}(s,t) = \text{Cov}(\hat{\boldsymbol{\beta}}(s), \hat{\boldsymbol{\beta}}(t)) = \gamma(s,t)(\mathbf{X}^T\mathbf{X})^{-1},$$

as desired. By (6.9), we have

$$(n-p-1)\hat{\gamma}(s,t) = \hat{\mathbf{v}}(s)^T\hat{\mathbf{v}}(t) = \mathbf{v}(s)^T(\mathbf{I}_n - \mathbf{P}_{\mathbf{X}})\mathbf{v}(t),$$

where $\mathbf{v}(t) \sim \text{GP}_n(\mathbf{0}, \gamma\mathbf{I}_n)$ and $\mathbf{I}_n - \mathbf{P}_{\mathbf{X}}$ is an idempotent matrix of rank $n-p-1$. By Theorem 4.8 of Chapter 4, we have $(n-p-1)\hat{\gamma}(s,t) \sim \text{WP}(n-p-1, \gamma)$. The theorem is proved.

Proof of Theorem 6.2 Notice that $\hat{\boldsymbol{\beta}}(t) = (\mathbf{X}^T\mathbf{X})^{-1}\mathbf{X}^T\mathbf{y}(t) = \boldsymbol{\beta}(t) + (\mathbf{X}^T\mathbf{X})^{-1}\mathbf{X}^T\mathbf{v}(t)$. Under Assumption LMA3, we have $(\mathbf{X}^T\mathbf{X})^{-1} = n^{-1}\mathbf{\Omega}^{-1} +$

$o_p(1)$ so that $\hat{\boldsymbol{\beta}}(t) - \boldsymbol{\beta}(t) = n^{-1}\sum_{i=1}^n \boldsymbol{\Omega}^{-1}\mathbf{x}_i\mathbf{v}_i(t) + o_p(1)$. It follows that for any constant vector $\mathbf{a} \in \mathcal{R}^{p+1}$, we have

$$\sqrt{n}\mathbf{a}^T\left[\hat{\boldsymbol{\beta}}(t) - \boldsymbol{\beta}(t)\right] = n^{-1/2}\sum_{i=1}^n z_i(t) + o_p(1),$$

where $z_i(t) = \mathbf{a}^T\boldsymbol{\Omega}^{-1}\mathbf{x}_i\mathbf{v}_i(t)$. Under Assumptions LMA1, LMA3, and LMA4, we have $\mathrm{E}[z_i(t)] = 0$ and

$$\mathrm{E}\|z_1\|^2 = \mathrm{E}\int_{\mathcal{T}} z_1^2(t)dt = \mathbf{a}^T\boldsymbol{\Omega}^{-1}\mathbf{a}\mathrm{tr}(\gamma) < \infty.$$

In addition, we have

$$\mathrm{Cov}(z_1(s), z_1(t)) = \mathrm{E}\left[\mathbf{a}^T\boldsymbol{\Omega}^{-1}\mathbf{x}_1\mathbf{v}_1(s)\mathbf{a}^T\boldsymbol{\Omega}^{-1}\mathbf{x}_1\mathbf{v}_1(t)\right] = \mathbf{a}^T\boldsymbol{\Omega}^{-1}\mathbf{a}\gamma(s,t).$$

By the central limit theorem for i.i.d. stochastic processes, Theorem 4.12 of Chapter 4, as $n \to \infty$, we have

$$\sqrt{n}\mathbf{a}^T\left[\hat{\boldsymbol{\beta}}(t) - \boldsymbol{\beta}(t)\right] \xrightarrow{d} \mathrm{GP}(0, \mathbf{a}^T\boldsymbol{\Omega}^{-1}\mathbf{a}\gamma).$$

The theorem then follows immediately.

Proof of Theorem 6.3 Notice that

$$\begin{aligned}
\hat{\gamma}(s,t) &= (n-p-1)^{-1}\mathbf{y}(s)^T(\mathbf{I}_n - \mathbf{P}_{\mathbf{X}})\mathbf{y}(t) \\
&= (n-p-1)^{-1}\mathbf{v}(s)^T(\mathbf{I}_n - \mathbf{P}_{\mathbf{X}})\mathbf{v}(t) \\
&= (n-p-1)^{-1}\textstyle\sum_{i=1}^n v_i(s)v_i(t) - (n-p-1)^{-1}\mathbf{v}(s)^T\mathbf{P}_{\mathbf{X}}\mathbf{v}(t) \\
&= (n-p-1)^{-1}\textstyle\sum_{i=1}^n z_i(s,t) - (n-p-1)^{-1}w(s,t),
\end{aligned}$$

where $w(s,t) = \mathbf{v}(s)^T\mathbf{P}_{\mathbf{X}}\mathbf{v}(t)$, and $z_i(s,t) = v_i(s)v_i(t), i = 1, 2, \cdots, n$ are i.i.d. with $\mathrm{E}(z_1(s,t)) = \gamma(s,t)$ and

$$\begin{aligned}
\varpi[(s_1,t_1),(s_2,t_2)] &= \mathrm{Cov}(z_1(s_1,t_1), z_1(s_2,t_2)) \\
&= \mathrm{E}\left[v_1(s_1)v_1(t_1)v_1(s_2)v_1(t_2)\right] - \gamma(s_1,t_1)\gamma(s_2,t_2).
\end{aligned}$$

We first show that as $n \to \infty$, we have

$$\sqrt{n}\left[n^{-1}\sum_{i=1}^n z_i(s,t) - \gamma(s,t)\right] \xrightarrow{d} \mathrm{GP}(0, \varpi).$$

This actually follows directly from the central limit theorem for i.i.d. stochastic processes, Theorem 4.12, as by Assumption LMA5, we have

$$\mathrm{E}\|z_1\|^2 = \mathrm{E}\int_{\mathcal{T}^2}[v_1(s)v_1(t)]^2dsdt = \mathrm{E}\left[\int_{\mathcal{T}} v_1^2(t)dt\right]^2 = \mathrm{E}\|v_1\|^4 < \infty.$$

It remains to show that $w(s,t) = O_{UP}(1)$, that is, $w(s,t)$ is bounded in probability uniformly. By the Cauchy-Schwarz inequality, we have $\mathrm{E}|w(s,t)| \le \sqrt{\mathrm{E}w(s,s)\mathrm{E}w(t,t)}$; and under Assumption LMA6, we have

$$\mathrm{E}w(t,t) = \mathrm{E}\mathbf{v}(t)^T\mathbf{P_X}\mathbf{v}(t) = \mathrm{tr}(\mathbf{P_X}\gamma(t,t)) = (p+1)\gamma(t,t) \le (p+1)\rho.$$

It follows that $\mathrm{E}|w(s,t)| \le (p+1)\rho < \infty$. Therefore, $w(s,t) = O_{UP}(1)$. The theorem is proved based on the above results.

Proof of Theorem 6.4 Under the given assumptions, we have

$$\mathrm{SSH}_n(t) = \mathbf{v}(t)^T(\mathbf{P_X} - \mathbf{P_{X_1}})\mathbf{v}(t), \quad \mathrm{SSE}_n(t) = \mathbf{v}(t)^T(\mathbf{I}_n - \mathbf{P_X})\mathbf{v}(t),$$

where $\mathbf{v}(t) \sim \mathrm{GP}_n(\mathbf{0}, \gamma\mathbf{I}_n)$. The theorem follows from Theorems 4.9 and 4.10 of Chapter 4 as $\mathbf{P_X} - \mathbf{P_{X_1}}$ and $\mathbf{I}_n - \mathbf{P_X}$ are idempotent matrices of rank q and $n-p-1$, respectively, and $[\mathbf{P_X} - \mathbf{P_{X_1}}][\mathbf{I}_n - \mathbf{P_X}] = \mathbf{0}$.

Proof of Theorem 6.5 Notice that $\mathrm{SSH}_n(t) = \mathbf{z}(t)^T\mathbf{z}(t)$, where $\mathbf{z}(t) = (\mathbf{P_X} - \mathbf{P_{X_1}})\mathbf{y}(t) = \mathbf{X}\hat{\boldsymbol{\beta}}(t) - \mathbf{X}_1\hat{\boldsymbol{\beta}}_1(t)$. Under the given assumptions, by Theorem 6.2, we have that both $\hat{\boldsymbol{\beta}}(t)$ and $\hat{\boldsymbol{\beta}}_1(t)$ are asymptotically Gaussian. Then given $\mathbf{X}$, we have that $\mathbf{z}(t)$ is also asymptotically Gaussian. In addition, under the null hypothesis, we have

$$\mathrm{E}[\mathbf{z}(t)] = \mathbf{0}, \quad \mathrm{Cov}(\mathbf{z}(s), \mathbf{z}(t)) = \mathrm{E}\mathbf{z}(s)\mathbf{z}(t)^T = \gamma(s,t)\left[\mathbf{P_X} - \mathbf{P_{X_1}}\right].$$

Thus, when n is sufficiently large, we can write $\mathbf{z}(t) \sim \mathrm{GP}_n\left(\mathbf{0}, \gamma[\mathbf{P_X} - \mathbf{P_{X_1}}]\right)$ approximately. Let the idempotent matrix $\mathbf{P_X} - \mathbf{P_{X_1}}$ have the singular value decomposition

$$\mathbf{P_X} - \mathbf{P_{X_1}} = \mathbf{U}\begin{pmatrix} \mathbf{I}_q & \mathbf{0} \\ \mathbf{0} & \mathbf{0} \end{pmatrix}\mathbf{U}^T,$$

where $\mathbf{U}$ is the orthonormal matrix whose columns are the eigenvectors of $\mathbf{P_X} - \mathbf{P_{X_1}}$. Set $\mathbf{w}(t) = \mathbf{U}^T\mathbf{z}(t) = [w_1(t), w_2(t), \cdots, w_n(t)]^T$. Then when n is sufficiently large, we can write

$$\mathbf{w}(t) \sim \mathrm{GP}_n\left[\mathbf{0}, \gamma\begin{pmatrix} \mathbf{I}_q & \mathbf{0} \\ \mathbf{0} & \mathbf{0} \end{pmatrix}\right] \text{ approximately.}$$

It follows that we have $\mathbf{w}_0(t) \equiv [w_1(t), \cdots, w_q(t)]^T \sim \mathrm{GP}_q(\mathbf{0}, \gamma\mathbf{I}_q)$ approximately when n is sufficiently large. Thus, as $n \to \infty$, we have

$$\int_{\mathcal{T}} \mathrm{SSH}_n(t)dt = \int_{\mathcal{T}} \mathbf{z}(t)^T\mathbf{z}(t)dt \xrightarrow{d} \int_{\mathcal{T}} \mathbf{w}_0(t)^T\mathbf{w}_0(t)dt.$$

The theorem follows from the continuous mapping theorem for random elements taking values in a Hilbert space (Billingsley, 1968, p. 34; Cuevas, Febrero, and Fraiman 2004) and Theorem 4.10 of Chapter 4.

Proof of Theorem 6.6 Under the given conditions, by Theorem 6.3, as $n \to \infty$, we have $\mathrm{E}[\hat{\gamma}(s,t) - \gamma(s,t)]^2 = \frac{\omega[(s,t),(s,t)]}{n}[1 + o(1)]$. By Assumptions LMA6 and LMA7, we have

$$|\omega[(s,t),(s,t)]| \le \mathrm{E}[v_1^2(s)v_1^2(t)] + \gamma^2(s,t) \le C + \rho, \quad \text{for all } (s,t) \in \mathcal{T}^2.$$

Therefore, as $n \to \infty$, we have $\hat{\gamma}(s,t) = \gamma(s,t) + O_{UP}(n^{-1/2}),\ (s,t) \in \mathcal{T}^2$. It follows that $\hat{\gamma}(s,t) \xrightarrow{p} \gamma(s,t)$ uniformly over $\mathcal{T}^2$. It follows that

$$\begin{aligned}
\lim_{n\to\infty} \mathrm{tr}(\hat{\gamma}) &= \int_{\mathcal{T}} \lim_{n\to\infty} \hat{\gamma}(t,t)dt = \int_{\mathcal{T}} \gamma(t,t)dt = \mathrm{tr}(\gamma), \\
\lim_{n\to\infty} \mathrm{tr}(\hat{\gamma}^{\otimes 2}) &= \int_{\mathcal{T}} \int_{\mathcal{T}} \lim_{n\to\infty} \hat{\gamma}^2(s,t)dsdt \\
&= \int_{\mathcal{T}} \int_{\mathcal{T}} \gamma^2(s,t)dsdt = \mathrm{tr}(\gamma^{\otimes 2}),
\end{aligned}$$

in probability. It follows from (6.26) and (6.29) that as $n \to \infty$, $\hat{\beta} \xrightarrow{p} \beta$ and $\hat{\kappa} \xrightarrow{p} \kappa$. The theorem is proved.

Proof of Theorem 6.7 The proof is obvious.

Proof of Theorem 6.8 Under the given assumptions and by Theorem 6.2, we have $\hat{\boldsymbol{\beta}}(t)$ is asymptotically a $(p+1)$-dimensional Gaussian process. It follows that $\mathbf{z}(t)$ is also asymptotically a Gaussian process. The theorem is proved by noting that $\mathrm{E}\mathbf{z}(t) = \mathbf{0}$ and $\mathrm{Cov}\,[\mathbf{z}(s), \mathbf{z}(t)] = \gamma(s,t)\mathbf{I}_q$.

Proof of Theorem 6.9 Under the given assumptions and by Theorem 6.7, we have

$$\int_{\mathcal{T}} \mathrm{SSH}_n(t)dt = \int_{\mathcal{T}} \mathbf{z}(t)^T\mathbf{z}(t)dt, \quad \text{where } \mathbf{z}(t) \sim \mathrm{GP}_q(\mathbf{0}, \gamma\mathbf{I}_q).$$

The theorem follows from Theorem 4.10 of Chapter 4 immediately.

Proof of Theorem 6.10 Under the given assumptions and by Theorem 6.8 and the continuous mapping theorem for random elements taking values in a Hilbert space (Billingsley, 1968, p. 34; Cuevas, Febrero, and Fraiman 2004), we have

$$\int_{\mathcal{T}} \mathrm{SSH}_n(t)dt \xrightarrow{d} \int_{\mathcal{T}} \mathbf{z}(t)^T\mathbf{z}(t)dt, \quad \text{where } \mathbf{z}(t) \sim \mathrm{GP}_q(\mathbf{0}, \gamma\mathbf{I}_q).$$

The theorem follows immediately from Theorem 4.10 of Chapter 4.

Proof of Theorem 6.11 Under the null hypothesis in (6.60), we have $\mathbf{y}(t) = \mathbf{X}_1(t)\boldsymbol{\beta}_1(t) + \mathbf{v}(t)$ so that we can further express T_n as

$$T_n = \sum_{j=1}^{M} \mathbf{v}(t_j)^T\mathbf{H}(t_j)\mathbf{v}(t_j) = \mathbf{v}^T\mathbf{H}\mathbf{v},$$

where $\mathbf{v} = [\mathbf{v}(t_1)^T, \mathbf{v}(t_2)^T, \cdots, \mathbf{v}(t_M)^T]^T$. Under Assumptions LMB1 and LMB2, we have $\mathbf{v} \sim N_{nM}(\mathbf{0}, \mathbf{\Gamma}_M \otimes \mathbf{I}_n)$. The theorem follows immediately from Lemma 6.1.

Proof of Theorem 6.12 The proof is along the same lines as those for the proof of Theorem 6.11 after replacing the matrix $\mathbf{H}$ in the proof of Theorem 6.11 with the matrix $\mathbf{H} - \rho\mathbf{G}$.

6.5 Concluding Remarks and Bibliographical Notes

In this chapter, we discuss various tests for linear models with functional responses. The covariates can be time independent or time dependent. We reviewed the pointwise, L^2-norm-based, F-type, and bootstrap tests. For FLMs with time independent covariates, the approximate null distributions of the L^2-norm-based and F-type tests are simple. They mainly depend on the underlying covariance function. For FLMs with time-dependent covariates, the approximate null distributions of the L^2-norm-based and F-type tests are much more involved. That is why we just studied the L^2-norm-based and F-type tests under the Gaussian assumption. Further study in this direction is interesting and warranted.

The pointwise F-test for FLMs was discussed in Ramsay and Silverman (2005, Section 13.2.2, Chapter 13). The L^2-norm-based test for comparing two nested FLMs with time independent covariates was first proposed by Faraway (1997), who approximated the null distribution of the L^2-norm-based test by a bootstrap approach. Zhang and Chen (2007) extended the L^2-norm-based test for the GLHT problem (6.45). They proposed several approaches for approximating the associated null distributions. Some of these techniques are described in Section 4.3 of Chapter 4. The F-type test for comparing two nested FLMs was proposed by Shen and Faraway (2004). They approximated the associated null distribution by the two-cumulant matched F-approximation method as described in Section 4.4 of Chapter 4. An application of the F-type test in longitudinal data analysis is given in Yang, Shen, Xu, and Shoptaw (2007). Zhang (2011a) further extended and studied this F-type test for the GLHT problem (6.45). He showed that the F-type test is root-n consistent. Recently, Xu, Shen, Yang, and Shoptaw (2011) proposed a so-called quasi F-test for comparing two nested FLMs with time-dependent covariates. This test is essentially the same as the F-type test described in Section 6.3. Some further investigation about FLMs with time-dependent covariates can be founded in Zhang (2013a).

The FLMs considered in this chapter belong to a type of FLM that establishes a regression relationship between a functional response and one or several scalar covariates. In the literature, other types of FLMs, generalized FLMs, and nonparametric functional regression models, among others, have also caught much attention for the past decade.

The FLMs with scalar responses (Ramsay and Silverman 2005, Chapter 15) are usually referred to FLMs with a scalar response and one functional predictor. A coefficient function is introduced to measure the effect of the functional predictor on the scalar response. Various methods for estimating this coefficient function have been proposed, including the functional PCA method (Cardot, Ferraty, and Sarda 1999), the B-spline method (Cardot, Ferraty, and Sarda 2003), the smoothing spline method (Crambes, Kneip, and Sarda 2009), and the wavelet-based LASSO method (Zhao, Ogden, and Reiss 2012), among others. James, Wang, and Zhu (2009) proposed a technique to estimate the coefficient function flexibly so that it is exactly zero over some region. This allows us to obtain interpretable, flexible, and accurate estimators for the coefficient function. Some interesting theoretical work has also been done. Cai and Hall (2006) and Li and Hsing (2007) investigated the rate of convergence of the estimated coefficient function. Cardot, Mas, and Sarda (2007) derived a central limit theory for some linear functions of the estimated coefficient function under mild conditions. It is often of interest to check if the coefficient function is actually zero everywhere. This problem is often referred to as the no-effect test. This has been done by a number of authors, including Cardot, Ferraty, Mas, and Sarda (2003) and Cardot, Goia, and Sarda (2004), among others. Most recently, Swihart, Goldsmith, and Crainiceanu (2012) provided a transparent, robust, and computationally feasible statistical approach for testing the necessity of functional effects against standard linear models. Horvath, Kokoszka, and Reimherr (2009) investigated a two-sample testing procedure.

Much attention has also foucused on the FLMs with a functional response and a functional predictor (Ramsay and Silverman 2005, Chapter 16). A coefficient function is used to model the effect of the functional predictor on the functional response. Various methods for estimating this coefficient function have also been proposed, for example, by Cuevas, Febrero, and Fraiman (2002), Chiou, Müller, and Wang (2004), Yao, Müller, and Wang (2005), Aguilera, Ocana, and Valderrama (2008), and Antoch, Prchal, Rosa, and Sarda (2008), among others. For such FLMs, Kokoszka, Maslova, Sojka, and Zhu (2008) proposed a simple no-effect test and Febrero, Galeano, and Gonzalez-Manteiga (2008) studied how to identify influential curves.

When FLMs are not appropriate for establishing the relationship between a scalar/functional response and a scalar/functional predictor, generalized FLMs are natural alternatives, extending the classical generalized linear models (Nelder and Wedderburn 1972, McCullagh and Nelder 1989). James (2002) proposed a technique to deal with linear, logistic, and censored regression models with functional predictors. He particularly focused on dealing with the problem when only fragments of each curve have been observed. Other work includes Müller and Stadtmuller (2005), Zhu and Cox (2009), Goldsmith, Bobb, Crainiceanu, Caffo, and Reich (2011), and Aguilera-Morillo, Aguilera, Escabias, and Valderrama (2012), among others. A quadratic regression model for functional data is proposed in Yao and Müller (2010). Quantile regression with functional covariates are investigated by Cardot, Crambes, and Sarda (2005), and Chen and Müller (2012), among others. Fast fitting methods for generalized FLMs were proposed by Goldsmith, Bobb, Crainiceanu, Caffo, and Reich (2011).

When parametric models are not appropriate for establishing the relationship between a scalar/functional response and a functional predictor, some nonparametric regression models may be used. Ferraty and Vieu (2002), Ferraty, Goia, and Vieu (2007), and Ferraty, Van Keilegom, and Vieu (2012) studied a kernel method. Benhenni, Ferraty, Rachdi, and Vieu (2007) and Amparo and Aurea (2009) proposed local linear methods. Burba, Ferraty, and Vieu (2009) investigated a k-nearest neighbor method.

As pointed out by Zhao, Marron, and Wells (2004), longitudinal data can be viewed as a type of functional data that are sparse. This functional viewpoint is not typical for most analysts of longitudinal data, but provides a route for powerful new insights. An interesting comparison between functional data and longitudinal data can be found in Rice (2004). Functional PCA for longitudinal data is investigated by James, Hastie, and Sugar (2000) and Kayano and Konishi (2010), among others. Various FLMs for longitudinal data analysis have been proposed. Yao, Müller, and Wang (2005a,b) proposed nonparametric methods for FLMs with sparse longitudinal data. Sentürk and Müller (2010), Müller and Yang (2010), and Wu, Fan, and Müller (2010) investigated functional varying coefficient models for longitudinal data. Chiou and Li (2007) proposed a functional k-center clustering method for longitudinal data. A review on FDA tools in longitudinal data analysis can be found in Müller (2005). An application of the F-type test in longitudinal data analysis can be found in Yang, Shen, Xu, and Shoptaw (2007).

FLMs have also been extended to other research areas. Müller, Sen, and Stadtmuller (2011) discussed some interesting applications of FLMs in finance while Malfait and Ramsay (2003) discussed a useful historical FLM. Mas and Pumo (2009) considered an FLM, taking into account the first-order derivative of the data.

6.6 Exercises

1. For the ergonomics data, the global multiple correlation coefficient of the quadratic FLM (6.1) is $R_n^2 = 85.22\%$ as given in (6.19). Test (6.39) using the approximate null distribution of the F-type test as given in (6.37) or in (6.38).
2. Under the FLM with time independent covariates (6.4), for some $\mathbf{C}$ and $\mathbf{c}(t), t \in \mathcal{T}$, the GLHT problem (6.45) is equivalent to the single coefficient function testing problem (6.41).
   (a) Identify $\mathbf{C}$ and $\mathbf{c}(t), t \in \mathcal{T}$.
   (b) Without using the pointwise relationship (6.42), show that the test statistic of the L^2-norm-based test, T_n, for (6.41) can be expressed as (6.43).

(c) Explain why one can give the significance of all the coefficient functions simultaneously, provided that the full FLM (6.4) is fitted.

3. For the ergonomics data, let the full FLM be the quadratic FLM (6.1). Of interest is to test the following null hypothesis;

$$H_0 : \beta_6(t) \equiv \beta_8(t) \equiv 0, t \in \mathcal{T}. \tag{6.82}$$

By some simple calculation, we have $\int_{\mathcal{T}} \text{SSH}_n(t)dt = 61.926$ and $\int_{\mathcal{T}} \text{SSE}_n(t)dt = 703.26$.

(a) Conduct an L^2-norm-based test for (6.82) using the naive method.
(b) Conduct an F-type test for (6.82) using the bias-reduced method.
(c) Are the conclusions by the L^2-norm-based test and the F-type test consistent?

4. Table 6.1 presents the significance of all the individual coefficient functions of the quadratic FLM (6.1) for the ergonomics data. The approximate null distributions of the L^2-norm-based test and the F-type tests were obtained using the naive method. Can you construct a similar table using the bias-reduced method based on the information presented in this chapter?

5. Under the FLM with time-dependent covariates (6.55), for some $\mathbf{C}$ and $\mathbf{c}(t), t \in \mathcal{T}$, the GLHT problem (6.45) is equivalent to the single coefficient function testing problem (6.41).

(a) Identify $\mathbf{C}$ and $\mathbf{c}(t), t \in \mathcal{T}$.
(b) Show that the test statistic of the L^2-norm-based test, T_n, for (6.41) can be expressed as

$$T_n = \sum_{l=1}^{M} \frac{\hat{\beta}_j^2(t_l)}{[\mathbf{X}(t_l)^T\mathbf{X}(t_l)]_{jj}^{-1}},$$

where $[\mathbf{X}(t_l)^T\mathbf{X}(t_l)]_{jj}^{-1}$ denotes the jth diagonal entry of $[\mathbf{X}(t_l)^T\mathbf{X}(t_l)]^{-1}$.

(c) Show that for the single coefficient function testing problem (6.41), the matrix $\mathbf{H}(t)$ defined in (6.80) can be expressed as

$$H(t) = \left\{\mathbf{G}(t)\mathbf{G}(t)^T\right\} / [\mathbf{X}(t)^T\mathbf{X}(t)]_{jj}^{-1},$$

where $\mathbf{G}(t) = \mathbf{X}(t)[\mathbf{X}(t)^T\mathbf{X}(t)]^{-1}\mathbf{e}_{j+1,p+1}$ and as usual, $\mathbf{e}_{r,q}$ denotes the q-dimensional vector whose rth entry is 1 and others 0.

(d) Show that the approximate null distributions of T_n given in (6.71) are different for different $j = 0, 1, \cdots, p$ unless $\mathbf{X}(t) \equiv \mathbf{X}$, that is, $\mathbf{X}(t)$ is time independent.

Chapter 7

Ill-Conditioned Functional Linear Models

7.1 Introduction

In Chapter 6 we have studied the functional linear model (FLM) (6.4) under the assumption that the time-independent design matrix $\mathbf{X}$ is a full-rank matrix. In that chapter, under some mild regularity conditions, various pointwise, L^2-norm-based, F-type, and bootstrap tests for the FLM (6.4) were discussed. As in classical linear models, in many situations, however, the design matrix $\mathbf{X}$ in an FLM may be ill-conditioned, that is, it is non-full-rank. The associated FLM is called an ill-conditioned or non-full-rank FLM, which can be defined as follows:

$$\mathbf{y}(t) = \mathbf{X}\boldsymbol{\beta}(t) + \mathbf{v}(t), \quad \mathbf{v}(t) \sim \mathrm{SP}_n(\mathbf{0}, \gamma\mathbf{I}_n), \; t \in \mathcal{T}, \tag{7.1}$$

where $\mathbf{y}(t) : n \times 1$ is a vector of response functions, $\mathbf{X} : n \times (p+1)$ is an ill-conditioned time-independent design matrix with rank $k < (p+1) < n$, $\mathbf{v}(t) : n \times 1$ is a vector of subject-effect functions, and $\boldsymbol{\beta}(t) : (p+1) \times 1$ is a vector of coefficient functions. When the first column of $\mathbf{X}$ is $\mathbf{1}_n$, an n-dimensional vector of ones, the first component of $\boldsymbol{\beta}(t)$ models the intercept function of the non-full-rank FLM (7.1). In this chapter, we aim to extend the methodologies developed for the full-rank FLM (6.4) in Chapter 6 for the above non-full-rank FLM (7.1).

Remark 7.1 *As the design matrix* $\mathbf{X}$ *is ill-conditioned, those methodologies developed for the full-rank FLM (6.4) in Chapter 6 are not directly applicable. In fact, unlike in Chapter 6, the vector of coefficient functions,* $\boldsymbol{\beta}(t)$*, of the non-full-rank FLM (7.1) is not estimable. That is, it does not have an unbiased estimator that is a linear function of* $\mathbf{y}(t)$.

Remark 7.2 *All the functional ANOVA models studied in Chapter 5 can be expressed in the form of the non-full-rank FLM (7.1) with the design matrix* $\mathbf{X}$ *being non-full-rank. As an illustrative example, we can re-express the one-way functional ANOVA model (5.26) of Chapter 5 in the form of (7.1).*

Example 7.1 *For easy reference, we rewrite the one-way functional ANOVA model (5.26) below:*

$$\begin{aligned} y_{ij}(t) = \eta(t) + \alpha_i(t) + v_{ij}(t),\ v_{ij}(t) \overset{i.i.d.}{\sim} SP(0,\gamma), \\ j = 1,2,\cdots,n_i; i = 1,2,\cdots,a, \end{aligned} \tag{7.2}$$

where $\eta(t)$ is the grand mean function, and $\alpha_i(t), i = 1,2,\cdots,a$ are the main-effect functions. Set

$$\begin{aligned} \mathbf{y}(t) &= [y_{11}(t),\cdots,y_{1n_1}(t),y_{21}(t),\cdots,y_{2n_2}(t),\cdots,y_{a1}(t),\cdots,y_{an_a}(t)]^T, \\ \mathbf{v}(t) &= [v_{11}(t),\cdots,v_{1n_1}(t),v_{21}(t),\cdots,v_{2n_2}(t),\cdots,v_{a1}(t),\cdots,v_{an_a}(t)]^T, \end{aligned}$$

$\boldsymbol{\beta}(t) = [\eta(t),\alpha_1(t),\cdots,\alpha_a(t)]^T$, and

$$\mathbf{X} = \begin{pmatrix} \mathbf{1}_{n_1} & \mathbf{1}_{n_1} & \mathbf{0} & \cdots & \mathbf{0} \\ \mathbf{1}_{n_2} & \mathbf{0} & \mathbf{1}_{n_1} & \cdots & \mathbf{0} \\ \vdots & \vdots & \vdots & \vdots & \vdots \\ \mathbf{1}_{n_a} & \mathbf{0} & \mathbf{0} & \dots & \mathbf{1}_{n_a} \end{pmatrix}, \tag{7.3}$$

where $\mathbf{0}$ denotes a zero vector of some length, and $\mathbf{1}_{n_i}$ a vector of ones of length n_i. Then the one-way functional ANOVA model (7.2) is expressed into the form of the non-full-rank FLM (7.1) with rank$(\mathbf{X}) = a < a+1$.

Therefore, this chapter also provides alternative methods for all the functional ANOVA models discussed in Chapter 5 in a unified manner, while in Chapter 5, the methodologies for one-way and two-way functional ANOVA models are discussed in separate sections.

Remark 7.3 *The functional ANOVA models discussed in Chapter 5 cannot accommodate many factors with more than two levels as it requires that each cell or combination of the factor levels has at least one functional observation. In practice, it is very expensive and time consuming. For example, suppose we have seven factors (which is not uncommon), each having three levels. Then the total number of cells is $3^7 = 2,187$ and hence the total number of runs in the experiment is at least $2,187$ when there is at least one run at each cell. This is really very expensive and time consuming if one wants to conduct all these runs of experiments.*

Remark 7.4 *In modern experimental designs, when several factors are involved, a fractional factorial design is often more favorable. In a fractional factorial design, each factor is allowed to have 2 levels, usually denoted as "+1" for high level and "–1" for low level, and it assigns experimental runs only at a fractional part of cells with each cell having one run. For example, a fractional factorial design 2^{7-2} involves seven factors, each factor having two levels, but only $25\% = 2^{-2}$ of the cells having one experimental run, with total thirty-two runs in the experiment. Therefore, the methodologies studied in Chapter 5 may not be directly applicable for functional data based on a fractional factorial design.*

$$\mathbf{X} = \begin{pmatrix}
1 & 1 & 0 & 1 & 0 & 1 & 0 & 1 & 0 & 1 & 0 & 0 & 1 & 0 & 1 \\
1 & 1 & 0 & 1 & 0 & 1 & 0 & 1 & 0 & 0 & 1 & 0 & 1 & 1 & 0 \\
1 & 1 & 0 & 1 & 0 & 1 & 0 & 0 & 1 & 1 & 0 & 1 & 0 & 1 & 0 \\
1 & 1 & 0 & 1 & 0 & 1 & 0 & 0 & 1 & 0 & 1 & 1 & 0 & 0 & 1 \\
1 & 1 & 0 & 1 & 0 & 0 & 1 & 1 & 0 & 1 & 0 & 1 & 0 & 0 & 1 \\
1 & 1 & 0 & 1 & 0 & 0 & 1 & 1 & 0 & 0 & 1 & 1 & 0 & 1 & 0 \\
1 & 1 & 0 & 1 & 0 & 0 & 1 & 0 & 1 & 1 & 0 & 0 & 1 & 1 & 0 \\
1 & 1 & 0 & 1 & 0 & 0 & 1 & 0 & 1 & 0 & 1 & 0 & 1 & 0 & 1 \\
1 & 1 & 0 & 0 & 1 & 1 & 0 & 1 & 0 & 1 & 0 & 1 & 0 & 1 & 0 \\
1 & 1 & 0 & 0 & 1 & 1 & 0 & 1 & 0 & 0 & 1 & 1 & 0 & 0 & 1 \\
1 & 1 & 0 & 0 & 1 & 1 & 0 & 0 & 1 & 1 & 0 & 0 & 1 & 0 & 1 \\
1 & 1 & 0 & 0 & 1 & 1 & 0 & 0 & 1 & 0 & 1 & 0 & 1 & 1 & 0 \\
1 & 1 & 0 & 0 & 1 & 0 & 1 & 1 & 0 & 1 & 0 & 0 & 1 & 1 & 0 \\
1 & 1 & 0 & 0 & 1 & 0 & 1 & 1 & 0 & 0 & 1 & 0 & 1 & 0 & 1 \\
1 & 1 & 0 & 0 & 1 & 0 & 1 & 0 & 1 & 1 & 0 & 1 & 0 & 0 & 1 \\
1 & 1 & 0 & 0 & 1 & 0 & 1 & 0 & 1 & 0 & 1 & 1 & 0 & 1 & 0 \\
1 & 0 & 1 & 1 & 0 & 1 & 0 & 1 & 0 & 1 & 0 & 1 & 0 & 1 & 0 \\
1 & 0 & 1 & 1 & 0 & 1 & 0 & 1 & 0 & 0 & 1 & 1 & 0 & 0 & 1 \\
1 & 0 & 1 & 1 & 0 & 1 & 0 & 0 & 1 & 1 & 0 & 0 & 1 & 0 & 1 \\
1 & 0 & 1 & 1 & 0 & 1 & 0 & 0 & 1 & 0 & 1 & 0 & 1 & 1 & 0 \\
1 & 0 & 1 & 1 & 0 & 0 & 1 & 1 & 0 & 1 & 0 & 0 & 1 & 1 & 0 \\
1 & 0 & 1 & 1 & 0 & 0 & 1 & 1 & 0 & 0 & 1 & 0 & 1 & 0 & 1 \\
1 & 0 & 1 & 1 & 0 & 0 & 1 & 0 & 1 & 1 & 0 & 1 & 0 & 0 & 1 \\
1 & 0 & 1 & 1 & 0 & 0 & 1 & 0 & 1 & 0 & 1 & 1 & 0 & 1 & 0 \\
1 & 0 & 1 & 0 & 1 & 1 & 0 & 1 & 0 & 1 & 0 & 0 & 1 & 0 & 1 \\
1 & 0 & 1 & 0 & 1 & 1 & 0 & 1 & 0 & 0 & 1 & 0 & 1 & 1 & 0 \\
1 & 0 & 1 & 0 & 1 & 1 & 0 & 0 & 1 & 1 & 0 & 1 & 0 & 1 & 0 \\
1 & 0 & 1 & 0 & 1 & 1 & 0 & 0 & 1 & 0 & 1 & 1 & 0 & 0 & 1 \\
1 & 0 & 1 & 0 & 1 & 0 & 1 & 1 & 0 & 1 & 0 & 1 & 0 & 0 & 1 \\
1 & 0 & 1 & 0 & 1 & 0 & 1 & 1 & 0 & 0 & 1 & 1 & 0 & 1 & 0 \\
1 & 0 & 1 & 0 & 1 & 0 & 1 & 0 & 1 & 1 & 0 & 0 & 1 & 1 & 0 \\
1 & 0 & 1 & 0 & 1 & 0 & 1 & 0 & 1 & 0 & 1 & 0 & 1 & 0 & 1 \\
1 & 0 & 1 & 0 & 1 & 0 & 1 & 0 & 1 & 0 & 1 & 0 & 1 & 0 & 1 \\
1 & 0 & 1 & 0 & 1 & 0 & 1 & 0 & 1 & 0 & 1 & 0 & 1 & 0 & 1 \\
1 & 0 & 1 & 0 & 1 & 0 & 1 & 0 & 1 & 0 & 1 & 0 & 1 & 0 & 1 \\
1 & 0 & 1 & 0 & 1 & 0 & 1 & 0 & 1 & 0 & 1 & 0 & 1 & 0 & 1
\end{pmatrix}. \tag{7.4}$$

Example 7.2 *Consider the audible noise data introduced in Section 1.2.5 of Chapter 1. There are seven factors, represented by A, B, C, D, E, F, and G, respectively, in which D is the noise factor while the others are control factors. Each factor has only two levels: low and high, denoted as "−1" and "+1" respectively. The associated study adopted a* 2_{IV}^{7-2} *design, supplemented by four additional replications at the high levels of all factors, resulting in an experiment with thirty-six runs. For each run, audible noise levels were mea-*

sured over a range of rotating speeds. The audible sound was recorded by the microphones located at several positions near the alternator. The response was a transformed pressure measurement, known as sound pressure level (SPL). For each response curve, forty-three measurements of sound pressure levels (in decibels) were recorded with rotating speeds ranging from $1,000$ *to* $2,500$ *revolutions per minute. Figure 1.8 in Chapter 1 displays the resulting thirty-six response curves. To model this audible noise data set using the non-full-rank FLM (7.1), we let*

$$\mathbf{y}(t) = \begin{pmatrix} y_1(t) \\ y_2(t) \\ y_3(t) \\ y_4(t) \\ y_5(t) \\ \vdots \\ y_{35}(t) \\ y_{36}(t) \end{pmatrix}, \quad \mathbf{v}(t) = \begin{pmatrix} v_1(t) \\ v_2(t) \\ v_3(t) \\ v_4(t) \\ v_5(t) \\ \vdots \\ v_{35}(t) \\ v_{36}(t) \end{pmatrix}, \quad \boldsymbol{\beta}(t) = \begin{pmatrix} \eta(t) \\ \alpha_{11}(t) \\ \alpha_{12}(t) \\ \alpha_{21}(t) \\ \alpha_{22}(t) \\ \vdots \\ \alpha_{71}(t) \\ \alpha_{72}(t) \end{pmatrix},$$

denote, respectively, the vectors of response functions, subject-effects functions, and coefficient functions. The vector of coefficient functions consists of the grand mean function $\eta(t)$ *and the main-effect functions*

$$\alpha_{ij}(t), j = 1, 2;\ i = 1, 2, \cdots, 7,$$

of the seven factors A, B, C, D, E, F, and G with each having two levels. The associated design matrix $\mathbf{X}$ *is given by (7.4), where except in the first column,* 0 *and* 1 *represent levels "*-1*" and "*$+1$*", respectively. Notice that* $\mathbf{X}$ *is of size* 36×15 *and we found that rank*$(\mathbf{X}) = 8 < 15 < 36$*. Therefore, the resulting FLM (7.1) is indeed non-full-rank or ill-conditioned. The methodologies developed in Chapters 5 and 6 cannot be directly applied to the analysis of this data set.*

In this chapter we study how to conduct statistical inferences for the non-full-rank FLM (7.1). The key is how to address the problems caused by the singularity of the non-full-rank design matrix $\mathbf{X}$. The main problem is that not all the linear functions of the vector of the coefficient function $\boldsymbol{\beta}(t)$ are estimable as indicated in Remark 7.1. The related problems include: (a) How to identify the estimable linear functions of $\boldsymbol{\beta}(t)$, (b) how to estimate these estimable linear functions, and (c) how to test these estimable linear functions. To solve these problems, in this chapter, we mainly discuss three methods, namely, the generalized inverse method, the reparameterization method, and the side-condition method. They are studied in Sections 7.2, 7.3, and 7.4, respectively. These three methods are essentially equivalent but are developed from three different views for conducting statistical inferences about the non-full-rank FLM (7.1). In fact, the generalized inverse method uses the generalized inverses of an ill-conditioned design matrix, the reparameterization method reparameterizes the non-full-rank FLM (7.1) into a full-rank FLM in the form of (6.4) that can be solved using the methodologies developed in Chapter 6, while the side-condition method solves the non-full-rank FLM (7.1) by adding some extra side conditions so that the solution to (7.1) is unique. Technical proofs of the main results are given in Section 7.5. Some concluding remarks and bibliographical notes are given in Section 7.6. Exercises for this chapter are listed in Section 7.7.

7.2 Generalized Inverse Method

7.2.1 Estimability of Regression Coefficient Functions

To solve the non-full-rank FLM (7.1), we use the ordinary pointwise least squares approach. The pointwise least squares approach aims to find $\boldsymbol{\beta}(t)$ to minimize the following integrated squared error:

$$Q(\boldsymbol{\beta}) = \int_{\mathcal{T}} \|\mathbf{y}(t) - \mathbf{X}\boldsymbol{\beta}(t)\|^2 dt. \tag{7.5}$$

This leads to solving the following normal equation system:

$$\mathbf{X}^T\mathbf{X}\boldsymbol{\beta}(t) = \mathbf{X}^T\mathbf{y}(t),\ t \in \mathcal{T}. \tag{7.6}$$

As $\mathbf{X} : n \times (p+1)$ has rank $k < (p+1) < n$, it has the following singular value decomposition (SVD):

$$\mathbf{X} = \mathbf{U}\begin{pmatrix} \mathbf{D}_k & \mathbf{0} \\ \mathbf{0} & \mathbf{0} \end{pmatrix}\mathbf{V}^T, \tag{7.7}$$

where $\mathbf{U} : n \times n$ is the orthonormal matrix with columns being the eigenvectors of $\mathbf{X}\mathbf{X}^T$, $\mathbf{V} : (p+1) \times (p+1)$ is the orthonormal matrix with columns being the eigenvectors of $\mathbf{X}^T\mathbf{X}$, and $\mathbf{D}_k$ is a real diagonal matrix with diagonal entries being the square roots of the non-zero eigenvalues of $\mathbf{X}\mathbf{X}^T$ and $\mathbf{X}^T\mathbf{X}$. In fact, we have

$$\mathbf{X}\mathbf{X}^T = \mathbf{U}\begin{pmatrix} \mathbf{D}_k^2 & \mathbf{0} \\ \mathbf{0} & \mathbf{0} \end{pmatrix}\mathbf{U}^T,\ \mathbf{X}^T\mathbf{X} = \mathbf{V}\begin{pmatrix} \mathbf{D}_k^2 & \mathbf{0} \\ \mathbf{0} & \mathbf{0} \end{pmatrix}\mathbf{V}^T. \tag{7.8}$$

To solve the normal equation system (7.6), we here introduce the concept of "generalized inverse of a real matrix." In mathematics, when a real matrix is non-full-rank, for example, the aforementioned $\mathbf{X}$, it is not invertible but we can define its generalized inverses. A real matrix $\mathbf{X}^-$ is called a generalized inverse of $\mathbf{X}$ if and only if it satisfies the following two equations:

$$\mathbf{X}\mathbf{X}^-\mathbf{X} = \mathbf{X},\ \mathbf{X}^-\mathbf{X}\mathbf{X}^- = \mathbf{X}^-. \tag{7.9}$$

Remark 7.5 *Based on the SVD (7.7), it is easy to show that* $\mathbf{X}^-$ *can be expressed as*

$$\mathbf{X}^- = \mathbf{V}\begin{pmatrix} \mathbf{D}_k^{-1} & \mathbf{D}_{12} \\ \mathbf{D}_{21} & \mathbf{D}_{22} \end{pmatrix}\mathbf{U}^T, \tag{7.10}$$

where $\mathbf{D}_{12}, \mathbf{D}_{21}$, and $\mathbf{D}_{22}$ can be any real matrices of proper sizes. It is easy to show that $\mathbf{X}^-$ defined in (7.10) satisfies (7.9). A real matrix has many generalized inverses unless it is a full-rank matrix, which has only one inverse.

Let $(\mathbf{X}^T\mathbf{X})^-$ denote any generalized inverse of $\mathbf{X}^T\mathbf{X}$. Then by (7.8) and (7.10), we have

$$(\mathbf{X}^T\mathbf{X})^- = \mathbf{V}\begin{pmatrix} \mathbf{D}_k^{-2} & \mathbf{S}_{12} \\ \mathbf{S}_{21} & \mathbf{S}_{22} \end{pmatrix}\mathbf{V}^T, \tag{7.11}$$

where the matrices $\mathbf{S}_{12}, \mathbf{S}_{21}$, and $\mathbf{S}_{22}$ are any real matrices of proper sizes. Set $\mathbf{P}_\mathbf{X} = \mathbf{X}(\mathbf{X}^T\mathbf{X})^-\mathbf{X}^T$. Then by (7.7) and (7.11), we have

$$\mathbf{P}_\mathbf{X} = \mathbf{X}(\mathbf{X}^T\mathbf{X})^-\mathbf{X}^T = \mathbf{U}\begin{pmatrix} \mathbf{I}_k & \mathbf{0} \\ \mathbf{0} & \mathbf{0} \end{pmatrix}\mathbf{U}^T. \tag{7.12}$$

This shows that $\mathbf{P}_\mathbf{X}$ is invariant to the choice of $(\mathbf{X}^T\mathbf{X})^-$ and it is an idempotent matrix of rank k. Simple algebra then shows that

$$\mathbf{P}_\mathbf{X}\mathbf{X} = \mathbf{X}, \;\; \mathbf{X}^T\mathbf{P}_\mathbf{X} = \mathbf{X}^T, \tag{7.13}$$

indicating that $\mathbf{P}_\mathbf{X}$ is a projection matrix of $\mathbf{X}$. It follows that

$$\mathbf{X}^T\mathbf{X}(\mathbf{X}^T\mathbf{X})^-\mathbf{X}^T\mathbf{y}(t) = \mathbf{X}^T\mathbf{P}_\mathbf{X}\mathbf{y}(t) = \mathbf{X}^T\mathbf{y}(t),$$

showing that $(\mathbf{X}^T\mathbf{X})^-\mathbf{X}^T\mathbf{y}(t)$ is a solution to the normal equation system (7.6). That is to say that the normal equation system (7.6) is consistent, that is, it has solutions. Denote any one of the solutions to (7.6) as

$$\hat{\boldsymbol{\beta}}(t) = (\mathbf{X}^T\mathbf{X})^-\mathbf{X}^T\mathbf{y}(t). \tag{7.14}$$

Let $\mathbf{C} : q \times (p+1)$ be a constant matrix. With respect to the non-full-rank FLM (7.1), a linear function of $\boldsymbol{\beta}(t)$, $\mathbf{C}\boldsymbol{\beta}(t)$, is said to be estimable if $\mathbf{C}\boldsymbol{\beta}(t)$ has an unbiased estimator that is a linear function of $\mathbf{y}(t)$. That is, when there is a constant matrix $\mathbf{A} : q \times n$ such that $\mathrm{E}[\mathbf{A}\mathbf{y}(t)] = \mathbf{C}\boldsymbol{\beta}(t)$, then $\mathbf{C}\boldsymbol{\beta}(t)$ is estimable. Notice that not all linear functions of $\boldsymbol{\beta}(t)$ are estimable. This shows that the non-full-rank FLM (7.1) is very different from the full-rank FLM (6.4) discussed in Chapter 6. In fact, $\boldsymbol{\beta}(t)$ itself is not estimable, as shown by the following theorem.

Theorem 7.1 *Under the non-full-rank FLM (7.1), we have (a) $\hat{\boldsymbol{\beta}}(t)$ is a linear function of $\mathbf{y}(t)$, (b) $\hat{\boldsymbol{\beta}}(t)$ is biased for $\boldsymbol{\beta}(t)$, and (c) $\boldsymbol{\beta}(t)$ is not estimable.*

The above theorem claims that under the non-full-rank FLM (7.1), there are no unbiased estimators for $\boldsymbol{\beta}(t)$. However, we will show that some of its linear functions are actually estimable, and we will describe some methods to find them in next subsection.

7.2.2 Methods for Finding Estimable Linear Functions

A question then arises naturally. What kind of linear functions of $\beta(t)$ are estimable? The following theorem answers this question.

Theorem 7.2 *Let* $\mathbf{C}$ *be a constant matrix of size* $q \times (p+1)$*. Under the non-full-rank FLM (7.1),* $\mathbf{C}\beta(t)$ *is estimable if and only if there exists a constant matrix* $\mathbf{A} : q \times n$ *such that* $\mathbf{C} = \mathbf{AX}$*.*

Theorem 7.2 says that $\mathbf{C}\beta(t)$ is estimable if and only if the rows of $\mathbf{C}$ are linear combinations of the rows of $\mathbf{X}$. In particular, $\mathbf{X}\beta(t)$ is estimable. As $\text{rank}(\mathbf{X}) = k$, there are k rows of $\mathbf{X}$ that are linearly independent. Notice that m real column vectors $\mathbf{a}_1, \mathbf{a}_2, \cdots, \mathbf{a}_m$ are linearly independent if and only if the matrix $[\mathbf{a}_1, \mathbf{a}_2, \cdots, \mathbf{a}_m]$ has rank m. Denote any k linearly independent rows of $\mathbf{X}$ as

$$\mathbf{x}_{i_l}^T, l = 1, 2, \cdots, k. \tag{7.15}$$

They form a linear basis for the linear space

$$\mathcal{L}(\mathbf{X}) = \left\{\mathbf{a}^T\mathbf{X} | \mathbf{a} \in \mathcal{R}^n\right\}, \tag{7.16}$$

spanned by the rows of $\mathbf{X}$, where $\mathcal{R}^n$ denotes the usual n-dimensional Euclidean linear space. Set

$$\mathbf{R} = [\mathbf{x}_{i_1}, \mathbf{x}_{i_2}, \cdots, \mathbf{x}_{i_k}]^T : k \times (p+1). \tag{7.17}$$

Then $\mathbf{R}$ is a full-rank matrix with rank k, $\mathcal{L}(\mathbf{R}) = \mathcal{L}(\mathbf{X})$, and $\mathbf{R}\beta(t)$ is estimable. As before, let $\mathbf{e}_{r,m}$ denote an m-dimensional column vector whose rth entry is 1 and others 0.

Remark 7.6 *Let* $\mathbf{E} = [\mathbf{e}_{i_1,n}, \mathbf{e}_{i_2,n}, \cdots, \mathbf{e}_{i_k,n}]^T$*. Then by (7.17), we have* $\mathbf{R} = \mathbf{EX}$*.*

In practice, the k linearly independent rows (7.15) of $\mathbf{X}$ can be obtained by the following simple method. Notice that $\mathbf{X}$ has no rows whose entries are all 0; otherwise, just remove those rows. We first set the first row of $\mathbf{X}$ as $\mathbf{x}_{i_1}^T$. We then check if the second row of $\mathbf{X}$ is linearly independent with $\mathbf{x}_{i_1}^T$. If it is a "yes," we set the second row as $\mathbf{x}_{i_2}^T$; otherwise, we try the next row of $\mathbf{X}$. Continue this process until all the rows of $\mathbf{X}$ have been checked. We then have the k linearly independent rows of $\mathbf{X}$ as listed in (7.15). In this way we actually obtain the first k linearly independent rows of $\mathbf{X}$.

Remark 7.7 *The design matrix* $\mathbf{X}$ *can be expressed as a linear function of* $\mathbf{R}$ *as the rows of* $\mathbf{R}$ *span the same linear space (7.16) as the rows of* $\mathbf{X}$*. That is, there is a real matrix* $\mathbf{Z} : n \times k$ *such that*

$$\mathbf{X} = \mathbf{ZR}. \tag{7.18}$$

In fact, as $\mathbf{RR}^T : k \times k$ *is full-rank, we have* $\mathbf{XR}^T = \mathbf{ZRR}^T$*. This gives that*

$$\mathbf{Z} = \mathbf{XR}^T(\mathbf{RR}^T)^{-1}. \tag{7.19}$$

By Theorem 7.2, we have the following theorem.

Theorem 7.3 *Let* $\mathbf{C}$ *be a constant matrix of size* $q \times (p+1)$. *Under the non-full-rank FLM (7.1),* $\mathbf{C}\boldsymbol{\beta}(t)$ *is estimable if and only if there exists a constant matrix* $\mathbf{B}: q \times k$ *such that* $\mathbf{C} = \mathbf{B}\mathbf{R}$, *where* $\mathbf{R}$ *is given in (7.17). In addition, when* $\mathbf{C}\boldsymbol{\beta}(t)$ *is estimable, we have*

$$\mathbf{B} = \mathbf{C}\mathbf{R}^T(\mathbf{R}\mathbf{R}^T)^{-1}, \tag{7.20}$$

and when $\mathbf{C}\boldsymbol{\beta}(t)$ *is nonestimable,* $\mathbf{C}\mathbf{R}^T(\mathbf{R}\mathbf{R}^T)^{-1}\mathbf{R} \neq \mathbf{C}$.

Theorem 7.3 is important as it can be used to check if a given linear function $\mathbf{C}\boldsymbol{\beta}(t)$ is estimable by checking if $\mathbf{C}\mathbf{S}$ is equal or not equal to $\mathbf{C}$ where

$$\mathbf{S} = \mathbf{R}^T(\mathbf{R}\mathbf{R}^T)^{-1}\mathbf{R}, \tag{7.21}$$

a projection matrix based on $\mathbf{R}^T$.

Example 7.3 *Consider the one-way functional ANOVA model (7.2). From the design matrix* $\mathbf{X}$ *given in (7.3), it is easy to see that* $\text{rank}(\mathbf{X}) = a$ *and the first* a *linearly independent rows of* $\mathbf{X}$ *form*

$$\mathbf{R} = [\mathbf{1}_a, \mathbf{I}_a]. \tag{7.22}$$

Therefore,

$$\boldsymbol{\phi}(t) = \mathbf{R}\boldsymbol{\beta}(t) = [\eta(t) + \alpha_1(t), \eta(t) + \alpha_2(t), \cdots, \eta(t) + \alpha_a(t)]^T \tag{7.23}$$

is estimable. That is, $\eta(t) + \alpha_i(t), i = 1, 2, \cdots, a$ *are all estimable.*

We now show that all the main-effect contrast functions of the one-way functional ANOVA model (7.2) are also estimable. To this end, let $\boldsymbol{\alpha}(t) = [\alpha_1(t), \cdots, \alpha_a(t)]^T$ *be the vector of the main-effect functions. The linear function* $\mathbf{c}^T\boldsymbol{\alpha}(t)$ *is called a main-effect contrast function if* $\mathbf{c}: a \times 1$ *is a real vector such that* $\mathbf{c}^T\mathbf{1}_a = 0$. *For example,* $\alpha_1(t) - \alpha_2(t), \alpha_3(t) - 2\alpha_4(t) + \alpha_a(t)$, *and* $2\alpha_1(t) - 3\alpha_3(t) + \alpha_5(t)$ *are all main-effect contrast functions. We claim "any main-effect contrast function* $\mathbf{c}^T\boldsymbol{\alpha}(t)$ *is estimable" as by (7.22),* $\mathbf{c}^T\mathbf{1}_a = 0$ *implies that* $[0, \mathbf{c}^T] = [0, \mathbf{c}^T]\mathbf{R}$. *That is,*

$$\mathbf{c}^T\boldsymbol{\alpha}(t) = [0, \mathbf{c}^T]\boldsymbol{\beta}(t) = [0, \mathbf{c}^T]\mathbf{R}\boldsymbol{\beta}(t) = [0, \mathbf{c}^T]\boldsymbol{\phi}(t),$$

which is estimable.

However, we can show that the grand mean function $\eta(t)$ *and the main-effect functions* $\alpha_i(t), i = 1, 2, \cdots, a$, *are nonestimable. To this end, we first obtain that* $\mathbf{R}\mathbf{R}^T = \mathbf{I}_a + \mathbf{J}_a$ *and* $(\mathbf{I}_a + \mathbf{J}_a)^{-1} = \mathbf{I}_a - (a+1)^{-1}\mathbf{J}_a$, *where* $\mathbf{J}_a = \mathbf{1}_a\mathbf{1}_a^T$, *the* $a \times a$ *matrix of ones. We then have*

$$\mathbf{S} = \mathbf{R}^T(\mathbf{R}\mathbf{R}^T)^{-1}\mathbf{R} = (a+1)^{-1}\begin{bmatrix} a & \mathbf{1}_a^T \\ \mathbf{1}_a & (a+1)\mathbf{I}_a - \mathbf{J}_a \end{bmatrix}. \tag{7.24}$$

It follows that $\mathbf{e}_{1,a+1}^T\mathbf{S} = (a+1)^{-1}[a, \mathbf{1}_a^T] \neq \mathbf{e}_{1,a+1}^T$, *implying that* $\eta(t) = \mathbf{e}_{1,a+1}^T\boldsymbol{\beta}(t)$ *is nonestimable, and* $\mathbf{e}_{i+1,a+1}^T\mathbf{S} = (a+1)^{-1}[1, (a+1)\mathbf{e}_{i,a}^T - \mathbf{1}_a^T] \neq \mathbf{e}_{i+1,a+1}^T$, *implying that* $\alpha_i(t) = \mathbf{e}_{i+1,a+1}^T\boldsymbol{\beta}(t)$ *is nonestimable for* $i = 1, 2, \cdots, a$.

Sometimes it is very challenging to find $\mathbf{R}$ manually, especially when the design matrix has a complicated structure. In this case, a calculator or a computer may be helpful for finding $\mathbf{R}, \mathbf{Z}, \mathbf{S}$, or $\mathbf{B}$, among others.

Example 7.4 *For the audible noise data, the non-full-rank design matrix* $\mathbf{X} : 36 \times 15$ *is given in (7.4). By some calculation, we have rank*$(\mathbf{X}) = 8$. *In addition, the matrix* $\mathbf{R}$ *(7.17) formed by the first* 8 *linearly independent rows of* $\mathbf{X}$ *is given by (7.25). Then the associated matrix* $\mathbf{Z}$ *computed using (7.19) such that* $\mathbf{X} = \mathbf{ZR}$ *is found to be given by (7.27). To check if the grand mean function* $\eta(t)$ *and the main-effect functions* $\alpha_{ij}(t), i = 1, 2, \cdots, 7, j = 1, 2$ *of Factors A, B, C, D, E, F, and G and the associated main-effect contrast functions are estimable, we compute the matrix* $\mathbf{S} = \mathbf{R}^T(\mathbf{RR}^T)^{-1}\mathbf{R}$ *(7.21), which is given by (7.29). We first notice that the grand mean function* $\eta(t)$ *and the main-effect functions* $\alpha_{ij}(t), i = 1, 2, \cdots, 7; j = 1, 2$, *are nonestimable because by (7.29), it is easy to check that* $\mathbf{e}_{r,15}^T\mathbf{S} \neq \mathbf{e}_{r,15}^T, r = 1, 2, \cdots, 15$. *We now check if the main-effect contrast functions of the seven factors are estimable. To this end, set* $\mathbf{C}$ *as given in (7.28). Then each row of* $\mathbf{C}$ *can be used to specify the main-effect contrast functions of a factor. For example, any main-effect contrast function of Factor C can be expressed as* $c[0, 0, 0, 0, 0, 1, -1, 0, 0, 0, 0, 0, 0, 0, 0]\boldsymbol{\beta}(t)$, *where* c *is any nonzero real number. By some calculation, we will find that we do have* $\mathbf{C} = \mathbf{CS}$. *Therefore, all the main-effect contrast functions of the seven factors are estimable. In fact, the matrix* $\mathbf{B} = \mathbf{CR}^T(\mathbf{RR}^T)^{-1}$ *such that* $\mathbf{C} = \mathbf{BR}$ *is given by (7.26).*

$$\mathbf{R} = \begin{pmatrix} 1 & 1 & 0 & 1 & 0 & 1 & 0 & 1 & 0 & 1 & 0 & 0 & 1 & 0 & 1 \\ 1 & 1 & 0 & 1 & 0 & 1 & 0 & 1 & 0 & 0 & 1 & 0 & 1 & 1 & 0 \\ 1 & 1 & 0 & 1 & 0 & 1 & 0 & 0 & 1 & 1 & 0 & 1 & 0 & 1 & 0 \\ 1 & 1 & 0 & 1 & 0 & 1 & 0 & 0 & 1 & 0 & 1 & 1 & 0 & 0 & 1 \\ 1 & 1 & 0 & 1 & 0 & 0 & 1 & 1 & 0 & 1 & 0 & 1 & 0 & 0 & 1 \\ 1 & 1 & 0 & 1 & 0 & 0 & 1 & 0 & 1 & 1 & 0 & 0 & 1 & 1 & 0 \\ 1 & 1 & 0 & 0 & 1 & 1 & 0 & 1 & 0 & 1 & 0 & 1 & 0 & 1 & 0 \\ 1 & 0 & 1 & 1 & 0 & 1 & 0 & 1 & 0 & 1 & 0 & 1 & 0 & 1 & 0 \end{pmatrix}. \tag{7.25}$$

$$\mathbf{B} = 1/2 \begin{pmatrix} 0 & 1 & 2 & -1 & 1 & -1 & 0 & -2 \\ 0 & 1 & 2 & -1 & 1 & -1 & -2 & 0 \\ 1 & 0 & 1 & 0 & -1 & -1 & 0 & 0 \\ 0 & 1 & 0 & -1 & 1 & -1 & 0 & 0 \\ 1 & -1 & 1 & -1 & 0 & 0 & 0 & 0 \\ -1 & 0 & 1 & 0 & 1 & -1 & 0 & 0 \\ -1 & 1 & 1 & -1 & 0 & 0 & 0 & 0 \end{pmatrix}. \tag{7.26}$$

$$
\mathbf{Z} = \begin{pmatrix}
1 & 0 & 0 & 0 & 0 & 0 & 0 & 0 \\
0 & 1 & 0 & 0 & 0 & 0 & 0 & 0 \\
0 & 0 & 1 & 0 & 0 & 0 & 0 & 0 \\
0 & 0 & 0 & 1 & 0 & 0 & 0 & 0 \\
0 & 0 & 0 & 0 & 1 & 0 & 0 & 0 \\
-1 & 1 & 0 & 0 & 1 & 0 & 0 & 0 \\
0 & 0 & 0 & 0 & 0 & 1 & 0 & 0 \\
0 & 0 & -1 & 1 & 0 & 1 & 0 & 0 \\
0 & 0 & 0 & 0 & 0 & 0 & 1 & 0 \\
0 & 0 & -1 & 1 & 0 & 0 & 1 & 0 \\
1 & -1 & -1 & 1 & -1 & 1 & 1 & 0 \\
0 & 0 & -1 & 1 & -1 & 1 & 1 & 0 \\
0 & 0 & -1 & 0 & 0 & 1 & 1 & 0 \\
0 & 0 & -2 & 1 & 0 & 1 & 1 & 0 \\
0 & -1 & -1 & 1 & 0 & 1 & 1 & 0 \\
-1 & 0 & -1 & 1 & 0 & 1 & 1 & 0 \\
0 & 0 & 0 & 0 & 0 & 0 & 0 & 1 \\
0 & 0 & -1 & 1 & 0 & 0 & 0 & 1 \\
1 & -1 & -1 & 1 & -1 & 1 & 0 & 1 \\
0 & 0 & -1 & 1 & -1 & 1 & 0 & 1 \\
0 & 0 & -1 & 0 & 0 & 1 & 0 & 1 \\
0 & 0 & -2 & 1 & 0 & 1 & 0 & 1 \\
0 & -1 & -1 & 1 & 0 & 1 & 0 & 1 \\
-1 & 0 & -1 & 1 & 0 & 1 & 0 & 1 \\
1 & -1 & -2 & 1 & -1 & 1 & 1 & 1 \\
0 & 0 & -2 & 1 & -1 & 1 & 1 & 1 \\
0 & -1 & -1 & 1 & -1 & 1 & 1 & 1 \\
0 & -1 & -2 & 2 & -1 & 1 & 1 & 1 \\
0 & -1 & -2 & 1 & 0 & 1 & 1 & 1 \\
-1 & 0 & -2 & 1 & 0 & 1 & 1 & 1 \\
0 & -1 & -2 & 1 & -1 & 2 & 1 & 1 \\
0 & -1 & -3 & 2 & -1 & 2 & 1 & 1 \\
0 & -1 & -3 & 2 & -1 & 2 & 1 & 1 \\
0 & -1 & -3 & 2 & -1 & 2 & 1 & 1 \\
0 & -1 & -3 & 2 & -1 & 2 & 1 & 1 \\
0 & -1 & -3 & 2 & -1 & 2 & 1 & 1
\end{pmatrix}. \tag{7.27}
$$

$$
\mathbf{C} = (\mathbf{C}_1, \mathbf{C}_2), \tag{7.28}
$$

with $\mathbf{C}_1$ and $\mathbf{C}_2$ being

$$\begin{pmatrix} 0 & 1 & -1 & 0 & 0 & 0 & 0 & 0 \\ 0 & 0 & 0 & 1 & -1 & 0 & 0 & 0 \\ 0 & 0 & 0 & 0 & 0 & 1 & -1 & 0 \\ 0 & 0 & 0 & 0 & 0 & 0 & 0 & 1 \\ 0 & 0 & 0 & 0 & 0 & 0 & 0 & 0 \\ 0 & 0 & 0 & 0 & 0 & 0 & 0 & 0 \\ 0 & 0 & 0 & 0 & 0 & 0 & 0 & 0 \end{pmatrix},$$

and

$$\begin{pmatrix} 0 & 0 & 0 & 0 & 0 & 0 & 0 \\ 0 & 0 & 0 & 0 & 0 & 0 & 0 \\ 0 & 0 & 0 & 0 & 0 & 0 & 0 \\ -1 & 0 & 0 & 0 & 0 & 0 & 0 \\ 0 & 1 & -1 & 0 & 0 & 0 & 0 \\ 0 & 0 & 0 & 1 & -1 & 0 & 0 \\ 0 & 0 & 0 & 0 & 0 & 1 & -1 \end{pmatrix}.$$

$$\mathbf{S} = \frac{1}{18}(\mathbf{S}_1, \mathbf{S}_2), \tag{7.29}$$

with $\mathbf{S}_1$ and $\mathbf{S}_2$ being

$$\begin{pmatrix} 4 & 2 & 2 & 2 & 2 & 2 & 2 & 2 \\ 2 & 10 & -8 & 1 & 1 & 1 & 1 & 1 \\ 2 & -8 & 10 & 1 & 1 & 1 & 1 & 1 \\ 2 & 1 & 1 & 10 & -8 & 1 & 1 & 1 \\ 2 & 1 & 1 & -8 & 10 & 1 & 1 & 1 \\ 2 & 1 & 1 & 1 & 1 & 10 & -8 & 1 \\ 2 & 1 & 1 & 1 & 1 & -8 & 10 & 1 \\ 2 & 1 & 1 & 1 & 1 & 1 & 1 & 10 \\ 2 & 1 & 1 & 1 & 1 & 1 & 1 & -8 \\ 2 & 1 & 1 & 1 & 1 & 1 & 1 & 1 \\ 2 & 1 & 1 & 1 & 1 & 1 & 1 & 1 \\ 2 & 1 & 1 & 1 & 1 & 1 & 1 & 1 \\ 2 & 1 & 1 & 1 & 1 & 1 & 1 & 1 \\ 2 & 1 & 1 & 1 & 1 & 1 & 1 & 1 \\ 2 & 1 & 1 & 1 & 1 & 1 & 1 & 1 \end{pmatrix}$$

and

$$
\begin{pmatrix}
2 & 2 & 2 & 2 & 2 & 2 & 2\\
1 & 1 & 1 & 1 & 1 & 1 & 1\\
1 & 1 & 1 & 1 & 1 & 1 & 1\\
1 & 1 & 1 & 1 & 1 & 1 & 1\\
1 & 1 & 1 & 1 & 1 & 1 & 1\\
1 & 1 & 1 & 1 & 1 & 1 & 1\\
1 & 1 & 1 & 1 & 1 & 1 & 1\\
-8 & 1 & 1 & 1 & 1 & 1 & 1\\
10 & 1 & 1 & 1 & 1 & 1 & 1\\
1 & 10 & -8 & 1 & 1 & 1 & 1\\
1 & -8 & 10 & 1 & 1 & 1 & 1\\
1 & 1 & 1 & 10 & -8 & 1 & 1\\
1 & 1 & 1 & -8 & 10 & 1 & 1\\
1 & 1 & 1 & 1 & 1 & 10 & -8\\
1 & 1 & 1 & 1 & 1 & -8 & 10
\end{pmatrix}.
$$

7.2.3 Estimation of Estimable Linear Functions

By Theorem 7.2, we know that $\mathbf{X}\boldsymbol{\beta}(t)$ and all its linear functions $\mathbf{AX}\boldsymbol{\beta}(t)$ are estimable; all other linear functions of $\boldsymbol{\beta}(t)$ are nonestimable. Now another question arises. If $\mathbf{C}\boldsymbol{\beta}(t)$ is estimable, how do we find its unbiased estimator? We have the following theorem.

Theorem 7.4 *Let $\hat{\boldsymbol{\beta}}(t)$ be as defined in (7.14) and $\mathbf{C}: q \times (p+1)$ be a full-rank matrix with rank($\mathbf{C}$) $= q \le k < p+1 < n$ such that $\mathbf{C}\boldsymbol{\beta}(t)$ is estimable. Then under the non-full-rank FLM (7.1), we have: (a) $\mathbf{C}\hat{\boldsymbol{\beta}}(t)$ is an unbiased linear estimator of $\mathbf{C}\boldsymbol{\beta}(t)$ with $E\left[\mathbf{C}\hat{\boldsymbol{\beta}}(t)\right] = \mathbf{C}\boldsymbol{\beta}(t)$ and $Cov\left[\mathbf{C}\hat{\boldsymbol{\beta}}(s), \mathbf{C}\hat{\boldsymbol{\beta}}(t)\right] = \gamma(s,t)\mathbf{C}(\mathbf{X}^T\mathbf{X})^-\mathbf{C}^T$; (b) $\mathbf{C}\hat{\boldsymbol{\beta}}(t)$ and $\mathbf{C}(\mathbf{X}^T\mathbf{X})^-\mathbf{C}^T$ are invariant to the choice of $(\mathbf{X}^T\mathbf{X})^-$; and (c) $\mathbf{C}(\mathbf{X}^T\mathbf{X})^-\mathbf{C}^T$ is a full-rank matrix of rank q.*

The above theorem indicates that provided $\mathbf{C}\boldsymbol{\beta}(t)$ is estimable, it is very easy to obtain its unbiased estimator, which is $\mathbf{C}\hat{\boldsymbol{\beta}}(t)$. As $\mathbf{X}\boldsymbol{\beta}(t)$ is estimable, by the above theorem, its unbiased estimator is $\mathbf{X}\hat{\boldsymbol{\beta}}(t) = \mathbf{P_X}\mathbf{y}(t)$, which is the vector of the fitted response functions. It follows that the vector of the fitted subject-effect functions is

$$\hat{\mathbf{v}}(t) = \mathbf{y}(t) - \mathbf{X}\hat{\boldsymbol{\beta}}(t) = \mathbf{y}(t) - \mathbf{P_X}\mathbf{y}(t) = (\mathbf{I}_n - \mathbf{P_X})\mathbf{y}(t), \tag{7.30}$$

which is invariant to the choice of $\hat{\boldsymbol{\beta}}(t)$. Set $W(s,t) = \sum_{i=1}^n \hat{v}_i(s)\hat{v}_i(t)$. Then by (7.13), we have

$$W(s,t) = \hat{\mathbf{v}}(s)^T\hat{\mathbf{v}}(t) = \mathbf{y}(s)^T(\mathbf{I}_n - \mathbf{P_X})\mathbf{y}(t) = \mathbf{v}(s)^T(\mathbf{I}_n - \mathbf{P_X})\mathbf{v}(t).$$

It follows that $\text{E}\{W(s,t)\} = (n-k)\gamma(s,t)$ as $\mathbf{P_X}$ is an idempotent matrix with rank k; see (7.12). It follows that the unbiased estimator of the covariance function $\gamma(s,t)$ is

$$\hat{\gamma}(s,t) = (n-k)^{-1}\sum_{i=1}^{n} \hat{v}_i(s)\hat{v}_i(t) = (n-k)^{-1}\mathbf{y}(s)^T\left(\mathbf{I}_n - \mathbf{P_X}\right)\mathbf{y}(t). \quad (7.31)$$

Obviously, $\hat{\gamma}(s,t)$ is invariant to the choice of $\hat{\boldsymbol{\beta}}(t)$.

7.2.4 Tests of Testable Linear Hypotheses

Under the non-full-rank FLM (7.1) with $\text{rank}(\mathbf{X}) = k < (p+1) < n$, consider the following GLHT problem:

$$\begin{array}{lll} & H_0 & : \mathbf{C}\boldsymbol{\beta}(t) \equiv \mathbf{c}(t), \ t \in \mathcal{T}, \\ \text{versus} & H_1 & : \mathbf{C}\boldsymbol{\beta}(t) \neq \mathbf{c}(t), \ \text{ for some } t \in \mathcal{T}, \end{array} \quad (7.32)$$

where $\mathbf{c}(t) = [c_1(t), \cdots, c_q(t)]^T$ is a vector of given functions and $\mathbf{C}$ is a given $q\times(p+1)$ full-rank matrix with rank $q \le k < p+1 < n$ and $\mathbf{C}\boldsymbol{\beta}(t)$ is estimable. We would like to emphasize that in the above GLHT problem, it is required that $\mathbf{C}\boldsymbol{\beta}(t)$ be estimable. As $\mathbf{C}\boldsymbol{\beta}(t)$ is estimable, the GLHT problem (7.32) is testable.

Remark 7.8 *For a testable hypothesis testing problem that can be written in the form of the GLHT problem (7.32), the matrix* $\mathbf{C}$ *and the known function* $\mathbf{c}(t)$ *are not unique. In fact, the GLHT problem, obtained by replacing the matrix* $\mathbf{C}$ *and the function* $\mathbf{c}(t)$ *in the GLHT problem (7.32) with*

$$\tilde{\mathbf{C}} = \mathbf{PC}, \quad \text{and } \tilde{\mathbf{c}}(t) = \mathbf{Pc}(t), \quad (7.33)$$

where $\mathbf{P}$ *is a full-rank matrix of size* $q \times q$, *is also testable and is equivalent to the original GLHT problem (7.32).*

We now describe how to construct a proper test statistic for the GLHT problem (7.32). First of all, by Theorem 7.4, we have $\text{E}\left[\mathbf{C}\hat{\boldsymbol{\beta}}(t) - \mathbf{c}(t)\right] = \mathbf{C}\boldsymbol{\beta}(t) - \mathbf{c}(t)$ and

$$\text{Cov}\left[\mathbf{C}\hat{\boldsymbol{\beta}}(s) - \mathbf{c}(s), \mathbf{C}\hat{\boldsymbol{\beta}}(t) - \mathbf{c}(t)\right] = \gamma(s,t)\mathbf{C}(\mathbf{X}^T\mathbf{X})^-\mathbf{C}^T.$$

By Theorem 7.4 again, $\mathbf{C}(\mathbf{X}^T\mathbf{X})^-\mathbf{C}^T$ is a full-rank matrix of size $q\times q$. Thus, we have the pivotal test function for the GLHT problem (7.32) as

$$\mathbf{z}(t) = \left[\mathbf{C}(\mathbf{X}^T\mathbf{X})^-\mathbf{C}^T\right]^{-1/2}\left[\mathbf{C}\hat{\boldsymbol{\beta}}(t) - \mathbf{c}(t)\right]. \quad (7.34)$$

Simple calculation gives the vector of mean functions and the matrix of covariance functions of $\mathbf{z}(t)$ as

$$\begin{array}{c} \boldsymbol{\eta}_z(t) = \left[\mathbf{C}(\mathbf{X}^T\mathbf{X})^-\mathbf{C}^T\right]^{-1/2}\left[\mathbf{C}\boldsymbol{\beta}(t) - \mathbf{c}(t)\right], \\ \boldsymbol{\Gamma}_z(s,t) = \gamma(s,t)\mathbf{I}_q. \end{array} \quad (7.35)$$

That is, the components of $\mathbf{z}(t)$ are uncorrelated. Under the Gaussian assumption, the components of $\mathbf{z}(t)$ are independent and have the common covariance function $\gamma(s,t)$. In addition, $(n-k)\hat{\gamma}(s,t)$ is a Wishart process with $n-k$ degrees of freedom and covariance function $\gamma(s,t)$. Notice that the degree of freedom of $(n-k)\hat{\gamma}(s,t)$ depends on the rank of the non-full-rank design matrix $\mathbf{X}$. We summarize these results as the following theorem.

Theorem 7.5 *Assume that in the non-full-rank FLM (7.1), $\mathbf{v}(t) \sim GP_n(\mathbf{0}, \gamma\mathbf{I}_n)$ and $tr(\gamma) < \infty$. Then we have*

$$\mathbf{z}(t) \sim GP_q(\boldsymbol{\eta}_z, \gamma\mathbf{I}_q), \quad (n-k)\hat{\gamma}(s,t) \sim WP(n-k, \gamma). \tag{7.36}$$

Based on the above theorem, we can describe various tests for the GLHT problem (7.32). For this purpose, we define the following pointwise SSH (sum of squares due to hypothesis) and SSE (sum of squares due to errors) as

$$\begin{array}{l} \text{SSH}_n(t) = \left[\mathbf{C}\hat{\boldsymbol{\beta}}(t) - \mathbf{c}(t)\right]^T \left[\mathbf{C}(\mathbf{X}^T\mathbf{X})^-\mathbf{C}^T\right]^{-1} \left[\mathbf{C}\hat{\boldsymbol{\beta}}(t) - \mathbf{c}(t)\right], \\ \text{SSE}_n(t) = (n-k)\hat{\gamma}(t,t) = \mathbf{y}(t)^T(\mathbf{I}_n - \mathbf{P}_{\mathbf{X}})\mathbf{y}(t). \end{array} \tag{7.37}$$

Remark 7.9 *The pivotal test function $\mathbf{z}(t)$ (7.34) is constructed in a way such that each component of $\mathbf{z}(t)$ is a process with the same covariance function $\gamma(s,t)$ as indicated in (7.35). In addition, the squared L^2-norm $SSH_n(t)$ (7.37) of $\mathbf{z}(t)$ is invariant under the transformation (7.33). This is a nice property for the GLHT problem (7.32) as for a testable hypothesis testing problem that can be written in the form of (7.32), the associated $\mathbf{C}$ and $\mathbf{c}(t)$ are not unique, as mentioned in Remark 7.8.*

Under the Gaussian assumption, we can show that the integrated SSH and SSE are χ^2-type mixtures, as stated in the following theorem. Notice that the degree of freedom of the χ^2-variates in the random expression of the integrated SSE depends on the rank of the non-full-rank design matrix $\mathbf{X}$.

Theorem 7.6 *Assume that in the non-full-rank FLM (7.1), $\mathbf{v}(t) \sim GP_n(\mathbf{0}, \gamma\mathbf{I}_n)$ and $tr(\gamma) < \infty$. Under the null hypothesis in (7.32), we have*

$$\begin{array}{lcl} \int_{\mathcal{T}} SSH_n(t)dt & \stackrel{d}{=} & \sum_{r=1}^m \lambda_r A_r, \ A_r \stackrel{i.i.d.}{\sim} \chi_q^2, \\ \int_{\mathcal{T}} SSE_n(t)dt & \stackrel{d}{=} & \sum_{r=1}^m \lambda_r E_r, \ E_r \stackrel{i.i.d.}{\sim} \chi_{n-k}^2, \end{array}$$

where $\lambda_1, \cdots, \lambda_m$ are all the positive eigenvalues of $\gamma(s,t)$.

Using $\text{SSH}_n(t)$ and $\text{SSE}_n(t)$, we are now ready to present various tests for the GLHT problem (7.32).

Pointwise F-Test The pointwise F-test for the testable GLHT problem (7.32) is conducted using the test statistic $F_n(t) = \frac{\text{SSH}_n(t)}{\text{SSE}_n(t)}, t \in \mathcal{T}$. Under the Gaussian assumption and the null hypothesis in (7.32), we have

$$F_n(t) \sim F_{q,n-k}, t \in \mathcal{T}.$$

The pointwise F-test can then be conducted accordingly.

L^2-Norm-Based Test As usual, the L^2-norm-based test for (7.32) uses the test statistic

$$T_n = \int_{\mathcal{T}} \mathrm{SSH}_n(t)dt. \tag{7.38}$$

By Theorem 7.6, we can approximate the null distribution of T_n by the Welch-Satterthwaite χ^2-approximation method described in Section 4.3 of Chapter 4 and obtain

$$T_n \sim \beta\chi^2_d \text{ approximately, where } \beta = \frac{\mathrm{tr}(\gamma^{\otimes 2})}{\mathrm{tr}(\gamma)},\ \kappa = \frac{\mathrm{tr}^2(\gamma)}{\mathrm{tr}(\gamma^{\otimes 2})},\ d = q\kappa. \tag{7.39}$$

As in the previous chapters, the parameters β and κ can be estimated by a naive method and a bias-reduced method. By the naive method, we replace $\gamma(s,t)$ in (7.39) with its unbiased estimator $\hat{\gamma}(s,t)$ given in (7.31) so that

$$\hat{\beta} = \frac{\mathrm{tr}(\hat{\gamma}^{\otimes 2})}{\mathrm{tr}(\hat{\gamma})},\ \hat{\kappa} = \frac{\mathrm{tr}^2(\hat{\gamma})}{\mathrm{tr}(\hat{\gamma}^{\otimes 2})},\ \hat{d} = q\hat{\kappa}, \tag{7.40}$$

and hence

$$T_n \sim \hat{\beta}\chi^2_{\hat{d}} \text{ approximately.} \tag{7.41}$$

By the bias-reduced method, we replace $\mathrm{tr}^2(\gamma)$ and $\mathrm{tr}(\gamma^{\otimes 2})$ in (7.39) with their unbiased estimators to obtain $\hat{\beta}$ and $\hat{\kappa}$. By Theorems 4.6 and 7.5 we can obtain the unbiased estimators of $\mathrm{tr}^2(\gamma)$ and $\mathrm{tr}(\gamma^{\otimes 2})$ as

$$\begin{aligned}\widehat{\mathrm{tr}^2(\gamma)} &= \frac{(n-k)(n-k+1)}{(n-k-1)(n-k+2)}\left[\mathrm{tr}^2(\hat{\gamma}) - \frac{2\mathrm{tr}(\hat{\gamma}^{\otimes 2})}{n-k+1}\right],\\ \widehat{\mathrm{tr}(\gamma^{\otimes 2})} &= \frac{(n-k)^2}{(n-k-1)(n-k+2)}\left[\mathrm{tr}(\hat{\gamma}^{\otimes 2}) - \frac{\mathrm{tr}^2(\hat{\gamma})}{n-k}\right].\end{aligned} \tag{7.42}$$

The L^2-norm-based test for (7.32) can then be conducted accordingly.

F-Type Test As usual, the F-type test statistic is defined as

$$F_n = \frac{\int_{\mathcal{T}} \mathrm{SSH}_n(t)dt/q}{\int_{\mathcal{T}} \mathrm{SSE}_n(t)dt/(n-k)}. \tag{7.43}$$

Under the Gaussian assumption and the null hypothesis in (7.32), by Theorem 7.6, we have

$$F_n \stackrel{d}{=} \frac{\sum_{r=1}^m \lambda_r A_r/q}{\sum_{r=1}^m \lambda_r E_r/(n-k)},$$

where $A_r \stackrel{i.i.d.}{\sim} \chi^2_q$, $E_r \stackrel{i.i.d.}{\sim} \chi^2_{n-k}$, and they are all independent; $\lambda_1, \lambda_2, \cdots, \lambda_m$ are all the positive eigenvalues of $\gamma(s,t)$. It follows that the null distribution

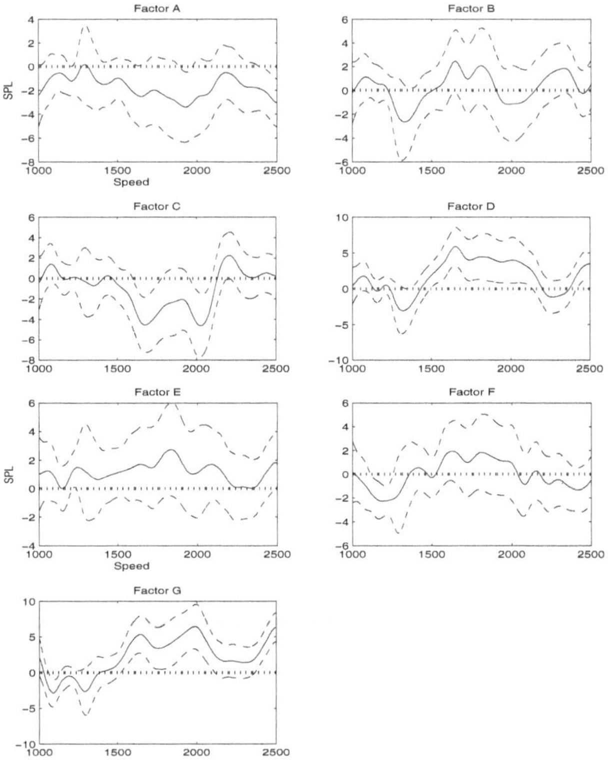

Figure 7.1 *Estimated main-effect contrast functions* $\hat{\alpha}_{i1}(t) - \hat{\alpha}_{i2}(t), i = 1, 2, \cdots, 7$ *(solid) of Factors A, B, C, D, E, F, and G of the audible noise data with their 95% pointwise confidence bands (dashed). The main-effect functions of Factors B, E, and F may be not significant or less significant as their* 95% *pointwise confidence bands cover the zero horizontal line (dotted) most of time while the other four main-effect functions are likely to be highly significant.*

of F_n can be approximated by the two-cumulant matched F-approximation method described in Section 4.4 of Chapter 4. We then have

$$F_n \sim F_{q\hat{\kappa},(n-k)\hat{\kappa}} \text{ approximately,}$$

where $\hat{\kappa}$ is obtained by the naive method or the bias-reduced method as described above for the L^2-norm-based test. The F-type test for (7.32) is then conducted accordingly.

Bootstrap Tests Under the Gaussian assumption, we can use the pointwise F-test, the L^2-norm-based test, and the F-type test described previously for the GLHT problem (7.32). When the Gaussian assumption is not valid, we can use some nonparametric bootstrap tests instead. Let $v_i^*(t), i = 1, 2, \cdots, n$, be a bootstrap sample randomly generated from the estimated subject-effect functions $\hat{v}_i(t), i = 1, 2, \cdots, n$, the n-components of $\hat{\mathbf{v}}(t), t \in \mathcal{T}$ as defined in (7.30). Set $\mathbf{v}^*(t) = [v_1^*(t), \cdots, v_n^*(t)]^T$. Compute $\mathbf{y}^*(t) = \mathbf{X}\hat{\boldsymbol{\beta}}(t) + \mathbf{v}^*(t)$ and $\hat{\boldsymbol{\beta}}^*(t) = (\mathbf{X}^T\mathbf{X})^- \mathbf{X}^T \mathbf{y}^*(t)$. Then we compute $\mathrm{SSH}_n^*(t) = \left[\hat{\boldsymbol{\beta}}^*(t) - \hat{\boldsymbol{\beta}}(t)\right]^T \mathbf{C}^T[\mathbf{C}(\mathbf{X}^T\mathbf{X})^-\mathbf{C}^T]^{-1}\mathbf{C}\left[\hat{\boldsymbol{\beta}}^*(t) - \hat{\boldsymbol{\beta}}(t)\right]$ and $\mathrm{SSE}_n^*(t) = \mathbf{y}^*(t)^T(\mathbf{I}_n - \mathbf{P}_\mathbf{X})\mathbf{y}^*(t)$. Notice that the values of $\mathrm{SSH}_n^*(t)$ and $\mathrm{SSE}_n^*(t)$ are invariant to the choice of $(\mathbf{X}^T\mathbf{X})^-$. The pointwise bootstrap test statistic, the L^2-norm-based bootstrap test statistic, and the F-type bootstrap test statistic can then be computed as

$$F_n^*(t) = \frac{\mathrm{SSH}_n^*(t)/q}{\mathrm{SSE}_n^*(t)/(n-k)}, \; T_n^* = \int_\mathcal{T} \mathrm{SSH}_n^*(t)dt, \text{ or}$$
$$F_n^* = \frac{\int_\mathcal{T} \mathrm{SSH}_n^*(t)dt/q}{\int_\mathcal{T} \mathrm{SSE}_n^*(t)dt/(n-k)}.$$

Repeat the above bootstrapping process a large number of times, calculate its $100(1-\alpha)$-percentile, and then conduct the pointwise bootstrap F-test, the L^2-norm-based bootstrap test, or the F-type bootstrap test accordingly.

Example 7.5 *For the audible noise data, we showed that all the main-effect contrast functions of the seven factors are estimable in Example 7.4. Therefore, the following problem about the main-effect contrast function of Factor i is testable:*

$$\begin{array}{lll} & H_0 & : \alpha_{i1}(t) - \alpha_{i2}(t) \equiv 0, \\ \text{versus} & H_1 & : \alpha_{i1}(t) - \alpha_{i2}(t) \neq 0 \textit{ for some } t \in \mathcal{T}, \end{array} \tag{7.44}$$

where $i = 1, 2, \cdots, 6$ or 7. This problem can be written in the form of the testable GLHT problem (7.32) by setting $\mathbf{C}$ *being the ith row $\mathbf{c}_i^T$ of the matrix* $\mathbf{C}$ *as defined in (7.28) and $\mathbf{c}(t) \equiv 0$. For example, when $i = 3$, we have $\mathbf{c}_i^T = [0, 0, 0, 0, 0, 1, -1, 0, 0, 0, 0, 0, 0, 0, 0]$. The unbiased estimators of the main-effect contrast functions are*

$$\mathbf{c}_i^T \hat{\boldsymbol{\beta}}(t), i = 1, 2, \cdots, 7.$$

These estimated main-effect contrast functions with their 95% pointwise confidence bands are depicted in Figure 7.1 as solid and dashed curves, respectively. One can use the raw audible noise data where each curve has forty-three measurement records while in this example, the audible noise data were pre-smoothed using the regression spline method described in Chapter 3 and the resolution number $M = 1,000$. It is seen that (a) the main-effect contrast functions of Factors B and E may not be significant as their 95% pointwise confidence bands contain the zero horizontal line (dashed) all the times; (b) the main-effect contrast functions of Factors D and G must be highly significant as their 95% pointwise confidence bands do not contain the zero horizontal line

Table 7.1 *L^2-norm-based test for the audible noise data with resolution* $M = 1,000$.

Factor	$T_n/10^4$	Naive P-value	Bias-reduced P-value
A	3.0665	0.0330	0.0221
B	1.4291	0.3168	0.3071
C	3.9031	0.0092	0.0049
D	7.3724	0.0000	0.0000
E	1.5365	0.2774	0.2643
F	1.2732	0.3818	0.3784
G	9.7866	0.0000	0.0000

Note: For the naive method, $\hat{\beta} = 2,745.6, \hat{d} = \hat{\kappa} = 4.4095$ *and for the bias-reduced method,* $\hat{\beta} = 2,239.0, \hat{d} = \hat{\kappa} = 5.3359$*, as given in (7.46) and (7.47), respectively. It is seen that the P-values by the bias-reduced method are generally smaller than those by the naive method.*

(dashed) over a big range; and (c) the main-effect contrast functions of Factors A, C and F may be significant or not significant as their 95% *pointwise confidence bands contain the zero horizontal line (dashed) only over a small range. The pointwise F-test for this data set resulted in similar conclusions.*

To apply the aforementioned L^2-norm-based test for (7.44), we first computed the sample covariance function of the audible noise data using (7.31). We then obtained the following two important quantities:

$$tr(\hat{\gamma}) = 1.2107e4, \quad tr(\hat{\gamma}^{\otimes 2}) = 3.3242e7, \tag{7.45}$$

where and throughout 1.2107e4 *means* 1.2107×10^4. *By the naive method, we obtained*

$$\hat{\beta} = 2,745.6, \quad \hat{d} = \hat{\kappa} = 4.4095. \tag{7.46}$$

By the bias-reduced method, we obtained

$$\hat{\beta} = 2,239.0, \quad \hat{d} = \hat{\kappa} = 5.3359. \tag{7.47}$$

Then applying the L^2-norm-based test for (7.44) for the seven factors resulted in Table 7.1. It is seen that (a) the main-effect contrast functions of Factors B, E, and F are not significant; (b) the main-effect contrast functions of Factors D and G are highly significant; and (c) the main-effect contrast functions of Factors A and C are significant at nominal levels 5% *and* 1%, *respectively. Applying the F-type test resulted in similar results. These results are also consistent with those obtained by Shen and Xu (2007). We also notice that the significance of the main-effect contrast function of Factor F over a small range (see Figure 7.1) is not detected by the L^2-norm-based and F-type tests. This is probably due to the fact that the P-values of the pointwise F-test used here were not corrected using some multiple comparison methods (for example, Cox and Lee 2008), or the L^2-norm-based test and the F-type test are not very powerful in such situations.*

7.3 Reparameterization Method

7.3.1 The Methodology

The key idea of the reparameterization method is to transform the non-full-rank FLM (7.1) into a full-rank FLM with a vector of new coefficient functions so that the methodologies developed in Chapter 6 may be applicable. Recall that $\mathbf{X}$ is a non-full-rank design matrix of size $n \times (p+1)$ with rank $k < p+1 \leq n$. Assume that we can write

$$\mathbf{X} = \mathbf{Z}\mathbf{R}, \tag{7.48}$$

where $\mathbf{Z} : n \times k$ and $\mathbf{R} : k \times (p+1)$ are full-rank matrices with rank k. For convenience, we can call $\mathbf{Z}$ and $\mathbf{R}$ the reparameterization matrices. See (7.18) for such a decomposition of $\mathbf{X}$. Set $\boldsymbol{\phi}(t) = \mathbf{R}\boldsymbol{\beta}(t)$ be the vector of new coefficient functions that are estimable. Then the non-full-rank FLM (7.1) is transformed into the following full-rank FLM:

$$\mathbf{y}(t) = \mathbf{Z}\boldsymbol{\phi}(t) + \mathbf{v}(t), \quad \mathbf{v}(t) \sim \mathrm{SP}_n(\mathbf{0}, \gamma\mathbf{I}_n). \tag{7.49}$$

It is often easy to specify the matrix $\mathbf{R}$ based on the relationship $\boldsymbol{\phi}(t) = \mathbf{R}\boldsymbol{\beta}(t)$. Then by (7.19), we can easily obtain $\mathbf{Z} = \mathbf{X}\mathbf{R}^T(\mathbf{R}\mathbf{R}^T)^{-1}$.

As (7.49) is a full-rank FLM, the methodologies developed in Chapter 6 can be applied. In particular, we have

$$\begin{aligned} \hat{\boldsymbol{\phi}}(t) &= (\mathbf{Z}^T\mathbf{Z})^{-1}\mathbf{Z}^T\mathbf{y}(t), \\ \hat{\gamma}(s,t) &= (n-k)^{-1}\mathbf{y}(t)^T(\mathbf{I}_n - \mathbf{P}_{\mathbf{Z}})\mathbf{y}(t), \end{aligned}$$

where $\mathbf{P}_{\mathbf{Z}} = \mathbf{Z}(\mathbf{Z}^T\mathbf{Z})^{-1}\mathbf{Z}^T$ is the usual projection matrix based on $\mathbf{Z}$.

7.3.2 Determining the Reparameterization Matrices

We now describe how to determine the reparameterization matrices $\mathbf{R}$ and $\mathbf{Z}$ for (7.48). Sometimes, $\mathbf{R}$ can be determined easily by the given reparameterization transformation $\boldsymbol{\phi}(t) = \mathbf{R}\boldsymbol{\beta}(t)$, which is estimable. In this case, $\mathbf{Z} = \mathbf{X}\mathbf{R}^T(\mathbf{R}\mathbf{R}^T)^{-1}$ (7.19) can also be computed easily. When the reparameterization transformation $\boldsymbol{\phi}(t) = \mathbf{R}\boldsymbol{\beta}(t)$ is not obvious, we can determine $\mathbf{R}$ by finding the k linearly independent rows of $\mathbf{X}$ as in Section 7.2.2. See (7.17) and (7.19) for the associated $\mathbf{R}$ and $\mathbf{Z}$ so that $\mathbf{X} = \mathbf{Z}\mathbf{R}$ is the desired decomposition of $\mathbf{X}$.

When n and p are large, it may be difficult to find the k linearly independent rows of $\mathbf{X}$. In this case, we may choose $\mathbf{R}$ and $\mathbf{Z}$ using the SVD (7.7). In fact, based on (7.7), we can write $\mathbf{X} = \mathbf{Z}\mathbf{R}$ with

$$\mathbf{Z} = \mathbf{U}\begin{pmatrix} \mathbf{D}_k^{1/2} \\ \mathbf{0} \end{pmatrix} : n \times k, \quad \mathbf{R} = \left(\mathbf{D}_k^{1/2}, \mathbf{0}\right)\mathbf{V}^T : k \times (p+1), \tag{7.50}$$

where $\mathbf{U} : n \times n$ and $\mathbf{V} : (p+1) \times (p+1)$ are the orthonormal matrices, and $\mathbf{D}_k : k \times k$ is the diagonal matrix of rank k as defined in (7.7).

Example 7.6 *For the one-way functional ANOVA model (7.2),* $\boldsymbol{\phi}(t) = [\eta(t)+\alpha_1(t), \eta(t)+\alpha_2(t), \cdots, \eta(t)+\alpha_a(t)]^T$ *is estimable as shown in Example 7.3. As* $\boldsymbol{\beta}(t) = [\eta(t), \alpha_1(t), \cdots, \alpha_a(t)]^T$, $\boldsymbol{\phi}(t) = \mathbf{R}\boldsymbol{\beta}(t)$ *implies that* $\mathbf{R} = [\mathbf{1}_a, \mathbf{I}_a]$ *as given in (7.22). From the design matrix* $\mathbf{X}$ *given in (7.3), by some simple linear algebra, we have* $\mathbf{Z} = \mathbf{X}\mathbf{R}^T(\mathbf{R}\mathbf{R}^T)^{-1} = diag(\mathbf{1}_{n_1}, \mathbf{1}_{n_2}, \cdots, \mathbf{1}_{n_a})$*. Alternatively, we can identify the matrix* $\mathbf{R}$ *from the design matrix* $\mathbf{X}$ *directly. In fact, from (7.3), we have that rank(*$\mathbf{X}$*)* $= a$ *and the* a *linearly independent rows of* $\mathbf{X}$ *forms* $\mathbf{R} = [\mathbf{1}_a, \mathbf{I}_a]$ *and hence* $\mathbf{Z} = \mathbf{X}\mathbf{R}^T(\mathbf{R}\mathbf{R}^T)^{-1} = diag(\mathbf{1}_{n_1}, \mathbf{1}_{n_2}, \cdots, \mathbf{1}_{n_a})$*.*

Example 7.7 *For the audible noise data, the matrix* $\mathbf{R}$ *formed by the first 8 linearly independent rows of* $\mathbf{X}$ *is given by (7.25) and the associated matrix* $\mathbf{Z}$ *is given in (7.27).*

7.3.3 Invariance of the Reparameterization

It is obvious that the decomposition $\mathbf{X} = \mathbf{Z}\mathbf{R}$ is not unique as we can always write $\mathbf{X} = (\mathbf{Z}\mathbf{Q})(\mathbf{Q}^{-1}\mathbf{R})$, where $\mathbf{Q} : k \times k$ is any full-rank real matrix. A question arises naturally. Are the statistical inferences based on the reparameterization invariant to the choice of the decompositions of $\mathbf{X}$? To answer this question, we need to establish a relationship between any two decompositions of $\mathbf{X}$ in the form of (7.48).

Theorem 7.7 *Let* $\mathbf{X} = \mathbf{Z}\mathbf{R} = \tilde{\mathbf{Z}}\tilde{\mathbf{R}}$ *be two different decompositions of* $\mathbf{X}$ *such that* $\mathbf{Z} : n\times k, \mathbf{R} : k\times(p+1), \tilde{\mathbf{Z}} : n\times k$, *and* $\tilde{\mathbf{R}} : k\times(p+1)$ *are full-rank matrices with rank* $k < p+1 < n$*. Then, (a) there is a unique full-rank matrix* $\mathbf{Q} : k\times k$ *such that* $\tilde{\mathbf{Z}} = \mathbf{Z}\mathbf{Q}$ *and* $\tilde{\mathbf{R}} = \mathbf{Q}^{-1}\mathbf{R}$*; and (b)* $\mathbf{P}_{\mathbf{Z}} = \mathbf{P}_{\mathbf{X}} = \mathbf{X}(\mathbf{X}^T\mathbf{X})^{-}\mathbf{X}^T$*.*

Let $\mathbf{C}\boldsymbol{\beta}(t)$ be estimable. Then by Theorem 7.2, there is a real matrix $\mathbf{A}$ such that $\mathbf{C} = \mathbf{A}\mathbf{X}$. It follows that

$$\begin{aligned} \mathbf{C}\boldsymbol{\beta}(t) &= \mathbf{A}\mathbf{X}\boldsymbol{\beta}(t) = \mathbf{A}\mathbf{Z}\boldsymbol{\phi}(t) = \mathbf{A}\mathbf{X}\mathbf{R}^T(\mathbf{R}\mathbf{R}^T)^{-1}\boldsymbol{\phi}(t) \\ &= \mathbf{C}\mathbf{R}^T(\mathbf{R}\mathbf{R}^T)^{-1}\boldsymbol{\phi}(t) = \mathbf{B}\boldsymbol{\phi}(t), \end{aligned} \tag{7.51}$$

where $\mathbf{B} = \mathbf{C}\mathbf{R}^T(\mathbf{R}\mathbf{R}^T)^{-1}$. Notice that (7.51) holds for any full-rank $\mathbf{R}$ such that the decomposition (7.48) is valid. The matrix $\mathbf{R}$ mentioned in Theorem 7.3 is a special case. Let $\widehat{\mathbf{C}\boldsymbol{\beta}(t)}$ denote the estimator of $\mathbf{C}\boldsymbol{\beta}(t)$ obtained by the relationship (7.51):

$$\widehat{\mathbf{C}\boldsymbol{\beta}(t)} = \mathbf{B}\hat{\boldsymbol{\phi}}(t) = \mathbf{C}\mathbf{R}^T(\mathbf{R}\mathbf{R}^T)^{-1}(\mathbf{Z}^T\mathbf{Z})^{-1}\mathbf{Z}^T\mathbf{y}(t). \tag{7.52}$$

Then we have the following theorem to identify the mean and covariance functions of $\widehat{\mathbf{C}\boldsymbol{\beta}(t)}$.

Theorem 7.8 *Let* $\mathbf{C} : q\times(p+1)$ *be a full-rank matrix with rank(*$\mathbf{C}$*)* $= q \le k < p+1 < n$ *such that* $\mathbf{C}\boldsymbol{\beta}(t)$ *is estimable. Let* $\widehat{\mathbf{C}\boldsymbol{\beta}(t)}$ *be as defined in (7.52).*

Then under the non-full-rank FLM (7.1) and with $\mathbf{B} = \mathbf{C}\mathbf{R}^T(\mathbf{R}\mathbf{R}^T)^{-1}$, *we have: (a)* $\widehat{\mathbf{C}\boldsymbol{\beta}(t)}$ *is an unbiased estimator of* $\mathbf{C}\boldsymbol{\beta}(t)$ *with* $Cov\left[\widehat{\mathbf{C}\boldsymbol{\beta}(s)}, \widehat{\mathbf{C}\boldsymbol{\beta}(t)}\right] = \gamma(s,t)\mathbf{B}(\mathbf{Z}^T\mathbf{Z})^{-1}\mathbf{B}^T$; *and (b)* $\widehat{\mathbf{C}\boldsymbol{\beta}(t)}$ *and* $\mathbf{B}(\mathbf{Z}^T\mathbf{Z})^{-1}\mathbf{B}^T$ *are invariant to the choice of the decompositions of* $\mathbf{X}$ *as defined in (7.48).*

7.3.4 Tests of Testable Linear Hypotheses

Consider the testable GLHT problem (7.32). By (7.51), it can be expressed as

$$H_0 : \mathbf{B}\boldsymbol{\phi}(t) \equiv \mathbf{c}(t), \quad \text{versus} \quad H_1 : \mathbf{B}\boldsymbol{\phi}(t) \neq \mathbf{c}(t), \text{ for some } t \in \mathcal{T}, \tag{7.53}$$

where $\mathbf{B} = \mathbf{C}\mathbf{R}^T(\mathbf{R}\mathbf{R}^T)^{-1}$. The above problem can be tested using the methodologies developed in Chapter 6 based on the full-rank FLM (7.49).

Notice that by Theorem 7.8, $\widehat{\mathbf{C}\boldsymbol{\beta}(t)} = \mathbf{B}\hat{\boldsymbol{\phi}}(t)$ and $\mathbf{P}_\mathbf{Z} = \mathbf{P}_\mathbf{X} = \mathbf{X}(\mathbf{X}^T\mathbf{X})^{-}\mathbf{X}^T$ are invariant to the choice of the reparameterization (7.48). It follows that

$$\begin{aligned} \text{SSH}_n(t) &= \left[\mathbf{B}\hat{\boldsymbol{\phi}}(t) - \mathbf{c}(t)\right]^T \left[\mathbf{B}(\mathbf{Z}^T\mathbf{Z})^{-1}\mathbf{B}^T\right]^{-1} \left[\mathbf{B}\hat{\boldsymbol{\phi}}(t) - \mathbf{c}(t)\right], \\ \text{SSE}_n(t) &= (n-k)\hat{\gamma}(t,t) = \mathbf{y}(t)^T(\mathbf{I}_n - \mathbf{P}_\mathbf{Z})\mathbf{y}(t), \end{aligned}$$

are also invariant. Therefore, the test results will be invariant to the choice of the reparameterization specified by (7.48).

An application of the reparameterization method to the audible noise data resulted in exactly the same results as those shown in Example 7.5.

7.4 Side-Condition Method

7.4.1 The Methodology

Under the non-full-rank FLM (7.1), the side-condition method aims to impose some side-conditions on $\boldsymbol{\beta}(t)$ so that the normal equation system (7.6) has a unique solution. Those side-conditions must be nonestimable functions of $\boldsymbol{\beta}(t)$.

Under the non-full-rank FLM (7.1), the design matrix $\mathbf{X} : n \times (p+1)$ is ill-conditioned with rank $k < p+1 < n$. Therefore, the deficiency in the rank of $\mathbf{X}$ is $(p+1-k)$. In order to make the solution to the normal equation system (7.6) unique, we must define the side-conditions that make up this deficiency in rank. Accordingly, we define the following side-conditions:

$$\mathbf{W}\boldsymbol{\beta}(t) = \mathbf{0}, \tag{7.54}$$

where $\mathbf{W} = [\mathbf{w}_1, \mathbf{w}_2, \cdots, \mathbf{w}_{p+1-k}]^T : (p+1-k) \times (p+1)$ is a full-rank matrix such that each of $\mathbf{w}_i^T\boldsymbol{\beta}(t), i = 1, 2, \cdots, p+1-k$ is nonestimable.

Theorem 7.9 *Under the non-full-rank FLM (7.1) and the side-condition (7.54), the normal equation system (7.6) has a unique solution* $\hat{\boldsymbol{\beta}}_\mathbf{W}(t) = (\mathbf{X}^T\mathbf{X} + \mathbf{W}^T\mathbf{W})^{-1}\mathbf{X}^T\mathbf{y}(t)$.

Based on $\hat{\boldsymbol{\beta}}_{\mathbf{W}}(t)$, the vector of the fitted subject-effect functions is given by

$$\hat{\mathbf{v}}_{\mathbf{W}}(t) = \mathbf{y}(t) - \mathbf{X}\hat{\boldsymbol{\beta}}_{\mathbf{W}}(t) = \mathbf{y}(t) - \mathbf{P}_{\mathbf{X}|\mathbf{W}}\mathbf{y}(t) = (\mathbf{I}_n - \mathbf{P}_{\mathbf{X}|\mathbf{W}})\mathbf{y}(t), \tag{7.55}$$

where

$$\mathbf{P}_{\mathbf{X}|\mathbf{W}} = \mathbf{X}(\mathbf{X}^T\mathbf{X} + \mathbf{W}^T\mathbf{W})^{-1}\mathbf{X}^T. \tag{7.56}$$

Theorem 7.10 *Under the non-full-rank FLM (7.1), we have* $\mathbf{P}_{\mathbf{X}|\mathbf{W}} = \mathbf{P}_{\mathbf{X}} = \mathbf{X}(\mathbf{X}^T\mathbf{X})^{-}\mathbf{X}^T$.

Theorem 7.10 says that $\mathbf{P}_{\mathbf{X}|\mathbf{W}}$ is a rank-k projection matrix of $\mathbf{X}$ and it is invariant to the choice of $\mathbf{W}$. From the proof of Theorem 7.10, we see that $\hat{\boldsymbol{\beta}}_{\mathbf{W}}(t)$ essentially specifies a particular solution to the normal equation system (7.6) by imposing the side-condition (7.54). Theorem 7.10 also indicates that the vector of the fitted subject-effect functions (7.55) is invariant to the choice of $\mathbf{W}$. Therefore,

$$\begin{aligned} & \hat{\mathbf{v}}_{\mathbf{W}}(s)^T\hat{\mathbf{v}}_{\mathbf{W}}(t) = \mathbf{y}(s)^T\left(\mathbf{I}_n - \mathbf{P}_{\mathbf{X}|\mathbf{W}}\right)\mathbf{y}(t) \\ = \ & \mathbf{y}(s)^T\left(\mathbf{I}_n - \mathbf{P}_{\mathbf{X}}\right)\mathbf{y}(t) = \mathbf{v}(s)^T\left(\mathbf{I}_n - \mathbf{P}_{\mathbf{X}}\right)\mathbf{v}(t). \end{aligned}$$

It follows that the unbiased estimator of the covariance function $\gamma(s,t)$ is given by

$$\hat{\gamma}_{\mathbf{W}}(s,t) = (n-k)^{-1}\hat{\mathbf{v}}_{\mathbf{W}}(s)^T\hat{\mathbf{v}}_{\mathbf{W}}(t) = (n-k)^{-1}\mathbf{y}(s)^T\left(\mathbf{I}_n - \mathbf{P}_{\mathbf{X}|\mathbf{W}}\right)\mathbf{y}(t), \tag{7.57}$$

which is invariant to the choice of $\mathbf{W}$.

7.4.2 Methods for Specifying the Side-Conditions

We now describe some methods for specifying the side-condition (7.54). When there are more than $(p+1-k)$ components of $\boldsymbol{\beta}(t) : (p+1)\times 1$ that are not estimable, we can find $(p+1-k)$ nonestimable linearly independent components from $\boldsymbol{\beta}(t)$ directly to form $\mathbf{W}\boldsymbol{\beta}(t)$ and to determine $\mathbf{W}$.

When it is not easy to see which components of $\boldsymbol{\beta}(t)$ are nonestimable, we can determine $\mathbf{W}$ by using the k linearly independent rows (7.15) of $\mathbf{X}$ or equivalently using the matrix $\mathbf{R}$ (7.17). Let the k linearly independent rows of $\mathbf{X}$ be $\mathbf{x}_{i_1}^T, \cdots, \mathbf{x}_{i_k}^T$ as given in (7.15). Find

$$\mathbf{w}_1, \mathbf{w}_2, \cdots, \mathbf{w}_{p+1-k} \tag{7.58}$$

from $\mathcal{R}^{p+1}$ so that $\mathbf{x}_{i_1}, \cdots, \mathbf{x}_{i_k}, \mathbf{w}_1, \cdots, \mathbf{w}_{p+1-k}$ are linearly independent. Set $\mathbf{W} = [\mathbf{w}_1, \mathbf{w}_2, \cdots, \mathbf{w}_{p+1-k}]^T$. Then $\mathbf{W}\boldsymbol{\beta}(t) = 0$ is the desired side-condition.

In practice, the $[(p+1)-k]$ linearly independent vectors (7.58) of $\mathcal{R}^{p+1}$ can be obtained using a simple method similar to the one for finding the k linearly independent rows of $\mathbf{X}$ as described in Section 7.2. We can use the rows of the identity matrix $\mathbf{I}_{p+1}$ that form a natural linear basis of the $(p+1)$-dimensional Euclidean space $\mathcal{R}^{p+1}$. We first check if the first row of $\mathbf{I}_{p+1}$ is linearly independent with $\mathbf{x}_{i_1}^T, \mathbf{x}_{i_2}^T, \cdots, \mathbf{x}_{i_k}^T$. If it is a "yes," we set it as $\mathbf{w}_1^T$; otherwise, we try the next row of $\mathbf{I}_{p+1}$. We continue this process until we find $[(p+1)-k]$ rows of $\mathbf{I}_{p+1}$ to obtain (7.58).

We can also use the SVD (7.7) of $\mathbf{X}$ to specify $\mathbf{W}$. Based on (7.7), we can write $\mathbf{X} = \mathbf{ZR}$ with $\mathbf{Z}$ and $\mathbf{R}$ as defined in (7.50) so that $\mathbf{R}\boldsymbol{\beta}(t)$ is estimable. Set $\mathbf{W} = (\mathbf{0}, \mathbf{I}_{p+1-k})\mathbf{V}^T : [(p+1)-k] \times (p+1)$, where $\mathbf{V} : (p+1) \times (p+1)$ is the orthonormal matrix defined in (7.7). Then

$$\begin{pmatrix} \mathbf{R} \\ \mathbf{W} \end{pmatrix} = \begin{pmatrix} \mathbf{D}_k^{1/2} & \mathbf{0} \\ \mathbf{0} & \mathbf{I}_{p+1-k} \end{pmatrix} \mathbf{V}^T$$

is a full-rank matrix. It follows that each component of $\mathbf{W}\boldsymbol{\beta}(t)$ is nonestimable and hence $\mathbf{W}\boldsymbol{\beta}(t) = \mathbf{0}$ gives a desired side-condition.

Example 7.8 *For the one-way functional ANOVA model (7.2), each of the components of* $\boldsymbol{\beta}(t) = [\eta(t), \alpha_1(t), \cdots, \alpha_a(t)]^T : (a+1) \times 1$ *is nonestimable as shown in Example 7.3 and we have rank*$(\mathbf{X}) = a$. *In this case, by the first method mentioned above, we only need to find the* $(a+1)-a = 1$ *nonestimable component from* $\boldsymbol{\beta}(t)$. *When we take* $\mathbf{W}\boldsymbol{\beta}(t) = \eta(t)$, *we have* $\mathbf{W} = \mathbf{e}_{1,a+1}^T$, *and when we take* $\mathbf{W}\boldsymbol{\beta}(t) = \alpha_i(t)$ *for* $1 \le i \le a$, *we have* $\mathbf{W} = \mathbf{e}_{i+1,a+1}^T$.

Example 7.9 *For the audible noise data, based on the matrix* $\mathbf{R}$ *given in (7.25), by the second method mentioned above, the side-condition matrix* $\mathbf{W}$ *is found to be*

$$\mathbf{W} = \begin{pmatrix} 1 & 0 & 0 & 0 & 0 & 0 & 0 & 0 & 0 & 0 & 0 & 0 & 0 & 0 & 0 \\ 0 & 1 & 0 & 0 & 0 & 0 & 0 & 0 & 0 & 0 & 0 & 0 & 0 & 0 & 0 \\ 0 & 0 & 0 & 1 & 0 & 0 & 0 & 0 & 0 & 0 & 0 & 0 & 0 & 0 & 0 \\ 0 & 0 & 0 & 0 & 0 & 1 & 0 & 0 & 0 & 0 & 0 & 0 & 0 & 0 & 0 \\ 0 & 0 & 0 & 0 & 0 & 0 & 0 & 1 & 0 & 0 & 0 & 0 & 0 & 0 & 0 \\ 0 & 0 & 0 & 0 & 0 & 0 & 0 & 0 & 0 & 1 & 0 & 0 & 0 & 0 & 0 \\ 0 & 0 & 0 & 0 & 0 & 0 & 0 & 0 & 0 & 0 & 0 & 1 & 0 & 0 & 0 \end{pmatrix}. \quad (7.59)$$

Then a desired side-condition is given by $\mathbf{W}\boldsymbol{\beta}(t) = \mathbf{0}$.

7.4.3 Invariance of the Side-Condition Method

It is obvious that $\hat{\boldsymbol{\beta}}_{\mathbf{W}}(t)$ depends on the choice of $\mathbf{W}$. However, it is not so obvious if $\mathbf{C}\hat{\boldsymbol{\beta}}_{\mathbf{W}}(t)$ depends on the choice of $\mathbf{W}$ when $\mathbf{C}\boldsymbol{\beta}(t)$ is estimable. The following theorem gives an answer.

Theorem 7.11 *Let* $\mathbf{C} : q \times (p+1)$ *be a full-rank matrix with rank*$(\mathbf{C}) = q \le k < p+1 < n$ *such that* $\mathbf{C}\boldsymbol{\beta}(t)$ *is estimable. Then under the non-full-rank FLM (7.1), we have that (a)* $\mathbf{C}\hat{\boldsymbol{\beta}}_{\mathbf{W}}(t)$ *is an unbiased linear estimator of* $\mathbf{C}\boldsymbol{\beta}(t)$ *with Cov*$\left[\mathbf{C}\hat{\boldsymbol{\beta}}_{\mathbf{W}}(s), \mathbf{C}\hat{\boldsymbol{\beta}}_{\mathbf{W}}(t)\right] = \gamma(s,t)\mathbf{C}(\mathbf{X}^T\mathbf{X} + \mathbf{W}^T\mathbf{W})^{-1}\mathbf{C}^T$; *and (b)* $\mathbf{C}\hat{\boldsymbol{\beta}}_{\mathbf{W}}(t)$ *and* $\mathbf{C}(\mathbf{X}^T\mathbf{X} + \mathbf{W}^T\mathbf{W})^{-1}\mathbf{C}^T$ *are invariant to the choice of* $\mathbf{W}$.

Theorem 7.11 says that provided $\mathbf{C}\boldsymbol{\beta}(t)$ is estimable, $\mathbf{C}\hat{\boldsymbol{\beta}}_{\mathbf{w}}(t)$ will be the same for different choices of $\mathbf{W}$. This invariance property is important for the side-condition method as different ways can be used to specify the side-condition (7.54) as shown in the previous subsection, and the final inference results will not be affected by different choices of the side-condition.

7.4.4   Tests of Testable Linear Hypotheses

Under the non-full-rank FLM (7.1) with rank$(\mathbf{X}) = k < (p+1) < n$, consider the testable GLHT problem (7.32) by the side-condition method. From Theorem 7.9, we have $\hat{\boldsymbol{\beta}}_{\mathbf{w}}(t) = (\mathbf{X}^T\mathbf{X} + \mathbf{W}^T\mathbf{W})^{-1}\mathbf{X}^T\mathbf{y}(t)$. By Theorem 7.11, we have $\mathrm{E}\left[\mathbf{C}\hat{\boldsymbol{\beta}}_{\mathbf{w}}(t) - \mathbf{c}(t)\right] = \mathbf{C}\boldsymbol{\beta}(t) - \mathbf{c}(t)$ and

$$\mathrm{Cov}\left[\mathbf{C}\hat{\boldsymbol{\beta}}_{\mathbf{w}}(s) - \mathbf{c}(s), \mathbf{C}\hat{\boldsymbol{\beta}}_{\mathbf{w}}(t) - \mathbf{c}(t)\right] = \gamma(s,t)\mathbf{C}(\mathbf{X}^T\mathbf{X} + \mathbf{W}^T\mathbf{W})^{-1}\mathbf{C}^T.$$

By Theorem 7.11 again, $\mathbf{C}(\mathbf{X}^T\mathbf{X} + \mathbf{W}^T\mathbf{W})^{-1}\mathbf{C}^T$ is a full-rank matrix of rank q. Thus, we can set the following pivotal test function for (7.32):

$$\mathbf{z}_{\mathbf{w}}(t) = \left[\mathbf{C}(\mathbf{X}^T\mathbf{X} + \mathbf{W}^T\mathbf{W})^{-1}\mathbf{C}^T\right]^{-1/2}\left[\mathbf{C}\hat{\boldsymbol{\beta}}_{\mathbf{w}}(t) - \mathbf{c}(t)\right]. \qquad (7.60)$$

The mean function vector and the covariance function matrix of $\mathbf{z}_{\mathbf{w}}(t)$ are given by

$$\begin{aligned} \boldsymbol{\eta}_{z|\mathbf{w}}(t) &= \left[\mathbf{C}(\mathbf{X}^T\mathbf{X} + \mathbf{W}^T\mathbf{W})^{-1}\mathbf{C}^T\right]^{-1/2}[\mathbf{C}\boldsymbol{\beta}(t) - \mathbf{c}(t)], \\ \boldsymbol{\Gamma}_{z|\mathbf{w}}(s,t) &= \gamma(s,t)\mathbf{I}_q. \end{aligned} \qquad (7.61)$$

It says that the components of $\mathbf{z}_{\mathbf{w}}(t)$ are uncorrelated. Under the Gaussian assumption, we have the following result.

Theorem 7.12 *Assume that in the non-full-rank FLM (7.1),* $\mathbf{v}(t) \sim GP_n(\mathbf{0}, \gamma\mathbf{I}_n)$ *and* $tr(\gamma) < \infty$. *Then we have*

$$\mathbf{z}_{\mathbf{w}}(t) \sim GP_q(\boldsymbol{\eta}_{z|\mathbf{w}}, \gamma\mathbf{I}_q), \quad (n-k)\hat{\gamma}_{\mathbf{w}}(s,t) \sim WP(n-k, \gamma).$$

Based on the above theorem, we can describe various tests for the testable GLHT problem (7.32). Based on $\hat{\boldsymbol{\beta}}_{\mathbf{w}}(t)$, we define the following pointwise SSH and SSE as

$$\begin{aligned} \mathrm{SSH}_{n\mathbf{w}}(t) &= \|\mathbf{z}_{\mathbf{w}}(t)\|^2 = \left[\mathbf{C}\hat{\boldsymbol{\beta}}_{\mathbf{w}}(t) - \mathbf{c}(t)\right]^T \\ &\quad \times \left[\mathbf{C}(\mathbf{X}^T\mathbf{X} + \mathbf{W}^T\mathbf{W})^{-1}\mathbf{C}^T\right]^{-1}\left[\mathbf{C}\hat{\boldsymbol{\beta}}_{\mathbf{w}}(t) - \mathbf{c}(t)\right], \\ \mathrm{SSE}_{n\mathbf{w}}(t) &= (n-k)\hat{\gamma}_{\mathbf{w}}(t,t) = \|\hat{\mathbf{v}}_{\mathbf{w}}(t)\|^2 = \mathbf{y}(t)^T(\mathbf{I}_n - \mathbf{P}_{\mathbf{x}|\mathbf{w}})\mathbf{y}(t). \end{aligned}$$

We have the following theorem stating that the associated integrated SSH and SSE are χ^2-type mixtures.

Theorem 7.13 *Assume that in the non-full-rank FLM (7.1),* $\mathbf{v}(t) \sim GP_n(\mathbf{0}, \gamma \mathbf{I}_n)$ *and* $tr(\gamma) < \infty$. *Under the null hypothesis in (7.32), we have*

$$\begin{aligned} \int_{\mathcal{T}} SSH_{n\mathbf{w}}(t)dt &\stackrel{d}{=} \sum_{r=1}^{m} \lambda_r A_r, \; A_r \stackrel{i.i.d.}{\sim} \chi_q^2, \\ \int_{\mathcal{T}} SSE_{n\mathbf{w}}(t)dt &\stackrel{d}{=} \sum_{r=1}^{m} \lambda_r E_r, \; E_r \stackrel{i.i.d.}{\sim} \chi_{n-k}^2, \end{aligned}$$

where $\lambda_1, \cdots, \lambda_m$ *are all the positive eigenvalues of* $\gamma(s,t)$.

Using $\text{SSH}_{n\mathbf{w}}(t)$ and $\text{SSE}_{n\mathbf{w}}(t)$, we can now present various tests for the testable GLHT problem (7.32). They are nearly the same as those presented in Section 7.2.4.

Pointwise F-Test The pointwise F-test for (7.32) is conducted using the test statistic $F_{n\mathbf{w}}(t) = \frac{\text{SSH}_{n\mathbf{w}}(t)}{\text{SSE}_{n\mathbf{w}}(t)}, t \in \mathcal{T}$. Under the Gaussian assumption and the null hypothesis in (7.32), we have

$$F_{n\mathbf{w}}(t) \sim F_{q,n-k}, t \in \mathcal{T}.$$

The pointwise F-test can then be conducted accordingly.

L^2-Norm-Based Test The L^2-norm-based test for (7.32) uses the test statistic

$$T_{n\mathbf{w}} = \int_{\mathcal{T}} \text{SSH}_{n\mathbf{w}}(t)dt.$$

Under the Gaussian assumption and the null hypothesis in (7.32), by Theorem 7.13, we can approximate the null distribution of $T_{n\mathbf{w}}$ by the Welch-Satterthwaite χ^2-approximation method described in Section 4.3 of Chapter 4 and obtain

$$T_{n\mathbf{w}} \sim \beta \chi_d^2 \text{ approximately, where } \beta = \frac{\text{tr}(\gamma^{\otimes 2})}{\text{tr}(\gamma)}, \; \kappa = \frac{\text{tr}^2(\gamma)}{\text{tr}(\gamma^{\otimes 2})}, \; d = q\kappa.$$

The parameters β and κ can be estimated using the naive method or the bias-reduced method as described in Section 7.2.4 but replacing $\hat{\gamma}(s,t)$ in the expressions in (7.40) and (7.42) by $\hat{\gamma}_{\mathbf{w}}(s,t)$ given in (7.57). The L^2-norm-based test for (7.32) can then be conducted accordingly.

F-Type Test The F-type test statistic is now defined as

$$F_{n\mathbf{w}} = \frac{\int_{\mathcal{T}} \text{SSH}_{n\mathbf{w}}(t)dt/q}{\int_{\mathcal{T}} \text{SSE}_{n\mathbf{w}}(t)dt/(n-k)}.$$

Under the Gaussian assumption and the null hypothesis in (7.32) and by Theorem 7.13, we have

$$F_{n\mathbf{w}} \stackrel{d}{=} \frac{\sum_{r=1}^{m} \lambda_r A_r/q}{\sum_{r=1}^{m} \lambda_r E_r/(n-k)},$$

where $A_r \overset{i.i.d.}{\sim} \chi_q^2$, $E_r \overset{i.i.d.}{\sim} \chi_{n-k}^2$ and they are all independent, and $\lambda_1, \lambda_2, \cdots, \lambda_m$ are all the positive eigenvalues of $\gamma(s,t)$. It follows that the null distribution of $F_{n\mathbf{w}}$ can be approximated by the two-cumulant matched F-approximation method described in Section 4.4 of Chapter 4. By that method, we have

$$F_{n\mathbf{w}} \sim F_{q\hat{\kappa},(n-k)\hat{\kappa}} \text{ approximately,}$$

where $\hat{\kappa}$ is obtained by the naive method or the bias-reduced method as described above for the L^2-norm-based test. The F-type test for (7.32) is then conducted accordingly.

Bootstrap Tests As for the generalized inverse method, when the Gaussian assumption is not valid, we can use some nonparametric bootstrap tests, which are briefly described as follows. Let $v_{i\mathbf{w}}^*(t), i = 1, 2, \cdots, n$, be a bootstrap sample randomly generated from the estimated subject-effect functions $\hat{v}_{i\mathbf{w}}(t), i = 1, 2, \cdots, n$, the n-components of $\hat{\mathbf{v}}_{\mathbf{w}}(t), t \in \mathcal{T}$ as defined in (7.55). Set $\mathbf{v}_{\mathbf{w}}^*(t) = [v_{1\mathbf{w}}^*(t), \cdots, v_{n\mathbf{w}}^*(t)]^T$. Compute $\mathbf{y}_{\mathbf{w}}^*(t) = \mathbf{X}\hat{\boldsymbol{\beta}}_{\mathbf{w}}(t) + \mathbf{v}_{\mathbf{w}}^*(t)$ and $\hat{\boldsymbol{\beta}}_{\mathbf{w}}^*(t) = (\mathbf{X}^T\mathbf{X} + \mathbf{W}^T\mathbf{W})^{-1}\mathbf{X}^T\mathbf{y}_{\mathbf{w}}^*(t)$. Then compute $\text{SSH}_{n\mathbf{w}}^*(t) = \left[\hat{\boldsymbol{\beta}}_{\mathbf{w}}^*(t) - \hat{\boldsymbol{\beta}}_{\mathbf{w}}(t)\right]^T \mathbf{C}^T[\mathbf{C}(\mathbf{X}^T\mathbf{X} + \mathbf{W}^T\mathbf{W})^{-1}\mathbf{C}^T]^{-1}\mathbf{C}\left[\hat{\boldsymbol{\beta}}_{\mathbf{w}}^*(t) - \hat{\boldsymbol{\beta}}_{\mathbf{w}}(t)\right]$ and $\text{SSE}_{n\mathbf{w}}^*(t) = \mathbf{y}_{\mathbf{w}}^*(t)^T(\mathbf{I}_n - \mathbf{P}_{\mathbf{x}|\mathbf{w}})\mathbf{y}_{\mathbf{w}}^*(t)$. Notice that the values of $\text{SSH}_{n\mathbf{w}}^*(t)$ and $\text{SSE}_{n\mathbf{w}}^*(t)$ are invariant to the choice of $\mathbf{W}$. The pointwise bootstrap test statistic, the L^2-norm-based bootstrap test statistic, or the F-type bootstrap test statistic can then be computed as

$$F_{n\mathbf{w}}^*(t) = \frac{\text{SSH}_{n\mathbf{w}}^*(t)/q}{\text{SSE}_{n\mathbf{w}}^*(t)/(n-k)}, \; T_{n\mathbf{w}}^* = \int_{\mathcal{T}} \text{SSH}_{n\mathbf{w}}^*(t)dt, \text{ or}$$
$$F_{n\mathbf{w}}^* = \frac{\int_{\mathcal{T}} \text{SSH}_{n\mathbf{w}}^*(t)dt/q}{\int_{\mathcal{T}} \text{SSE}_{n\mathbf{w}}^*(t)dt/(n-k)}.$$

Repeat the above bootstrapping process a large number of times, calculate its $100(1-\alpha)$-percentile, and then conduct the pointwise bootstrap test, the L^2-norm-based bootstrap test, or the F-type bootstrap test accordingly.

An application of the side-condition method to the audible noise data resulted in exactly the same results as those presented in Example 7.5.

7.5 Technical Proofs

In this section, we outline the proofs of the main results described in this chapter.

Proof of Theorem 7.1 Assertion (a) is obvious. Assertion (b) follows from

$$\text{E}\hat{\boldsymbol{\beta}}(t) = (\mathbf{X}^T\mathbf{X})^-\mathbf{X}^T\text{E}[\mathbf{y}(t)] = (\mathbf{X}^T\mathbf{X})^-\mathbf{X}^T\mathbf{X}\boldsymbol{\beta}(t),$$

which depends on $(\mathbf{X}^T\mathbf{X})^-$ and is not equal to $\boldsymbol{\beta}(t)$. We now show Assertion (c). Suppose we have a linear function $\mathbf{Ay}(t)$ that estimates $\boldsymbol{\beta}(t)$ unbiasedly. Then we have

$$\boldsymbol{\beta}(t) = \mathrm{E}[\mathbf{Ay}(t)] = \mathrm{E}\left[\mathbf{AX}\boldsymbol{\beta}(t) + \mathbf{Av}(t)\right] = \mathbf{AX}\boldsymbol{\beta}(t).$$

As this must hold for all possible $\boldsymbol{\beta}(t)$, we must have $\mathbf{I}_{p+1} = \mathbf{AX}$. Then $p+1 = \mathrm{rank}(\mathbf{I}_{p+1}) = \mathrm{rank}(\mathbf{AX}) \le \mathrm{rank}(\mathbf{X}) = k < p+1$. This contradiction implies that there does not exist $\mathbf{A}$ such that $\mathrm{E}\left[\mathbf{Ay}(t)\right] = \boldsymbol{\beta}(t)$. Therefore, $\boldsymbol{\beta}(t)$ is not estimable. The theorem is proved.

Proof of Theorem 7.2 If there exists a constant matrix $\mathbf{A}$ such that $\mathbf{C} = \mathbf{AX}$, then $\mathrm{E}\left[\mathbf{Ay}(t)\right] = \mathbf{AX}\boldsymbol{\beta}(t) = \mathbf{C}\boldsymbol{\beta}(t)$, showing that $\mathbf{C}\boldsymbol{\beta}(t)$ is estimable. Conversely, if $\mathbf{C}\boldsymbol{\beta}(t)$ is estimable, then there is a constant matrix $\mathbf{A} : q \times n$ such that $\mathrm{E}\left[\mathbf{Ay}(t)\right] = \mathbf{C}\boldsymbol{\beta}(t)$. Under the non-full-rank FLM (7.1), we have $\mathrm{E}\left[\mathbf{Ay}(t)\right] = \mathbf{AX}\boldsymbol{\beta}(t)$. It follows that $\mathbf{C}\boldsymbol{\beta}(t) = \mathbf{AX}\boldsymbol{\beta}(t)$ holds for any $\boldsymbol{\beta}(t)$ so that $\mathbf{C} = \mathbf{AX}$.

Proof of Theorem 7.3 For the first assertion, the "if" part is obvious. We now prove its "only if" part. By Theorem 7.2, if $\mathbf{C}\boldsymbol{\beta}(t)$ is estimable, there exists a constant matrix $\mathbf{A}$ such that $\mathbf{C} = \mathbf{AX}$. As $\mathbf{X} = \mathbf{ZR}$, we have $\mathbf{C} = \mathbf{AZR} = \mathbf{BR}$ as desired where $\mathbf{B} = \mathbf{AZ}$. To show the remaining assertions, notice that when $\mathbf{C}\boldsymbol{\beta}(t)$ is estimable, we have $\mathbf{CR}^T = \mathbf{BRR}^T$. The expression (7.20) follows immediately. However, when $\mathbf{C}\boldsymbol{\beta}(t)$ is nonestimable, we must have $\mathbf{C} \neq \mathbf{CR}^T(\mathbf{RR}^T)^{-1}\mathbf{R}$, as otherwise, by setting $\mathbf{B} = \mathbf{CR}^T(\mathbf{RR}^T)^{-1}$, we have $\mathbf{C} = \mathbf{BR}$ which implies that $\mathbf{C}\boldsymbol{\beta}(t)$ is estimable. This contradicts the assumption that $\mathbf{C}\boldsymbol{\beta}(t)$ is nonestimable. The theorem is proved.

Proof of Theorem 7.4 By Theorem 7.2, if $\mathbf{C}\boldsymbol{\beta}(t)$ is estimable, there exists a constant matrix $\mathbf{A}$ such that $\mathbf{C} = \mathbf{AX}$. By (7.13), we have

$$\begin{aligned}
\mathrm{E}\left[\mathbf{C}\hat{\boldsymbol{\beta}}(t)\right] &= \mathbf{C}(\mathbf{X}^T\mathbf{X})^-\mathbf{X}^T\mathbf{X}\boldsymbol{\beta}(t) = \mathbf{AP_x X}\boldsymbol{\beta}(t)\\
&= \mathbf{AX}\boldsymbol{\beta}(t) = \mathbf{C}\boldsymbol{\beta}(t),\\
\mathrm{Cov}\left[\mathbf{C}\hat{\boldsymbol{\beta}}(s), \mathbf{C}\hat{\boldsymbol{\beta}}(t)\right] &= \gamma(s,t)\mathbf{C}(\mathbf{X}^T\mathbf{X})^-\mathbf{X}^T\mathbf{X}(\mathbf{X}^T\mathbf{X})^-\mathbf{C}^T\\
&= \gamma(s,t)\mathbf{AX}(\mathbf{X}^T\mathbf{X})^-\mathbf{X}^T\mathbf{X}(\mathbf{X}^T\mathbf{X})^-\mathbf{X}^T\mathbf{A}^T\\
&= \gamma(s,t)\mathbf{AX}(\mathbf{X}^T\mathbf{X})^-\mathbf{X}^T\mathbf{A}^T\\
&= \gamma(s,t)\mathbf{C}(\mathbf{X}^T\mathbf{X})^-\mathbf{C}^T.
\end{aligned}$$

This proves (a). To show (b), let $\mathbf{G}_1$ and $\mathbf{G}_2$ be any two generalized inverses of $\mathbf{X}^T\mathbf{X}$ and set $\hat{\boldsymbol{\beta}}_1(t) = \mathbf{G}_1\mathbf{X}^T\mathbf{y}(t)$ and $\hat{\boldsymbol{\beta}}_2(t) = \mathbf{G}_2\mathbf{X}^T\mathbf{y}(t)$. Then by (7.12), we have $\mathbf{XG}_1\mathbf{X}^T = \mathbf{XG}_2\mathbf{X}^T = \mathbf{P_x}$. It follows that we have

$$\begin{gathered}
\mathbf{C}\hat{\boldsymbol{\beta}}_1(t) - \mathbf{C}\hat{\boldsymbol{\beta}}_2(t) = \mathbf{C}(\mathbf{G}_1 - \mathbf{G}_2)\mathbf{X}^T\mathbf{y}(t) = \mathbf{A}(\mathbf{XG}_1\mathbf{X}^T - \mathbf{XG}_2\mathbf{X}^T)\mathbf{y}(t) = 0,\\
\mathbf{CG}_1\mathbf{C}^T - \mathbf{CG}_2\mathbf{C}^T = \mathbf{C}(\mathbf{G}_1 - \mathbf{G}_2)\mathbf{C}^T = \mathbf{A}(\mathbf{XG}_1\mathbf{X}^T - \mathbf{XG}_2\mathbf{X}^T)\mathbf{A}^T = 0.
\end{gathered}$$

This shows (b). We now show (c). As $\mathbf{C} = \mathbf{AX}$ and $\mathbf{P_x X} = \mathbf{X}$, we have $\mathbf{C} =$

$\mathbf{C}(\mathbf{X}^T\mathbf{X})^-\mathbf{X}^T\mathbf{X}$. Using the fact that $\text{rank}(\mathbf{S}_1\mathbf{S}_2) \le \min[\text{rank}(\mathbf{S}_1), \text{rank}(\mathbf{S}_2)]$, we have

$$\text{rank}(\mathbf{C}) \le \text{rank}\left[(\mathbf{C}\mathbf{X}^T\mathbf{X})^-\mathbf{X}^T\right] \le \text{rank}(\mathbf{C}),$$

implying that $\text{rank}\left[\mathbf{C}(\mathbf{X}^T\mathbf{X})^-\mathbf{X}^T\right] = \text{rank}(\mathbf{C})$. Using the fact that $\text{rank}(\mathbf{S}) = \text{rank}(\mathbf{S}\mathbf{S}^T)$, we have

$$\begin{aligned} q &= \text{rank}(\mathbf{C}) = \text{rank}(\mathbf{C}(\mathbf{X}^T\mathbf{X})^-\mathbf{X}^T) \\ &= \text{rank}\left[\mathbf{C}(\mathbf{X}^T\mathbf{X})^-\mathbf{X}^T\mathbf{X}(\mathbf{X}^T\mathbf{X})^-\mathbf{C}^T\right] \\ &= \text{rank}\left[\mathbf{A}\mathbf{P}_\mathbf{X}\mathbf{A}^T\right] = \text{rank}\left[\mathbf{C}(\mathbf{X}^T\mathbf{X})^-\mathbf{C}^T\right]. \end{aligned}$$

This shows (c). The whole theorem is proved.

Proof of Theorem 7.5 It is easy to show that $\mathbf{C}\hat{\boldsymbol{\beta}}(t) = \mathbf{C}\boldsymbol{\beta}(t) + \mathbf{C}(\mathbf{X}^T\mathbf{X})^-\mathbf{X}^T\mathbf{v}(t)$. As $\mathbf{v}(t) \sim \text{GP}_n(\mathbf{0}, \gamma\mathbf{I}_n)$, we have $\mathbf{C}[\hat{\boldsymbol{\beta}}(t) - \boldsymbol{\beta}(t)] \sim \text{GP}_q(\mathbf{0}, \gamma\mathbf{C}(\mathbf{X}^T\mathbf{X})^-\mathbf{C}^T)$. We then have $\mathbf{z}(t) \sim \text{GP}_q(\boldsymbol{\eta}_z, \gamma\mathbf{I}_q)$. By (7.31), we have $(n-k)\hat{\gamma}(s,t) = \mathbf{v}(s)^T(\mathbf{I}_n - \mathbf{P}_\mathbf{X})\mathbf{v}(t)$. As $\mathbf{I}_n - \mathbf{P}_\mathbf{X}$ is an idempotent matrix with rank $n-k$, by Theorem 4.8, we have $(n-k)\hat{\gamma}(s,t) \sim \text{WP}(n-k,\gamma)$. The theorem is proved.

Proof of Theorem 7.6 Under the given conditions and by Theorem 7.5, we have $\mathbf{z}(t) \sim \text{GP}_q(\mathbf{0}, \gamma\mathbf{I}_n)$ and $(n-k)\hat{\gamma}(s,t) \sim \text{WP}(n-k,\gamma)$. Then by Theorem 4.10 we have

$$\int_\mathcal{T} \text{SSH}_n(t)dt = \int_\mathcal{T} \|\mathbf{z}(t)\|^2 dt \stackrel{d}{=} \sum_{r=1}^m \lambda_r A_r, \quad A_r \stackrel{i.i.d.}{\sim} \chi_q^2,$$

where $\lambda_r, r = 1, 2, \cdots, m$ are all the positive eigenvalues of $\gamma(s,t)$. Similarly, by Theorem 4.5, we have $\int_\mathcal{T} \text{SSE}_n(t)dt = \text{tr}[(n-k)\hat{\gamma}] \stackrel{d}{=} \sum_{r=1}^m \lambda_r E_r, E_r \stackrel{i.i.d.}{\sim} \chi_{n-k}^2$ as desired. The theorem is proved.

Proof of Theorem 7.7 As $\mathbf{X} = \mathbf{Z}\mathbf{R} = \tilde{\mathbf{Z}}\tilde{\mathbf{R}}$ with $\mathbf{Z} : n \times k, \mathbf{R} : k \times (p+1), \tilde{\mathbf{Z}} : n \times k$ and $\tilde{\mathbf{R}} : k \times (p+1)$ are full-rank matrices with rank k, we have that $\tilde{\mathbf{R}}\tilde{\mathbf{R}}^T$ is a full-rank matrix and $\mathbf{Z}\mathbf{R}\tilde{\mathbf{R}}^T = \tilde{\mathbf{Z}}\tilde{\mathbf{R}}\tilde{\mathbf{R}}^T$. It follows that $\tilde{\mathbf{Z}} = \mathbf{Z}\mathbf{Q}$ where $\mathbf{Q} = \mathbf{R}\tilde{\mathbf{R}}^T(\tilde{\mathbf{R}}\tilde{\mathbf{R}}^T)^{-1}$ is unique and full-rank. We then have $\mathbf{Z}\mathbf{R} = \mathbf{Z}\mathbf{Q}\tilde{\mathbf{R}}$ and hence $\mathbf{Z}^T\mathbf{Z}\mathbf{R} = \mathbf{Z}^T\mathbf{Z}\mathbf{Q}\tilde{\mathbf{R}}$. As $\mathbf{Z}^T\mathbf{Z}$ is a full-rank matrix, we have $\tilde{\mathbf{R}} = \mathbf{Q}^{-1}\mathbf{R}$. Assertion (a) is proved.

To show (b), let $\mathbf{Z}^*$ and $\mathbf{R}^*$ denote the matrices $\mathbf{Z}$ and $\mathbf{R}$ defined in (7.50) obtained using the SVD (7.7). For this particular decomposition, by (7.12) and (7.50), it is straightforward to show that

$$\mathbf{P}_{\mathbf{Z}^*} = \mathbf{Z}^*(\mathbf{Z}^{*T}\mathbf{Z}^*)^{-1}\mathbf{Z}^{*T} = \mathbf{U}\begin{pmatrix} \mathbf{I}_k & \mathbf{0} \\ \mathbf{0} & \mathbf{0} \end{pmatrix}\mathbf{U}^T = \mathbf{P}_\mathbf{X}. \tag{7.62}$$

Let $\mathbf{X} = \mathbf{Z}\mathbf{R}$ be any decomposition of $\mathbf{X}$ such that $\mathbf{Z} : n \times k$ and $\mathbf{R} : k \times (p+1)$

are full-rank matrices with rank $k < p+1 < n$. By (a), there is a unique full-rank matrix $\mathbf{Q} : k \times k$ such that $\mathbf{Z} = \mathbf{Z}^*\mathbf{Q}$ and $\mathbf{R} = \mathbf{Q}^{-1}\mathbf{R}^*$. It follows that

$$\begin{aligned}\mathbf{P}_{\mathbf{Z}} &= \mathbf{Z}(\mathbf{Z}^T\mathbf{Z})^{-1}\mathbf{Z}^T = \mathbf{Z}^*\mathbf{Q}\left[(\mathbf{Z}^*\mathbf{Q})^T(\mathbf{Z}^*\mathbf{Q})\right]^{-1}(\mathbf{Z}^*\mathbf{Q})^T \\ &= \mathbf{Z}^*(\mathbf{Z}^{*T}\mathbf{Z}^*)^{-1}\mathbf{Z}^{*T} = \mathbf{P}_{\mathbf{Z}^*} = \mathbf{P}_{\mathbf{X}}.\end{aligned}$$

The theorem is proved.

Proof of Theorem 7.8 First of all, we have $\mathrm{E}\hat{\boldsymbol{\phi}}(t) = \boldsymbol{\phi}(t)$ and $\mathrm{Cov}(\hat{\boldsymbol{\phi}}(s), \hat{\boldsymbol{\phi}}(t)) = \gamma(s,t)(\mathbf{Z}^T\mathbf{Z})^{-1}$. It follows from (7.52) that

$$\begin{aligned}\mathrm{E}\widehat{\mathbf{C}\boldsymbol{\beta}(t)} &= \mathbf{B}\boldsymbol{\phi}(t) = \mathbf{C}\boldsymbol{\beta}(t), \\ \mathrm{Cov}\left[\widehat{\mathbf{C}\boldsymbol{\beta}(s)}, \widehat{\mathbf{C}\boldsymbol{\beta}(t)}\right] &= \gamma(s,t)\mathbf{B}(\mathbf{Z}^T\mathbf{Z})^{-1}\mathbf{B}^T,\end{aligned}$$

where $\mathbf{B} = \mathbf{C}\mathbf{R}^T(\mathbf{R}\mathbf{R}^T)^{-1}$. Assertion (a) is proved. To show (b), let $\mathbf{Z}^*$ and $\mathbf{R}^*$ be as defined in the proof of Theorem 7.7. Then by Theorem 7.7, there exists a unique full-rank matrix $\mathbf{Q} : k \times k$ such that $\mathbf{Z} = \mathbf{Z}^*\mathbf{Q}$ and $\mathbf{R} = \mathbf{Q}^{-1}\mathbf{R}^*$. It follows that

$$\begin{aligned}&\mathbf{R}^T(\mathbf{R}\mathbf{R}^T)^{-1}(\mathbf{Z}^T\mathbf{Z})^{-1}\mathbf{Z}^T = \mathbf{R}^{*T}(\mathbf{R}^*\mathbf{R}^{*T})^{-1}(\mathbf{Z}^{*T}\mathbf{Z}^*)^{-1}\mathbf{Z}^{*T}, \\ &\mathbf{B} = \mathbf{C}\mathbf{R}^T(\mathbf{R}\mathbf{R}^T)^{-1} = \mathbf{C}\mathbf{R}^{*T}(\mathbf{R}^*\mathbf{R}^{*T})^{-1} = \mathbf{B}^*\mathbf{Q},\end{aligned}$$

Therefore,

$$\begin{aligned}\widehat{\mathbf{C}\boldsymbol{\beta}(t)} &= \mathbf{C}\mathbf{R}^T(\mathbf{R}\mathbf{R}^T)^{-1}(\mathbf{Z}^T\mathbf{Z})^{-1}\mathbf{Z}^T\mathbf{y}(t) \\ &= \mathbf{C}\mathbf{R}^{*T}(\mathbf{R}^*\mathbf{R}^{*T})^{-1}(\mathbf{Z}^{*T}\mathbf{Z}^*)^{-1}\mathbf{Z}^{*T}\mathbf{y}(t), \\ \mathbf{B}(\mathbf{Z}^T\mathbf{Z})^{-1}\mathbf{B}^T &= \mathbf{B}^*\mathbf{Q}(\mathbf{Q}^T\mathbf{Z}^{*T}\mathbf{Z}^*\mathbf{Q})^{-1}(\mathbf{B}^*\mathbf{Q})^T \\ &= \mathbf{B}^*(\mathbf{Z}^{*T}\mathbf{Z}^*)^{-1}\mathbf{B}^{*T}.\end{aligned}$$

The theorem is proved.

Proof of Theorem 7.9 As each of $\mathbf{w}_i^T\boldsymbol{\beta}(t), i = 1, 2, \cdots, p+1-k$ is nonestimable, the rows of $\mathbf{W}$ and the rows of $\mathbf{X}$ must be linearly independent because otherwise, there exists one $\mathbf{w}_i$ such that $\mathbf{w}_i^T = \mathbf{a}_i^T\mathbf{X}$ for some $a_i : n \times 1$, meaning that $\mathbf{w}_i^T\boldsymbol{\beta}(t)$ is estimable. It follows that the matrix

$$\begin{pmatrix}\mathbf{X} \\ \mathbf{W}\end{pmatrix} : (p+1) \times (n+p+1-k)$$

is a full-rank matrix with rank $p+1$. Therefore,

$$\begin{pmatrix}\mathbf{X} \\ \mathbf{W}\end{pmatrix}^T\begin{pmatrix}\mathbf{X} \\ \mathbf{W}\end{pmatrix} = \mathbf{X}^T\mathbf{X} + \mathbf{W}^T\mathbf{W}$$

is a full-rank matrix with rank $p+1$ and hence is nonsingular. As $\mathbf{W}\mathbf{W}^T : (p+1-k) \times (p+1-k)$ is full-rank, $\mathbf{W}\boldsymbol{\beta}(t) = \mathbf{0}$ if and only if $\mathbf{W}^T\mathbf{W}\boldsymbol{\beta}(t) = 0$.

It follows that under the side-condition (7.54), the normal equation system (7.6) is equivalent to the following equation system:

$$(\mathbf{X}^T\mathbf{X} + \mathbf{W}^T\mathbf{W})\boldsymbol{\beta}(t) = \mathbf{X}^T\mathbf{y}(t),$$

which has a unique solution $\hat{\boldsymbol{\beta}}_{\mathbf{W}}(t) = (\mathbf{X}^T\mathbf{X}+\mathbf{W}^T\mathbf{W})^{-1}\mathbf{X}^T\mathbf{y}(t)$. The theorem is proved.

Proof of Theorem 7.10 By (7.14), any solution to the normal equation system (7.6) can be expressed as $\mathbf{G}\mathbf{X}^T\mathbf{y}(t)$, where $\mathbf{G}$ denotes any one of the generalized inverses of $\mathbf{X}^T\mathbf{X}$. By Theorem 7.9, $\hat{\boldsymbol{\beta}}_{\mathbf{W}}(t) = (\mathbf{X}^T\mathbf{X}+\mathbf{W}^T\mathbf{W})^{-1}\mathbf{X}^T\mathbf{y}(t)$ is one of the solutions to the normal equation system (7.6), which also satisfies the side-condition (7.54). Therefore, there must exist a particular generalized inverse of $\mathbf{X}^T\mathbf{X}$, namely $\mathbf{G}_{\mathbf{W}}$, such that

$$\hat{\boldsymbol{\beta}}_{\mathbf{W}}(t) = (\mathbf{X}^T\mathbf{X} + \mathbf{W}^T\mathbf{W})^{-1}\mathbf{X}^T\mathbf{y}(t) = \mathbf{G}_{\mathbf{W}}\mathbf{X}^T\mathbf{y}(t),$$

which holds for any $\mathbf{y}(t)$. It follows that we must have $(\mathbf{X}^T\mathbf{X}+\mathbf{W}^T\mathbf{W})^{-1}\mathbf{X}^T = \mathbf{G}_{\mathbf{W}}\mathbf{X}^T$. We then have $\mathbf{P}_{\mathbf{X}|\mathbf{W}} = \mathbf{X}(\mathbf{X}^T\mathbf{X} + \mathbf{W}^T\mathbf{W})^{-1}\mathbf{X}^T = \mathbf{X}\mathbf{G}_{\mathbf{W}}\mathbf{X}^T = \mathbf{P}_{\mathbf{X}}$ due to the fact that $\mathbf{P}_{\mathbf{X}} = \mathbf{X}(\mathbf{X}^T\mathbf{X})^{-}\mathbf{X}^T = \mathbf{X}\mathbf{G}\mathbf{X}^T$ is invariant to the choice of any generalized inverse $\mathbf{G}$ of $\mathbf{X}^T\mathbf{X}$. The theorem is proved.

Proof of Theorem 7.11 By Theorem 7.2, if $\mathbf{C}\boldsymbol{\beta}(t)$ is estimable, there exists a constant matrix $\mathbf{A}$ such that $\mathbf{C} = \mathbf{A}\mathbf{X}$. Then by Theorem 7.10, we have

$$\begin{aligned}\mathbf{C}\hat{\boldsymbol{\beta}}_{\mathbf{W}}(t) &= \mathbf{A}\mathbf{X}(\mathbf{X}^T\mathbf{X} + \mathbf{W}^T\mathbf{W})^{-1}\mathbf{X}^T\mathbf{y}(t)\\ &= \mathbf{A}\mathbf{P}_{\mathbf{X}|\mathbf{W}}\mathbf{y}(t) = \mathbf{A}\mathbf{P}_{\mathbf{X}}\mathbf{y}(t) = \mathbf{C}\hat{\boldsymbol{\beta}}(t),\\ \mathbf{C}(\mathbf{X}^T\mathbf{X} + \mathbf{W}^T\mathbf{W})^{-1}\mathbf{C}^T &= \mathbf{A}\mathbf{X}(\mathbf{X}^T\mathbf{X} + \mathbf{W}^T\mathbf{W})^{-1}\mathbf{X}^T\mathbf{A}^T\\ &= \mathbf{A}\mathbf{P}_{\mathbf{X}|\mathbf{W}}\mathbf{A} = \mathbf{A}\mathbf{P}_{\mathbf{X}}\mathbf{A}^T = \mathbf{C}(\mathbf{X}^T\mathbf{X})^{-}\mathbf{C}^T,\end{aligned} \tag{7.63}$$

where $\hat{\boldsymbol{\beta}}(t) = \mathbf{X}(\mathbf{X}^T\mathbf{X})^{-}\mathbf{X}^T\mathbf{y}(t)$ as defined in (7.14). It follows from (7.63) and Theorem 7.4 that (a) $\mathrm{E}\left[\mathbf{C}\hat{\boldsymbol{\beta}}_{\mathbf{W}}(t)\right] = \mathrm{E}\left[\mathbf{C}\hat{\boldsymbol{\beta}}(t)\right] = \mathbf{C}\boldsymbol{\beta}(t)$ and

$$\begin{aligned}&\mathrm{Cov}\left[\mathbf{C}\hat{\boldsymbol{\beta}}_{\mathbf{W}}(s), \mathbf{C}\hat{\boldsymbol{\beta}}_{\mathbf{W}}(t)\right] = \mathrm{Cov}\left[\mathbf{C}\hat{\boldsymbol{\beta}}(s), \mathbf{C}\hat{\boldsymbol{\beta}}(t)\right]\\ &= \gamma(s,t)\mathbf{C}(\mathbf{X}^T\mathbf{X})^{-}\mathbf{C}^T = \gamma(s,t)\mathbf{C}(\mathbf{X}^T\mathbf{X} + \mathbf{W}^T\mathbf{W})^{-1}\mathbf{C}^T;\end{aligned}$$

and (b) $\mathbf{C}\hat{\boldsymbol{\beta}}_{\mathbf{W}}(t)$ and $\mathbf{C}(\mathbf{X}^T\mathbf{X} + \mathbf{W}^T\mathbf{W})^{-1}\mathbf{C}^T$ are invariant to the choice of $\mathbf{W}$. The theorem is proved.

Proof of Theorem 7.12 From the proof of Theorem 7.11, we have $\mathbf{C}\hat{\boldsymbol{\beta}}_{\mathbf{W}}(t) = \mathbf{C}\boldsymbol{\beta}(t) + \mathbf{C}(\mathbf{X}^T\mathbf{X} + \mathbf{W}^T\mathbf{W})^{-1}\mathbf{X}^T\mathbf{v}(t)$. As $\mathbf{v}(t) \sim \mathrm{GP}_n(\mathbf{0}, \gamma\mathbf{I}_n)$, we have $\mathbf{C}[\hat{\boldsymbol{\beta}}_{\mathbf{W}}(t) - \boldsymbol{\beta}(t)] \sim \mathrm{GP}_q(\mathbf{0}, \gamma\mathbf{C}(\mathbf{X}^T\mathbf{X} + \mathbf{W}^T\mathbf{W})^{-1}\mathbf{C}^T)$. We then have $\mathbf{z}_{\mathbf{W}}(t) \sim \mathrm{GP}_q(\boldsymbol{\eta}_{z|\mathbf{W}}, \gamma\mathbf{I}_q)$. By (7.57), we have $(n-k)\hat{\gamma}_{\mathbf{W}}(s,t) = \mathbf{v}(s)^T(\mathbf{I}_n - \mathbf{P}_{\mathbf{X}})\mathbf{v}(t)$. As $\mathbf{I}_n - \mathbf{P}_{\mathbf{X}}$ is an idempotent matrix with rank $(n-k)$, by Theorem 4.8, we have $(n-k)\hat{\gamma}_{\mathbf{W}}(s,t) \sim \mathrm{WP}(n-k, \gamma)$. The theorem is proved.

Proof of Theorem 7.13 Under the given conditions and by Theorem 7.12, we have $\mathbf{z}_{\mathbf{w}}(t) \sim \mathrm{GP}_q(\mathbf{0}, \gamma \mathbf{I}_n)$ and $(n-k)\hat{\gamma}_{\mathbf{w}}(s,t) \sim \mathrm{WP}(n-k, \gamma)$. Then by Theorem 4.10, we have

$$\int_{\mathcal{T}} \mathrm{SSH}_{n\mathbf{w}}(t)dt = \int_{\mathcal{T}} \|\mathbf{z}_{\mathbf{w}}(t)\|^2 dt \stackrel{d}{=} \sum_{r=1}^{m} \lambda_r A_r, \quad A_r \stackrel{i.i.d.}{\sim} \chi_q^2,$$

where $\lambda_r, r = 1, 2, \cdots, m$ are all the positive eigenvalues of $\gamma(s,t)$. Similarly, by Theorem 4.5, we have $\int_{\mathcal{T}} \mathrm{SSE}_{n\mathbf{w}}(t)dt = \mathrm{tr}[(n-k)\hat{\gamma}] \stackrel{d}{=} \sum_{r=1}^{m} \lambda_r E_r, E_r \stackrel{i.i.d.}{\sim} \chi_{n-k}^2$ as desired. The theorem is proved.

7.6 Concluding Remarks and Bibliographical Notes

In this chapter, we discussed the generalized inverse, reparameterization, and side-condition methods for ill-conditioned functional linear models. These methods can be used to handle the functional ANOVA problems discussed in Chapter 5 in a unified manner provided that the non-full-rank design matrix can be efficiently specified.

The non-full-rank FLM (7.1) may be handled by deleting some columns of the design matrix so that it becomes a full-rank FLM. This method is similar to the side-condition method in spirit by setting some coefficient functions to be 0. When the functional data were collected from a fractional factorial design with each factor having two levels, one can just delete all the columns in the design matrix that are associated with the second levels of the factors. Nair et al. (2002) adopted this method when they analyzed the audible noise data using a pointwise F-test. Shen and Xu (2007) adopted this method when they analyzed the audible noise data using an F-type test.

7.7 Exercises

1. Consider the non-full-rank FLM $\mathbf{y}(t) = \mathbf{X}\boldsymbol{\beta}(t) + \mathbf{v}(t)$, where

$$\mathbf{X} = \begin{pmatrix} 1 & 1 & 0 & 1 & 0 \\ 1 & 1 & 0 & 0 & 1 \\ 1 & 0 & 1 & 1 & 0 \\ 1 & 0 & 1 & 0 & 1 \end{pmatrix}, \quad \boldsymbol{\beta}(t) = \begin{pmatrix} \eta(t) \\ \alpha_1(t) \\ \alpha_2(t) \\ \beta_1(t) \\ \beta_2(t) \end{pmatrix}.$$

   (a) Show that $\mathrm{rank}(\mathbf{X}) = 3$ and find three linearly independent rows of $\mathbf{X}$.
   (b) Show that $\alpha_1(t) - \alpha_2(t)$ and $\beta_1(t) - \beta_2(t)$ are estimable.
   (c) Show that $\eta(t), \alpha_1(t), \alpha_2(t), \beta_1(t)$ and $\beta_2(t)$ are nonestimable.

2. Consider the following one-way functional ANOVA model:

$$y_{ij}(t) = \eta(t) + \alpha_i(t) + v_{ij}(t), t \in \mathcal{T}, j = 1, 2; i = 1, 2. \qquad (7.64)$$

Set $\mathbf{y}(t) = [y_{11}(t), y_{12}(t), y_{21}(t), y_{22}(t)]^T$, $\boldsymbol{\beta}(t) = [\eta(t), \alpha_1(t), \alpha_2(t)]^T$, and $\mathbf{v}(t) = [v_{11}(t), v_{12}(t), v_{21}(t), v_{22}(t)]^T$.

   (a) Identify the design matrix $\mathbf{X}$ such that the one-way functional ANOVA model (7.64) can be written as $\mathbf{y}(t) = \mathbf{X}\boldsymbol{\beta}(t) + \mathbf{v}(t)$.
   (b) Show that $\mathbf{X}$ has rank 2 and hence it is a non-full-rank matrix.
   (c) Find two linearly independent rows of $\mathbf{X}$ to form the matrix $\mathbf{R}$ and then determine the matrix $\mathbf{Z}$ so that $\mathbf{X} = \mathbf{ZR}$.
   (d) Determine the reparameterization matrix $\mathbf{R}$ for the reparameterization transformation $\boldsymbol{\phi}(t) = [\eta(t)+\alpha_1(t), \eta(t)+\alpha_2(t)] = \mathbf{R}\boldsymbol{\beta}(t)$ and then determine $\mathbf{Z}$ such that $\mathbf{X} = \mathbf{ZR}$.

3. Consider the non-full-rank FLM defined in Exercise 1.

   (a) Given $\eta(t) = 0$, find $\hat{\boldsymbol{\beta}}_{\mathbf{W}}(t)$ and $[0, 1, -1, 0, 0]\hat{\boldsymbol{\beta}}_{\mathbf{W}}(t)$.
   (b) Given $\alpha_1(t) = 0$, find $\hat{\boldsymbol{\beta}}_{\mathbf{W}}(t)$ and $[0, 1, -1, 0, 0]\hat{\boldsymbol{\beta}}_{\mathbf{W}}(t)$.
   (c) Given $\beta_1(t) = 0$, find $\hat{\boldsymbol{\beta}}_{\mathbf{W}}(t)$ and $[0, 1, -1, 0, 0]\hat{\boldsymbol{\beta}}_{\mathbf{W}}(t)$.

4. Consider the one-way functional ANOVA model defined in Exercise 2.

   (a) Given $\eta(t) = 0$, find $\hat{\boldsymbol{\beta}}_{\mathbf{W}}(t)$ and $[0, 1, -1]\hat{\boldsymbol{\beta}}_{\mathbf{W}}(t)$.
   (b) Given $\alpha_1(t) = 0$, find $\hat{\boldsymbol{\beta}}_{\mathbf{W}}(t)$ and $[0, 1, -1]\hat{\boldsymbol{\beta}}_{\mathbf{W}}(t)$.
   (c) Given $\alpha_2(t) = 0$, find $\hat{\boldsymbol{\beta}}_{\mathbf{W}}(t)$ and $[0, 1, -1]\hat{\boldsymbol{\beta}}_{\mathbf{W}}(t)$.
   (d) Given $\alpha_1(t) + \alpha_2(t) = 0$, find $\hat{\boldsymbol{\beta}}_{\mathbf{W}}(t)$ and $[0, 1, -1]\hat{\boldsymbol{\beta}}_{\mathbf{W}}(t)$.

5. For the one-way functional ANOVA model (7.2), we have $\mathbf{R} = [\mathbf{1}_a, \mathbf{I}_a]$. Find $\mathbf{W}$ such that $\mathbf{W}\boldsymbol{\beta}(t) = \mathbf{0}$ is a desired side-condition.

6. For the audible noise data, consider conducting the F-type test for the main-effect contrast functions of the seven factors using only the data information presented in this chapter.

   (a) Compute the test statistics F_n (7.43) for the testing problem (7.44) with $i = 1, 2, \cdots, 7$, respectively.
   (b) Compute the associate approximate degrees of freedom by the naive method and the bias-reduced method.
   (c) Compute the associate P-values using computer software, for example, S-PLUS, R, or MATLAB®.
   (d) Compare the results with those listed in Example 7.5.

Chapter 8

Diagnostics of Functional Observations

8.1 Introduction

Suppose we have a functional data set with n cases:

$$[\mathbf{x}_i^T, y_i(t)],\ t \in \mathcal{T},\ i = 1, 2, \cdots, n, \tag{8.1}$$

where $\mathbf{x}_i, i = 1, 2, \cdots, n$ denote the observed $(p+1)$-dimensional covariates or predictors, and $y_i(t), i = 1, 2, \cdots, n$ are the associated response functions over $\mathcal{T}$, a compact support of t. As in Chapter 6, in many situations, the relationship between the response functions and the covariates can be modeled using the following functional linear model (FLM):

$$y_i(t) = \mathbf{x}_i^T\boldsymbol{\beta}(t) + v_i(t),\ \ v_i(t) \overset{i.i.d.}{\sim} \text{SP}(0, \gamma), t \in \mathcal{T}, \tag{8.2}$$

where the vector of coefficient functions $\boldsymbol{\beta}(t) = [\beta_0(t), \beta_1(t), \cdots, \beta_p(t)]^T$ models the effects of covariates and $v_i(t), i = 1, 2, \cdots, n$ denote the subject-effect functions that are not explained by the linear term $\mathbf{x}_i^T\boldsymbol{\beta}(t)$. In the studies of Chapter 6, we made an implicit assumption that each case $[\mathbf{x}_i^T, y_i(t)]$ in (8.1) follows the FLM (8.2) well. In practice, however, this assumption may be not satisfied by all the cases in (8.1); rather, there are some unusual cases in (8.1) that do not follow the FLM (8.2) closely.

Example 8.1 *Figure 8.1 displays $n = 60$ reconstructed right elbow angle curves of the ergonomics data introduced in Section 1.2.8 of Chapter 1 and analyzed in Chapter 6 using the FLM (6.1) with $p+1 = 10$ covariates formed by the coordinates of the targets that the mobile driver wanted to reach. Curve 37 is highlighted as a wide dashed curve as it is quite different from other curves in shape. It is suspected to be an unusual curve that does not follow the FLM (6.1) well. Figure 8.2 displays the first four fitted coefficient functions and their 95% confidence bands, computed using all the data (solid) or with Curve 37 excluded (dashed). The effects of deleting Curve 37 are easily spotted, especially when t takes values around 0.5. That is why in the statistical analysis conducted in Chapter 6, Curve 37 was removed to guarantee that the associated inferences are reliable.*

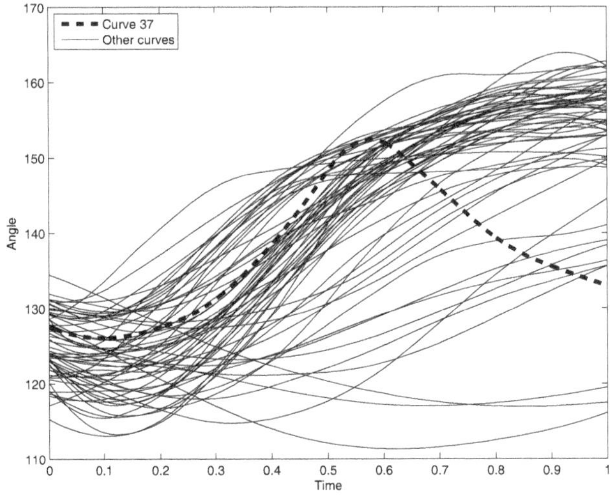

Figure 8.1 *Reconstructed right elbow angle curves of the ergonomics data. Curve 37 is highlighted as a wide dashed curve as it is very different from other curves in shape.*

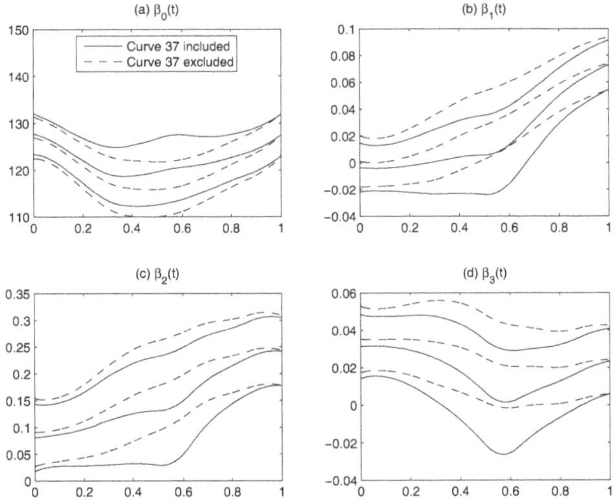

Figure 8.2 *First four fitted coefficient functions with their* 95% *confidence bands, computed using all the ergonomics data (solid) or with Curve 37 excluded (dashed). The effects of deleting Curve 37 are easily spotted, especially when the time* t *takes values around* 0.5.

As in the above example, analysis with unusual cases may lead to misleading conclusions. In this chapter, we aim to develop some diagnostics methodologies to detect and examine these unusual cases.

For further discussion, we write the FLM (8.2) in matrix notation as follows:

$$\mathbf{y}(t) = \mathbf{X}\boldsymbol{\beta}(t) + \mathbf{v}(t), \quad \mathbf{v}(t) \sim \mathrm{SP}_n(\mathbf{0}, \gamma\mathbf{I}_n), t \in \mathcal{T}, \tag{8.3}$$

where

$$\begin{aligned} \mathbf{y}(t) &= [y_1(t), y_2(t), \cdots, y_n(t)]^T, \\ \mathbf{v}(t) &= [v_1(t), v_2(t), \cdots, v_n(t)]^T, \\ \mathbf{X} &= [\mathbf{x}_1, \mathbf{x}_2, \cdots, \mathbf{x}_n]^T, \end{aligned}$$

are known as the vector of response functions, the vector of subject-effect functions, and the design matrix, respectively. For simplicity, throughout this chapter, we assume that the design matrix $\mathbf{X} : n \times (p+1)$ is full rank and time independent. It would be straightforward to extend the methodologies studied in this chapter to the cases when $\mathbf{X}$ is non-full rank and time dependent. See Chapter 6 for FLMs with time-dependent design matrices and Chapter 7 for FLMs with non-full-rank design matrices.

As in classical linear regression models, unusual cases in a functional data set (8.1) for the FLM (8.3) may be classified as high-leverage, outlying, or influential cases. A high-leverage case is a case $[\mathbf{x}_i^T, y_i(t)]$ with high leverage, that is, whose covariates take extreme values. The leverages of the cases in (8.1) are defined as the diagonal entries

$$h_{ii}, i = 1, 2, \cdots, n, \tag{8.4}$$

of the hat or projection matrix

$$\mathbf{H} = (h_{ij}) = \mathbf{P}_{\mathbf{X}} = \mathbf{X}(\mathbf{X}^T\mathbf{X})^{-1}\mathbf{X}^T, \tag{8.5}$$

associated with the FLM (8.3). Thus, as in classical linear regression models, high-leverage cases are usually related to the design matrix $\mathbf{X}$ only. They can be easily identified by a scatterplot of the leverages (8.4). Furthermore, the high-leverage problem can be avoided by properly designing the associated experiment or the functional observation collection process for the functional data set (8.1). Thus, in this chapter, we do not discuss how to detect high-leverage cases for the FLM (8.3).

An outlying case in a functional data set (8.1), on the other hand, is a case $[\mathbf{x}_i^T, y_i(t)]$ that does not follow the FLM (8.2) well. The associated response function $y_i(t)$ may take extreme values over $\mathcal{T}$ or may be very different from other response functions in shape or in other geometrical structures. For example, in the ergonomics data set, the case associated with Curve 37 (see Figure 8.1) looks like an outlying case as Curve 37 is very different from other curves in shape. The response function of an outlying case is also known as a functional outlier or simply an outlier. Notice that on the one hand, functional outliers may be functions with gross errors resulting from recording or typing mistakes. These gross errors should be identified and corrected whenever possible. They, on the other hand, may be "true" functions in the sense that they are not due to gross errors but are somehow suspicious or surprising as they look like they do not to follow the same FLM (8.2) as the rest of functions. For this kind of functional outlier, some further analysis is needed to discover what source generates these so-called outliers.

A case in a functional data set (8.1) is considered an influential case if its deletion will cause a big change in the fit to the FLM (8.3). Later we will know, as in classical linear regression models, that an influential case $[\mathbf{x}_i^T, y_i(t)]$ in a functional data set (8.1) can be a high-leverage case (when $\mathbf{x}_i$ is an outlier), an outlying case (when $y_i(t)$ is an outlier), or both (when both $\mathbf{x}_i$ and $y_i(t)$ are outliers). As in classical linear regression models, when we fit the FLM (8.3), we should avoid situations where the fit to the FLM (8.3) is overly determined by one or a few influential cases.

In this chapter, we study how to identify outlying and influential cases in the functional data set (8.1) with respect to the FLM (8.3). We first describe three different kinds of residual functions in Section 8.2. In Section 8.3, we discuss several methods for functional outlier detection, including a standardized residual-based method, a jackknife residual-based method, and some functional depth-based methods. Influential case detection is studied in Section 8.4. How to obtain the robust estimators of the coefficient and covariance functions for the FLM (8.3) is briefly discussed in Section 8.5. Technical proofs of two main results are given in Section 8.7. Some concluding remarks and bibliographical notes are given in Section 8.8. Exercises for this chapter are listed in Section 8.9.

8.2 Residual Functions

In this section, we introduce three kinds of residual functions with respect to the FLM (8.3) and the functional data set (8.1). These residual functions will be used to construct various measures for detecting if there are some unusual cases in the functional data set (8.1). Notice that, throughout this chapter, when we say "under the FLM (8.3)," we mean that "there are no unusual cases in the functional data set (8.1) with respect to the FLM (8.3)."

8.2.1 Raw Residual Functions

By Chapter 6, the vector of fitted response functions for the FLM (8.3) can be expressed as

$$\hat{\mathbf{y}}(t) = \mathbf{X}\hat{\boldsymbol{\beta}}(t) = \mathbf{P}_{\mathbf{X}}\mathbf{y}(t), \tag{8.6}$$

where $\mathbf{P}_{\mathbf{X}}$ is the associated projection matrix as defined in (8.5). It follows that the vector of raw residual functions can be expressed as

$$\hat{\mathbf{v}}(t) = [\hat{v}_1(t), \hat{v}_2(t), \cdots, \hat{v}_n(t)]^T = (\mathbf{I}_n - \mathbf{P}_{\mathbf{X}})\mathbf{y}(t). \tag{8.7}$$

These raw residual functions measure how close the FLM (8.3) fits the functional data set (8.1). It is easy to see that under the FLM (8.3), we have $\mathrm{E}\hat{\mathbf{v}}(t) \equiv \mathbf{0}$ and

$$\mathrm{Cov}\left[\hat{\mathbf{v}}(s), \hat{\mathbf{v}}(t)\right] = (\mathbf{I}_n - \mathbf{P_X})\gamma(s,t). \tag{8.8}$$

That is, under the above assumption, for $i = 1, 2, \cdots, n$, we have $\mathrm{E}\hat{v}_i(t) = 0$ and

$$\mathrm{Cov}\left(\hat{v}_i(s), \hat{v}_i(t)\right) = (1 - h_{ii})\gamma(s,t), \tag{8.9}$$

where h_{ii} denotes the ith leverage as given in (8.4). It is seen that for $i = 1, 2, \cdots, n$, the ith raw residual function $\hat{v}_i(s)$ follows $\mathrm{SP}(0, (1 - h_{ii})\gamma)$, a stochastic process with mean function 0 and covariance function $(1-h_{ii})\gamma(s,t)$ that depends on the design matrix $\mathbf{X}$ through the ith leverage. As usual, the unbiased estimator of $\gamma(s,t)$ is given by

$$\begin{aligned} \hat{\gamma}(s,t) &= (n-p-1)^{-1}\textstyle\sum_{i=1}^n \hat{v}_i(s)\hat{v}_i(t) \\ &= (n-p-1)^{-1}\mathbf{y}(s)^T(\mathbf{I}_n - \mathbf{P_X})\mathbf{y}(t). \end{aligned} \tag{8.10}$$

8.2.2 *Standardized Residual Functions*

It is not a good idea to use the raw residual functions $\hat{v}_i(t), i = 1, 2, \cdots, n$ directly to construct measures for detecting unusual cases as these raw residual functions have different distributions. To overcome this difficulty, we define the following standardized residual functions:

$$z_i(t) = \hat{v}_i(t)/\sqrt{1 - h_{ii}}, i = 1, 2, \cdots, n, \tag{8.11}$$

so that these standardized residual functions have the same mean and covariance functions. In particular, under the FLM (8.3) and the Gaussian assumption:

$$v_1(t), v_2(t), \cdots, v_n(t) \overset{i.i.d.}{\sim} \mathrm{GP}(0, \gamma), \tag{8.12}$$

we have

$$z_i(t) \sim \mathrm{GP}(0, \gamma), i = 1, 2, \cdots, n. \tag{8.13}$$

Notice that even under the Gaussian assumption (8.12), $z_i(t), i = 1, 2, \cdots, n$ are not mutually independent. In the next section we will use these standardized residual functions to define various measures for detecting outlying and influential cases in the functional data set (8.1).

8.2.3 *Jackknife Residual Functions*

We can also define jackknife residual functions. Let $\mathbf{X}^{(-i)}$, $\mathbf{y}^{(-i)}(t)$, and $\mathbf{v}^{(-i)}(t)$, respectively, denote the design matrix $\mathbf{X}$, the vector of response functions $\mathbf{y}(t)$, and the vector of subject-effect functions $\mathbf{v}(t)$ in the FLM (8.3) after the ith rows are deleted. In this case, the FLM (8.3) can be written as

$$\begin{aligned} \mathbf{y}^{(-i)}(t) &= \mathbf{X}^{(-i)}\boldsymbol{\beta}(t) + \mathbf{v}^{(-i)}(t), \\ \mathbf{v}^{(-i)}(t) &\sim \mathrm{SP}_{n-1}(\mathbf{0}, \gamma\mathbf{I}_{n-1}), t \in \mathcal{T}. \end{aligned} \tag{8.14}$$

Let $\hat{\boldsymbol{\beta}}^{(-i)}(t)$ be the estimator of $\boldsymbol{\beta}(t)$ computed under the FLM (8.14):

$$\hat{\boldsymbol{\beta}}^{(-i)}(t) = \left(\mathbf{X}^{(-i)T}\mathbf{X}^{(-i)}\right)^{-1}\mathbf{X}^{(-i)T}\mathbf{y}^{(-i)}(t).$$

Then the prediction of the ith response function $y_i(t)$ using all the cases except $[\mathbf{x}_i^T, y_i(t)]$ is given by

$$\hat{y}_i^{(-i)}(t) = \mathbf{x}_i^T\hat{\boldsymbol{\beta}}^{(-i)}(t) = \mathbf{x}_i^T(\mathbf{X}^{(-i)T}\mathbf{X}^{(-i)})^{-1}\mathbf{X}^{(i)T}\mathbf{y}^{(-i)}(t).$$

The prediction error function is then given by

$$\hat{v}_i^{(-i)}(t) = y_i(t) - \hat{y}_i^{(-i)}(t). \tag{8.15}$$

Obviously, we have $\text{E}\left[\hat{v}_i^{(-i)}(t)\right] \equiv 0$. Notice that $y_i(t)$ and $\hat{y}_i^{(-i)}(t)$ are independent as computation of $\hat{y}_i^{(-i)}(t)$ does not involve the ith case $[\mathbf{x}_i^T, y_i(t)]$. It follows that the covariance function between $\hat{v}_i^{(-i)}(s)$ and $\hat{v}_i^{(-i)}(t)$ is given by

$$\begin{aligned} \rho^{(-i)}(s,t) &= \text{Cov}\left[\hat{v}_i^{(-i)}(s), \hat{v}_i^{(-i)}(t)\right] \\ &= \left[1 + \mathbf{x}_i^T(\mathbf{X}^{(-i)T}\mathbf{X}^{(-i)})^{-1}\mathbf{x}_i\right]\gamma(s,t). \end{aligned} \tag{8.16}$$

Therefore, $\hat{v}_i^{(-i)}(t)$ follows $\text{SP}(0, \rho^{(-i)})$, a stochastic process having mean function 0 and the covariance function $\rho^{(-i)}(s,t)$ that depends on the ith case and the whole design matrix $\mathbf{X}$. Notice that under the FLM (8.14), the unbiased estimator of $\gamma(s,t)$ is given by

$$\hat{\gamma}^{(-i)}(s,t) = (n-p-2)^{-1}\mathbf{y}^{(-i)}(s)^T(\mathbf{I}_{n-1} - \mathbf{P}_{\mathbf{X}^{(-i)}})\mathbf{y}^{(-i)}(t), \tag{8.17}$$

obtained using all the data except the ith case $[\mathbf{x}_i^T, y_i(t)]$.

As for the raw residual functions, it is not a good idea to use the prediction error functions $\hat{v}_i^{(-i)}(t), i = 1, 2, \cdots, n$ directly to construct measures for detecting unusual cases as these prediction error functions have different covariance functions. To overcome this difficulty, using (8.16), we define the following jackknife residual functions:

$$z_i^{(-i)}(t) = \frac{\hat{v}_i^{(-i)}(t)}{\sqrt{1 + \mathbf{x}_i^T\left[\mathbf{X}^{(-i)T}\mathbf{X}^{(-i)}\right]^{-1}\mathbf{x}_i}}, \quad i = 1, 2, \cdots, n, \tag{8.18}$$

so that they have the same mean and covariance functions. In particular, under the FLM (8.3) and the Gaussian assumption (8.12), we have

$$z_i^{(-i)}(t) \sim \text{GP}(0, \gamma), i = 1, 2, \cdots, n. \tag{8.19}$$

8.3 Functional Outlier Detection

In this section, we introduce several methods for functional outlier detection based on the standardized and jackknife residual functions defined in the previous section. As in outlier detection in classical linear regression models, functional outlier detection depends on an ordering of the response functions in the functional data set (8.1) with respect to the FLM (8.3). To order these response functions, we need to introduce some measure for the outlyingness of these response functions with respect to the FLM (8.3). An outlyingness measure defines a way to order the response functions in the functional data set (8.1) from center to outward so that those response functions near the center have smaller outlyingness measures and those response functions far from the center have greater outlyingness measures. For the FLM (8.3), an outlyingness measure determines how well a case $[\mathbf{x}_i^T, y_i(t)]$ in the functional data set (8.1) satisfies the FLM (8.3). In other words, for a case $[\mathbf{x}_i^T, y_i(t)]$, the center is the conditional expectation of the FLM (8.3), which is given by

$$\mathrm{E}\left[y_i(t)|\mathbf{x}_i\right] = \mathbf{x}_i^T\boldsymbol{\beta}(t).$$

Therefore, measuring how well a case $[\mathbf{x}_i^T, y_i(t)]$ fits to the FLM (8.3) is equivalent to measuring how close the standardized residual functions

$$z_i(t), i = 1, 2, \cdots, n, \tag{8.20}$$

or the jackknife residual functions

$$z_i^{(-i)}(t), i = 1, 2, \cdots, n, \tag{8.21}$$

to a zero function that equals to 0 for any $t \in \mathcal{T}$. In what follows we use these standardized and jackknife residual functions to construct some outlyingness measures for functional outlier detection for the FLM (8.3).

8.3.1 Standardized Residual-Based Method

The standardized residual-based method uses the squared L^2-norms of the standardized residual functions (8.20) as the outlyingness measures of the response functions $y_i(t), i = 1, 2, \cdots, n$:

$$S_i^2 = \|z_i - 0\|^2 = \int_{\mathcal{T}} z_i^2(t)dt, i = 1, 2, \cdots, n. \tag{8.22}$$

For easy reference, we call (8.22) the standardized residual outlyingness (SRO) measures.

Remark 8.1 *The underlying rationale for using the SRO measures for functional outlier detection is as follows. When $y_i(t)$ is a functional outlier, the case $[\mathbf{x}_i^T, y_i(t)]$ will not fit the FLM (8.2) well so that the ith standardized* *residual function $z_i(t)$ will be very large in absolute values over $\mathcal{T}$, compared with the majority of the standardized residual functions (8.20). In this case, the value of S_i^2 will be very large, compared with the majority of $S_j^2, j = 1, 2, \cdots, n$. Therefore, we can use the SRO measures (8.22) to assess how well the FLM (8.2) fits the cases of the functional data set (8.1). The smaller the S_i^2 is, the better the FLM (8.2) fits the case $[\mathbf{x}_i^T, y_i(t)]$.*

A question arises naturally. How large is "large" for SRO measures (8.22)? As in classical linear regression models, we can solve this problem by drawing and analyzing the scatterplot of these SRO measures (8.22). Alternatively, we can find the cutoff values or the P-values of these SRO measures (8.22) to determine if they are "too large" under some criterion. The response functions associated with "too large" SRO measures may be declared as functional outliers. For this purpose, we need to derive the distribution of the SRO measures (8.22).

Theorem 8.1 *Under the FLM (8.3) with tr(γ) $< \infty$ and the Gaussian assumption (8.12), for $i = 1, 2, \cdots, n$, we have*

$$S_i^2 \stackrel{d}{=} \sum_{r=1}^{m} \lambda_r A_r, \quad A_r \stackrel{i.i.d.}{\sim} \chi_1^2,$$

where $\lambda_r, r = 1, 2, \cdots, m$ are all the positive eigenvalues of $\gamma(s,t)$.

Theorem 8.1 shows that under the FLM (8.3) and the Gaussian assumption (8.12), the SRO measures (8.22) follow a common χ^2-type mixture. Thus, we can approximate their common distribution using the methods described in Section 4.3 of Chapter 4. In particular, by the Welch-Satterthwaite χ^2-approximation method, we have

$$S_i^2 \sim \beta\chi_\kappa^2 \text{ approximately}, i = 1, 2, \cdots, n,$$

where

$$\beta = \frac{\text{tr}(\gamma^{\otimes 2})}{\text{tr}(\gamma)} \text{ and } \kappa = \frac{\text{tr}^2(\gamma)}{\text{tr}(\gamma^{\otimes 2})}.$$

In practice, the parameters β and κ may be estimated by the naive method and the bias-reduced method as described in Section 6.2.4 of Chapter 6. In fact, the naive and bias-reduced estimators, namely $\hat{\beta}$ and $\hat{\kappa}$, of β and κ are given in (6.26) and (6.29), respectively, so that we have

$$S_i^2 \sim \hat{\beta}\chi_{\hat{\kappa}}^2 \text{ approximately}, i = 1, 2, \cdots, n. \tag{8.23}$$

Using the above approximate distribution, for a given nominal significance level α, the cutoff value for the SRO measures (8.22) is given by

$$\hat{\beta}\chi_{\hat{\kappa}}^2(1-\alpha), \tag{8.24}$$

where as usual $\chi_d^2(1-\alpha)$ denotes the $100(1-\alpha)$ percentile of the χ^2-distribution with d degrees of freedom. When a SRO measure in (8.22) is larger than the cutoff value (8.24), the associated response function may be classified as an α-outlier.

Remark 8.2 *In practice, $\hat{\kappa}$ may be not an integer. In this case, we can truncate $\hat{\kappa}$ into its nearest integer when it is very big; otherwise, we treat $\chi_{\hat{\kappa}}^2$ as a gamma distribution with scale parameter $1/2$ and shape parameter $\hat{\kappa}/2$ to compute the cutoff value (8.24).*

Remark 8.3 *As in classical linear regression models, in the context of outlier detection, the nominal significance level α is very subjective. It is used to denote the maximum false positive rate allowed for classifying a non-outlier as an outlier. This number should not be too large and it often takes 5%, 1%, or even 0.1% so that the false positive rate can be well controlled.*

Alternatively, we can use the approximate distribution (8.23) to compute the approximate P-values of the SRO measures (8.22). Let $X \sim \chi_{\hat{\kappa}}^2$. Then the approximate P-values of the SRO measures (8.22) are given by

$$P(X \geq S_i^2/\hat{\beta}), i = 1, 2, \cdots, n, \tag{8.25}$$

which can be used to quantify how close the FLM (8.3) fits the cases in the functional data set (8.1).

In practice, the SRO measures (8.22) may be very large values. To overcome this difficulty, we can rescale them by $\text{tr}(\hat{\gamma})$, resulting in the following rescaled SRO measures:

$$\tilde{S}_i^2 = S_i^2/\text{tr}(\hat{\gamma}), i = 1, 2, \cdots, n. \tag{8.26}$$

We use $\text{tr}(\hat{\gamma})$ in the denominator due to the fact that under the FLM (8.3), we have $\text{E}\left(S_i^2\right) = \text{tr}(\gamma)$. In this case, the cutoff value for the rescaled SRO measures (8.26) is now given by

$$\hat{\beta}\chi_{\hat{\kappa}}^2(1-\alpha)/\text{tr}(\hat{\gamma}), \tag{8.27}$$

while the P-values of the rescaled SRO measures (8.26) are still given by (8.25). Admittedly, in the above computation of the cutoff value and the P-values of the rescaled SRO measures (8.26), the variation of $\text{tr}(\hat{\gamma})$ has not been taken into account.

Example 8.2 *For the ergonomics data presented in Example 8.1, we obtained $tr(\hat{\gamma}) = 1.7874e4$ and $tr(\hat{\gamma}^{\otimes 2}) = 1.9746e8$. By the bias-reduced method, we obtained $\hat{\beta} = 1.0488e4$ and $\hat{\kappa} = 1.6642$. The upper panel of Figure 8.3 displays the scatterplot of the rescaled SRO measures $\tilde{S}_i^2, i = 1, 2, \cdots, n$ for the ergonomics angle curves under the FLM (6.1). The associated cutoff values for*

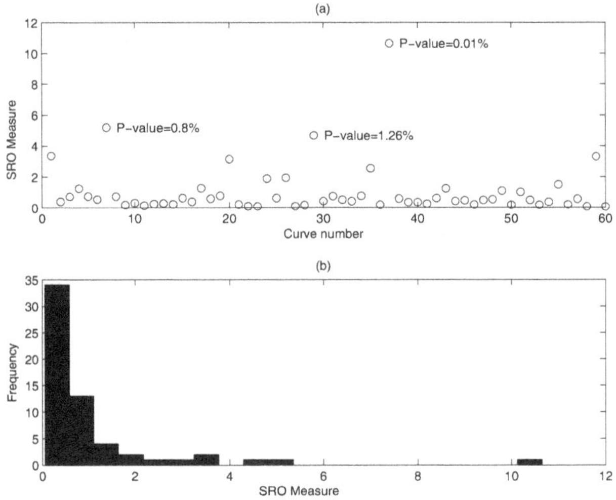

Figure 8.3 *Outlier detection for the ergonomics data using the standardized residual-based method. Panel (a) shows the scatterplot of SRO measure against curve number. Panel (b) displays the histogram of the SRO measures. Both panels indicate that Curve 37 is a functional outlier.*

$\alpha = 5\%, 1\%,$ *and* 0.1%, *computed using (8.27), are* $3.1233, 4.9393,$ *and* 7.5692, *respectively. It follows that when* $\alpha = 5\%$, *there are more than three outliers; when* $\alpha = 1\%$, *Curves 7 and 37 are outliers, and when* $\alpha = 0.1\%$, *only Curve 37 is an outlier. Therefore, the outliers detected based on the cutoff values are* α*-dependent and hence is very subjective. In the upper panel of Figure 8.3, we also labeled the approximate P-values of the three most outlying curves, computed using (8.25). It is seen that these three most outlying curves are very significant with their approximate P-values being* $1.26\%, 0.80\%,$ *and* 0.01%, *respectively.*

The lower panel of Figure 8.3 displays the histogram of the rescaled SRO measures $\tilde{S}_i^2, i = 1, 2, \cdots, n$. *This histogram seems to suggest that only Curve 37 is an outlier as the rescaled SRO measures of Curves 7 and 29 are much closer to the majority of the rescaled SRO measures than that of Curve 37. To further clarify if Curves 7 and 29 are outliers, we redrew all the right elbow angle curves of the ergonomics data in Figure 8.4 with Curves 37, 7, and 29 highlighted in wide solid, dashed, and dotted curves. It is seen that both Curves 7 and 29 have very similar shapes as the majority of the ergonomics angle curves but Curve 37 is very different from other curves in shape. Shen and Faraway (2004) pointed out that Curve 37 is an outlier as the mobile driver changed his mind about the target location in mid-reach. This is in agreement with our detection.*

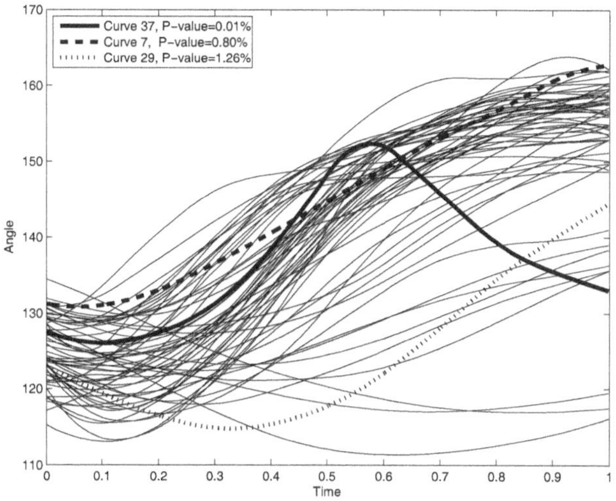

Figure 8.4 *Reconstructed right elbow angle curves of the ergonomics data. Curves 37, 7, and 29 are highlighted as wide solid, dashed, and dotted curves, respectively. Note that both Curves 7 and 29 have very similar shapes as the majority of the angle curves while Curve 37 is very different from other curves in shape.*

8.3.2 Jackknife Residual-Based Method

As the standardized residual-based method presented in the previous subsection, we can use the squared L^2-norms $\|z_i^{(-i)}\|^2, i = 1, 2, \cdots, n$ of the jackknife residual functions (8.21) as the outlyingness measures of the response functions $y_i(t), i = 1, 2, \cdots, n$. Under the FLM (8.3) and the Gaussian assumption (8.12), it is easy to show that for any $i = 1, 2, \cdots, n$, $\|z_i^{(-i)}\|^2$ follows a common χ^2-type mixture as described in Theorem 8.1 for the SRO measures (8.22). That is, we have

$$\|z_i^{(-i)}\|^2 \stackrel{d}{=} \sum_{r-1}^{m} \lambda_r A_r, \ A_r \stackrel{i.i.d.}{\sim} \chi_1^2, \tag{8.28}$$

where λ_r are the eigenvalues of $\gamma(s,t)$. Therefore, the cutoff value (8.24) can also be used for $\|z_i^{(-i)}\|^2$, and the P-values of $\|z_i^{(-i)}\|^2, i = 1, 2, \cdots, n$ can also be computed using the formula (8.25). In other words, the standardized residual-based method presented in the previous subsection works perfectly if we replace the standardized residual functions $z_i(t), i = 1, 2, \cdots, n$ there with the jackknife residual functions $z_i^{(-i)}(t), i = 1, 2, \cdots, n$.

The jackknife residual-based method we discuss in this subsection uses the following jackknife residual outlyingness (JRO) measures:

$$J_i^2 = \frac{\|z_i^{(-i)}\|^2}{\text{tr}(\hat{\gamma}^{(-i)})}, i = 1, 2, \cdots, n, \tag{8.29}$$

where $\hat{\gamma}^{(-i)}(s,t)$ denotes the unbiased sample covariance function (8.17) computed based on the FLM (8.14) after the ith case $[\mathbf{x}_i^T, y_i(t)]$ is excluded.

Remark 8.4 *The underlying rationale for using the JRO measures for functional outlier detection is along the same lines as that for using the SRO measures (8.22) as stated in Remark 8.1. When $y_i(t)$ is a functional outlier, it will not fit the FLM (8.2) well so that the ith jackknife residual function $z_i^{(-i)}(t)$ will be very large in absolute values over $\mathcal{T}$, compared with the majority of the jackknife residual functions (8.21). In this case, the value of J_i^2 will be very large, compared with the majority of $J_j^2, j = 1, 2, \cdots, n$, too. Therefore, we can use the JRO measures (8.29) to assess how well the FLM (8.2) fits the cases of the functional data set (8.1). The smaller the J_i^2 is, the better the FLM (8.2) fits the case $[\mathbf{x}_i^T, y_i(t)]$.*

As for the SRO measures, to detect outlying cases in the functional data set (8.1) with respect to the FLM (8.3), we can draw and analyze the scatterplot of the JRO measures (8.29). Alternatively, we can compute the cutoff values or the P-values of the JRO measures (8.29). The response functions with JRO measures larger than some cutoff value or with P-values smaller than some nominal false positive rate α may be declared functional outliers. Theorem 8.2 gives the distribution of the JRO measures (8.22) when the FLM (8.3) fits the functional data set (8.1) well.

Theorem 8.2 *Under the FLM (8.3) with $tr(\gamma) < \infty$ and the Gaussian assumption (8.12), we have*

$$J_i^2 \stackrel{d}{=} \frac{\sum_{r=1}^m \lambda_r A_r}{(\sum_{r=1}^m \lambda_r B_r)/(n-p-2)},$$

where $\lambda_r, r = 1, 2, \cdots, m$ are all the m positive eigenvalues of $\gamma(s,t)$ and $A_r \sim \chi_1^2, B_r \sim \chi_{n-p-2}^2, r = 1, 2, \cdots, m$ are independent.

Theorem 8.2 is due to Shen and Xu (2007), although their theorem is stated in a very different manner from ours. Theorem 8.2 states that $J_i^2, i = 1, 2, \cdots, n$ all follow the distribution of an F-type mixture. Therefore, the distribution of these JRO measures can be approximated using the methods discussed in Section 4.4 of Chapter 4. By the two-cumulant matched F-approximation method described there, we have

$$J_i^2 \sim F_{\kappa,(n-p-2)\kappa} \text{ approximately,} \tag{8.30}$$

where $\kappa = \frac{\text{tr}^2(\gamma)}{\text{tr}(\gamma^{\otimes 2})}$. In practice, we can estimate the parameter κ by the naive method or the bias-reduced method as described in Section 6.2.4 of Chapter 6. The associated naive and bias-reduced estimators of κ are given in (6.26) and (6.29), respectively, so that we have

$$J_i^2 \sim F_{\hat{\kappa},(n-p-2)\hat{\kappa}} \text{ approximately}, i = 1, 2, \cdots, n. \tag{8.31}$$

Using the above approximate distribution, for a given α, the cutoff value for the JRO measures (8.29) is then given by

$$F_{\hat{\kappa},(n-p-2)\hat{\kappa}}(1-\alpha), \tag{8.32}$$

where $F_{d_1,d_2}(1-\alpha)$ denotes the $100(1-\alpha)$ percentile of the F-distribution with d_1 and d_2 degrees of freedom. When a JRO measure in (8.29) is larger than the cutoff value (8.32), the associated response function may be classified as an α-outlier.

Alternatively, we can use the approximate distribution (8.31) to compute the approximate P-values of the JRO measures (8.29). Let $X \sim F_{\hat{\kappa},(n-p-2)\hat{\kappa}}$. Then the approximate P-values of the JRO measures (8.29) are given by

$$P(X \geq J_i^2), i = 1, 2, \cdots, n, \tag{8.33}$$

which can be used to quantify how close the FLM (8.3) fits the cases in the functional data set (8.1).

Notice that when n is very large, it is time consuming to compute the JRO measures $J_i^2, i = 1, 2, \cdots, n$ using (8.29) directly as we have to fit the FLM (8.14) n times, each with one case deleted from the functional data set (8.1). Fortunately, this can be avoided by a simple trick. By some simple calculation, Shen and Xu (2007) showed that

$$J_i^2 = \frac{(n-p-2)\tilde{S}_i^2}{(n-p-1)-\tilde{S}_i^2}, \; i = 1, 2, \cdots, n, \tag{8.34}$$

where $\tilde{S}_i^2 = S_i^2/\text{tr}(\hat{\gamma}), i = 1, 2, \cdots, n$ are the rescaled SRO measures (8.26).

An application of the jackknife residual-based method to the ergonomics data results in the same results as those shown in Example 8.2. We leave this as an exercise problem at the end of this chapter.

8.3.3 Functional Depth-Based Method

In functional data analysis, depth measures have been introduced by several authors (Fraiman and Muniz 2001; Cuevas, Febrero, and Fraiman 2006, 2007) for measuring the centrality of a function with respect to a sample of functions generated from a stochastic process. These measures provide a way to order functions in a functional sample from center to outward so that functions near the center have greater depth measures and functions far from the center have smaller depth measures. Thus, functional depth measures can be used to detect outlying functions in a sample of functions. Obviously, the concept of depth measures introduced here is opposite that of outlying measures introduced in the previous subsection but they work in a similar manner in detecting functional outliers by ordering functions in a functional data set properly.

In this subsection, we describe three functional depth measures that can be used to detect outlying cases in the functional data set (8.1) with respect to the FLM (8.2). These three functional depth measures are closely related to the three functional depths described by Febrero, Galeano, and Gonzalez-Manteiga (2008) and proposed by Fraiman and Muniz (2001) and Cuevas, Febrero, and Fraiman (2006, 2007), respectively. Notice that the three functional depth measures described here are based on the standardized residual functions $z_i(t), i = 1, 2, \cdots, n$ given in (8.11). The methodologies should work quite similarly if we replace these standardized residual functions by the jackknife residual functions $z_i^{(-i)}(t), i = 1, 2, \cdots, n$ given in (8.18).

Empirical Distribution-Based Depth This depth was originally proposed by Fraiman and Muniz (2001) which can be used for detecting outliers from a sample of functions (Febrero, Galeano, and Gonzalez-Manteiga 2008). It is defined as the integration of some pointwise depth measures of a sample of functions over a given range. For the FLM (8.2) based on the functional data set (8.1), it can be defined based on the standardized residual functions $z_i(t), t \in \mathcal{T}, i = 1, 2, \cdots, n$, which carry almost all the information about how well the FLM (8.2) fits all the cases in the functional data set (8.1). For any fixed $t \in \mathcal{T}$, it is well known that the empirical distribution function of $z_i(t), i = 1, 2, \cdots, n$ is given by

$$F_n\left(z_i(t)\right) = n^{-1} \sum_{j=1}^{n} I\left\{z_j(t) \leq z_i(t)\right\},$$

where $I\{A\}$ is the indicator of a set A. Set

$$D(z_i(t)) = 1 - \left|0.5 - F_n\left(z_i(t)\right)\right|, \tag{8.35}$$

which is a depth measure for assessing the centrality of $z_i(t)$ among the univariate sample $z_1(t), z_2(t), \cdots, z_n(t)$. Then the associated pointwise depth measures for the standardized residual functions are given by

$$D(z_i(t)), \; i = 1, 2, \cdots, n.$$

An integration of these pointwise depth measures gives the following empirical distribution (ED) based depth measures:

$$D_{\text{ED}}(z_i) = \int_{\mathcal{T}} D(z_i(t))dt, i = 1, 2, \cdots, n. \tag{8.36}$$

Notice that in the examples presented in this chapter, the depth defined in (8.35) for a univariate sample is generally adopted but it can be replaced, for example, with the halfspace depth (Tukey 1975) $\text{HD}(x) = \min\{F_n(x), 1 - F_n(x)\}$ or the simplicial depth (Liu 1990) $\text{SD}(x) = 2F_n(x)(1 - F_n(x))$, among others, if desired.

Functional KDE-Based Depth This depth was originally proposed by

Cuevas, Febrero, and Fraiman (2006) and can be used to detect functional outliers from a sample of functions. Its key idea is to use the probability density function (pdf) of "a random function" surrounded by the rest of "random functions" in a functional data set as a depth measure. In practice, this pdf is unknown and may be replaced by a kernel density estimation (KDE) of the function. That is why this depth is called a functional KDE-based depth.

For the FLM (8.2) based on the functional data set (8.1), the functional KDE-based depth measures are defined based on the standardized residual functions $z_i(t), t \in \mathcal{T}, i = 1, 2, \cdots, n$:

$$D_{\text{KDE}}(z_i) = n^{-1} \sum_{j=1}^{n} K_h \left(\|z_j - z_i\| \right), \; i = 1, 2, \cdots, n, \tag{8.37}$$

where $\|\cdot\|$ denotes a functional norm, $K_h(t) = K(t/h)/h$ with $K(t)$ a kernel function, and $h > 0$ being a bandwidth.

Remark 8.5 *To use (8.37), a functional norm $\|\cdot\|$, a kernel function $K(t)$, and a bandwidth h must be chosen. Following Cuevas, Febrero, and Fraiman (2006), we can use the L^2-norm, the Gaussian kernel $K(t) = \exp(-t^2/2)/\sqrt{\pi/2}, t > 0$, and the bandwidth h being taken as the 15th percentile of the empirical distribution of $\{\|z_i - z_j\|^2, i, j = 1, 2, \cdots, n\}$. Febrero, Galeano, and Gonzalez-Manteiga (2008) pointed out that the functional KDE (8.37) is not very sensitive to the choice of the bandwidth and a wide range of values of h are appropriate with the only requirement that the bandwidth h be not very small.*

Remark 8.6 *The Gaussian kernel $K(t) = \exp(-t^2/2)/\sqrt{\pi/2}, t > 0$ used here is different from the Gaussian kernel $K(t) = \exp(-t^2/2)/\sqrt{2\pi}, t \in (-\infty, \infty)$ used in the local polynomial kernel smoothing technique described in Chapters 2 and 3.*

Remark 8.7 *The underlying rationale for us to use the functional KDE-based depth measures (8.37) for functional outlier detection is that the probability density of a functional outlier surrounded by the rest of the functions is expected to be very small compared with those of the majority of the remaining functions.*

Projection-Based Depth Cuevas, Febrero, and Fraiman (2007) proposed a random projection depth for functional data classification. Febrero, Galeano, and Gonzalez-Manteiga (2008) adopted this random projection depth for functional outlier detection. The random projection depth measures for a functional data set depend on a sequence of random projection directions generated from some stochastic process. Use of a sequence of random projection directions, from a practical point of view, is not very convenient as it may lead to different conclusions. To overcome this problem, here we propose a projection-based depth measure for outlier detection for the functional data set (8.1) with respect to the FLM (8.2) based on a sequence of fixed projection directions. The projection-based depth measures are defined based on the standardized residual functions $z_i(t), i = 1, 2, \cdots, n$ given in (8.11).

Given a sequence of fixed projection directions

$$\phi_l(t), l = 1, 2, \cdots, L, \tag{8.38}$$

the projections of the standardized residual functions $z_i(t), i = 1, 2, \cdots, n$ are given by the inner product between $z_i(t)$ and $\phi_l(t)$:

$$x_{il} = \int_{\mathcal{T}} z_i(t)\phi_l(t)dt, i = 1, 2, \cdots, n; l = 1, 2, \cdots, L.$$

For $l = 1, 2, \cdots, L$, let $D(x_{il}), i = 1, 2, \cdots, n$ denote the depth measures of the univariate sample $x_{il}, i = 1, 2, \cdots, n$. Then the projection depth measures of $z_i(t), i = 1, 2, \cdots, n$ are defined as

$$D_P(z_i) = L^{-1} \sum_{l=1}^{L} D(x_{il}),\ i = 1, 2, \cdots, n. \tag{8.39}$$

In practice, the projection directions (8.38) and the univariate depth D must be chosen. To choose the projection directions (8.38), a simple method is to take the first L eigenfunctions of the unbiased sample covariance function $\hat{\gamma}(s, t)$ given in (8.10) as the sequence of the projection directions for some L, say, $L = n$. When the resolution M for computing $\hat{\gamma}(s, t)$ is too large, however, this method may be time consuming as in this case it is not easy to compute the first L eigenfunctions of $\hat{\gamma}(s, t)$. To overcome this difficulty, alternatively we can take $z_l(t), l = 1, 2, \cdots, n$ as the projection directions and let $L = n$, especially when n is much smaller than the resolution M. This method works reasonably well. As for the univariate depth D, we generally adopt the depth defined in (8.35) for our examples presented in this chapter but the halfspace depth (Tukey 1975) $\text{HD}(x) = \min\{F_n(x), 1 - F_n(x)\}$, or the simplicial depth (Liu 1990) $\text{SD}(x) = 2F_n(x)(1 - F_n(x))$, among others, can also be used if desired.

We are now ready to use the above functional depth measures for functional outlier detection for the functional data set (8.1) with respect to the FLM (8.2). This may be done by drawing and analyzing the scatterplots of the functional depth measures against the curve number. Those cases $[\mathbf{x}_i^T, y_i(t)]$ with functional depth measures much smaller than the majority of the functional depth measures for the whole functional data set (8.1) may be claimed as outlying cases. Alternatively, one can conduct functional outlier detection using some cutoff values for a functional depth measure. For any given false positive rate α, Febrero, Galeano, and Gonzalez-Manteiga (2008) proposed two bootstrap procedures for determining the cutoff values for a functional depth measure.

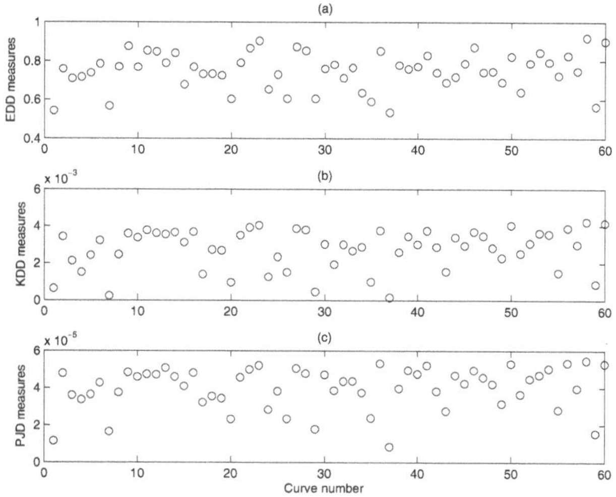

Figure 8.5 *Outlier detection for the ergonomics data using the (a) ED-, (b) functional KDE-, and (c) projection-based functional depth measures. At least visually, the ED-based functional depth measure does not work very well.*

Example 8.3 *For the ergonomics data with the FLM (6.1), the above three functional depth measures were calculated and are presented in Figure 8.5. Panel (a) presents the ED-based depth measures computed based on the univariate depth (8.35). It is seen that Curve 37 is not very special because its ED-based depth measure is very close to those of Curves 1 and 59. At the same time, Curves 7 and 29 are also not very special as their ED-based depth measures are close to the depth measures of Curves 20 and 26. Thus, compared with the standardized residual-based outlyingness measure, this ED-based functional depth measure does not work well in detecting functional outliers for the ergonomics data. This discovery is consistent with that observed from the simulation studies and real data applications presented in Febrero, Galeano, and Gonzalez-Manteiga (2008). Panels (b) and (c) present the associated functional KDE- and projection-based depth measures. They suggest that Curve 37 has the smallest depth measures, followed by Curves 7 and 29, and then by Curves 1, 59, 20, and 35, etc. From these two depth measures, we are sure that Curve 37 is a functional outlier, and Curves 7 and 29 are suspected functional outliers. One can also apply the two bootstrap procedures of Febrero, Galeano, and Gonzalez-Manteiga (2008) to determine the cutoff values of these depth measures.*

Alternatively, we can convert the above three functional depth measures to the following three functional outlyingness measures for functional outlier

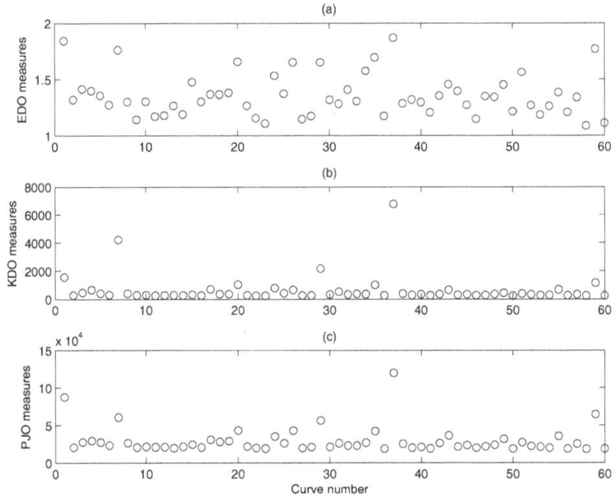

Figure 8.6 *Outlier detection for the ergonomics data based on the ED-, functional KDE-, and projection-based functional outlyingness measures. Visually, it appears that the functional KDE- and projection-based functional outlyingness measures work better than their associated depth measures in detecting functional outliers for the ergonomics data. The ED-based functional outlyingness measure still does not work well.*

detection:

$$\begin{aligned} O_{\text{ED}}(z_i) &= 1/D_{\text{ED}}(z_i), & i = 1, 2, \cdots, n, \\ O_{\text{KDE}}(z_i) &= 1/D_{\text{KDE}}(z_i), & i = 1, 2, \cdots, n, \\ O_P(z_i) &= 1/D_P(z_i), & i = 1, 2, \cdots, n, \end{aligned} \quad (8.40)$$

which may be called the ED-based, functional KDE-based, and projection-based outlyingness measures, respectively. They can be used to detect the functional outliers in the functional data set (8.1) with respect to the FLM (8.2) by drawing and analyzing the scatterplots of these functional outlyingness measures.

Example 8.4 *For the ergonomics data with the FLM (6.1), the ED-, functional KDE-, and projection-based outlyingness measures were calculated using (8.40) and are presented in panels (a), (b), and (c) of Figure 8.6, respectively. As expected, based on the ED-based outlying measure, it is not easy to detect Curve 37 as a functional outlier although it does have the largest outlyingness. However, based on the functional KDE- and projection-based outlyingness measures in panels (b) and (c), we can easily detect that Curve 37 is a functional outlier, and Curves 7 and 29 are suspected functional outliers. Compared with Figure 8.5, it seems that the functional KDE- and projection-based outlying measures outperform the functional KDE- and projection-based depth measures in functional outlier detection, at least visually.*

8.4 Influential Case Detection

In classical linear regression models, Cook's distance is a well-known tool for identifying influential cases. In this section, we study how to identify influential cases in the functional data set (8.1) with respect to the FLM (8.2). Let $\hat{\boldsymbol{\beta}}(t)$ and $\hat{\boldsymbol{\beta}}^{(-i)}(t)$ denote the estimators of $\boldsymbol{\beta}(t)$ obtained with and without the ith case $[\mathbf{x}_i^T, y_i(t)]$. If the ith case is influential, the change from $\hat{\boldsymbol{\beta}}(t)$ to $\hat{\boldsymbol{\beta}}^{(-i)}(t)$ will be big. Therefore, we can measure the influence of the ith case by the following generalized Cook's distance (Shen and Xu 2007):

$$C_i = \frac{\int_{\mathcal{T}} \left[\hat{\boldsymbol{\beta}}(t) - \hat{\boldsymbol{\beta}}^{(-i)}(t)\right]^T (\mathbf{X}^T\mathbf{X}) \left[\hat{\boldsymbol{\beta}}(t) - \hat{\boldsymbol{\beta}}^{(-i)}(t)\right] dt}{(p+1)\mathrm{tr}(\hat{\gamma})}. \tag{8.41}$$

Notice that for any fixed $t \in \mathcal{T}$, the FLM (8.3) is a classical linear regression model; it is then natural to define the pointwise Cook's distance at t for the ith case $[\mathbf{x}_i^T, y_i(t)]$ as

$$C_i(t) = \frac{\left[\hat{\boldsymbol{\beta}}(t) - \hat{\boldsymbol{\beta}}^{(-i)}(t)\right]^T (\mathbf{X}^T\mathbf{X}) \left[\hat{\boldsymbol{\beta}}(t) - \hat{\boldsymbol{\beta}}^{(-i)}(t)\right]}{(p+1)\hat{\gamma}(t,t)}. \tag{8.42}$$

It is easy to see that the generalized Cook's distance (8.41) can be expressed as a weighting integration of the pointwise Cook's distance (8.42). That is,

$$C_i = \int_{\mathcal{T}} w(t) C_i(t) dt,$$

where $w(t) = \frac{\hat{\gamma}(t,t)}{\mathrm{tr}(\hat{\gamma})}, t \in \mathcal{T}$ is a weight function such that $\int_{\mathcal{T}} w(t)dt = 1$. Notice that we can express the generalized Cook's distance (8.41) in terms of the L^2-distance between $\hat{\mathbf{y}}(t) = \mathbf{X}\hat{\boldsymbol{\beta}}(t)$ and $\hat{\mathbf{y}}^{(-i)}(t) = \mathbf{X}\hat{\boldsymbol{\beta}}^{(-i)}(t)$, the vectors of fitted response functions with and without the ith case. That is,

$$C_i = \frac{\int_{\mathcal{T}} \|\hat{\mathbf{y}}(t) - \hat{\mathbf{y}}^{(-i)}(t)\|^2 dt}{(p+1)\mathrm{tr}(\hat{\gamma})}, \tag{8.43}$$

where $\|\mathbf{a}\|$ denotes the usual L^2-norm of the vector $\mathbf{a}$. Cases with large C_i have a substantial influence on the estimated coefficient functions in $\hat{\boldsymbol{\beta}}(t)$ and on the fitted response functions in $\hat{\mathbf{y}}(t)$. Thus, deletion of these influential cases may result in important changes in conclusions.

It is very time consuming to compute the generalized Cook's distance using (8.41) as one needs to fit the FLM (8.3) n times, each with a case deleted. Fortunately, the generalized Cook's distance (8.41) can be computed directly from the rescaled SRO measure $\tilde{S}_i^2$ as defined in (8.26) and the leverage h_{ii} as we can express C_i as

$$C_i = \frac{h_{ii}}{1-h_{ii}} \frac{\tilde{S}_i^2}{p+1}. \tag{8.44}$$

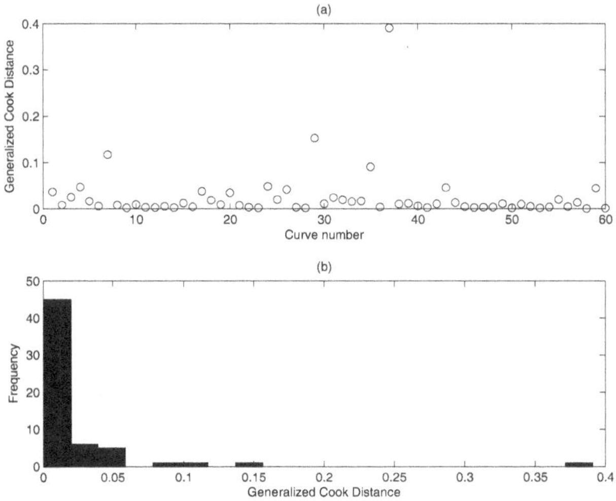

Figure 8.7 *Influential case detection for the ergonomics data using the generalized Cook's distance. Panel (a) shows the scatterplot of generalized Cook's distance against curve number. Panel (b) displays the histogram of the generalized Cook's distances. Both figures indicate that Curve 37 is highly influential.*

It is seen that a highly influential case must have a large leverage h_{ii}, a large rescaled SRO measure $\tilde{S}_i^2$, or both. That is, an influential case must be a high-leverage case, an outlying case, or both.

Example 8.5 *For the ergonomics data presented in Example 8.1, we computed the generalized Cook's distance for the cases in the ergonomics data with the FLM (6.1) using (8.44). Figure 8.7 displays the scatterplot (panel (a)) and histogram (panel (b)) of these generalized Cook's distances. It is seen that Case 37 is highly influential while Cases 29 and 7 are suspected to be influential. This is consistent with what we observed from Figure 8.4 where Curves 37, 7, and 29 are highlighted.*

8.5 Robust Estimation of Coefficient Functions

The functional outlyingness measures or the generalized Cook's distances can be used to order the cases $[\mathbf{x}_i^T, y_i(t)]$ in the functional data set (8.1) for fitting the FLM (8.3). Let the sorted cases be given by

$$[\mathbf{x}_{(i)}^T, y_{(i)}(t)], i = 1, 2, \cdots, n, \tag{8.45}$$

so that $[\mathbf{x}_{(1)}^T, y_{(1)}(t)]$ has the smallest outlyingness measure or the smallest generalized Cook's distance, followed by $[\mathbf{x}_{(2)}^T, y_{(2)}(t)]$, $[\mathbf{x}_{(3)}^T, y_{(3)}(t)]$, and so on, and finally $[\mathbf{x}_{(n)}^T, y_{(n)}(t)]$ has the largest outlyingness measure or the largest generalized Cook's distance. Let the first L cases in (8.45) be not outlying or influential cases. Alternatively, we can determine L using the given nominal false positive rate α so that $L = [(1-\alpha)n]$ where $[a]$ denotes the integer part of a. That is, the last $(n-L)$ cases are deleted from the functional data set (8.45). We then fit the FLM (8.2) to the first L cases given in (8.45) so that we can get the robust estimation (when the $(n-L)$ outlying cases are removed) or α-trimmed estimation (when the last $(n-L)$ most outlying cases are removed) of the coefficient function vector $\boldsymbol{\beta}(t)$ and the covariance function $\gamma(s,t)$. See Figure 8.2 for the first four fitted coefficient functions in the FLM (6.1) based on the ergonomics data with and without the functional outlying case associated with Curve 37.

8.6 Outlier Detection for a Sample of Functions

Detecting functional outliers from a sample of n functions

$$y_1(t), y_2(t), \cdots, y_n(t) \overset{i.i.d.}{\sim} \text{SP}(\eta, \gamma), t \in \mathcal{T}, \tag{8.46}$$

is equivalent to detecting functional outliers from the following mean function model:

$$y_i(t) = \eta(t) + v_i(t), \quad v_i(t) \overset{i.i.d.}{\sim} \text{SP}(0, \gamma), t \in \mathcal{T},\ i = 1, 2, \cdots, n. \tag{8.47}$$

This mean function model is a special case of the FLM (8.3) with the design matrix being $\mathbf{1}_n$. Therefore, all the methods developed in the previous sections can be used to detect functional outliers from (8.46), keeping in mind that the design matrix $\mathbf{X} = \mathbf{1}_n$. For readers' convenience, in what follows, we briefly describe all the related terms and methods for detecting functional outliers from (8.46). For this purpose, we list the following sample mean and covariance functions:

$$\begin{array}{c} \hat{\eta}(t) = \bar{y}(t) = n^{-1} \sum_{i=1}^n y_i(t), \\ \hat{\gamma}(s,t) = (n-1)^{-1} \sum_{i=1}^n [y_i(s) - \bar{y}(s)][y_i(t) - \bar{y}(t)]. \end{array} \tag{8.48}$$

8.6.1 Residual Functions

First of all, the raw residual functions are now given by

$$\hat{v}_i(t) = y_i(t) - \bar{y}(t), i = 1, 2, \cdots, n,$$

and the leverages are given by $h_{ii} = 1/n, i = 1, 2, \cdots, n$. It follows that the standardized residual functions are given by

$$z_i(t) = \frac{\hat{v}_i(t)}{\sqrt{1 - h_{ii}}} = \sqrt{n/(n-1)}[y_i(t) - \bar{y}(t)], i = 1, 2, \cdots, n. \tag{8.49}$$

We have $z_i(t) \sim \text{SP}(0, \gamma), i = 1, 2, \cdots, n$. When the sample (8.46) is Gaussian, the standardized residual functions (8.49) are also Gaussian. The raw jackknife residual functions are given by

$$\hat{v}_i^{(-i)}(t) = y_i(t) - \bar{y}^{(-i)}(t) = \frac{n}{n-1}[y_i(t) - \bar{y}(t)], \ i = 1, 2, \cdots, n,$$

where $\bar{y}^{(-i)}(t)$ denotes the usual sample mean function using all the functions except function i. In this case, we have $\mathbf{x}_i = 1$ and $1 + \mathbf{x}_i^T[\mathbf{X}^{(-i)T}\mathbf{X}^{(-i)}]^{-1}\mathbf{x}_i = n/(n-1)$. It follows that the associated jackknife residual functions are given by

$$\begin{aligned} z_i^{(-i)}(t) &= \frac{\hat{v}_i^{(-i)}(t)}{\sqrt{n/(n-1)}} \\ &= \sqrt{n/(n-1)}[y_i(t) - \bar{y}(t)] = z_i(t), i = 1, 2, \cdots, n. \end{aligned} \tag{8.50}$$

Therefore, under the mean function model (8.47), the jackknife residual functions and the standardized residual functions are exactly the same.

8.6.2 *Functional Outlier Detection*

In this subsection, we outline the three functional outlier detection methods for the functional sample (8.46).

Standardized Residual-Based Method Under the mean function model (8.47), the SRO outlyingness measures of the functions $y_i(t), i = 1, 2, \cdots, n$ are now given by

$$S_i^2 = \|z_i\|^2 = \frac{n}{n-1}\int_{\mathcal{T}}[y_i(t) - \bar{y}(t)]^2 dt, i = 1, 2, \cdots, n. \tag{8.51}$$

Under the mean function model (8.47) and the Gaussian assumption (8.12), Theorem 8.1 is still valid. The approximate distribution given in (8.23) is still valid with the parameters $\hat{\boldsymbol{\beta}}$ and $\hat{\kappa}$ given in (6.26) or (6.29). We can still use the cutoff value specified in (8.24) and the P-values specified in (8.25) for functional outlier detection. The rescaled SRO measures (8.26) and their cutoff values (8.27) can still be used. Please keep in mind, however, that in all these calculations, the associated sample covariance function $\hat{\gamma}(s, t)$ is given in (8.48).

Jackknife Residual-Based Method Under the mean function model (8.47), the JRO outlyingness measures of the functions $y_i(t), i = 1, 2, \cdots, n$, are now given by

$$J_i^2 = \frac{\|z_i^{(-i)}\|^2}{\text{tr}(\hat{\gamma}^{(-i)})}, i = 1, 2, \cdots, n, \tag{8.52}$$

where $\text{tr}(\hat{\gamma}^{(-i)})$ is the usual sample covariance function using all the functions except function i. Under the mean function model (8.47) and the Gaussian assumption (8.12), Theorem 8.2 is still valid. The approximate distribution given in (8.31) is still valid with the parameter $\hat{\kappa}$ given in (6.26) or (6.29). We can still use the cutoff value specified in (8.32) and the P-values specified in (8.33) for functional outlier detection. In addition, the JRO measures $J_i^2, i = 1, 2, \cdots, n$ can be computed using the relationship (8.34) with the rescaled SRO measures $\tilde{S}_i^2, i = 1, 2, \cdots, n$. Again, in all these calculations, the sample covariance function $\hat{\gamma}(s,t)$ is given in (8.48).

Functional Depth-Based Method The three depth measures described in Section 8.3.3 were originally proposed for functional outlier detection for a sample of functions (Fraiman and Muniz 2001; Cuevas, Febrero, and Fraiman 2006, 2007; Febrero, Galeano, and Gonzalez-Manteiga 2008). These depth measures can now be directly applied to the functions from the functional sample (8.46).

The ED-based depth measures are now given by

$$D_{\text{ED}}(y_i) = \int_{\mathcal{T}} D(y_i(t))dt, i = 1, 2, \cdots, n,$$

where the univariate depth D is the depth defined in (8.35) in our examples presented in this section but can be replaced with the halfspace depth (Tukey 1975) $\text{HD}(x) = \min\{F_n(x), 1 - F_n(x)\}$ or the simplicial depth (Liu 1990) $\text{SD}(x) = 2F_n(x)(1 - F_n(x))$, among others, if desired.

The functional KDE-based depth measures are now given by

$$D_{\text{KDE}}(y_i) = n^{-1} \sum_{j=1}^{n} K_h (\|y_j - y_i\|), \ i = 1, 2, \cdots, n,$$

where we can use the L^2-norm for $\|\cdot\|$, the Gaussian kernel $K(t) = \exp(-t^2/2)/\sqrt{\pi/2}, t > 0$, and the bandwidth h being taken as the 15th percentile of the empirical distribution of $\{\|y_i - y_j\|^2, i, j = 1, 2, \cdots, n\}$ as in Cuevas, Febrero, and Fraiman (2006) and Febrero, Galeano, and Gonzalez-Manteiga (2008).

Given a sequence of fixed projection directions (8.38), the projections of $y_i(t), i = 1, 2, \cdots, n$ are given by the inner product between $y_i(t)$ and $\phi_l(t)$:

$$x_{il} = \int_{\mathcal{T}} y_i(t)\phi_l(t)dt, i = 1, 2, \cdots, n; l = 1, 2, \cdots, L.$$

Then the projection depth measures of $y_i(t), i = 1, 2, \cdots, n$ are given by

$$D_P(y_i) = L^{-1} \sum_{l=1}^{L} D(x_{il}), \ i = 1, 2, \cdots, n,$$

where $D(x_{il}), i = 1, 2, \cdots, n$ denote the depth measures of the one-dimensional projections $x_{il}, i = 1, 2, \cdots, n$. Again, we can take the first L eigenfunctions of the unbiased sample covariance function $\hat{\gamma}(s,t)$ given in (8.48) or directly take $y_i(t), i = 1, 2, \cdots, n$ as the projection directions. In addition, we adopt the depth defined in (8.35) for the univariate depth D for our examples presented in this section but it can be replaced with the halfspace depth (Tukey 1975) $\text{HD}(x) = \min\{F_n(x), 1 - F_n(x)\}$, or the simplicial depth (Liu 1990) $\text{SD}(x) = 2F_n(x)(1 - F_n(x))$, among others, if desired.

Under the mean function model (8.47), it is not difficult to notice that in all the above calculations, using the standardized residual functions (8.49), the jackknife residual functions (8.50), or the functions in (8.46) are equivalent. In addition, as in Section 8.3.3, one can directly use the above ED-, functional KDE-, or projection-based depth measures or use the associated ED-, functional KDE-, or projection-based outlyingness measures converted using (8.40) for functional outlier detection from the functional sample (8.46).

Example 8.6 *The nitrogen oxides (NOx) emission level data were introduced in Section 1.2.3 of Chapter 1. In this example, we consider the standardized residual-based method for detecting the functional outliers in the NOx emission level curves for* $n = 76$ *working days. These curves were depicted in the upper panel of Figure 1.5. By some calculations, we obtained* $tr(\hat{\gamma}) = 4.4377\text{e}4$ *and* $tr(\hat{\gamma}^{\otimes 2}) = 6.0163\text{e}8$. *Based on these quantities and by the bias-reduced method, we obtained* $\hat{\beta} = 1.2800\text{e}4$ *and* $\hat{\kappa} = 3.4404$. *The upper panel of Figure 8.8 displays the scatterplot of the rescaled SRO measures* $\tilde{S}_i^2, i = 1, 2, \cdots, n$ *for the* 76 *NOx emission level curves under the FLM (8.47). The associated cutoff values for* $\alpha = 5\%, 1\%$, *and* 0.1%, *computed using (8.27) are* $2.4706, 3.5231$, *and* 4.9780, *respectively. It follows that when* $\alpha = 5\%$, *there are more than three outliers; when* $\alpha = 1\%$, *Curves 14, 16, and 37 are outliers; and when* $\alpha = 0.1\%$, *Curves 16 and 37 are outliers. In the upper panel of Figure 8.8, the approximate P-values of the three most outlying curves, computed using (8.25), were depicted. It is seen that these three most outlying curves are very significant with their approximate P-values being* $0.38\%, 0.01\%$, *and* 0.00%, *respectively.*

The lower panel of Figure 8.8 displays the histogram of the rescaled SRO measures $\tilde{S}_i^2, i = 1, 2, \cdots, n$. *This histogram seems to suggest that Curves 14, 16, and 37 are all functional outliers. To further clarify if they are outliers, we redrew all the NOx emission level curves in Figure 8.9 with Curves 14, 16, and 37 highlighted in wide dot-dashed, solid, and dashed curves, and we also identified their associated dates as "16/03/2005, 18/03/2005, 29/04/2005," respectively. According to Febrero, Galeano, and Gonzalez-Manteiga (2008), the day "18/03/2005" was a Friday corresponding with the beginning of the Eastern vacation in Spain in the year 2005. The day "29/04/2005" was another Friday associated with the beginning of a long weekend. Therefore, the outliers associated with these two days "18/03/2005" and "29/04/2005" were related to small vacation periods that produced a large traffic concentration in specific periods due to the fact that people were hurriedly heading home to cel-*

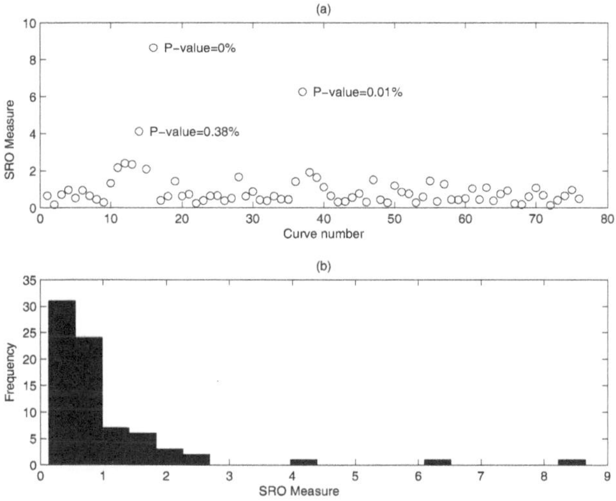

Figure 8.8 *Outlier detection for the NOx emission level curves for seventy-six working days using the standardized residual-based method. Panel (a) shows the scatterplot of SRO measure against curve number. Panel (b) displays the histogram of the SRO measures. Both the figures suggest that there are three functional outliers.*

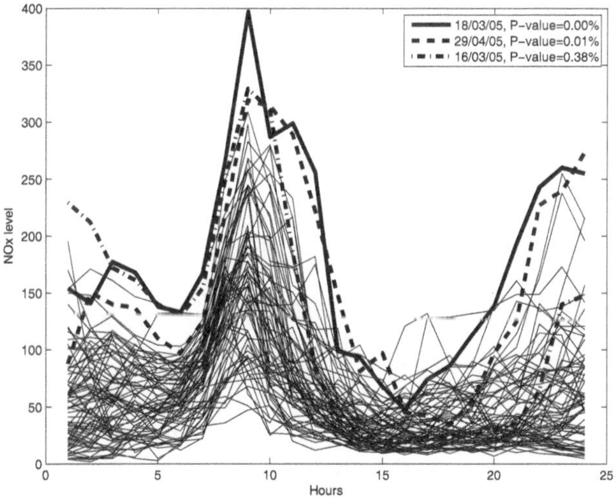

Figure 8.9 *NOx emission level curves for seventy-six working days. Curves 14, 16, and 37 are detected as functional outliers and they are highlighted in wide dot-dashed, solid, and dashed curves, respectively. Their associated dates are identified as "16/03/2005, 18/03/2005, 29/04/2005," respectively.*

ebrate the holidays. However, it is not clear why the day "16/03/2005" was so special although it is not difficult to notice that the NOx emission levels before 10 am that day were very high compared with those of other working days.

8.7 Technical Proofs

In this section, we outline the proofs of the two theorems described in this chapter.

Proof of Theorem 8.1 Under the FLM (8.3) and the Gaussian assumption (8.12), for $i = 1, 2, \cdots, n$, by (8.13), we have $z_i(t) \sim \text{GP}(0, \gamma)$. The theorem is proved by applying Theorem 4.2 of Chapter 4.

Proof of Theorem 8.2 Under the FLM (8.3) and the Gaussian assumption (8.12), $y_i(t), \hat{y}_i^{(-i)}(t)$ and $\hat{\gamma}^{(-i)}(s,t)$ are independent. It follows that $z_i^{(-i)}(t)$ and $\text{tr}(\hat{\gamma}^{(-i)})$ are independent. By (8.19) and Theorem 4.2 of Chapter 4, we have $\|z_i^{(-i)}\|^2 \stackrel{d}{=} \sum_{r=1}^m \lambda_r A_r, A_r \stackrel{i.i.d.}{\sim} \chi_1^2$, where $\lambda_r, r = 1, 2, \cdots, m$ are all the m positive eigenvalues of $\gamma(s,t)$. In addition, under the FLM (8.3) and the Gaussian assumption (8.12), we have $(n-p-2)\hat{\gamma}^{(-i)}(s,t) \sim \text{WP}(n-p-2, \gamma)$. By Theorem 4.5 of Chapter 4, we have $\text{tr}(\hat{\gamma}^{(-i)}) \stackrel{d}{=} \sum_{r=1}^m \lambda_r B_r/(n-p-2), B_r \stackrel{i.i.d.}{\sim} \chi_{n-p-2}^2$. The theorem is then proved.

8.8 Concluding Remarks and Bibliographical Notes

In this chapter, we reviewed a number of methods for unusual functional observation detection. The functional outlier detection methods are based on either functional outlyingness or functional depth measures while functional influential observation detection is based on the generalized Cook's distance proposed by Shen and Xu (2007).

The standardized residual and jackknife residual methods were originally studied by Shen and Xu (2007) although in a quite different manner. The empirical distribution-based functional depth was first proposed and studied by Fraiman and Muniz (2001) for functional outlier detection of a sample of functions. The functional KDE-based functional depth was studied by Cuevas, Febrero, and Fraiman (2006), and the random projection depth was investigated by Cuevas, Febrero, and Fraiman (2007). These three functional depths were reviewed and adopted by Febrero, Galeano, and Gonzalez-Manteiga (2008), and two bootstrap procedures for cutoff value determination were proposed.

We now give a brief review of other related literature. Based on residual process analysis, Chiou and Müller (2007) proposed a diagnostics method for functional regression models that relates a functional response to predictor variables that can be multivariate vectors or random functions. Cuesta-Albertos and Nieto-Reyes (2008) proposed a random depth to approximate

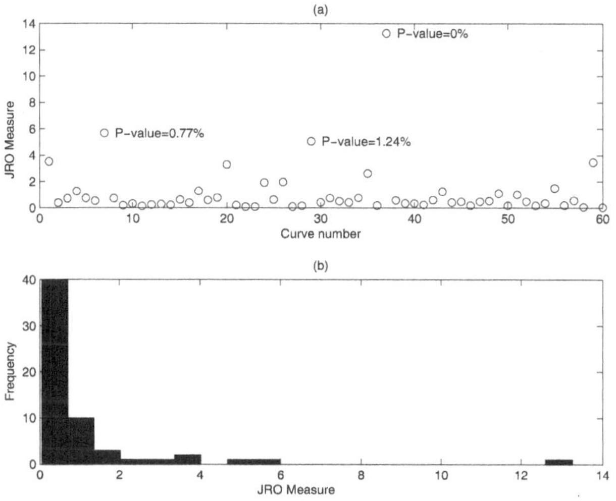

Figure 8.10 *Outlier detection for the ergonomics data using the jackknife residual-based method. Panel (a) shows the scatterplot of JRO measure against curve number. Panel (b) displays the histogram of the JRO measures.*

the Tukey depth (Tukey 1974) to overcome the computational difficulty when it is applied in the context of functional data analysis. Most recently, Gervini (2012) introduced trimmed estimators for the mean and covariance functions of general functional data. The estimators are based on a new measure of "outlyingness" or data depth that is well defined on any metric space, although he focused on Euclidean spaces. He computed the breakdown point of the estimators and showed that the optimal breakdown point is attainable for the appropriate choice of tuning parameters. The small-sample behavior of the estimators was studied by simulation, and it was shown to have better outlier-resistance properties than alternative estimators. Yu, and Zou, and Wang (2012), on the other hand, proposed a principal component-based testing procedure for outlier detection in functional data.

8.9 Exercises

1. Consider the jackknife residual-based method for the ergonomics data presented in Example 8.1. Figure 8.10 displays the scatterplot (upper panel) and histogram (lower panel) of the JRO measures $J_i^2, i = 1, 2, \cdots, n$ of the right elbow angle curves of the ergonomics data with respect to the FLM (6.1). The approximate P-values of the three most outlying curves, computed using (8.33), are also depicted.

   (a) Using the information given in Example 8.2, find the bias-

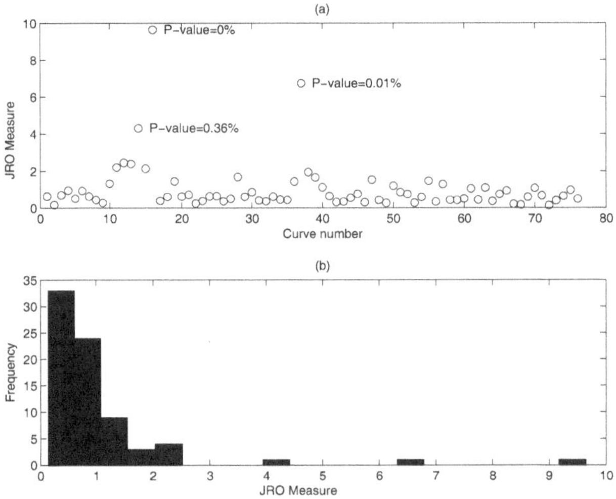

Figure 8.11 *Outlier detection for the NOx emission level data for seventy-six working days using the jackknife residual-based method. Panel (a) shows the scatterplot of JRO measure against curve number. Panel (b) displays the histogram of the JRO measures.*

reduced estimator of κ and then specify the approximation distribution of the JRO measures (8.29) using (8.31).

(b) Compute the approximate cutoff values of the JRO measures (8.29) for $\alpha = 5\%, 1\%$, and 0.1%. For these three values of α, find the associated outliers. Are the results consistent with those obtained by the standardized residual-based method presented in Example 8.2?

(c) Compare the approximate P-values of the three most outlying curves with their approximate P-values depicted in the upper panel of Figure 8.3.

(d) Based on the histogram showed in the lower panel of Figure 8.10, can we claim that Curve 37 is an outlier while Curves 7 and 29 are not?

2. Consider the jackknife residual-based method for the NOx emission level data for seventy-six working days as in Example 8.6. Figure 8.11 displays the scatterplot (upper panel) and histogram (lower panel) of the JRO measures $J_i^2, i = 1, 2, \cdots, n$ of the NOx emission level curves with respect to the FLM (8.47). The approximate P-values of the three most outlying curves, computed using (8.33), are also depicted.

(a) Using the information given in Example 8.6, find the bias-

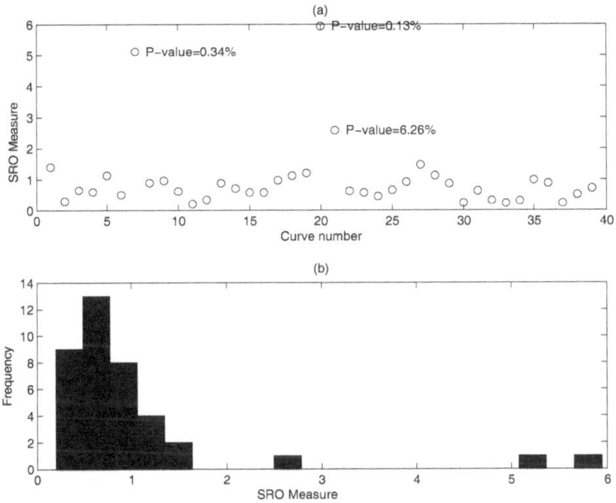

Figure 8.12 *Outlier detection for the NOx emission level curves for thirty-nine non-working days using the standardized residual-based method. Panel (a) shows the scatterplot of SRO measures against curve number. Panel (b) displays the histogram of the SRO measures.*

reduced estimator of κ and then specify the approximation distribution of the JRO measures (8.52) using (8.31).

(b) Compute the approximate cutoff values of the JRO measures (8.52) for $\alpha = 5\%, 1\%$, and 0.1%. For these three values of α, find the associated outliers. Are the results consistent with those obtained by the standardized residual-based method presented in Example 8.6?

(c) Compare the approximate P-values of the three most outlying curves with their approximate P-values depicted in the upper panel of Figure 8.8.

(d) Based on the histogram showed in the lower panel of Figure 8.11, can we claim that Curves 14, 16, and 37 are functional outliers?

3. Consider the standardized residual-based method for detecting the functional outliers in the NOx emission level curves for thirty-nine non-working days. These curves were depicted in the lower panel of Figure 1.5. By some calculations, we obtained $\text{tr}(\hat{\gamma}) = 3.0640e4$ and $\text{tr}(\hat{\gamma}^{\otimes 2}) = 4.2976e8$. The scatterplot of the rescaled SRO measures $\tilde{S}_i^2, i = 1, 2, \cdots, n$, for the NOx emission level curves for thirty-nine non-working days under the FLM (8.47) is displayed in the upper panel of Figure 8.12 where the approximate P-values of the three most outlying curves, computed using (8.25), are depicted. The histogram

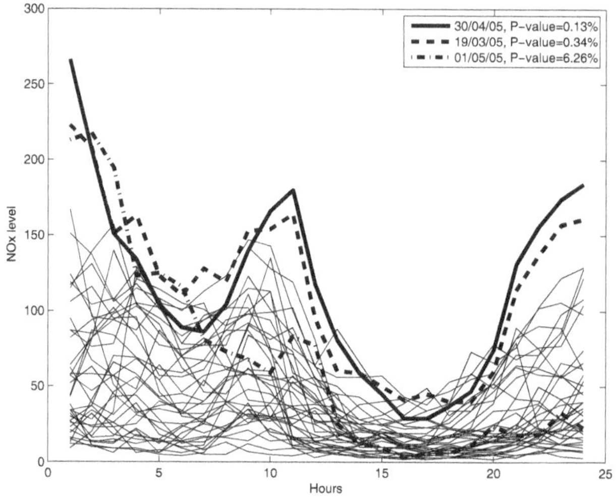

Figure 8.13 *NOx emission level curves of non-working days with the three detected functional outliers highlighted. Curves 7, 20, and 21 are detected as functional outliers and they are highlighted as wide dashed, solid, and dot-dashed curves. Their associated dates are identified as "19/03/2005, 30/04/2005, 01/05/2005," respectively.*

of the rescaled SRO measures $\tilde{S}_i^2, i = 1, 2, \cdots, n$ is displayed in the lower panel of Figure 8.8. The scatterplot suggests that Curves 7, 20, and 21 possibly are functional outliers. We have identified these dates as "19/03/2005, 30/04/2005, 01/05/2005" and highlighted them as wide dashed, solid, and dot-dashed curves in Figure 8.12 where all the NOx emission level curves for thirty-nine non-working days are displayed.

(a) Using the bias-reduced method, compute $\hat{\beta}$ and $\hat{\kappa}$ for the standardized residual-based method for detecting functional outliers from the NOx emission level curves for thirty-nine non-working days. Write down the associated approximation pdf of the SRO measures (8.51) using (8.23).

(b) Compute the approximate cutoff values of the rescaled SRO measures $\tilde{S}_i^2, i = 1, 2, \cdots, n$ for $\alpha = 5\%, 1\%$, and 0.1%. For these three values of α, find the associated functional outliers.

(c) Based on the histogram showed in the lower panel of Figure 8.11, can we claim that Curves 7, 20, and 21 are functional outliers?

(d) Based on some comments given in Example 8.6, comment on why Curves 7, 20, and 21 are so special compared with the rest of NOx emission level curves for thirty-nine non-working days.

4. For the NOx emission level data of thirty-nine non-working days with the FLM (8.47), the ED-, functional KDE-, and projection-based functional depth measures are calculated and are presented in Figure 8.14.

   (a) Identify the functional outliers of the NOx emission level curves for thirty-nine non-working days using the ED-, functional KDE-, and projection-based functional depth measures, respectively.
   (b) Compare the three functional depth measures in terms of functional outlier detection capability.

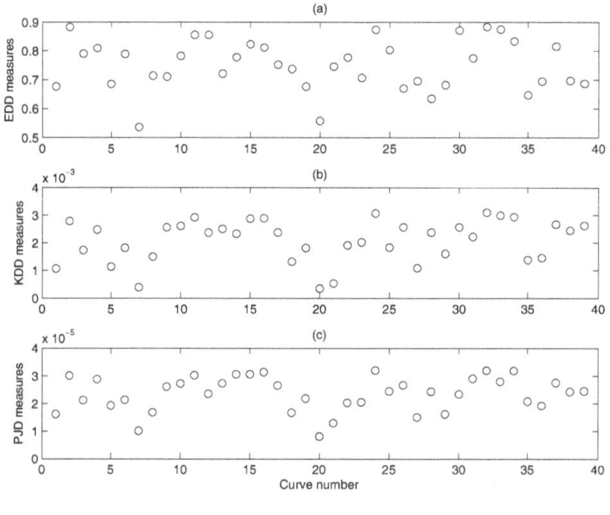

Figure 8.14 *Outlier detection for the NOx emission level data for thirty-nine non-working days. The scatterplots of ED-, functional KDE-, and projection-based functional depth measures against curve number are displayed in panels (a), (b), and (c), respectively.*

5. For the NOx emission level data for thirty-nine non-working days with the FLM (8.47), the ED-, functional KDE-, and projection-based functional outlyingness measures are calculated and are presented in Figure 8.15.

   (a) Identify the functional outliers of the NOx emission level curves for thirty-nine non-working days using the ED-, functional KDE-, and projection-based functional outlyingness measures, respectively.

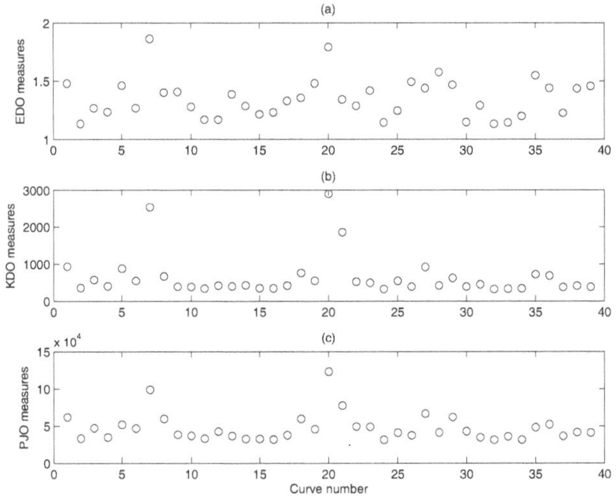

Figure 8.15 *Outlier detection for the NOx emission level data for thirty-nine non-working days. The scatterplots of ED-, functional KDE-, and projection-based functional outlyingness measures against curve number are displayed in panels (a), (b), and (c), respectively.*

(b) Compare the three functional outlyingness measures in terms of functional outlier detection capability.

(c) Compare the three functional outlyingness measures and the three functional depth measures in terms of functional outlier detection capability based on Figures 8.14 and 8.15.

6. Notice that for any fixed $t \in \mathcal{T}$, the FLM (8.3) is a classical linear regression model with the design matrix $\mathbf{X}$ and the response vector $\mathbf{y}(t)$. Then, based on the theory of classical linear regression models (Montgomery 2005, p. 398), we have

$$1 + \mathbf{x}_i^T[\mathbf{X}^{(-i)T}\mathbf{X}^{(-i)}]^{-1}\mathbf{x}_i = (1 - h_{ii})^{-1},$$
$$\hat{v}_i^{(-i)}(t) = y_i(t) - \hat{y}_i^{(-i)}(t) = \frac{\hat{v}_i(t)}{1-h_{ii}},\ t \in \mathcal{T},$$

and

$$(n-p-2)\hat{\gamma}^{(-i)}(t,t) = (n-p-1)\hat{\gamma}(t,t) - \frac{\hat{v}_i^2(t)}{1-h_{ii}},$$

keeping in mind that here $\gamma(t,t)$ is the common variance of $y_i(t), i = 1, 2, \cdots, n$ at $t \in \mathcal{T}$.

(a) Show that $z_i(t) = z_i^{(-i)}(t), i = 1, 2, \cdots, n$. That is, the standardized residual functions $z_i(t), i = 1, 2, \cdots, n$ and the jackknife residual functions $z_i^{(-i)}(t), i = 1, 2, \cdots, n$ are exactly the same.

(b) Show that $(n-p-2)\mathrm{tr}(\hat{\gamma}^{(-i)}) = (n-p-1)\mathrm{tr}(\hat{\gamma}) - S_i^2,\ i = 1, 2, \cdots, n$.

(c) Show that the JRO measures (8.29) satisfy (8.34). That is, show that for $i = 1, 2, \cdots, n$,

$$J_i^2 = \frac{(n-p-2)S_i^2}{(n-p-1)\mathrm{tr}(\hat{\gamma}) - S_i^2} = \frac{(n-p-2)\tilde{S}_i^2}{(n-p-1) - \tilde{S}_i^2}.$$

Chapter 9

Heteroscedastic ANOVA for Functional Data

9.1 Introduction

In Chapter 5 we studied how to conduct various ANOVA inferences for functional data under the assumption that different groups have a common covariance function. In this chapter, this assumption is dropped as in practice, this assumption may not always be satisfied. From this point of view, the ANOVA models studied in Chapter 5 can be referred to as homogeneous ANOVA models while the ANOVA models considered in this chapter are called heteroscedastic ANOVA models. When there are only two groups of functions involved, the associated problem may be referred to as a two-sample Behrens-Fisher problem for functional data.

As in classical heteroscedastic ANOVA/MANOVA (Behrens 1929, Fisher 1935, Welch 1947, James 1954, Johansen 1980, Kim 1992, Krishnamoorthy and Yu 2004, Zhang and Xu 2009, Zhang 2011b, etc.), we shall see that the key for heteroscedastic ANOVA for functional data is how to take into account the heteroscedasticity of the group covariance functions in constructing the test statistics and in deriving and approximating the null distributions of the test statistics. The possible effects of the group covariance function heteroscedasticity are as follows. First of all, the methodologies proposed under the covariance function homogeneity assumption may not be applicable and the associated results may be misleading when the homogeneity assumption is seriously violated (Zhang, Liang, and Xiao 2010). Second, the group covariance functions must be estimated separately based on the associated group sample only so that the convergence rates of the sample covariance functions are limited by $O_p(n_{\min}^{-1/2})$ which is much slower than the convergence rate $O_p(n^{-1/2})$ of the pooled sample covariance function constructed under the covariance function homogeneity assumption where $n_{\min}$ and n denote the minimum sample size and the total sample size, respectively. Third, deriving and approximating the null distributions of the test statistics under heteroscedastic ANOVA models are more complicated than under homogeneous ANOVA models.

The remaining part of this chapter is organized as follows. We start with

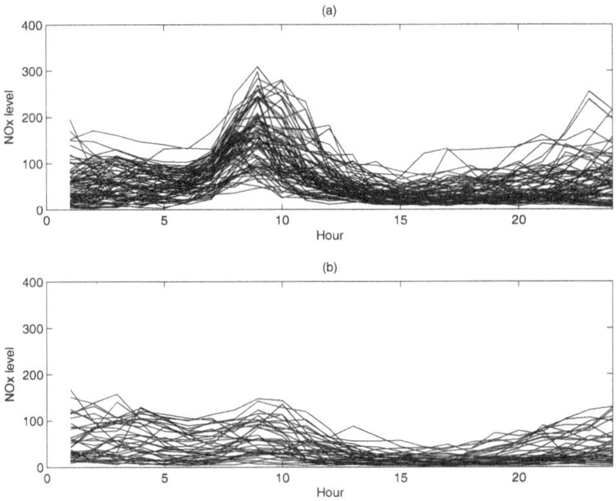

Figure 9.1 *NOx emission level curves for (a) seventy-three working days and (b) thirty-six non-working days with outlying curves removed. No smoothing has been applied to reconstruct these NOx emission level curves. It appears that the NOx emission level curves of working days are more variable than those of non-working days.*

the two-sample Behrens-Fisher problem for functional data in Section 9.2. Studies about heteroscedastic one-way ANOVA and two-way ANOVA for functional data are then described in Sections 9.3 and 9.4. Technical proofs of the main results are outlined in Section 9.5. Some concluding remarks and bibliographical notes are given in Section 9.6. Section 9.7 is devoted to some exercise problems related to this chapter.

9.2 Two-Sample Behrens-Fisher Problems

We use the following example to motivate a two-sample Behrens-Fisher (BF) problem for functional data. This BF problem was discussed in Zhang, Liang, and Xiao (2010).

Example 9.1 *The nitrogen oxides (NOx) emission level data were introduced in Section 1.2.3 of Chapter 1. Figure 9.1 displays the NOx emission level curves for (a) seventy-three working days and (b) thirty-six non-working days after six outlying curves were removed using one of the methods described in Chapter 8. The NOx emission levels for each day were evenly recorded twenty-four times daily, once an hour and the data were not smoothed. It appears that the NOx emission level curves of working days are more variable than those of non-working days. The sample covariance functions of the NOx emission level*

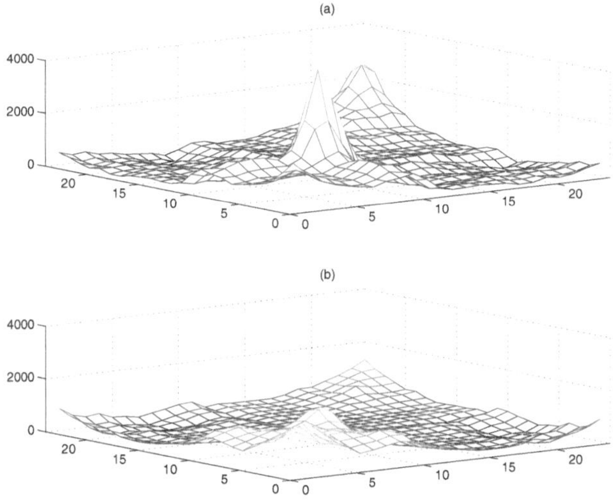

Figure 9.2 *Sample covariance functions of NOx emission level curves for (a) seventy-three working days and (b) thirty-six non-working days with outlying curves removed. It appears that there are some differences between these two sample covariance functions.*

curves of working days and non-working days displayed in Figure 9.2 seem to verify this point of view to some degree. Of interest is to test if the mean NOx emission level curves of working and non-working days are significantly different without assuming that the covariance functions of the NOx emission level curves of working and non-working days are equal. This two-sample problem for functional data cannot be handled directly using those methods for two-sample problems for functional data described in Section 5.2 of Chapter 5 as those methods assume that the covariance functions of the two functional samples are equal. This motivates a heteroscedastic two-sample problem for functional data or a two-sample BF problem for functional data.

A general two-sample BF problem for functional data can be defined as follows. Suppose we have two functional samples

$$y_{11}(t), \cdots, y_{1n_1}(t) \stackrel{i.i.d.}{\sim} \text{SP}(\eta_1, \gamma_1), \quad y_{21}(t), \cdots, y_{2n_2}(t) \stackrel{i.i.d.}{\sim} \text{SP}(\eta_2, \gamma_2), \quad (9.1)$$

where $\eta_1(t)$ and $\eta_2(t)$ are two unknown mean functions, and $\gamma_1(s,t)$ and $\gamma_2(s,t)$ are two unknown covariance functions that may be unequal. We wish to test the following hypotheses:

$$\begin{array}{lll} & H_0 & : \eta_1(t) \equiv \eta_2(t), \; t \in \mathcal{T}, \\ \text{versus} & H_1 & : \eta_1(t) \neq \eta_2(t), \; \text{ for some } t \in \mathcal{T}. \end{array} \quad (9.2)$$

When $\gamma_1(s,t) \equiv \gamma_2(s,t), \; s,t \in \mathcal{T}$ is satisfied, the above problem (9.2) reduces to the two-sample problem (5.2) discussed in Section 5.2 in Chapter 5. In practice, however, the equality of $\gamma_1(s,t)$ and $\gamma_2(s,t)$ may not be satisfied and is often not easy to check, as indicated in Example 9.1. Therefore, it is of practical interest to propose some methodologies for testing (9.2) without assuming the equality of the covariance functions of the two functional samples (9.1).

9.2.1 Estimation of Mean and Covariance Functions

To test (9.2), we first give the sample mean and covariance functions of the two samples (9.1) as

$$\begin{aligned} \hat{\eta}_i(t) &= \bar{y}_i(t) = n_i^{-1}\sum_{j=1}^{n_i} y_{ij}(t),\ i=1,2, \\ \hat{\gamma}_i(s,t) &= (n_i-1)^{-1}\sum_{j=1}^{n_i}\left[y_{ij}(s)-\bar{y}_i(s)\right]\left[y_{ij}(t)-\bar{y}_i(t)\right], \\ & \quad i=1,2, \end{aligned} \tag{9.3}$$

which are the unbiased estimators of $\eta_1(t), \eta_2(t)$ and $\gamma_1(s,t), \gamma_2(s,t)$, respectively. Notice that to take the covariance function heteroscedasticity into account, we have to estimate $\gamma_1(s,t)$ and $\gamma_2(s,t)$ separately, using the data in the associated sample only. This shows that the convergence rates of the covariance function estimators are limited by $O_p(n_{\min}^{-1/2})$, much slower than the convergence rate $O_p(n^{-1/2})$ of the pooled sample covariance function

$$\hat{\gamma}(s,t) = [(n_1-1)\hat{\gamma}_1(s,t) + (n_2-1)\hat{\gamma}_2(s,t)]/(n-2), \tag{9.4}$$

constructed under the equal-covariance function assumption where $n_{\min} = \min(n_1,n_2)$ and $n = n_1+n_2$ denote the minimum sample size and the total sample size of the two samples, respectively.

Set $\Delta_n(t) = \sqrt{n_1n_2/n}[\bar{y}_1(t)-\bar{y}_2(t)]$ as the pivotal test function for the two-sample BF problem (9.2). We have

$$\begin{aligned} \eta_{\Delta_n}(t) &= \mathrm{E}\Delta_n(t) = \sqrt{n_1n_2/n}[\eta_1(t)-\eta_2(t)], \\ \gamma_{\Delta_n}(s,t) &= \mathrm{Cov}\left[\Delta_n(s),\Delta_n(t)\right] = \tfrac{n_2}{n}\gamma_1(s,t)+\tfrac{n_1}{n}\gamma_2(s,t). \end{aligned} \tag{9.5}$$

A natural unbiased estimator of $\gamma_{\Delta_n}(s,t)$ is given by

$$\hat{\gamma}_{\Delta_n}(s,t) = \frac{n_2}{n}\hat{\gamma}_1(s,t) + \frac{n_1}{n}\hat{\gamma}_2(s,t), \tag{9.6}$$

where $\hat{\gamma}_1(s,t)$ and $\hat{\gamma}_2(s,t)$ are given in (9.3).

To study various procedures for testing the two-sample BF problem (9.2), we need to investigate the properties of $\Delta_n(t)$ and $\hat{\gamma}_1(s,t), \hat{\gamma}_2(s,t)$ under various conditions. For this purpose, we list the following assumptions:

Heteroscedastic Two-Sample Problem Assumptions (HTS)

1. The two samples (9.1) are with $\eta_i(t) \in \mathcal{L}^2(\mathcal{T})$ and $\mathrm{tr}(\gamma_i) < \infty$ for $i=1,2$.
2. The two samples (9.1) are Gaussian.

3. As $n_{\min} \to \infty$, we have $n_1/n \to \tau > 0$.
4. For each $i = 1, 2$, the subject-effect functions $v_{ij}(t) = y_{ij}(t) - \eta_i(t), j = 1, 2, \cdots, n_i$ are i.i.d..
5. For each $i = 1, 2$, the subject-effect function $v_{i1}(t)$ satisfies $\mathrm{E}\|v_{i1}\|^4 < \infty$.
6. For each $i = 1, 2$, the maximum variance $\rho_i = \max_{t \in \mathcal{T}} \gamma_i(t, t) < \infty$.
7. For each $i = 1, 2$, the expectation $\mathrm{E}[v_{i1}^2(s)v_{i1}^2(t)]$ is uniformly bounded. That is, for any $(s, t) \in \mathcal{T}^2$, we have $\mathrm{E}[v_{i1}^2(s)v_{i1}^2(t)] < C_i < \infty$, $i = 1, 2$, where C_1, C_2 are two constants independent of (s, t).

The above assumptions are similar to those assumptions for homogeneous two-sample problems for functional data as listed in Section 5.2 of Chapter 5, keeping in mind that the two functional samples (9.1) now have different covariance functions.

First of all, we have the following simple result under the Gaussian assumption. This result holds for the two samples (9.1) with finite sample sizes.

Theorem 9.1 *Under Assumptions HTS1, HTS2, and HTS4, we have*

$$\Delta_n(t) - \eta_{\Delta_n}(t) \sim GP(0, \gamma_{\Delta_n}), \quad (n_i - 1)\hat{\gamma}_i(s, t) \sim WP(n_i - 1, \gamma_i), \; i = 1, 2.$$

Theorem 9.1 states that under the Gaussian assumption HTS2, $\Delta_n(t)$ is a Gaussian process while $(n_i - 1)\hat{\gamma}_i(s, t), i = 1, 2$ are Wishart processes. These results are expected. They follow from Theorem 4.14 of Chapter 4 immediately. When the Gaussian assumption HTS2 is not valid, provided some additional assumptions, we have the following two theorems.

Theorem 9.2 *Under Assumptions HTS1, HTS3, and HTS4, as $n_{\min} \to \infty$, we have*

$$\Delta_n(t) - \eta_{\Delta_n}(t) \xrightarrow{d} GP(0, \gamma_\Delta),$$

where with τ defined in Assumption HTS3, we have

$$\gamma_\Delta(s, t) = \lim_{n_{\min} \to \infty} \gamma_{\Delta_n}(s, t) = (1 - \tau)\gamma_1(s, t) + \tau\gamma_2(s, t).$$

Theorem 9.3 *Under Assumptions HTS1, HTS3 through HTS6, and HTS7, as $n_{\min} \to \infty$, we have*

$$\sqrt{n_i}\,\{\hat{\gamma}_i(s, t) - \gamma_i(s, t)\} \xrightarrow{d} GP(0, \varpi_i),$$

where for $i = 1, 2$,

$$\varpi_i\{(s_1, t_1), (s_2, t_2)\} = E\{v_{i1}(s_1)v_{i1}(t_1)v_{i1}(s_2)v_{i1}(t_2)\} - \gamma_i(s_1, t_1)\gamma_i(s_2, t_2).$$

In addition, we have $\hat{\gamma}_i(s, t) = \gamma_i(s, t) + O_{UP}(n_{\min}^{-1/2})$, $i = 1, 2$, where, as before, O_{UP} means "bounded in probability uniformly."

Theorems 9.2 and 9.3 show that $\Delta_n(t), \hat{\gamma}_1(s,t)$ and $\hat{\gamma}_2(s,t)$ are all asymptotically Gaussian. From Theorem 9.2, we can see that if $\gamma_1(s,t) \equiv \gamma_2(s,t) \equiv \gamma(s,t), s,t \in \mathcal{T}$, that is, the covariance function homogeneity assumption is satisfied, we have $\gamma_\Delta(s,t) \equiv \gamma(s,t)$. From Theorem 9.3, we can see that the convergence rates of $\hat{\gamma}_1(s,t), \hat{\gamma}_2(s,t)$ are limited by $O_{UP}(n_{\min}^{-1/2})$.

The above three theorems are important for constructing various tests for the two-sample BF problem (9.2) for Gaussian and non-Gaussian functional data.

9.2.2 *Testing Methods*

We are now ready to describe an L^2-norm-based test, an F-type test, and two nonparametric bootstrap tests for (9.2).

L^2-Norm-Based Test The test statistic of the L^2-norm-based test for testing (9.2) is defined as the squared L^2-norm of $\Delta_n(t)$:

$$T_n = \|\Delta_n\|^2 = \frac{n_1 n_2}{n} \int_{\mathcal{T}} [\bar{y}_1(t) - \bar{y}_2(t)]^2 dt. \tag{9.7}$$

The null distributions of T_n are given below when the two functional samples (9.1) are Gaussian and non-Gaussian.

Theorem 9.4 *Under Assumptions HTS1 and HTS2 and the null hypothesis in (9.2), we have*

$$T_n \stackrel{d}{=} \sum_{r=1}^{m} \lambda_r A_r, \; A_r \stackrel{i.i.d.}{\sim} \chi_1^2,$$

where $\lambda_r, r = 1, 2, \cdots, m$ are all the positive eigenvalues of $\gamma_{\Delta_n}(s,t)$.

Theorem 9.5 *Under Assumptions HTS1, HTS3, HTS4, and HTS5, and the null hypothesis in (9.2), as $n_{\min} \to \infty$, we have*

$$T_n \xrightarrow{d} \sum_{r=1}^{m} \lambda_r A_r, \; A_r \stackrel{i.i.d.}{\sim} \chi_1^2,$$

where $\lambda_r, r = 1, 2, \cdots, m$ are all the positive eigenvalues of $\gamma_\Delta(s,t)$ defined in Theorem 9.2.

Theorem 9.4 shows that under the Gaussian assumption HTS2, the test statistic T_n is a χ^2-type mixture and Theorem 9.5 shows that this is also asymptotically true under some proper assumptions even when the Gaussian assumption HTS2 is not valid.

Remark 9.1 *In Theorem 9.4, the eigenvalues $\lambda_r, r = 1, 2, \cdots, m$ depend on the sample sizes n_1 and n_2 of the two samples (9.1) while in Theorem 9.5, the eigenvalues $\lambda_r, r = 1, 2, \cdots, m$ do not depend on the sample sizes n_1 and n_2.*

By Theorems 9.4 and 9.5, the null distribution of T_n can be approximated using the methods described in Section 4.3 of Chapter 4. By the Welch-Satterthwaite χ^2-approximation method described there, we have

$$T_n \sim \beta_1 \chi^2_{d_1} \quad \text{approximately}, \tag{9.8}$$

where

$$\beta_1 = \frac{B_1}{A_1}, \quad d_1 = \frac{A_1^2}{B_1}. \tag{9.9}$$

In the above expressions, under the conditions of Theorem 9.4, we have

$$A_1 = \mathrm{tr}(\gamma_{\Delta_n}), \quad B_1 = \mathrm{tr}(\gamma_{\Delta_n}^{\otimes 2}), \tag{9.10}$$

where $\gamma_{\Delta_n}(s,t)$ is given in (9.5) and under the conditions of Theorem 9.5, we have

$$A_1 = \mathrm{tr}(\gamma_{\Delta}), \quad B_1 = \mathrm{tr}(\gamma_{\Delta}^{\otimes 2}), \tag{9.11}$$

where $\gamma_{\Delta}(s,t)$ is given in Theorem 9.2.

Let $\hat{\beta}_1$ and $\hat{d}_1$ be the estimators of β_1 and d_1 based on the two samples (9.1) under the null hypothesis in (9.2). The approximate null distribution of T_n is then given by

$$T_n \sim \hat{\beta}_1 \chi^2_{\hat{d}_1} \quad \text{approximately}. \tag{9.12}$$

The L^2-norm-based test is then conducted accordingly.

As usual, we can obtain $\hat{\beta}_1$ and $\hat{d}_1$ by a naive method and a bias-reduced method. However, we shall see that deriving the naive method and the bias-reduced method under the two-sample BF problem (9.2) is more involved than under the homogeneous two-sample problem (5.2).

First of all, the naive estimators $\hat{\beta}_1$ and $\hat{d}_1$ are obtained by replacing $\gamma_{\Delta_n}(s,t)$ in (9.10) or $\gamma_{\Delta}(s,t)$ in (9.11) by their natural estimator $\hat{\gamma}_{\Delta_n}(s,t)$ as given in (9.6), resulting in the following expressions:

$$\begin{aligned}
\hat{\beta}_1 = \hat{B}_1/\hat{A}_1, \quad \hat{d}_1 = \hat{A}_1^2/\hat{B}_1,\\
\hat{A}_1 = \mathrm{tr}(\hat{\gamma}_{\Delta_n}) = \tfrac{n_2}{n}\mathrm{tr}(\hat{\gamma}_1) + \tfrac{n_1}{n}\mathrm{tr}(\hat{\gamma}_2),\\
\hat{B}_1 = \mathrm{tr}(\hat{\gamma}_{\Delta_n}^{\otimes 2}) = \tfrac{n_2^2}{n^2}\mathrm{tr}(\hat{\gamma}_1^{\otimes 2}) + \tfrac{n_1^2}{n^2}\mathrm{tr}(\hat{\gamma}_2^{\otimes 2}) + \tfrac{2n_1 n_2}{n^2}\mathrm{tr}(\hat{\gamma}_1 \otimes \hat{\gamma}_2),
\end{aligned} \tag{9.13}$$

where $\hat{\gamma}_1(s,t)$ and $\hat{\gamma}_2(s,t)$ are given in (9.3). It is seen that these formulas are more complicated than those listed in (5.16) under the homogeneous two-sample problem (5.2).

Second, it is also more involved to obtain the bias-reduced estimators of β_1 and d_1 under the two-sample BF problem (9.2) than under the homogeneous two-sample problem (5.2). We first need to replace A_1, A_1^2, and B_1 in (9.9) with their unbiased estimators under the Gaussian assumption HTS2. The unbiased estimator of A_1 is already given in (9.13). With some calculations, we obtain the unbiased estimators of A_1^2 and B_1 as

$$\begin{aligned} \widehat{A_1^2} &= \tfrac{n_2^2}{n^2}\widehat{\mathrm{tr}^2(\gamma_1)} + \tfrac{n_1^2}{n^2}\widehat{\mathrm{tr}^2(\gamma_2)} + \tfrac{2n_1n_2}{n^2}\mathrm{tr}(\hat{\gamma}_1)\mathrm{tr}(\hat{\gamma}_2), \\ \hat{B}_1 &= \tfrac{n_2^2}{n^2}\widehat{\mathrm{tr}(\gamma_1^{\otimes 2})} + \tfrac{n_1^2}{n^2}\widehat{\mathrm{tr}(\gamma_2^{\otimes 2})} + \tfrac{2n_1n_2}{n^2}\mathrm{tr}(\hat{\gamma}_1 \otimes \hat{\gamma}_2), \end{aligned} \tag{9.14}$$

where for each $i = 1, 2$, $\widehat{\mathrm{tr}^2(\gamma_i)}$ and $\widehat{\mathrm{tr}(\gamma_i^{\otimes 2})}$ denote the unbiased estimators of $\mathrm{tr}^2(\gamma_i)$ and $\mathrm{tr}(\gamma_i^{\otimes 2})$, respectively; and as the two functional samples (9.1) are independent, $\mathrm{tr}(\hat{\gamma}_1)\mathrm{tr}(\hat{\gamma}_2)$ and $\mathrm{tr}(\hat{\gamma}_1 \otimes \hat{\gamma}_2)$ are unbiased for $\mathrm{tr}(\gamma_1)\mathrm{tr}(\gamma_2)$ and $\mathrm{tr}(\gamma_1 \otimes \gamma_2)$, respectively. When the two samples (9.1) are Gaussian, by Theorem 9.1 and Theorem 4.6 of Chapter 4, the unbiased estimators of $\mathrm{tr}^2(\gamma_i)$ and $\mathrm{tr}(\gamma_i^{\otimes 2})$ are, respectively, given by

$$\begin{aligned} \widehat{\mathrm{tr}^2(\gamma_i)} &= \tfrac{(n_i-1)n_i}{(n_i-2)(n_i+1)}\left[\mathrm{tr}^2(\hat{\gamma}_i) - 2\mathrm{tr}(\hat{\gamma}_i^{\otimes 2})/n_i\right], \ i = 1, 2, \\ \widehat{\mathrm{tr}(\gamma_i^{\otimes 2})} &= \tfrac{(n_i-1)^2}{(n_i-2)(n_i+1)}\left[\mathrm{tr}(\hat{\gamma}_i^{\otimes 2}) - \mathrm{tr}^2(\hat{\gamma}_i)/(n_i - 1)\right], \ i = 1, 2. \end{aligned} \tag{9.15}$$

It follows that the bias-reduced estimators of β_1 and d_1 are given by

$$\hat{\beta}_1 = \hat{B}_1/\hat{A}_1, \ \ \hat{d}_1 = \widehat{A_1^2}/\hat{B}_1. \tag{9.16}$$

From the above, it is seen that the bias-reduced method is more complicated than the bias-reduced method described in Section 5.2 of Chapter 5 under the homogeneous two-sample problem (5.2).

Overall, it is seen that the L^2-norm-based test for a two-sample BF problem is quite similar to that for a homogeneous two-sample problem except now we need to estimate and use the two covariance functions $\gamma_i(s,t), i = 1, 2$, separately.

Example 9.2 *As an illustrative example, we now apply the L^2-norm-based test for two-sample BF problems to the NOx emission level data for testing if the mean NOx emission levels of working days and non-working days are the same over time. The calculated L^2-norm-based test statistic is $T_n = 6.7128e5$. To compute the approximate null distribution, we first obtain the following quantities:*

$$\begin{array}{ll} tr(\hat{\gamma}_1) = 3.4236e4, & tr(\hat{\gamma}_2) = 2.0841e4, \\ tr(\hat{\gamma}_1^{\otimes 2}) = 2.7298e8, & tr(\hat{\gamma}_2^{\otimes 2}) = 1.5896e8, \\ tr(\hat{\gamma}_1)tr(\hat{\gamma}_2) = 7.1350e8, & tr(\hat{\gamma}_1 \otimes \hat{\gamma}_2) = 1.8204e8. \end{array} \tag{9.17}$$

It is seen that there are some differences between $\hat{\gamma}_1(s,t)$ and $\hat{\gamma}_2(s,t)$ although we are still not sure if the differences between $\gamma_1(s,t)$ and $\gamma_2(s,t)$ are significant. Plugging these quantities into (9.13), it is easy to obtain $\hat{A}_1 = 2.5265e4, \hat{A}_1^2 = 6.3831e8, \hat{B}_1 = 3.6322e8$. Then the naive estimators of β_1 and d_1 are given by $\hat{\beta}_1 = 7.1883e3, \hat{d}_1 = 3.5147$. The approximate P-value for the L^2-norm-based test is then computed as $P(\hat{\beta}_1\chi^2_{\hat{d}_1} \geq T_n) = P(\chi^2_{3.5147} \geq$

93.8354), *which is essentially* 0, *showing that the mean NOx emission levels of working days and non-working days over time are unlikely the same.*

Remark 9.2 *We have also applied the L^2-norm-based test for homogeneous two-sample problems developed in Section 5.2 of Chapter 5 to the NOx emission level data and the conclusion is the same as that obtained in Example 9.2. This is expected as the evidence from the observed data is very strong that different testing methods can still produce the same testing result.*

F-Type Test To partially take into account the variations of the sample covariance functions $\hat{\gamma}_i(s,t), i = 1,2$, we can consider the so-called F-type test for the two-sample BF problem (9.2) under the Gaussian assumption. The Gaussian assumption is an adequate condition for the F-type test to work. The test statistic of the F-type test is defined as

$$F_n = \frac{T_n}{S_n} = \frac{\frac{n_1 n_2}{n} \int_{\mathcal{T}} [\bar{y}_1(t) - \bar{y}_2(t)]^2 dt}{\text{tr}(\hat{\gamma}_{\Delta_n})}, \tag{9.18}$$

where

$$S_n = \text{tr}(\hat{\gamma}_{\Delta_n}) = \frac{n_2}{n}\text{tr}(\hat{\gamma}_1) + \frac{n_1}{n}\text{tr}(\hat{\gamma}_2). \tag{9.19}$$

Remark 9.3 *The F-type test statistic (9.18) is constructed in a way such that we have*

$$E(T_n) = tr(\gamma_{\Delta_n}) = \frac{n_2}{n} tr(\gamma_1) + \frac{n_1}{n} tr(\gamma_2) = E(S_n).$$

That is, the numerator and denominator of F_n have the same expectation. This allows, by the two-cumulant matched F-approximation method described in Chapter 4, that the null distribution of F_n can be approximated by a usual F-distribution with degrees of freedom estimated from the data. The F-type tests for heteroscedastic one-way and two-way ANOVA developed later will be constructed in a similar way.

To derive the null distribution of F_n, we first need to find out the distribution of S_n. The following theorem shows that under the Gaussian assumption HTS2, S_n is a χ^2-type mixture.

Theorem 9.6 *Under the Gaussian assumption HTS2, we have*

$$S_n \stackrel{d}{=} \frac{n_2}{n(n_1-1)} \sum_{r=1}^{m_1} \lambda_{1r} A_{1r} + \frac{n_1}{n(n_2-1)} \sum_{r=1}^{m_2} \lambda_{2r} A_{2r}, \tag{9.20}$$

where all $A_{ir}, r = 1, 2, \cdots, m_i; i = 1, 2$ are independent, and for each $i = 1, 2$, $A_{i1}, A_{i2}, \cdots, A_{im_i} \stackrel{i.i.d.}{\sim} \chi^2_{n_i-1}$ and $\lambda_{i1}, \lambda_{i2}, \cdots, \lambda_{im_i}$ are all the m_i positive eigenvalues of $\gamma_i(s,t)$. Furthermore, we have

$$Var(S_n) = 2\left[\frac{n_2^2}{n^2(n_1-1)} tr(\gamma_1^{\otimes 2}) + \frac{n_1^2}{n^2(n_2-1)} tr(\gamma_2^{\otimes 2})\right]. \tag{9.21}$$

Theorem 9.6 shows that the random expression of $S_n = \text{tr}(\hat{\gamma}_{\Delta_n})$ is more complicated than the random expression (5.22) of $\text{tr}(\hat{\gamma})$ under the homogeneous two-sample problem (5.2). This is the effect of the covariance function heteroscedasticity. Fortunately, the result of Theorem 9.6 also indicates that S_n is a χ^2-type mixture and hence its distribution can be approximated using the methods described in Section 4.3 of Chapter 4. Again, by the Welch-Satterthwaite χ^2-approximation method described there, we have

$$S_n \sim \beta_2 \chi^2_{d_2} \text{ approximately,} \tag{9.22}$$

where

$$\beta_2 = B_2/A_1, \quad d_2 = A_1^2/B_2, \tag{9.23}$$

with $A_1 = \text{E}(S_n) = \text{E}(T_n)$ given in (9.9) and

$$B_2 = \text{Var}(S_n)/2 = \frac{n_2^2}{n^2(n_1-1)}\text{tr}(\gamma_1^{\otimes 2}) + \frac{n_1^2}{n^2(n_2-1)}\text{tr}(\gamma_2^{\otimes 2}).$$

Under the Gaussian assumption HTS2, it is easy to see that T_n and S_n are independent. In addition, by (9.8) and (9.22) and Remark 9.3, we have $\beta_1 d_1 = A_1 = \beta_2 d_2$ and approximately have

$$T_n \sim (\beta_1 d_1)\frac{\chi^2_{d_1}}{d_1}, \quad S_n \sim (\beta_2 d_2)\frac{\chi^2_{d_2}}{d_2}.$$

It follows that we have

$$F_n = \frac{T_n}{S_n} \sim F_{d_1, d_2} \text{ approximately,} \tag{9.24}$$

where d_1 and d_2 are defined in (9.9) and (9.23), respectively.

Let $\hat{d}_1$ and $\hat{d}_2$ be the estimators of d_1 and d_2 based on the two samples (9.1) and under the null hypothesis in (9.2). Then under the null hypothesis in (9.2), we have

$$F_n \sim F_{\hat{d}_1, \hat{d}_2} \text{ approximately.} \tag{9.25}$$

The F-type test is then conducted accordingly.

From (9.25), it is seen that we only need to obtain $\hat{d}_2$ as $\hat{d}_1$ is already given in (9.13) and (9.16) obtained by the naive and bias-reduced methods, respectively. We can also obtain $\hat{d}_2$ by a naive method or a bias-reduced method. Again, we shall see that the naive method and the bias-reduced method for d_2 are more involved than those under the homogeneous two-sample problem (5.2).

Based on (9.23), the naive estimators of d_2 are obtained by replacing

$\gamma_1(s,t)$ and $\gamma_2(s,t)$ in (9.23) directly with their estimators $\hat{\gamma}_1(s,t)$ and $\hat{\gamma}_2(s,t)$ given in (9.6):

$$\begin{array}{c} \hat{d}_2 = \hat{A}_1^2/\hat{B}_2, \quad \text{where} \\ \hat{B}_2 = \frac{n_2^2}{n^2(n_1-1)}\text{tr}(\hat{\gamma}_1^{\otimes 2}) + \frac{n_1^2}{n^2(n_2-1)}\text{tr}(\hat{\gamma}_2^{\otimes 2}), \end{array} \tag{9.26}$$

and $\hat{A}_1$ is given in (9.13).

To obtain the bias-reduced estimator of d_2, we replace A_1^2 and B_2 in the expression of d_2 in (9.23) with their unbiased estimators under the Gaussian assumption HTS2. The unbiased estimator of A_1^2 is already given (9.16) while the unbiased estimator of B_2 is given by

$$\hat{B}_2 = \frac{n_2^2}{n^2(n_1-1)}\widehat{\text{tr}(\gamma_1^{\otimes 2})} + \frac{n_1^2}{n^2(n_2-1)}\widehat{\text{tr}(\gamma_2^{\otimes 2})},$$

where $\widehat{\text{tr}(\gamma_1^{\otimes 2})}$ and $\widehat{\text{tr}(\gamma_2^{\otimes 2})}$ are the unbiased estimators of $\text{tr}(\gamma_1^{\otimes 2})$ and $\text{tr}(\gamma_2^{\otimes 2})$ as given in (9.15). The bias-reduced estimator of d_2 is then given by

$$\hat{d}_2 = \widehat{A_1^2}/\hat{B}_2. \tag{9.27}$$

Example 9.3 *We now apply the F-type test for two-sample BF problems to the NOx emission level data for testing if the mean NOx emission levels of working days and non-working days are the same over time. The calculated test statistic is $F_n = 2.6570$. We already obtained $\hat{d}_1 = 3.5147$ by the naive method as in Example 9.2 and we now need to compute the naive estimator $\hat{d}_2$ using (9.26). By (9.13) and (9.26), we obtained the naive estimates of A_1^2 and B_2 as $\hat{A}_1^2 = 6.3831e8$ and $\hat{B}_2 = 1.6495e3$. It follows that the naive estimate of d_2 is $\hat{d}_2 = \hat{A}_1^2/\hat{B}_2 = 3.8697e5$. Therefore, we have $F_n \sim F_{3.5147,3.8697e5}$ approximately. The approximate P-value for the F-type test is then computed as $P(F_{\hat{d}_1,\hat{d}_2} \geq F_n) = P(F_{3.5147,3.8697e5} \geq 2.6570)$, which is essentially 0, showing that the mean NOx emission levels of working days and non-working days over time are unlikely the same. Notice that $\hat{d}_2$ is very large; it is no surprise that the P-value of the F-type test is the same as the P-value of the L^2-norm-based test.*

Bootstrap Tests When the sample sizes n_1 and n_2 are small, the Welch-Satterthwaite χ^2-approximation for the null distribution of the L^2-norm-based test statistic T_n is not recommended as in this case, Theorem 9.5 may not be valid. When the two samples (9.1) are not Gaussian, the two-cumulant matched F-approximation method for the null distribution of the F-type test statistic F_n is not recommended as in this case, Theorem 9.6 may not be valid. In these cases, a simple nonparametric bootstrap method can be used to bootstrap the critical values of the L^2-norm-based test statistic or the F-type test statistic. This nonparametric bootstrap method can be described as follows.

Let $y_{ij}^*(t), j = 1, 2, \cdots, n_i; i = 1, 2$ be two bootstrap samples randomly generated from the two functional samples in (9.1), respectively. Then we can construct the two bootstrap sample mean functions as $\bar{y}_1^*(t), \bar{y}_2^*(t)$ and the two bootstrap sample covariance functions $\hat{\gamma}_1^*(s,t), \hat{\gamma}_2^*(s,t)$, computed as in (9.3) but based on the two bootstrap samples. Notice that given the two original samples (9.1), the two bootstrap samples have the mean functions $\bar{y}_1(t), \bar{y}_2(t)$ and the covariance functions $\hat{\gamma}_1(s,t), \hat{\gamma}_2(s,t)$, respectively. We can then compute $\Delta_n^*(t) = \sqrt{n_1 n_2/n}\,[(\bar{y}_1^*(t) - \bar{y}_2^*(t)) - (\bar{y}_1(t) - \bar{y}_2(t))]$ and $\gamma_{\Delta_n^*}^* = \frac{n_2}{n}\hat{\gamma}_1^*(s,t) + \frac{n_1}{n}\hat{\gamma}_2^*(s,t)$.

For the L^2-norm-based bootstrap test or the F-type bootstrap test, one computes

$$T_n^* = \|\Delta_n^*\|^2, \quad \text{or } F_n^* = \frac{\|\Delta_n^*\|^2}{\text{tr}(\gamma_{\Delta_n^*}^*)}.$$

Repeat the above bootstrapping process a large number of times, calculate the $100(1-\alpha)$-percentile for T_n^* or F_n^*, and then conduct the L^2-norm-based bootstrap test or the F-type bootstrap test accordingly. The above nonparametric bootstrap tests do not require the Gaussian assumption or large sample sizes, but they are generally time consuming.

9.3 Heteroscedastic One-Way ANOVA

One-way functional ANOVA was defined in Section 5.3 of Chapter 5. It aims to test the equality of the mean functions of several functional samples having the same covariance function. In practice, these functional samples may have different covariance functions so that the methods described in Section 5.3 of Chapter 5 may not be directly applicable. In this section, we study how to test the equality of the mean functions of several functional samples having different covariance functions by extending the main ideas of the previous section developed for two-sample Behrens-Fisher problems. This problem may be referred to as heteroscedastic one-way ANOVA for functional data. We use the following example to motivate this heteroscedastic one-way ANOVA problem for functional data.

Example 9.4 *The Canadian temperature data (Canadian Climate Program 1982) were introduced in Section 1.2.4 of Chapter 1. They are the daily temperature records of thirty-five Canadian weather stations over a year (365 days), among which, fifteen are in Eastern, another fifteen in Western and the remaining five in Northern Canada. The reconstructed Canadian temperature curves were displayed in Figure 5.5 of Chapter 5. In Section 5.3 of Chapter 5, the equality of the mean temperature curves of the Eastern, Western, and Northern weather stations was tested under the assumption that these Eastern, Western, and Northern weather stations have a common covariance function. However, Figure 9.3 seems indicate that the three groups of the temperature curves may have quite different covariance functions. This is also* *partially verified by the quantities presented in Table 9.1. It is then desirable to drop this equal-covariance function assumption. This gives a motivation for a heteroscedastic one-way ANOVA problem for functional data.*

A heteroscedastic one-way functional ANOVA problem may be defined as follows. Suppose we have k independent functional samples

$$y_{i1}(t), \cdots, y_{in_i}(t) \overset{i.i.d.}{\sim} \text{SP}(\eta_i, \gamma_i), i = 1, \cdots, k, \tag{9.28}$$

where $\eta_i(t), i = 1, 2, \cdots, k$ are unknown group mean functions and $\gamma_i(s,t), i = 1, 2, \cdots, k$ are unknown group covariance functions and are possibly unequal. We wish to test the following null hypothesis:

$$H_0 : \eta_1(t) \equiv \eta_2(t) \equiv \cdots \equiv \eta_k(t), \ t \in \mathcal{T}. \tag{9.29}$$

This may be also called a k-sample BF problem for functional data, extending the two-sample BF problem for functional data discussed in the previous section.

When the k-sample BF problem (9.29) is rejected, we may also consider some post hoc or contrast tests as defined in Section 5.3 of Chapter 5 but in the current setup, that is, with the k functional samples (9.28) possibly having different covariance functions.

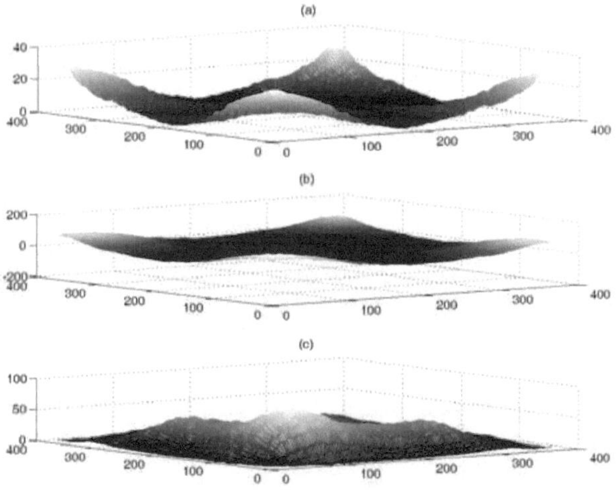

Figure 9.3 *Sample covariance functions of the Canadian temperature data for (a) fifteen Eastern weather stations, (b) fifteen Western weather stations, and (c) five Northern weather stations. The resolution number is $M = 1,000$. The first two sample covariance functions look quite similar to each other but they are somewhat different from the third sample covariance function.*

9.3.1 Estimation of Group Mean and Covariance Functions

The group mean and covariance functions of the k functional samples (9.28) are given by

$$\begin{array}{rcl} \hat{\eta}_i(t) & = & \bar{y}_{i\cdot}(t) = n_i^{-1}\sum_{j=1}^{n_i} y_{ij}(t), \\ \hat{\gamma}_i(s,t) & = & (n_i-1)^{-1}\sum_{j=1}^{n_i}(y_{ij}(s)-\bar{y}_{i\cdot}(s))(y_{ij}(t)-\bar{y}_{i\cdot}(t)), \\ & & i=1,2,\cdots,k. \end{array} \tag{9.30}$$

In the above, to take the covariance function heteroscedasticity into account, the covariance functions are estimated separately based on the associated sample only. Notice that $\hat{\eta}_i(t), i=1,2,\cdots,k$ are independent and

$$\mathrm{E}\hat{\eta}_i(t)=\eta_i(t), \quad \mathrm{Cov}\left[\hat{\eta}_i(s),\hat{\eta}_i(t)\right]=\gamma_i(s,t)/n_i, i=1,2,\cdots,k.$$

Based on the above, we can set

$$\begin{array}{c} \mathbf{z}(t)=[z_1(t),z_2(t),\cdots,z_k(t)]^T, \\ z_i(t)=\sqrt{n_i}[\hat{\eta}_i(t)-\eta_i(t)], \ i=1,2,\cdots,k. \end{array} \tag{9.31}$$

Then we have

$$\mathbf{z}(t)\sim \mathrm{SP}_k(\mathbf{0},\mathbf{\Gamma}_z), \quad \mathbf{\Gamma}_z(s,t)=\mathrm{diag}\left[\gamma_1(s,t),\gamma_2(s,t),\cdots,\gamma_k(s,t)\right], \tag{9.32}$$

where $\mathrm{SP}_k(\boldsymbol{\eta},\mathbf{\Gamma})$ denotes a k-dimensional stochastic process with parameters $\boldsymbol{\eta}(t)$ and $\mathbf{\Gamma}(s,t)$. When the covariance function homogeneity assumption is satisfied, we will have $\mathbf{\Gamma}_z(s,t)=\gamma(s,t)\mathbf{I}_k$ as in Chapter 5, where $\gamma(s,t)$ is the common covariance function.

To study various procedures for testing the heteroscedastic one-way

Table 9.1 *Various quantities of the Canadian temperature data, calculated with resolution $M=1{,}000$ over various seasons (1=Eastern weather stations, 2=Western weather stations, and 3=Northern weather stations).*

	Spring	Summer	Fall	Winter	Whole year
[a,b]	$[60,151]$	$[152,243]$	$[244,334]$	$[335,365]\&[1,59]$	$[1,365]$
$\mathrm{tr}(\hat{\gamma}_1)$	1.1105e3	5.4352e2	7.5718e2	2.2843e3	4.6955e3
$\mathrm{tr}(\hat{\gamma}_2)$	2.9930e3	6.0016e2	1.6832e3	7.9373e3	1.3214e4
$\mathrm{tr}(\hat{\gamma}_3)$	5.4837e3	2.8722e3	2.2024e3	7.9802e2	1.1356e4
$\mathrm{tr}(\hat{\gamma}_1^{\otimes 2})$	9.7813e5	2.7072e5	5.2033e5	5.1234e6	1.5441e7
$\mathrm{tr}(\hat{\gamma}_2^{\otimes 2})$	7.8196e6	3.2025e5	2.5980e6	6.2174e7	1.4505e8
$\mathrm{tr}(\hat{\gamma}_3^{\otimes 2})$	2.9227e7	8.0657e6	3.8508e6	4.2252e5	9.9106e7
$\mathrm{tr}(\hat{\gamma}_1)\mathrm{tr}(\hat{\gamma}_2)$	3.3224e6	3.2620e5	1.2745e6	1.8131e7	6.2045e7
$\mathrm{tr}(\hat{\gamma}_1)\mathrm{tr}(\hat{\gamma}_3)$	6.0899e6	1.5611e6	1.6676e6	1.8229e6	5.3324e7
$\mathrm{tr}(\hat{\gamma}_2)\mathrm{tr}(\hat{\gamma}_3)$	1.6413e7	1.7238e6	3.7072e6	6.3341e6	1.5006e8
$\mathrm{tr}(\hat{\gamma}_1\otimes\hat{\gamma}_2)$	2.5800e6	2.9245e5	1.1170e6	1.7739e7	4.4924e7
$\mathrm{tr}(\hat{\gamma}_1\otimes\hat{\gamma}_3)$	5.1651e6	1.4670e6	1.3315e6	1.2882e6	2.5237e7
$\mathrm{tr}(\hat{\gamma}_2\otimes\hat{\gamma}_3)$	1.2691e7	1.6024e6	2.6416e6	4.5688e6	5.4188e7

ANOVA problem (9.29), we need to establish the properties of $\mathbf{z}(t)$ and $\hat{\gamma}_i(s,t), i = 1, 2, \cdots, k$ under various conditions. For this purpose, we list the following assumptions:

Heteroscedastic One-Way ANOVA Assumptions (HKS)

1. The k samples (9.28) are with $\eta_i(t) \in \mathcal{L}^2(\mathcal{T})$ and $\text{tr}(\gamma_i) < \infty, i = 1, 2, \cdots, k$.
2. The k samples (9.28) are Gaussian.
3. As $n_{\min} = \min_{i=1}^k n_i \to \infty$, the k sample sizes satisfy $n_i/n \to \tau_i > 0,\ i = 1, 2, \cdots, k$ where $n = \sum_{i=1}^k n_i$ is the total sample size of the k samples.
4. For each $i = 1, 2, \cdots, k$, the subject-effect functions $v_{ij}(t) = y_{ij}(t) - \eta_i(t), j = 1, 2, \cdots, n_i$ are i.i.d..
5. For each $i = 1, 2, \cdots, k$, the subject-effect function $v_{i1}(t)$ satisfies $\text{E}\|v_{i1}\|^4 < \infty$.
6. For each $i = 1, 2, \cdots, k$, the maximum variance $\rho_i = \max_{t \in \mathcal{T}} \gamma_i(t,t) < \infty$.
7. For each $i = 1, 2, \cdots, k$, the expectation $\text{E}[v_{i1}^2(s)v_{i1}^2(t)]$ is uniformly bounded. That is, there is a constant $C_i < \infty$ such that $\text{E}[v_{i1}^2(s)v_{i1}^2(t)] \le C_i$.

The above assumptions are comparable with those assumptions for one-way functional ANOVA listed in Section 5.3 of Chapter 5 but now the k samples (9.28) have different covariance functions. We first have the following theorem stating that under the Gaussian assumption HKS2, $\mathbf{z}(t)$ is a k-dimensional Gaussian process and $(n_i - 1)\hat{\gamma}_i(s,t), i = 1, 2, \cdots, k$ are Wishart processes.

Theorem 9.7 *Under Assumptions HKS1 and HKS2, we have* $\mathbf{z}(t) \sim GP_k(\mathbf{0}, \Gamma_z)$ *and* $(n_i - 1)\hat{\gamma}_i(s,t) \sim WP(n_i - 1, \gamma_i), i = 1, 2, \cdots, k$.

Theorem 9.7 is a natural extension of Theorem 9.1. It is useful for constructing various tests for (9.29) when the k functional samples (9.28) are Gaussian. Notice that the components of $\mathbf{z}(t)$ are independent but are no longer i.i.d. due to the covariance function heteroscedasticity. When the Gaussian assumption HKS2 is not valid but the sample sizes $n_1, n_2, \cdots, n_k$ are large, by the central limit theorem of i.i.d. stochastic processes, Theorem 4.12 of Chapter 4, we have the following theorem stating that $\mathbf{z}(t)$ is asymptotically a k-dimensional Gaussian process.

Theorem 9.8 *Under Assumptions HKS1, HKS3, and HKS4, as* $n_{\min} \to \infty$, *we have* $\mathbf{z}(t) \xrightarrow{d} GP_k(\mathbf{0}, \Gamma_z)$.

Notice that we impose Assumptions HKS3 and HKS4 for Theorem 9.8 to guarantee that as $n_{\min} \to \infty$, each of the sample mean functions $\hat{\eta}_i(t) = \bar{y}_{i.}(t), i = 1, 2, \cdots, k$ converges to a Gaussian process weakly.

Theorem 9.9 *Under Assumptions HKS1, HKS3 through HKS6, and HKS7, as $n_{\min} \to \infty$, we have*

$$\sqrt{n_i}\left\{\hat{\gamma}_i(s,t) - \gamma_i(s,t)\right\} \xrightarrow{d} GP(0, \varpi_i),\ i = 1, 2, \cdots, k, \tag{9.33}$$

where

$$\varpi_i\left\{(s_1,t_1),(s_2,t_2)\right\} = E\{v_{i1}(s_1)v_{i1}(t_1)v_{i1}(s_2)v_{i1}(t_2)\} - \gamma_i(s_1,t_1)\gamma_i(s_2,t_2).$$

In addition, we have

$$\hat{\gamma}_i(s,t) = \gamma_i(s,t) + O_{UP}(n_{\min}^{-1/2}),\ i = 1, 2, \cdots, k. \tag{9.34}$$

The above theorem shows that under some proper assumptions, $\hat{\gamma}_i(s,t), i = 1, 2, \cdots, k$ are asymptotically Gaussian and are consistent uniformly over $\mathcal{T}^2$. From Theorem 9.9 we can see that the convergence rates of $\hat{\gamma}_i(s,t), i = 1, 2, \cdots, k$ are limited by $O_{UP}(n_{\min}^{-1/2})$.

As for the homogeneous one-way ANOVA model, for the main-effect, post hoc, or contrast tests of the heteroscedastic one-way ANOVA model, we do not need to identify the main-effect functions defined in (5.26) of Chapter 5. They are actually not identifiable unless some constraint is imposed. If we do want to estimate these main-effect functions, the most commonly used constraint is given in (5.35) of Chapter 5 so that the unbiased main-effect estimators are given in (5.36) of Chapter 5. From Section 5.3 of Chapter 5, the main-effect, post hoc, and contrast testing problems can be expressed as (5.25), (5.29), and (5.31), respectively, in terms of the group mean functions $\eta_i(t), i = 1, 2, \cdots, k$.

9.3.2 Heteroscedastic Main-Effect Test

For the heteroscedastic one-way functional ANOVA problem (9.29), we denote the pointwise between-subject variations as

$$\mathrm{SSH}_n(t) = \sum_{i=1}^{k} n_i[\bar{y}_{i\cdot}(t) - \bar{y}_{\cdot\cdot}(t)]^2, \tag{9.35}$$

where $\bar{y}_{\cdot\cdot}(t)$ denotes the sample grand mean function of the k functional samples (9.28). That is, $\bar{y}_{\cdot\cdot}(t) = n^{-1}\sum_{i=1}^{k}\sum_{j=1}^{n_i} y_{ij}(t) = n^{-1}\sum_{i=1}^{k} n_i\bar{y}_{i\cdot}(t)$. Under the null hypothesis (9.29), it is easy to show that

$$\mathrm{SSH}_n(t) = \mathbf{z}(t)^T\mathbf{A}_n\mathbf{z}(t),\ \ \mathbf{A}_n = \mathbf{I}_k - \mathbf{b}_n\mathbf{b}_n^T/n, \tag{9.36}$$

where $\mathbf{z}(t)$ is as defined in (9.31) and $\mathbf{b}_n = [n_1^{1/2}, n_2^{1/2}, \cdots, n_k^{1/2}]^T$. Notice that $\mathbf{A}_n : k \times k$ is an idempotent matrix such that $\mathbf{A}_n = \mathbf{A}_n^T, \mathbf{A}_n^2 = \mathbf{A}_n$. See (5.145) of Chapter 5. It is easy to identify the entries of $\mathbf{A}_n$. In fact, its (i,j)th entry is

$$a_{n,ij} = \begin{cases} 1 - n_i/n, & \text{if } i = j, \\ -\sqrt{n_i n_j}/n, & \text{if } i \neq j. \end{cases} \tag{9.37}$$

As $n_{\min} \to \infty$, we have

$$\mathbf{A}_n \to \mathbf{A} = (a_{ij}) : k \times k, \text{ where } a_{ij} = \begin{cases} 1 - \tau_i, & \text{if } i = j, \\ -\sqrt{\tau_i \tau_j}, & \text{if } i \neq j. \end{cases} \tag{9.38}$$

We now describe three tests for the heteroscedastic one-way ANOVA problem (9.29). They are an L^2-norm-based test, an F-type test, and some nonparametric bootstrap tests.

L^2-Norm-Based Test The test statistic of the L^2-norm-based test for (9.29) can be defined as

$$T_n = \int_{\mathcal{T}} \text{SSH}_n(t)dt = \sum_{i=1}^{k} n_i \int_{\mathcal{T}} [\bar{y}_{i\cdot}(t) - \bar{y}_{\cdot\cdot}(t)]^2 dt. \tag{9.39}$$

This test statistic T_n is exactly the same as the one defined in Chapter 5 for the homogeneous one-way main-effect test. However, it is much harder to find the null distribution of T_n under the current context than under the homogeneous one-way ANOVA problem discussed in Chapter 5. In fact, unlike for the two-sample BF problem (9.2) and for the homogeneous one-way ANOVA problem (5.25), it is not easy to see if the test statistic T_n is a χ^2-type mixture. This difficulty is mainly due to the covariance function heteroscedasticity. Nevertheless, as T_n is nonnegative and possibly skewed to the left or right, we can still approximate the null distribution of T_n with that of a random variable of form $R_1 \sim \beta_1 \chi^2_{d_1}$ by matching the means and variances of T_n and R_1. As the mean and variance of R_1 are known to be

$$\text{E}(R_1) = \beta_1 d_1, \text{Var}(R_1) = 2\beta_1^2 d_1, \tag{9.40}$$

what we need to do next is to find the mean and variance of T_n under the null hypothesis (9.29). The following two theorems give the mean and variance of T_n under the null hypothesis for Gaussian and non-Gaussian functional data.

Theorem 9.10 *Under Assumptions HKS1 and HKS2 and the null hypothesis (9.29), we have*

$$E(T_n) = \sum_{i=1}^{k} a_{n,ii} tr(\gamma_i), \quad Var(T_n) = 2\sum_{i=1}^{k}\sum_{j=1}^{k} a_{n,ij}^2 tr(\gamma_i \otimes \gamma_j).$$

Theorem 9.11 *Under Assumptions HKS1, HKS3, and HKS4 and the null hypothesis (9.29), as $n_{\min} \to \infty$, we have*

$$\begin{aligned} E(T_n) &= \textstyle\sum_{i=1}^{k} a_{n,ii} tr(\gamma_i) = \sum_{i=1}^{k} a_{ii} tr(\gamma_i) + o(1), \\ Var(T_n) &= 2\textstyle\sum_{i=1}^{k}\sum_{j=1}^{k} a_{ij}^2 tr(\gamma_i \otimes \gamma_j) + o(1), \end{aligned}$$

where $a_{ij}, i, j = 1, 2, \cdots, k$ *are given in (9.38).*

Based on the above two theorems, we are now ready to approximate the null distribution of T_n using that of $R_1 = \beta_1 \chi^2_{d_1}$. By matching the means and variances of T_n and R_1, we obtain

$$T_n \sim \beta_1 \chi^2_{d_1} \text{ approximately,} \tag{9.41}$$

where

$$\beta_1 = B_1/A_1,\ d_1 = A_1^2/B_1,\ A_1 = \text{E}(T_n) = \sum_{i=1}^{k} a_{n,ii}\text{tr}(\gamma_i), \tag{9.42}$$

and under Assumptions HKS1 and HKS2, we have

$$B_1 = \text{Var}(T_n)/2 = \sum_{i=1}^{k}\sum_{j=1}^{k} a_{n,ij}^2 \text{tr}(\gamma_i \otimes \gamma_j), \tag{9.43}$$

and under Assumptions HKS1, HKS3 and HKS4, we have

$$B_1 = \text{Var}(T_n)/2 = \sum_{i=1}^{k}\sum_{j=1}^{k} a_{ij}^2 \text{tr}(\gamma_i \otimes \gamma_j) + o(1). \tag{9.44}$$

Let $\hat{\beta}_1$ and $\hat{d}_1$ be the estimators of β_1 and d_1 based on the data under the null hypothesis (9.29). The approximate null distribution of T_n is then given by

$$T_n \sim \hat{\beta}_1 \chi^2_{\hat{d}_1} \quad \text{approximately.} \tag{9.45}$$

The L^2-norm-based test is then conducted accordingly.

As before, we can obtain $\hat{\beta}_1$ and $\hat{d}_1$ by a naive method and a bias-reduced method although these two methods are more involved than those for the homogenous one-way ANOVA problem (5.25) described in Chapter 5. As usual, the naive estimators $\hat{\beta}_1$ and $\hat{d}_1$ are obtained by replacing $\gamma_i(s,t), i = 1, 2, \cdots, k$ in (9.42) through (9.44) directly with their estimators $\hat{\gamma}_i(s,t), i = 1, 2, \cdots, k$, given in (9.30):

$$\begin{array}{c} \hat{\beta}_1 = \hat{B}_1/\hat{A}_1,\ \ \hat{d}_1 = \hat{A}_1^2/\hat{B}_1, \text{ where} \\ \hat{A}_1 = \sum_{i=1}^k a_{n,ii}\text{tr}(\hat{\gamma}_i),\ \ \hat{B}_1 = \sum_{i=1}^k \sum_{j=1}^k a_{n,ij}^2 \text{tr}(\hat{\gamma}_i \otimes \hat{\gamma}_j), \end{array} \tag{9.46}$$

and in (9.44), $a_{ij}, i, j = 1, 2, \cdots, k$ are also replaced with $a_{n,ij}, i, j = 1, 2, \cdots, k$.

To obtain the bias-reduced estimators of β_1 and d_1, we replace A_1, A_1^2, and B_1 in (9.42) through (9.44) with their unbiased estimators under the Gaussian assumption. The unbiased estimator of A_1 is already given in (9.46). By some calculations, we obtain the unbiased estimators of A_1^2 and B_1 as

$$\begin{array}{rcl} \widehat{A_1^2} &=& \sum_{i=1}^k a_{n,ii}^2 \widehat{\text{tr}^2(\gamma_i)} + \sum_{i \neq j} a_{n,ii}a_{n,jj}\text{tr}(\hat{\gamma}_i)\text{tr}(\hat{\gamma}_j), \\ \hat{B}_1 &=& \sum_{i=1}^k a_{n,ii}^2 \widehat{\text{tr}(\gamma_i^{\otimes 2})} + \sum_{i \neq j} a_{n,ij}^2 \text{tr}(\hat{\gamma}_i \otimes \hat{\gamma}_j), \end{array} \tag{9.47}$$

Table 9.2 *Heteroscedastic one-way ANOVA of the Canadian temperature data by the L^2-norm-based test, with resolution $M = 1,000$.*

Method	Time period	T_n	$\hat{\beta}_1$	$\hat{d}_1$	P-value
Naive	Spring	8.58e4	3.90e3	1.81	1.23e − 5
	Summer	1.87e4	2.12e3	1.47	6.18e − 3
	Fall	7.60e4	1.45e3	2.27	6.36e − 12
	Winter	1.22e5	4.52e3	1.44	5.43e − 7
	Whole year	3.02e5	7.58e3	2.63	6.50e − 9
Bias-reduced	Spring	8.58e4	1.96e3	3.06	1.79e − 9
	Summer	1.87e4	9.57e2	2.59	1.25e − 4
	Fall	7.60e4	8.22e2	3.63	0
	Winter	1.22e5	3.68e3	1.66	3.50e − 8
	Whole year	3.02e5	4.58e3	4.05	1.73e − 13

Note: The approximate degrees of freedom $\hat{d}_1$ by the naive method are generally smaller than those by the bias-reduced method while the P-values by the naive method are generally larger than those by the bias-reduced method.

where for each $i = 1, 2, \cdots, k$, $\widehat{\mathrm{tr}^2(\gamma_i)}$ and $\widehat{\mathrm{tr}(\gamma_i^{\otimes 2})}$ denote the unbiased estimators of $\mathrm{tr}^2(\gamma_i)$ and $\mathrm{tr}(\gamma_i^{\otimes 2})$, respectively, and as the k functional samples (9.28) are independent, for $i \neq j$, $\mathrm{tr}(\hat{\gamma}_i)\mathrm{tr}(\hat{\gamma}_j)$ and $\mathrm{tr}(\hat{\gamma}_i \otimes \hat{\gamma}_j)$ are unbiased for $\mathrm{tr}(\gamma_i)\mathrm{tr}(\gamma_j)$ and $\mathrm{tr}(\gamma_i \otimes \gamma_j)$, respectively. When the k functional samples (9.28) are Gaussian, by Theorem 9.7 and Theorem 4.6 of Chapter 4, we have

$$\begin{aligned} \widehat{\mathrm{tr}^2(\gamma_i)} &= \frac{(n_i-1)n_i}{(n_i-2)(n_i+1)}\left[\mathrm{tr}^2(\hat{\gamma}_i) - \frac{2\mathrm{tr}(\hat{\gamma}_i^{\otimes 2})}{n_i}\right], \quad i = 1, 2, \cdots, k, \\ \widehat{\mathrm{tr}(\gamma_i^{\otimes 2})} &= \frac{(n_i-1)^2}{(n_i-2)(n_i+1)}\left[\mathrm{tr}(\hat{\gamma}_i^{\otimes 2}) - \frac{\mathrm{tr}^2(\hat{\gamma}_i)}{n_i-1}\right], \quad i = 1, 2, \cdots, k. \end{aligned} \tag{9.48}$$

It follows that the bias-reduced estimators of β_1 and d_1 are given by

$$\hat{\beta}_1 = \hat{B}_1/\hat{A}_1, \quad \hat{d}_1 = \widehat{A_1^2}/\hat{B}_1. \tag{9.49}$$

Example 9.5 *We now apply the L^2-norm-based test to the heteroscedastic one-way ANOVA of the Canadian temperature data for testing if the mean temperature functions of the Eastern, Western, and Northern weather stations are the same over some time periods as specified in Table 9.1. Table 9.2 shows the test results with the null distributions of T_n approximated by the naive and bias-reduced methods. The estimated parameters $\hat{\beta}_1$ and $\hat{d}_1$ can be computed directly using the quantities listed in Table 9.1. Note that the approximate degrees of freedom $\hat{d}_1$ by the naive method are generally smaller than those by the bias-reduced methods and the P-values by the naive method are generally larger than those by the bias-reduced method. These P-values show that there is strong evidence showing that the mean functions of the Canadian temperature curves for the Eastern, Western, and Northern weather stations are not the same.*

Remark 9.4 *Table 9.2 is parallel to Table 5.7 which was obtained under the assumption that the temperature curves of the Eastern, Western, and Northern weather stations have the same covariance function. Compared with Table 5.7, the P-values of Table 9.2 are much larger than those of Table 5.7, showing that the possible effects of the heteroscedasticity of the covariance functions for different groups. As this heteroscedasticity has been carefully taken into account, the results of Table 9.2 are more reliable than those of Table 5.7.*

F-Type Test Under the Gaussian assumption (HKS2), we can consider the so-called F-type test to partially take into account the variations of the group sample covariance functions $\hat{\gamma}_i(s,t), i = 1, 2, \cdots, k$. Again, the Gaussian assumption is an adequate condition for the F-type test. The test statistic of the F-type test is given by

$$F_n = \frac{T_n}{S_n}, \tag{9.50}$$

where T_n is the test statistic of the L^2-norm-based test as given in (9.39) and S_n is given by

$$S_n = \sum_{i=1}^{k} a_{n,ii}\mathrm{tr}(\hat{\gamma}_i). \tag{9.51}$$

As pointed out in Remark 9.3, the test statistic F_n is constructed in this way as under the null hypothesis (9.29), we have

$$\mathrm{E}(T_n) = \sum_{i=1}^{k} a_{n,ii}\mathrm{tr}(\gamma_i) = \mathrm{E}(S_n), \tag{9.52}$$

so that the denominator and numerator of F_n have the same expectation. This allows, by the two-cumulant matched F-approximation method as described below, that the null distribution of F_n can be well approximated by a usual F-distribution with degrees of freedom estimated from the data. To derive this null distribution of F_n, we need to find the approximate distribution of S_n as the approximate null distribution of T_n has already been obtained.

Theorem 9.12 *Under the Gaussian assumption (HKS2), we have*

$$S_n \stackrel{d}{=} \sum_{i=1}^{k}\sum_{r=1}^{m_i} \frac{a_{n,ii}\lambda_{ir}}{n_i - 1} A_{ir}, \tag{9.53}$$

where all $A_{ir}, r = 1, 2, \cdots, m_i; i = 1, 2, \cdots, k$ are independent and for each $i = 1, 2, \cdots, k$, $A_{i1}, A_{i2}, \cdots, A_{im_i} \stackrel{i.i.d.}{\sim} \chi^2_{n_i - 1}$ and $\lambda_{i1}, \lambda_{i2}, \cdots, \lambda_{im_i}$ are all the m_i positive eigenvalues of $\gamma_i(s,t)$. Furthermore, we have

$$Var(S_n) = 2\sum_{i=1}^{k} \frac{a^2_{n,ii}}{n_i - 1} tr(\gamma_i^{\otimes 2}). \tag{9.54}$$

By Theorem 9.12, S_n is a χ^2-type mixture that is more complicated in shape than the denominator of the F-type test statistic F_n (5.53) for the homogeneous one-way ANOVA problem (5.25). Nevertheless, the distribution of S_n can be well approximated using the methods described in Section 4.3 of Chapter 4. In fact, by the Welch-Satterthwaite χ^2-approximation method described there, we have

$$S_n \sim \beta_2 \chi^2_{d_2} \text{ approximately,} \tag{9.55}$$

with

$$\beta_2 = B_2/A_1, \quad d_2 = A_1^2/B_2, \tag{9.56}$$

where $A_1 = \mathrm{E}(S_n) = \mathrm{E}(T_n)$ is given in (9.43) and

$$B_2 = \mathrm{Var}(S_n)/2 = \sum_{i=1}^k \frac{a_{n,ii}^2}{n_i - 1} \mathrm{tr}(\gamma_i^{\otimes 2}).$$

Under the Gaussian assumption HKS2, it is easy to see that T_n and S_n are independent. In addition, by (9.41), (9.52), and (9.55), we have $\beta_1 d_1 = A_1 = \beta_2 d_2$ and approximately have

$$T_n \sim (\beta_1 d_1)\frac{\chi^2_{d_1}}{d_1}, \quad S_n \sim (\beta_2 d_2)\frac{\chi^2_{d_2}}{d_2}.$$

It follows that we have

$$F_n = \frac{T_n}{S_n} \sim F_{d_1, d_2} \text{ approximately,} \tag{9.57}$$

where d_1 and d_2 are defined in (9.43) and (9.56), respectively.

Let $\hat{d}_1$ and $\hat{d}_2$ be the estimators of d_1 and d_2 based on the data. Then under the null hypothesis (9.29), we have

$$F_n \sim F_{\hat{d}_1, \hat{d}_2} \text{ approximately.} \tag{9.58}$$

The F-type test is then conducted accordingly.

From (9.58), it is seen that we only need to obtain $\hat{d}_2$ as $\hat{d}_1$ is already given in (9.46) and (9.49) obtained by the naive and bias-reduced methods, respectively. We can also obtain $\hat{d}_2$ by a naive method or a bias-reduced method. Based on (9.56), the naive estimator of d_2 is obtained by replacing $\gamma_i(s,t), i = 1, 2, \cdots, k$ in (9.56) directly with their estimators $\hat{\gamma}_i(s,t), i = 1, 2, \cdots, k$ given in (9.30):

$$\hat{d}_2 = \hat{A}_1^2/\hat{B}_2, \quad \text{where } \hat{B}_2 = \sum_{i=1}^k \frac{a_{n,ii}^2}{n_i - 1} \mathrm{tr}(\hat{\gamma}_i^{\otimes 2}), \tag{9.59}$$

Table 9.3 *Heteroscedastic one-way ANOVA of the Canadian temperature data by the F-type test, with resolution* $M = 1,000$.

Method	Time period	F_n	$\hat{d}_1$	$\hat{d}_2$	P-value
Naive	Spring	12.18	1.81	230	$2.03e-5$
	Summer	6.02	1.47	164	$7.03e-3$
	Fall	23.16	2.27	376	$3.36e-11$
	Winter	18.65	1.44	4224	$5.67e-7$
	Whole year	15.14	2.63	535	$1.29e-8$
Bias-reduced	Spring	12.18	3.06	492	$8.34e-8$
	Summer	6.02	2.59	328	$1.03e-3$
	Fall	23.16	3.63	925	$1.11e-16$
	Winter	18.65	1.66	6292	$1.11e-7$
	Whole year	15.14	4.05	1350	$3.14e-12$

Note: The approximate degrees of freedom $\hat{d}_2$ *are generally large so that the P-values of the F-type test are comparable to those of the* L^2*-norm-based test presented in Table 9.2.*

and $\hat{A}_1$ is given in (9.46).

To obtain the bias-reduced estimator of d_2, we replace A_1^2 and B_2 in the expression of d_2 in (9.56) with their unbiased estimators under the Gaussian assumption. The unbiased estimator of A_1^2 is already given (9.49) while the unbiased estimator of B_2 is given by

$$\hat{B}_2 = \sum_{i=1}^{k} \frac{a_{n,ii}^2}{n_i - 1} \widehat{\mathrm{tr}(\gamma_i^{\otimes 2})},$$

where $\widehat{\mathrm{tr}(\gamma_i^{\otimes 2})}, i = 1, 2, \cdots, k$ are the unbiased estimators of $\mathrm{tr}(\gamma_i^{\otimes 2}), i = 1, 2, \cdots, k$ as given in (9.48). The bias-reduced estimator of d_2 is then given by

$$\hat{d}_2 = \widehat{A_1^2}/\hat{B}_2. \tag{9.60}$$

Remark 9.5 *Unlike for the homogeneous one-way ANOVA problem (5.25), the approximate degrees of freedom* $\hat{d}_1$ *and* $\hat{d}_2$ *of* F_n *are no longer proportional to the degrees of freedom adjustment factor* $\hat{\kappa}$ *as in (5.54). This is of course due to the covariance function heteroscedasticity.*

Example 9.6 *We now apply the F-type test to the heteroscedastic one-way ANOVA of the Canadian temperature data. Table 9.3 shows the test results with the null distributions of* F_n *approximated by the naive and bias-reduced methods. The estimated degrees of freedom* $\hat{d}_1$ *and* $\hat{d}_2$ *can be computed directly using the quantities listed in Table 9.1. Notice that the values of* $\hat{d}_1$ *are exactly the same as those listed in Table 9.2. Table 9.3 is parallel to Table 5.8 and the P-values of the former table are generally larger than those of the latter table. As Table 5.8 was obtained under the assumption that the temperature curves of the Eastern, Western, and Northern weather stations have the same* *covariance functions while Table 9.3 was obtained without this assumption, the conclusions made based on Table 9.3 are more reliable.*

Bootstrap Tests As for the two-sample BF problems for functional data, when $n_{\min}$ is small, the Welch-Satterthwaite χ^2-approximation for the null distribution of the L^2-norm-based test statistic T_n is not preferred and when the k functional samples (9.28) are not Gaussian, the two-cumulant matched F-approximation method for the null distribution of the F-type test statistic F_n is also not preferred. In these cases, we can use a nonparametric bootstrap approach to bootstrap the critical values of the L^2-norm-based test statistic or the F-type test statistic. The details are as follows.

For each $i = 1, 2, \cdots, k$, let $y_{ij}^*(t), j = 1, 2, \cdots, n_i$ be a bootstrap sample randomly generated from $y_{ij}(t), j = 1, 2, \cdots, n_i$. Then we can construct the k bootstrap sample mean functions as $\bar{y}_{1\cdot}^*(t), \bar{y}_{2\cdot}^*(t), \cdots, \bar{y}_{k\cdot}^*(t)$, and the k bootstrap sample covariance function $\hat{\gamma}_1^*(s,t), \hat{\gamma}_2^*(s,t), \cdots, \hat{\gamma}_k^*(s,t)$, computed as in (9.30) but based on the k bootstrap samples. Notice that given the k original samples (9.28), the k bootstrap samples have the mean functions $\bar{y}_{i\cdot}(t), i = 1, 2, \cdots, k$ and the covariance functions $\hat{\gamma}_i(s,t), i = 1, 2, \cdots, k$, respectively. We then compute $\mathbf{z}^*(t) = [z_1^*(t), z_2^*(t), \cdots, z_k^*(t)]^T$, with $z_i^*(t) = \sqrt{n_i}\left[\bar{y}_{i\cdot}^*(t) - \bar{y}_{i\cdot}(t)\right], i = 1, 2, \cdots, k$.

For the L^2-norm-based bootstrap test or the F-type bootstrap test, one computes $T_n^* = \int_{\mathcal{T}} \mathbf{z}^*(t)^T \mathbf{A}_n \mathbf{z}^*(t) dt$, or $F_n^* = \frac{T_n^*}{S_n^*}$ with $S_n^* = \sum_{i=1}^k a_{n,ii} \text{tr}(\hat{\gamma}_i^*)$. Repeat the above bootstrapping process a large number of times, calculate the $100(1-\alpha)$-percentile for T_n^* or F_n^*, and then conduct the L^2-norm-based bootstrap test or the F-type bootstrap test accordingly. The bootstrap tests are generally time consuming although they are widely applicable.

9.3.3 *Tests of Linear Hypotheses under Heteroscedasticity*

We now consider the following heteroscedastic GLHT problem:

$$\begin{array}{lll} & H_0 & : \mathbf{C}\boldsymbol{\eta}(t) \equiv \mathbf{c}(t), \ t \in \mathcal{T}, \\ \text{versus} & H_1 & : \mathbf{C}\boldsymbol{\eta}(t) \neq \mathbf{c}(t), \ \text{for some } t \in \mathcal{T}, \end{array} \tag{9.61}$$

where $\boldsymbol{\eta}(t) = [\eta_1(t), \eta_2(t), \cdots, \eta_k(t)]^T$, $\mathbf{C} : q \times k$ is a known coefficient matrix with $\text{rank}(\mathbf{C}) = q$, and $\mathbf{c}(t) : q \times 1$ is a known constant function.

Remark 9.6 *The heteroscedastic GLHT problem (9.61) is very general. It reduces to the k-sample BF problem (9.29) if we set $\mathbf{c}(t) \equiv \mathbf{0}$ and $\mathbf{C} = [\mathbf{I}_{k-1}, -\mathbf{1}_{k-1}]$. In addition, any post hoc and contrast tests about the mean functions $\eta_i(t), i = 1, 2, \cdots, k$ can be written in the form of the heteroscedastic GLHT problem (9.61). For example, given the k functional samples (9.28), to conduct a post hoc test $H_0 : \eta_i(t) \equiv \eta_j(t), \ t \in \mathcal{T}$, we set $\mathbf{c}(t) \equiv \mathbf{0}$ and $\mathbf{C} = \mathbf{e}_{i,k} - \mathbf{e}_{j,k}$ where again $\mathbf{e}_{r,k}$ denotes a k-dimensional unit vector whose rth entry is 1 and others 0; to conduct a contrast test $H_0 : \mathbf{a}^T \boldsymbol{\eta}(t) \equiv 0, \ t \in \mathcal{T}$ where $\mathbf{a}^T \mathbf{1}_k = 0$, we set $\mathbf{c}(t) = 0$ and $\mathbf{C} = \mathbf{a}^T$.*

Notice that for a same hypothesis, the associated $\mathbf{C}$ is not unique and the test results should not be affected by the choice of $\mathbf{C}$. To this end, we define the pointwise between-subject variations due to hypothesis for the heteroscedastic GLHT problem (9.61) as

$$\text{SSH}_n(t) = [\mathbf{C}\hat{\boldsymbol{\eta}}(t) - \mathbf{c}(t)]^T (\mathbf{C}\mathbf{D}_n\mathbf{C}^T)^{-1} [\mathbf{C}\hat{\boldsymbol{\eta}}(t) - \mathbf{c}(t)], \tag{9.62}$$

where $\hat{\boldsymbol{\eta}}(t) = [\hat{\eta}_1(t), \hat{\eta}_2(t), \cdots, \hat{\eta}_k(t)]^T$ denotes the vector of the group sample mean functions of the k functional samples (9.28) and $\mathbf{D}_n = \text{diag}(1/n_1, 1/n_2, \cdots, 1/n_k)$. It is seen that the expression of the above $\text{SSH}_n(t)$ is exactly the same as that of the $\text{SSH}_n(t)$ for the homogeneous GLHT problem (5.62) defined in (5.67) of Chapter 5 but the null distribution of $\text{SSH}_n(t)$ is more involved in the current context.

Under the null hypothesis in (9.61), we have $\mathbf{C}\boldsymbol{\eta}(t) \equiv \mathbf{c}(t)$ so that we have

$$\text{SSH}_n(t) = \mathbf{z}(t)^T \mathbf{A}_n \mathbf{z}(t), \tag{9.63}$$

where $\mathbf{z}(t)$ is as defined in (9.31) and

$$\mathbf{A}_n = \mathbf{D}_n^{1/2}\mathbf{C}^T(\mathbf{C}\mathbf{D}_n\mathbf{C}^T)^{-1}\mathbf{C}\mathbf{D}_n^{1/2} = (a_{n,ij}) : k \times k. \tag{9.64}$$

It is easy to see that $\mathbf{A}_n$ is an idempotent matrix of rank q. Under Assumption HKS3, as $n_{\min} \to \infty$, we have

$$\mathbf{A}_n \to \mathbf{A} = \mathbf{D}^{1/2}\mathbf{C}^T(\mathbf{C}\mathbf{D}\mathbf{C}^T)^{-1}\mathbf{C}\mathbf{D}^{1/2} = (a_{ij}) : k \times k, \tag{9.65}$$

where

$$\mathbf{D} = \lim_{n_{\min}\to\infty} n\mathbf{D}_n = \text{diag}(1/\tau_1, 1/\tau_2, \cdots, 1/\tau_k).$$

A comparison (9.63) with (9.36) indicates that the L^2-norm-based, F-type, and bootstrap tests for the heteroscedastic one-way ANOVA problem (9.29) described in the previous section can now be extended straightforwardly for the heteroscedastic GLHT problem (9.61). Furthermore, Theorems 9.10, 9.11, and 9.12 are extended nicely in the current context. What we need to keep in mind is that the null hypothesis is now given in (9.61) and we now have $\text{SSH}_n(t), \mathbf{A}_n$, and $\mathbf{A}$ as defined in (9.62), (9.64), and (9.65), respectively.

9.4 Heteroscedastic Two-Way ANOVA

Two-way functional ANOVA was studied in Section 5.4 of Chapter 5. It aims to test the significance of the main and interaction effect functions of two factors under the assumption that all the cell covariance functions are equal. In practice, however, this assumption may be violated so that the methods described in Section 5.4 of Chapter 5 may not be directly applicable. In this section, we study how to conduct heteroscedastic two-way ANOVA for functional data by extending the main ideas of the previous two sections. We use the following example to motivate the heteroscedastic two-way ANOVA problem for functional data.

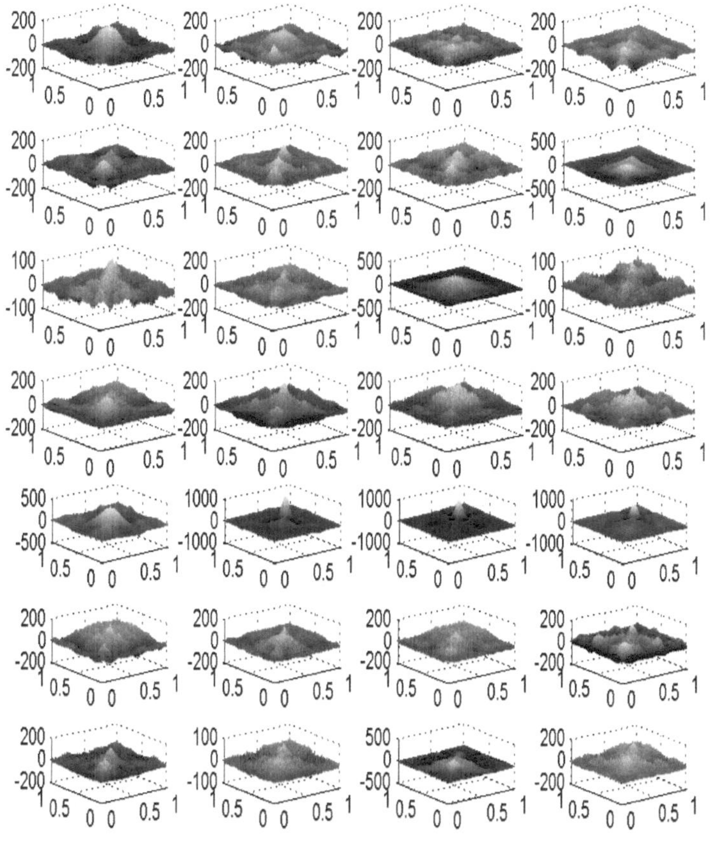

Figure 9.4 *Sample cell covariance functions of the orthosis data. The row panels are associated with the seven subjects and the column panels are associated with the four treatment conditions. It appears that the sample cell covariance functions may not be the same.*

Example 9.7 *The orthosis data were introduced in Section 1.2.7 of Chapter 1. They are the resultant moments of seven young male volunteers who wore a spring-loaded orthosis of adjustable stiffness under four experimental conditions: a control condition (without orthosis), an orthosis condition (with the orthosis only), and two spring conditions (spring 1, spring 2), each ten times, over a period of time (about ten seconds) that was equally divided into* 256 *time points. It motivates a functional ANOVA model with two factors (Subjects: seven young male volunteers and Treatments: four conditions) with ten observations per cell. Figure 1.12 in Chapter 1 displays the raw curves of the orthosis data, each panel of raw curves observed in the cell. To see if the*

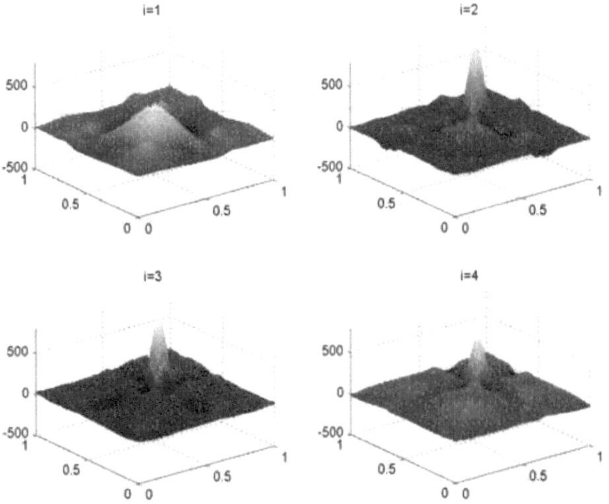

Figure 9.5 *Sample cell covariance functions of the orthosis data for Subject 5 at the four treatment conditions. It appears that the four sample cell covariance functions are not the same.*

covariance functions at different cells are the same, Figure 9.4 displays all the sample cell covariance functions. It appears that these twenty-eight sample cell covariance functions may not be the same. For example, the four sample cell covariance functions depicted in Row 5 (associated with Subject 5) are very different from the four sample cell covariance functions depicted in Row 6 (associated with Subject 6). To see more clearly how different some of the sample cell covariance functions are, Figure 9.5 displays the four sample cell covariance functions for Subject 5 *at all conditions. Based on this, it is clear that the four sample cell covariance functions are not the same. Chapter 10 will formally verify if this is the case. Nevertheless, it is desirable to assume that for this orthosis data, all the cell covariance functions are different for the two-way functional ANOVA model, resulting in a heteroscedastic two-way functional ANOVA model.*

More generally, consider a two-way experiment with two factors A and B having a and b levels, respectively, with a total of ab factorial combinations or cells. For $i = 1, 2, \cdots, a; j = 1, 2, \cdots, b$, suppose at the (i, j)th cell that we have a random functional sample:

$$y_{ijk}(t), k = 1, 2, \cdots, n_{ij}, \tag{9.66}$$

satisfying the following cell model:

$$y_{ijk}(t) = \eta_{ij}(t) + v_{ijk}(t), \;\; v_{ijk}(t) \sim \text{SP}(0, \gamma_{ij}), \;\; k = 1, \cdots, n_{ij}, \tag{9.67}$$

where $\eta_{ij}(t)$ and $\gamma_{ij}(s,t)$ are the associated mean and covariance functions. The covariance functions $\gamma_{ij}(s,t), i = 1,2,\cdots,a; j = 1,2,\cdots,b$ are unknown and may be different across the cells so that the model (9.67) is a heteroscedastic two-way ANOVA model. Notice that the ab functional samples (9.66) are assumed to be independent of each other.

As in Chapter 5, for a two-way ANOVA model, the cell mean functions $\eta_{ij}(t)$ are often decomposed into the form

$$\eta_{ij}(t) = \eta_0(t) + \alpha_i(t) + \beta_j(t) + \theta_{ij}(t), i = 1,2,\cdots,a; j = 1,2,\cdots,b, \quad (9.68)$$

where $\eta_0(t)$ is the grand mean function; $\alpha_i(t)$ and $\beta_j(t)$ are the ith and jth main-effect functions of factors A and B, respectively; and $\theta_{ij}(t)$ is the (i,j)th interaction-effect function between factors A and B so that the model (9.67) can be further written as the following heteroscedastic two-way ANOVA model:

$$\begin{array}{rcl} y_{ijk}(t) & = & \eta_0(t) + \alpha_i(t) + \beta_j(t) + \theta_{ij}(t) + v_{ijk}(t), \\ v_{ijk}(t) & \sim & \text{SP}(0, \gamma_{ij}), \ k = 1,\cdots,n_{ij}; \\ & & i = 1,\cdots,a; j = 1,\cdots,b. \end{array} \quad (9.69)$$

For this model, we are interested in the following three null hypotheses:

$$\begin{array}{lcl} H_{0A} & : & \alpha_i(t) \equiv 0, \ i = 1,2,\cdots,a, \ t \in \mathcal{T}, \\ H_{0B} & : & \beta_j(t) \equiv 0, \ j = 1,2,\cdots,b, \ t \in \mathcal{T}, \\ H_{0AB} & : & \theta_{ij}(t) \equiv 0, \ i = 1,2,\cdots,a; j = 1,2,\cdots,b, \ t \in \mathcal{T}. \end{array} \quad (9.70)$$

The first two null hypotheses aim to test if the main-effect functions of the two factors are significant, while the last one aims to test if the interaction-effect function between the two factors is significant.

The model (9.69) is not identifiable as the parameters $\eta_0(t), \alpha_i(t), \beta_j(t)$ and, $\theta_{ij}(t)$ are not uniquely defined unless further constraints are imposed. Given a sequence of nonnegative weights $w_{ij}, i = 1,2,\cdots,a; j = 1,2,\cdots,b$, as in Chapter 5, we can impose the following constraints:

$$\begin{array}{c} \sum_{i=1}^a w_{i\cdot}\alpha_i(t) \equiv 0, \ \sum_{i=1}^a w_{ij}\theta_{ij}(t) \equiv 0, \ j = 1,2,\cdots,b-1, \\ \sum_{j=1}^b w_{\cdot j}\beta_j(t) \equiv 0, \ \sum_{j=1}^b w_{ij}\theta_{ij}(t) \equiv 0, \ i = 1,2,\cdots,a-1, \\ \sum_{i=1}^a \sum_{j=1}^b w_{ij}\theta_{ij}(t) \equiv 0, \end{array} \quad (9.71)$$

where $w_{i\cdot} = \sum_{j=1}^b w_{ij}$ and $w_{\cdot j} = \sum_{i=1}^a w_{ij}$.

Remark 9.7 *We can continue to use the equal-weight method and the size-adapted-weight method described in Section 5.4 of Chapter 5 for homogeneous two-way ANOVA for specifying the weights $w_{ij}, i = 1,2,\cdots,a; j = 1,2,\cdots,b$. Concretely speaking, these two methods specify the weights as $w_{ij} = g_i h_j, i = 1,2,\cdots,a; j = 1,2,\cdots,b$ with the equal-weight method specifying $g_i = 1/a, h_j = 1/b, i = 1,2,\cdots,a; j = 1,2,\cdots,b$, while the size-adapted-weight method specifying $g_i = \sum_{j=1}^b n_{ij}/n, i = 1,2,\cdots,a$ and $h_j = \sum_{i=1}^a n_{ij}/n, j = 1,2,\cdots,b$, where and throughout, $n = \sum_{i=1}^a \sum_{j=1}^b n_{ij}$.*

Set

$$\begin{array}{rcl}\boldsymbol{\alpha}(t) &=& [\alpha_1(t), \cdots, \alpha_a(t)]^T, \ \boldsymbol{\beta}(t) = [\beta_1(t), \cdots, \beta_b(t)]^T, \\ \boldsymbol{\theta}(t) &=& [\theta_{11}(t), \cdots, \theta_{1b}(t), \cdots, \theta_{a1}(t), \cdots, \theta_{ab}(t)]^T.\end{array}$$

Then under constraints (9.71), the three null hypotheses (9.70) can be equivalently written as

$$\begin{array}{lll} H_{0A}: & \mathbf{H}_a\boldsymbol{\alpha}(t) \equiv 0, t \in \mathcal{T}, & \text{where } \mathbf{H}_a = \left(\mathbf{I}_{a-1}, -\mathbf{1}_{a-1}\right), \\ H_{0B}: & \mathbf{H}_b\boldsymbol{\beta}(t) \equiv 0, t \in \mathcal{T}, & \text{where } \mathbf{H}_b = \left(\mathbf{I}_{b-1}, -\mathbf{1}_{b-1}\right), \\ H_{0AB}: & \mathbf{H}_{ab}\boldsymbol{\theta}(t) \equiv 0, t \in \mathcal{T}, & \text{where } \mathbf{H}_{ab} = \mathbf{H}_a \otimes \mathbf{H}_b, \end{array} \tag{9.72}$$

where, as before, $\mathbf{I}_r$ and $\mathbf{1}_r$ denote an identity matrix of size r and a r-dimensional vector of ones, respectively, and $\otimes$ denotes the Kronecker product operation. The coefficient matrices $\mathbf{H}_a, \mathbf{H}_b$, and $\mathbf{H}_{ab}$ are full rank contrast matrices, having ranks $(a-1), (b-1)$, and $(a-1)(b-1)$, respectively.

When the weights can be written as $w_{ij} = g_i h_j, i = 1, 2, \cdots, a; j = 1, 2, \cdots, b$ as suggested in Remark 9.7 such that $g_i \geq 0$, $\sum_{i=1}^a g_i = 1$ and $h_j \geq 0$, $\sum_{j=1}^b h_j = 1$, we can easily identify the parameters $\eta_0(t), \alpha_i(t), \beta_j(t)$, and $\theta_{ij}(t)$ as

$$\begin{array}{rcl} \eta_0(t) &=& \sum_{i=1}^a \sum_{j=1}^b g_i h_j \eta_{ij}(t), \\ \alpha_i(t) &=& \sum_{j=1}^b h_j \eta_{ij}(t) - \eta_0(t), \\ \beta_j(t) &=& \sum_{i=1}^a g_i \eta_{ij}(t) - \eta_0(t), \\ \theta_{ij}(t) &=& \eta_{ij}(t) - \alpha_i(t) - \beta_j(t) - \eta_0(t). \end{array} \tag{9.73}$$

Let $\mathbf{g} = [g_1, \cdots, g_a]^T$ and $\mathbf{h} = [h_1, \cdots, h_b]^T$ collect the weights for the levels of Factors A and B, respectively, and let $\boldsymbol{\eta}(t) = [\eta_{11}(t), \cdots, \eta_{1b}(t), \cdots, \eta_{a1}(t), \cdots, \eta_{ab}(t)]^T$ collect all the cell mean functions. As before, denote a p-dimensional unit vector whose rth component is 1 and others are 0 as $\mathbf{e}_{r,p}$. Then we further have

$$\begin{array}{ll} \eta_0(t) = [\mathbf{g}^T \otimes \mathbf{h}^T]\boldsymbol{\eta}(t), & \alpha_i(t) = [(\mathbf{e}_{i,a} - \mathbf{g})^T \otimes \mathbf{h}^T]\boldsymbol{\eta}(t), \\ \beta_j(t) = [\mathbf{g}^T \otimes (\mathbf{e}_{j,b} - \mathbf{h})^T]\boldsymbol{\eta}(t), & \theta_{ij}(t) = [(\mathbf{e}_{i,a} - \mathbf{g})^T \otimes (\mathbf{e}_{j,b} - \mathbf{h})^T]\boldsymbol{\eta}(t), \\ & i = 1, 2, \cdots, a; j = 1, 2, \cdots, b. \end{array}$$

In matrix notation, we can further write

$$\begin{array}{rcl} \boldsymbol{\alpha}(t) &=& [(\mathbf{I}_a - \mathbf{1}_a\mathbf{g}^T) \otimes \mathbf{h}^T]\boldsymbol{\eta}(t) \equiv \mathbf{A}_a\boldsymbol{\eta}(t), \\ \boldsymbol{\beta}(t) &=& [\mathbf{g}^T \otimes (\mathbf{I}_b - \mathbf{1}_b\mathbf{h}^T)]\boldsymbol{\eta}(t) \equiv \mathbf{A}_b\boldsymbol{\eta}(t), \\ \boldsymbol{\theta}(t) &=& [(\mathbf{I}_a - \mathbf{1}_a\mathbf{g}^T) \otimes (\mathbf{I}_b - \mathbf{1}_b\mathbf{h}^T)]\boldsymbol{\eta}(t) \equiv \mathbf{A}_{ab}\boldsymbol{\eta}(t), \end{array} \tag{9.74}$$

where the matrices $\mathbf{A}_a, \mathbf{A}_b$, and $\mathbf{A}_{ab}$ are not full rank matrices, having ranks $(a-1), (b-1)$, and $(a-1)(b-1)$, respectively.

Notice that each of the testing problems associated with the three null hypotheses (9.72) can be equivalently expressed in the form of the following heteroscedastic GLHT problem:

$$\begin{array}{rcl} H_0 & : & \mathbf{C}\boldsymbol{\eta}(t) \equiv \mathbf{c}(t), \ t \in \mathcal{T}, \\ \text{versus} \quad H_1 & : & \mathbf{C}\boldsymbol{\eta}(t) \neq \mathbf{c}(t), \ \text{for some } t \in \mathcal{T}, \end{array} \tag{9.75}$$

where $\mathbf{C} : q \times (ab)$ is a known coefficient matrix of full rank with $\text{rank}(\mathbf{C}) = q$ and $\mathbf{c}(t) : q \times 1$ is a vector of known constant functions which is often specified as $\mathbf{0}$. In fact, it is easy to check that when $\mathbf{c}(t) \equiv \mathbf{0}$ and $\mathbf{C}$ equals one of the following matrices:

$$\begin{array}{rcl} \mathbf{C}_a = \mathbf{H}_a\mathbf{A}_a & = & \left(\mathbf{I}_{a-1}, -\mathbf{1}_{a-1}\right)\left[(\mathbf{I}_a - \mathbf{1}_a\mathbf{g}^T) \otimes \mathbf{h}^T\right], \\ \mathbf{C}_b = \mathbf{H}_b\mathbf{A}_b & = & \left(\mathbf{I}_{b-1}, -\mathbf{1}_{b-1}\right)\left[\mathbf{g}^T \otimes (\mathbf{I}_b - \mathbf{1}_b\mathbf{h}^T)\right], \\ \mathbf{C}_{ab} = \mathbf{H}_{ab}\mathbf{A}_{ab} & = & \left[(\mathbf{I}_{a-1}, -\mathbf{1}_{a-1})(\mathbf{I}_a - \mathbf{1}_a\mathbf{g}^T)\right] \\ & & \otimes\left[(\mathbf{I}_{b-1}, -\mathbf{1}_{b-1})(\mathbf{I}_b - \mathbf{1}_b\mathbf{h}^T)\right], \end{array}$$

the heteroscedastic GLHT problem (9.75) will reduce to one of the testing problems associated with the three null hypotheses (9.72).

Remark 9.8 *The post hoc and contrast tests about the main-effects of Factor A or Factor B can also be conducted using the heteroscedastic GLHT problem (9.75) provided that we specify $\mathbf{c}(t)$ and $\mathbf{C}$ properly. For example, to test $H_0 : \alpha_i(t) \equiv \alpha_j(t)$ for $1 \le i < j \le a$, we set $\mathbf{c}(t) \equiv 0$ and $\mathbf{C} = (\mathbf{e}_{i,a} - \mathbf{e}_{j,a})\mathbf{A}_a$; and to test $H_0 : \beta_i(t) - 3\beta_j(t) + 2\beta_k(t) \equiv 0$ for $1 \le i < j < k \le b$, we set $\mathbf{c}(t) \equiv 0$ and $\mathbf{C} = (\mathbf{e}_{i,b} - 3\mathbf{e}_{j,b} + 2\mathbf{e}_{k,b})\mathbf{A}_b$.*

9.4.1 Estimation of Cell Mean and Covariance Functions

To construct the test statistic for the heteroscedastic GLHT problem (9.75), we denote the usual unbiased estimators of the cell mean and covariance functions of the random samples (9.67) as

$$\begin{array}{rcl} \hat{\eta}_{ij}(t) & = & \bar{y}_{ij\cdot}(t) = n_{ij}^{-1}\sum_{k=1}^{n_{ij}} y_{ijk}(t), \\ \hat{\gamma}_{ij}(s,t) & = & (n_{ij}-1)^{-1}\sum_{k=1}^{n_{ij}}[y_{ijk}(s) - \bar{y}_{ij\cdot}(s)][y_{ijk}(t) - \bar{y}_{ij\cdot}(t)], \\ & & i = 1, 2, \cdots, a; j = 1, 2, \cdots, b. \end{array} \tag{9.76}$$

They are also known as the sample cell mean functions and the sample cell covariance functions respectively. Notice that $\hat{\eta}_{ij}(t), i = 1, 2, \cdots, a; j = 1, 2 \cdots, b$ are independent and

$$\begin{array}{c} \mathrm{E}\hat{\eta}_{ij}(t) = \eta_{ij}(t), \ \mathrm{Cov}\left[\hat{\eta}_{ij}(s), \hat{\eta}_{ij}(t)\right] = \gamma_{ij}(s,t)/n_{ij}, \\ i = 1, 2, \cdots, a; j = 1, 2, \cdots, b. \end{array}$$

Set

$$\begin{array}{rcl} \mathbf{z}(t) & = & [z_{11}(t), \cdots, z_{1b}(t), \cdots, z_{a1}(t), \cdots, z_{ab}(t)]^T, \\ z_{ij}(t) & = & \sqrt{n_{ij}}[\hat{\eta}_{ij}(t) - \eta_{ij}(t)], \ \ i = 1, 2, \cdots, a; j = 1, 2, \cdots, b. \end{array} \tag{9.77}$$

Then we have

$$\begin{aligned}\mathbf{z}(t) &\sim \mathrm{SP}_{ab}(\mathbf{0}, \mathbf{\Gamma}_z),\\ \mathbf{\Gamma}_z(s,t) &= \mathrm{diag}\left[\gamma_{11}(s,t), \cdots, \gamma_{1b}(s,t), \cdots, \gamma_{a1}(s,t), \cdots, \gamma_{ab}(s,t)\right].\end{aligned} \tag{9.78}$$

To study various procedures for testing the heteroscedastic two-way ANOVA problem (9.70), we need to establish the properties of $\mathbf{z}(t)$ and $\hat{\gamma}_{ij}(s,t), i = 1, 2, \cdots, a; j = 1, 2, \cdots, b$ under various conditions. For this purpose, we list the following assumptions:

Heteroscedastic Two-Way ANOVA Assumptions (HTW)

1. The ab samples (9.66) are with $\eta_{ij}(t) \in \mathcal{L}^2(\mathcal{T})$ and $\mathrm{tr}(\gamma_{ij}) < \infty$, $i = 1, 2, \cdots, a; j = 1, 2, \cdots, b$.
2. The ab samples (9.66) are Gaussian.
3. As $n_{\min} = \min_{1 \le i \le a, 1 \le j \le b} n_{ij} \to \infty$, the ab sample sizes satisfy $n_{ij}/n \to \tau_{ij}$ such that $\tau_{ij} \in (0, 1)$ $i = 1, 2, \cdots, a; j = 1, 2, \cdots, b$.
4. For $i = 1, 2, \cdots, a; j = 1, 2, \cdots, b$, the subject-effect functions $v_{ijk}(t) = y_{ijk}(t) - \eta_{ij}(t), k = 1, 2, \cdots, n_{ij}$ are i.i.d..
5. For $i = 1, 2, \cdots, a; j = 1, 2, \cdots, b$, the subject-effect function $v_{ij1}(t)$ satisfies $\mathrm{E}\|v_{ij1}\|^4 < \infty$.
6. For $i = 1, 2, \cdots, a; j = 1, 2, \cdots, b$, the maximum variance $\rho_{ij} = \max_{t \in \mathcal{T}} \gamma_{ij}(t,t) < \infty$.
7. For $i = 1, 2, \cdots, a; j = 1, 2, \cdots, b$, the expectation $\mathrm{E}[v_{ij1}^2(s) v_{ij1}^2(t)]$ is uniformly bounded. That is, there is a constant $C_{ij} < \infty$ such that $\mathrm{E}[v_{ij1}^2(s) v_{ij1}^2(t)] \le C_{ij}$.

The above assumptions are comparable with those assumptions for homogeneous two-way functional ANOVA listed in Section 5.4 of Chapter 5 but now the ab samples (9.66) have different covariance functions. We first have the following theorem stating that under the Gaussian assumption HTW2, $\mathbf{z}(t)$ is an (ab)-dimensional Gaussian process and $(n_{ij} - 1)\hat{\gamma}_{ij}(s,t), i = 1, 2, \cdots, a; j = 1, 2, \cdots, b$ are Wishart processes.

Theorem 9.13 *Under Assumptions HTW1 and HTW2, we have* $\mathbf{z}(t) \sim GP_{ab}(\mathbf{0}, \mathbf{\Gamma}_z)$ *and* $(n_{ij} - 1)\hat{\gamma}_{ij}(s,t) \sim WP(n_{ij} - 1, \gamma_{ij}), i = 1, 2, \cdots, a; j = 1, 2, \cdots, b$.

Theorem 9.13 is a natural extension of Theorem 9.7. It can be used for constructing various tests for (9.75) when the ab functional samples (9.66) are Gaussian. When the Gaussian assumption is not valid but the sample sizes n_{ij}, $i = 1, 2, \cdots, a; j = 1, 2, \cdots, b$ are large, by the central limit theorem of i.i.d. stochastic processes, Theorem 4.12 of Chapter 4, we have the following theorem stating that $\mathbf{z}(t)$ is asymptotically an (ab)-dimensional Gaussian process.

Theorem 9.14 *Under Assumptions HTW1, HTW3, and HTW4, as* $n_{\min} \to \infty$, *we have*

$$\mathbf{z}(t) \xrightarrow{d} GP_{ab}(\mathbf{0}, \mathbf{\Gamma}_z).$$

The following theorem shows that $\hat{\gamma}_{ij}(s,t), i = 1, 2, \cdots, a; j = 1, 2, \cdots, b$ are asymptotically Gaussian and are consistent uniformly under some mild assumptions.

Theorem 9.15 *Under Assumptions HTW1, HTW3 through HTW6, and HTW7, as* $n_{\min} \to \infty$, *we have*

$$\sqrt{n_{ij}}\left\{\hat{\gamma}_{ij}(s,t) - \gamma_{ij}(s,t)\right\} \xrightarrow{d} GP(0, \varpi_{ij}), \tag{9.79}$$

where

$$\varpi_{ij}\left\{(s_1,t_1),(s_2,t_2)\right\} = E\{v_{ij1}(s_1)v_{ij1}(t_1)v_{ij1}(s_2)v_{ij1}(t_2)\} - \gamma_{ij}(s_1,t_1)\gamma_{ij}(s_2,t_2),$$

for $i = 1, 2, \cdots, a; j = 1, 2, \cdots, b$. *In addition, we have*

$$\hat{\gamma}_{ij}(s,t) = \gamma_{ij}(s,t) + O_{UP}(n_{\min}^{-1/2}),\ i = 1, 2, \cdots, a; j = 1, 2, \cdots, b. \tag{9.80}$$

9.4.2 Tests of Linear Hypotheses under Heteroscedasticity

For the heteroscedastic GLHT problem (9.75), the pointwise between-subject variations due to hypothesis can be expressed as

$$\mathrm{SSH}_n(t) = \left[\mathbf{C}\hat{\boldsymbol{\eta}}(t) - \mathbf{c}(t)\right]^T (\mathbf{C}\mathbf{D}_n\mathbf{C}^T)^{-1}\left[\mathbf{C}\hat{\boldsymbol{\eta}}(t) - \mathbf{c}(t)\right], \tag{9.81}$$

where

$$\hat{\boldsymbol{\eta}}(t) = [\hat{\eta}_{11}(t), \hat{\eta}_{12}(t), \cdots, \hat{\eta}_{1b}(t), \cdots, \hat{\eta}_{a1}(t), \hat{\eta}_{a2}(t), \cdots, \hat{\eta}_{ab}(t)]^T$$

denotes the vector of the sample cell mean functions of the ab functional samples (9.66) and

$$\mathbf{D}_n = \mathrm{diag}(1/n_{11}, 1/n_{12}, \cdots, 1/n_{1b}, \cdots, 1/n_{a1}, 1/n_{a2}, \cdots, 1/n_{ab}).$$

Again, the expression of this $\mathrm{SSH}_n(t)$ is exactly the same as that of the $\mathrm{SSH}_n(t)$ for the homogeneous GLHT problem (5.91) defined in (5.96) of Chapter 5. However, the distribution of $\mathrm{SSH}_n(t)$ for the heteroscedastic GLHT problem (9.75) is more complicated than that of $\mathrm{SSH}_n(t)$ for the homogeneous GLHT problem (5.91). This is due to the cell covariance function heteroscedasticity.

Under the null hypothesis in (9.75), we have $\mathbf{C}\boldsymbol{\eta}(t) \equiv \mathbf{c}(t)$ so that we have

$$\mathrm{SSH}_n(t) = \mathbf{z}(t)^T\mathbf{A}_n\mathbf{z}(t), \tag{9.82}$$

where $\mathbf{z}(t)$ is as defined in (9.77) and

$$\mathbf{A}_n = \mathbf{D}_n^{1/2}\mathbf{C}^T(\mathbf{C}\mathbf{D}_n\mathbf{C}^T)^{-1}\mathbf{C}\mathbf{D}_n^{1/2} = \left(a_{n,ij,kl}\right) : (ab) \times (ab), \tag{9.83}$$

where for $i,k = 1,2,\cdots,a; j,l = 1,2,\cdots,b$; and $a_{n,ij,kl}$ denotes the (ij,kl)th entry of A_n whose entries are labeled associated with the entries of $\mathbf{z}(t)$ as defined in (9.77). It is easy to see that $\mathbf{A}_n$ is an idempotent matrix of rank q. Under Assumption HTW3, as $n_{\min} \to \infty$, we have

$$\mathbf{A}_n \to \mathbf{A} = \mathbf{D}^{1/2}\mathbf{C}^T(\mathbf{C}\mathbf{D}\mathbf{C}^T)^{-1}\mathbf{C}\mathbf{D}^{1/2} = \left(a_{ij,kl}\right) : (ab) \times (ab), \tag{9.84}$$

where $a_{ij,kl} = \lim_{n_{\min}\to\infty} a_{n,ij,kl}$ and

$$\begin{aligned} \mathbf{D} &= \lim_{n_{\min}\to\infty} n\mathbf{D}_n \\ &= \text{diag}(1/\tau_{11}, 1/\tau_{12}, \cdots, 1/\tau_{1b}, \cdots, 1/\tau_{a1}, 1/\tau_{a2}, \cdots, 1/\tau_{ab}). \end{aligned}$$

From the above, it is seen that the heteroscedastic GLHT problem (9.75) for two-way functional ANOVA is similar in shape to the heteroscedastic GLHT problem (9.61) and the expression (9.81) of $\text{SSH}_n(t)$ under heteroscedastic two-way ANOVA is nearly the same as the expression (9.62) of $\text{SSH}_n(t)$ under heteroscedastic one-way ANOVA. Therefore, it is expected that the various tests for the heteroscedastic GLHT problem (9.61) can be extended for the heteroscedastic GLHT problem (9.75) straightforwardly. We list them below purely for easy reference.

L^2-Norm-Based Test The test statistic of the L^2-norm-based test for the heteroscedastic GLHT problem (9.75) can be expressed as

$$\begin{array}{c} T_n = \int_{\mathcal{T}} \text{SSH}_n(t)dt = \\ \int_{\mathcal{T}} [\mathbf{C}\hat{\boldsymbol{\eta}}(t) - \mathbf{c}(t)]^T (\mathbf{C}\mathbf{D}_n\mathbf{C}^T)^{-1} [\mathbf{C}\hat{\boldsymbol{\eta}}(t) - \mathbf{c}(t)]\, dt. \end{array} \tag{9.85}$$

As T_n is nonnegative, we can approximate the distribution of T_n with that of a random variable of form $R_1 \sim \beta_1\chi^2_{d_1}$ by matching the means and variances of T_n and R_1. As the mean and variance of R_1 are given in (9.40), we only need to find the mean and variance of T_n under the null hypothesis (9.75). The following two theorems give the mean and variance of T_n for Gaussian and non-Gaussian functional data.

Theorem 9.16 *Under Assumptions HTW1 and HTW2 and the null hypothesis (9.75), we have*

$$\begin{aligned} E(T_n) &= \textstyle\sum_{i=1}^a \sum_{j=1}^b a_{n,ij,ij}\, tr(\gamma_{ij}), \\ Var(T_n) &= 2\textstyle\sum_{i,k=1}^a \sum_{j,l=1}^b a^2_{n,ij,kl}\, tr(\gamma_{ij} \otimes \gamma_{kl}). \end{aligned}$$

Theorem 9.17 *Under Assumptions HTW1, HTW3, HTW4, and the null hypothesis (9.75), as $n_{\min} \to \infty$, we have*

$$\begin{aligned} &E(T_n) = \textstyle\sum_{i=1}^a \sum_{j=1}^b a_{n,ij,ij}\, tr(\gamma_{ij}) = \sum_{i=1}^a \sum_{j=1}^b a_{ij,ij}\, tr(\gamma_{ij}) + o(1), \\ &Var(T_n) = 2\textstyle\sum_{i,k=1}^a \sum_{j,l=1}^b a^2_{ij,kl}\, tr(\gamma_{ij} \otimes \gamma_{kl}) + o(1). \end{aligned}$$

Based on the above two theorems, we are now ready to approximate the null distribution of T_n using that of $R_1 = \beta_1\chi^2_{d_1}$. By matching the means and variances of T_n and R_1, we obtain

$$T_n \sim \beta_1\chi^2_{d_1} \text{ approximately,} \tag{9.86}$$

where

$$\beta_1 = B_1/A_1,\ d_1 = A_1^2/B_1,\ A_1 = \mathrm{E}(T_n) = \sum_{i=1}^{a}\sum_{j=1}^{b} a_{n,ij,ij}\mathrm{tr}(\gamma_{ij}); \tag{9.87}$$

and under Assumptions HTW1 and HTW2, by Theorem 9.16, we have

$$B_1 = \mathrm{Var}(T_n)/2 = \sum_{i,k=1}^{a}\sum_{j,l=1}^{b} a^2_{n,ij,kl}\mathrm{tr}(\gamma_{ij}\otimes\gamma_{kl}); \tag{9.88}$$

and under Assumptions HTW1, HTW3, and HTW4, by Theorem 9.17, we have

$$B_1 = \mathrm{Var}(T_n)/2 = \sum_{i,k=1}^{a}\sum_{j,l=1}^{b} a^2_{ij,kl}\mathrm{tr}(\gamma_{ij}\otimes\gamma_{kl}) + o(1). \tag{9.89}$$

Let $\hat{\beta}_1$ and $\hat{d}_1$ be the estimators of β_1 and d_1 based on the data under the null hypothesis (9.75). The approximate null distribution of T_n is then given by

$$T_n \sim \hat{\beta}_1\chi^2_{\hat{d}_1} \quad \text{approximately.} \tag{9.90}$$

The L^2-norm-based test is then conducted accordingly.

As before, we can obtain $\hat{\beta}_1$ and $\hat{d}_1$ by a naive method and a bias-reduced method. The naive estimators $\hat{\beta}_1$ and $\hat{d}_1$ are obtained by replacing $\gamma_{ij}(s,t), i = 1,2,\cdots,a; j = 1,2,\cdots,b$ in (9.87) through (9.89) directly with their estimators $\hat{\gamma}_{ij}(s,t),\ i = 1,2,\cdots,a; j = 1,2,\cdots,b$ given in (9.76):

$$\begin{array}{c}\hat{\beta}_1 = \hat{B}_1/\hat{A}_1,\ \ \hat{d}_1 = \hat{A}_1^2/\hat{B}_1, \text{ where}\\ \hat{A}_1 = \sum_{i=1}^{a}\sum_{j=1}^{b} a_{n,ij,ij}\mathrm{tr}(\hat{\gamma}_{ij}),\\ \hat{B}_1 = \sum_{i,k=1}^{a}\sum_{j,l=1}^{b} a^2_{n,ij,kl}\mathrm{tr}(\hat{\gamma}_{ij}\otimes\hat{\gamma}_{kl}),\end{array} \tag{9.91}$$

and in (9.89), $a_{ij,kl}, i,k = 1,2,\cdots,a; j,l = 1,2,\cdots,b$ are also replaced with $a_{n,ij,kl}, i,k = 1,2,\cdots,a; j,l = 1,2,\cdots,b$.

To obtain the bias-reduced estimators of β_1 and d_1, we replace A_1, A_1^2 and B_1 in (9.87) through (9.89) with their unbiased estimators under the Gaussian assumption. The unbiased estimator of A_1 is already given in (9.91). By some calculations, we obtain the unbiased estimators of A_1^2 and B_1 as

$$\begin{array}{rcl}\widehat{A_1^2} &=& \sum_{i=1}^{a}\sum_{j=1}^{b} a^2_{n,ij,ij}\widehat{\mathrm{tr}^2(\gamma_{ij})}\\ && +\sum_{(ij)\neq(kl)} a_{n,ij,ij}a_{n,kl,kl}\mathrm{tr}(\hat{\gamma}_{ij})\mathrm{tr}(\hat{\gamma}_{kl}),\\ \hat{B}_1 &=& \sum_{i=1}^{a}\sum_{j=1}^{b} a^2_{n,ij,ij}\mathrm{tr}(\widehat{\gamma_{ij}^{\otimes 2}})\\ && +\sum_{(ij)\neq(kl)} a^2_{n,ij,kl}\mathrm{tr}(\hat{\gamma}_{ij}\otimes\hat{\gamma}_{kl}),\end{array} \tag{9.92}$$

Table 9.4 *Heteroscedastic two-way ANOVA for the orthosis data by the L^2-norm-based test.*

Method	Effect	T_n	$\hat{\beta}_1$	$\hat{d}_1$	P-value
Naive	Subject	1.08e7	3,599	26.40	0
	Treatment	3.46e6	2,859	16.62	0
	Subject×Treatment	2.23e6	5,146	55.38	0
	Control vs. Orthosis	1.04e5	2,394	6.39	$1.43e-7$
	Spring1 vs. Spring2	3.12e4	2,988	5.48	$8.29e-2$
	Cont.+Orth. vs. Spr1+Spr2	3.32e6	2,607	6.07	0
Bias-reduced	Subject	1.08e7	2,901	32.68	0
	Treatment	3.46e6	2,509	18.89	0
	Subject×Treatment	2.23e6	3,051	93.23	0
	Control vs. Orthosis	1.04e5	2,165	7.03	$3.59e-8$
	Spring1 vs. Spring2	3.12e4	2,751	5.93	$7.56e-2$
	Cont.+Orth. vs. Spr1+Spr2	3.32e6	2,490	6.35	0

Note: The P-values show that the main-effects of subjects and treatments and their interaction effects are highly significant.

where for $i = 1, 2, \cdots, a; j = 1, 2, \cdots, b$; $\widehat{\text{tr}^2(\gamma_{ij})}$ and $\widehat{\text{tr}(\gamma_{ij}^{\otimes 2})}$ denote the unbiased estimators of $\text{tr}^2(\gamma_{ij})$ and $\text{tr}(\gamma_{ij}^{\otimes 2})$, respectively; and as the ab functional samples (9.66) are independent, for $(ij) \neq (kl)$, $\text{tr}(\hat{\gamma}_{ij})\text{tr}(\hat{\gamma}_{kl})$ and $\text{tr}(\hat{\gamma}_{ij} \otimes \hat{\gamma}_{kl})$ are unbiased for $\text{tr}(\gamma_{ij})\text{tr}(\gamma_{kl})$ and $\text{tr}(\gamma_{ij} \otimes \gamma_{kl})$, respectively. When the ab samples (9.66) are Gaussian, by Theorem 9.13 and Theorem 4.6 of Chapter 4, we have

$$\begin{aligned}
\widehat{\text{tr}^2(\gamma_{ij})} &= \tfrac{(n_{ij}-1)n_{ij}}{(n_{ij}-2)(n_{ij}+1)}\left[\text{tr}^2(\hat{\gamma}_{ij}) - \frac{2\text{tr}(\hat{\gamma}_{ij}^{\otimes 2})}{n_{ij}}\right], \\
\widehat{\text{tr}(\gamma_{ij}^{\otimes 2})} &= \tfrac{(n_{ij}-1)^2}{(n_{ij}-2)(n_{ij}+1)}\left[\text{tr}(\hat{\gamma}_{ij}^{\otimes 2}) - \frac{\text{tr}^2(\hat{\gamma}_{ij})}{n_{ij}-1}\right], \\
& i = 1, 2, \cdots, a; j = 1, 2, \cdots, b.
\end{aligned} \tag{9.93}$$

It follows that the bias-reduced estimators of β_1 and d_1 are given by

$$\hat{\beta}_1 = \hat{B}_1/\hat{A}_1, \;\; \hat{d}_1 = \widehat{A_1^2}/\hat{B}_1. \tag{9.94}$$

Example 9.8 *Table 9.4 presents the test results of a heteroscedastic two-way ANOVA of the orthosis data by the L^2-norm-based test. The naive and bias-reduced methods for estimating the approximate null distributions of the L^2-norm-based test were considered. It is seen that the main-effects of subjects and treatments and their interaction effects are highly significant. Based on Remark 9.8, we found significant differences between Control and Orthosis and between Control+Orthosis and Spring 1+ Spring 2; and we also found that the difference between Spring 1 and Spring 2 is less significant [P-value=0.0829 (naive) and* 0.0756 *(bias-reduced)]. As each cell has the same number of observations, the equal-weight and size-adapted-weight methods produced exactly the same results.*

Remark 9.9 *From a homogeneous two-way ANOVA of the orthosis data by the L^2-norm-based test presented in Section 5.4 of Chapter 5, we found that the main-effects of Subject, Treatment, and their interaction are also highly significant (P-value=0). We also found significant differences between Control and Orthosis [P-value=5.10e − 7 (naive) and 3.62e − 7 (bias-reduced)] and between Control+Orthosis and Spring 1+Spring 2 (P-value=0). The difference between Spring 1 and Spring 2 is less significant [P-value=0.0652 (naive) and 0.0629 (bias-reduced)]. These results are similar to those obtained in the literature and under the equal cell covariance function assumption. Abramovich et al. (2004) and Abramovich and Angelini (2006) found significant differences between treatments, but not between Control and Orthosis, and not between Spring 1 and Spring 2 either. Antoniadis and Sapatinas (2007) treated Subject as a random-effect component and found significant differences between Control and Orthosis (P-value = 0.001), between Spring 1 and Spring 2 (P-value=0.020), and between Control + Orthosis and Spring 1 + Spring 2 (P-value=0). Cuesta-Albertos and Febrero-Bande (2010) analyzed the orthosis data by a random projection method and found significant main and interaction effects (P-value=0) and significant differences between Control and Orthosis (P-value = 0.0007), between Spring 1 and Spring 2 (P-value = 0.028), and between Control + Orthosis and Spring 1 + Spring 2 (P-value=0).*

F-Type Test Under the Gaussian assumption HTW2, we can consider the so-called F-type test for (9.75) to partially take into account the variations of the sample cell covariance functions $\hat{\gamma}_{ij}(s,t), i = 1, 2, \cdots, a; j = 1, 2, \cdots, b$. The associated test statistic is given by

$$F_n = \frac{T_n}{S_n}, \tag{9.95}$$

where T_n is the test statistic of the L^2-norm-based test given in (9.85) and S_n is given by

$$S_n = \sum_{i=1}^{a}\sum_{j=1}^{b} a_{n,ij,ij}\text{tr}(\hat{\gamma}_{ij}). \tag{9.96}$$

It is easy to see that

$$\text{E}(T_n) = \text{E}(S_n) = A_1, \tag{9.97}$$

where A_1 is given in (9.87). To approximate the null distribution of F_n, we need only to find the distribution of S_n as the null distribution of T_n has been given earlier.

Theorem 9.18 *Under the Gaussian assumption HTW2, we have*

$$S_n \stackrel{d}{=} \sum_{i=1}^{a}\sum_{j=1}^{b}\sum_{r=1}^{m_{ij}} \frac{a_{n,ij,ij}\lambda_{ijr}}{n_{ij}-1} A_{ijr}, \tag{9.98}$$

where all $A_{ijr}, r = 1, 2, \cdots, m_{ij}; i = 1, 2, \cdots, a; j = 1, 2, \cdots, b$ are independent *and for* $i = 1, 2, \cdots, a; j = 1, 2, \cdots, b$, $A_{ij1}, A_{ij2}, \cdots, A_{ijm_{ij}} \overset{i.i.d.}{\sim} \chi^2_{n_{ij}-1}$ *and* $\lambda_{ij1}, \lambda_{ij2}, \cdots, \lambda_{ijm_{ij}}$ *are all the* m_{ij} *positive eigenvalues of* $\gamma_{ij}(s,t)$. *Furthermore, we have*

$$Var(S_n) = 2\sum_{i=1}^{a}\sum_{j=1}^{b} \frac{a_{n,ij,ij}^2}{n_{ij}-1} tr(\gamma_{ij}^{\otimes 2}). \tag{9.99}$$

By Theorem 9.18, S_n is a χ^2-type mixture that is more complicated in shape than the denominator of the F-type test statistic F_n for the homogeneous two-way ANOVA problem described in Chapter 5. Therefore, the distribution of S_n can be approximated using the methods described in Section 4.3 of Chapter 4. In fact, by the Welch-Satterthwaite χ^2-approximation method described there, we have

$$S_n \sim \beta_2 \chi^2_{d_2} \text{ approximately}, \tag{9.100}$$

where

$$\beta_2 = B_2/A_1, \ d_2 = A_1^2/B_2, \tag{9.101}$$

where $A_1 = \mathrm{E}(S_n) = \mathrm{E}(T_n)$ as given in (9.88) and

$$B_2 = \mathrm{Var}(S_n)/2 = \sum_{i=1}^{a}\sum_{j=1}^{b} \frac{a_{n,ij,ij}^2}{n_{ij}-1} \mathrm{tr}(\gamma_{ij}^{\otimes 2}).$$

Under the Gaussian assumption, it is easy to see that T_n and S_n are independent. In addition, by (9.86), (9.97), and (9.100), we have $\beta_1 d_1 = A_1 = \beta_2 d_2$ and approximately have

$$T_n \sim (\beta_1 d_1)\frac{\chi^2_{d_1}}{d_1}, \ \ S_n \sim (\beta_2 d_2)\frac{\chi^2_{d_2}}{d_2}.$$

It follows that we have

$$F_n = \frac{T_n}{S_n} \sim F_{d_1, d_2} \text{ approximately}, \tag{9.102}$$

where d_1 and d_2 are defined in (9.88) and (9.101), respectively.

Let $\hat{d}_1$ and $\hat{d}_2$ be the estimators of d_1 and d_2 based on the data and under the null hypothesis (9.75). Then under the null hypothesis (9.75), we have

$$F_n \sim F_{\hat{d}_1, \hat{d}_2} \ \text{ approximately}. \tag{9.103}$$

The F-type test is then conducted accordingly.

From (9.103), it is seen that we only need to obtain $\hat{d}_2$ as $\hat{d}_1$ is already given in (9.91) and (9.94) obtained by the naive and bias-reduced methods, respectively. We can also obtain $\hat{d}_2$ by a naive method and a bias-reduced

Table 9.5 *Heteroscedastic two-way ANOVA of the orthosis data by the F-type test.*

Method	Effect	F_n	$\hat{d}_1$	$\hat{d}_2$	P-value
Naive	Subject	113.33	26.40	613	0
	Treatment	72.77	16.62	613	0
	Subject×Treatment	7.84	55.38	613	0
	Control vs. Orthosis	6.81	6.39	309	$1.45e-7$
	Spring1 vs. Spring2	1.90	5.48	305	$8.29e-2$
	Cont.+Orth. vs. Spr1+Spr2	209.77	6.07	613	0
Bias-reduced	Subject	113.33	32.68	1.23e5	0
	Treatment	72.77	18.89	1.23e5	0
	Subject×Treatment	7.84	93.23	1.23e5	0
	Control vs. Orthosis	6.81	7.03	6.34e4	$3.94e-8$
	Spring1 vs. Spring2	1.90	5.93	5.95e4	$7.69e-2$
	Cont.+Orth. vs. Spr1+Spr2	209.77	6.35	1.22e5	0

Note: It appears that the P-values here are comparable with those by the L^2-norm-based test presented in Table 9.4.

method. Based on (9.101), the naive estimators of d_2 is obtained by replacing $\gamma_{ij}(s,t), i = 1, 2, \cdots, a; j = 1, 2, \cdots, b$ in (9.101) directly with their estimators $\hat{\gamma}_{ij}(s,t), i = 1, 2, \cdots, a; j = 1, 2, \cdots, b$ given in (9.76) so that

$$\hat{d}_2 = \hat{A}_1^2/\hat{B}_2, \quad \text{where } \hat{B}_2 = \sum_{i=1}^{a}\sum_{j=1}^{b} \frac{a_{n,ij,ij}^2}{n_{ij}-1}\text{tr}(\hat{\gamma}_{ij}^{\otimes 2}), \tag{9.104}$$

and $\hat{A}_1$ is given in (9.91).

To obtain the bias-reduced estimator of d_2, we replace A_1^2 and B_2 in the expression of d_2 in (9.101) with their unbiased estimators obtained under the Gaussian assumption HTW2. The unbiased estimator of A_1^2 is already given (9.94) while the unbiased estimator of B_2 is given by

$$\hat{B}_2 = \sum_{i=1}^{a}\sum_{j=1}^{b} \frac{a_{n,ij,ij}^2}{n_{ij}-1}\widehat{\text{tr}(\gamma_{ij}^{\otimes 2})},$$

where $\widehat{\text{tr}(\gamma_{ij}^{\otimes 2})}, i = 1, 2, \cdots, a; j = 1, 2, \cdots, b$ are the unbiased estimators of $\text{tr}(\gamma_{ij}^{\otimes 2}), i = 1, 2, \cdots, a; j = 1, 2, \cdots, b$ as given in (9.93). The bias-reduced estimator of d_2 is then given by

$$\hat{d}_2 = \widehat{A_1^2}/\hat{B}_2. \tag{9.105}$$

Example 9.9 *Table 9.5 presents the testing results of a heteroscedastic two-way ANOVA of the orthosis data by the F-type test with the naive and bias-reduced methods. The P-values here are comparable with those by the L^2-norm-based test presented in Table 9.4.*

Remark 9.10 *A homogeneous two-way ANOVA of the orthosis data by the F-type test presented in Section 5.4 of Chapter 5 resulted in quite similar results as those presented in Remark 9.9 except now the P-values for the differences between Control and Orthosis and between Spring 1 and Spring 2 are* $6.27e-7$ *(naive),* $4.59e-7$ *(bias-reduced) and* 0.0659 *(naive),* 0.0639 *(bias-reduced), respectively.*

Bootstrap Tests When $n_{\min}$ is small, Theorem 9.17 may not be valid so that the Welch-Satterthwaite χ^2-approximation for the null distribution of the L^2-norm-based test statistic T_n may not be accurate. And when the ab samples (9.66) are not Gaussian, Theorems 9.16 and 9.18 may not be valid so that the two-cumulant matched F-approximation method for the null distribution of the F-type test statistic F_n may not be accurate. In these cases, some nonparametric bootstrap approaches may be used to bootstrap the critical values of the L^2-norm-based test statistic or the F-type test statistic. The bootstrap tests for the GLHT problem (9.75) are briefly described as follows.

For $i = 1, 2, \cdots, a; j = 1, 2, \cdots, b$, let $y^*_{ijk}(t), k = 1, 2, \cdots, n_{ij}$ be a bootstrap sample randomly generated from $y_{ijk}(t), k = 1, 2, \cdots, n_{ij}$. Then we can construct the ab bootstrap sample mean functions as $\bar{y}^*_{ij\cdot}(t),\ i = 1, 2, \cdots, a; j = 1, 2, \cdots, b$, and the ab bootstrap sample covariance function $\hat{\gamma}^*_{ij}(s, t),\ i = 1, 2, \cdots, a; j = 1, 2, \cdots, b$, computed as in (9.76) but based on the ab bootstrap samples. Notice that given the ab original samples (9.66), the ab bootstrap samples have the cell mean functions $\bar{y}_{ij\cdot}(t), i = 1, 2, \cdots, a; j = 1, 2, \cdots, b$ and the cell covariance functions $\hat{\gamma}_{ij}(s, t), i = 1, 2, \cdots, a; j = 1, 2, \cdots, b$, respectively. We then compute

$$\mathbf{z}^*(t) = [z^*_{11}(t), \cdots, z^*_{1b}(t), \cdots, z^*_{a1}(t), \cdots, z^*_{ab}(t)]^T,$$
$$z^*_{ij}(t) = \sqrt{n_{ij}}\left[\bar{y}^*_{ij\cdot}(t) - \bar{y}_{ij\cdot}(t)\right], i = 1, 2, \cdots, a; j = 1, 2, \cdots, b.$$

For the L^2-norm-based bootstrap test or the F-type bootstrap test, one computes

$$T^*_n = \int_{\mathcal{T}} \mathbf{z}^*(t)^T \mathbf{A}_n \mathbf{z}^*(t) dt, \quad \text{or}$$
$$F^*_n = \frac{T^*_n}{S^*_n} \text{ with } S^*_n = \sum_{i=1}^a \sum_{j=1}^b a_{n,ij,ij} \text{tr}(\hat{\gamma}^*_{ij}).$$

Repeat the above bootstrapping process a large number of times, calculate the $100(1-\alpha)$-percentile for T^*_n or F^*_n, and then conduct the L^2-norm-based bootstrap test or the F-type bootstrap test accordingly. Note that the bootstrap tests are generally time consuming.

9.5 Technical Proofs

In this section, we outline the proofs of the main results described in this chapter.

Proof of Theorem 9.1 Under the Gaussian assumption HTS2, the theorem is obvious.

Proof of Theorem 9.2 Under the given conditions, as $n_{\min} \to \infty$, by Theorem 4.15 of Chapter 4, we have

$$\sqrt{n_i}[\bar{y}_i(t) - \eta_i(t)] \xrightarrow{d} \mathrm{GP}(0, \gamma_i), i = 1, 2.$$

That is, $\sqrt{n_1 n_2/n}[\bar{y}_1(t) - \eta_1(t)] \xrightarrow{d} \mathrm{GP}(0, (1-\tau)\gamma_1)$ and $\sqrt{n_1 n_2/n}[\bar{y}_2(t) - \eta_2(t)] \xrightarrow{d} \mathrm{GP}(0, \tau\gamma_2)$. The theorem then follows from the independence of the two samples (9.1).

Proof of Theorem 9.3 The theorem is a special case of Theorem 9.9 and its proof is also a special case of the proof of Theorem 9.9.

Proof of Theorem 9.4 Under the given conditions, the theorem follows immediately from Theorem 9.1 and Theorem 4.2 of Chapter 4.

Proof of Theorem 9.5 Under the given conditions, the theorem follows immediately from Theorem 9.2, the continuous mapping theorem for random elements taking values in a Hilbert space (Billingsley, 1968, p. 34; Cuevas, Febrero, and Fraiman 2004), and Theorem 4.2 of Chapter 4.

Proof of Theorem 9.6 Notice that under the Gaussian assumption HTS2, we have $(n_i - 1)\hat{\gamma}_i(s,t) \sim \mathrm{WP}(n_i - 1, \gamma_i), i = 1, 2$, and they are independent with each other. It follows from Theorem 4.5 of Chapter 4 that

$$\mathrm{tr}((n_i - 1)\hat{\gamma}_i) \stackrel{d}{=} \sum_{r=1}^{m_i} \lambda_{ir} A_{ir},\ A_{i1}, A_{i2}, \cdots A_{im_i} \stackrel{i.i.d.}{\sim} \chi^2_{n_i - 1}; i = 1, 2.$$

The random expression (9.20) follows immediately. To show (9.21), by (9.20) and noticing that $\mathrm{Var}(A_{ir}) = 2(n_i - 1)$, we have

$$\begin{aligned} \mathrm{Var}(S_n) &= \tfrac{n_2^2}{n^2(n_1-1)^2} \textstyle\sum_{r=1}^{m_1} 2\lambda_{1r}^2(n_1 - 1) + \tfrac{n_1^2}{n^2(n_2-1)^2} \textstyle\sum_{r=1}^{m_2} 2\lambda_{2r}^2(n_2 - 1) \\ &= 2\left[\tfrac{n_2^2}{n^2(n_1-1)}\mathrm{tr}(\gamma_1^{\otimes 2}) + \tfrac{n_1^2}{n^2(n_2-1)}\mathrm{tr}(\gamma_2^{\otimes 2})\right], \end{aligned}$$

as desired. The theorem is proved.

Proof of Theorem 9.7 Under the given conditions, the k functional samples (9.28) are Gaussian and independent. The theorem follows immediately from Theorem 4.14 of Chapter 4.

Proof of Theorem 9.8 Notice that the components of $\mathbf{z}(t)$, $z_i(t) = \sqrt{n_i}[\hat{\eta}_i(t) - \eta_i(t)], i = 1, 2, \cdots, k$, are independent. Under the given conditions and by Theorem 4.15, as $n_{\min} \to \infty$, we have $z_i(t) \xrightarrow{d} \mathrm{GP}(0, \gamma_i), i = 1, 2, \cdots, k$. It follows that $\mathbf{z}(t) \xrightarrow{d} \mathrm{GP}_k(\mathbf{0}, \Gamma_z)$, as desired. The theorem is then proved.

Proof of Theorem 9.9 Under the given conditions, the expression (9.33) follows from Theorem 4.16 of Chapter 4 immediately. It follows that as $n_{\min} \to \infty$, we have

$$\mathrm{E}\left[\hat{\gamma}_i(s,t) - \gamma_i(s,t)\right]^2 = \frac{\omega_i\left[(s,t),(s,t)\right]}{n_i}[1+o(1)], i = 1,2,\cdots,k.$$

The expression (9.34) follows immediately from the fact that

$$\varpi_i\left[(s,t),(s,t)\right] \le \mathrm{E}[v_{i1}^2(s)v_{i1}^2(t)] + \gamma_i^2(s,t) \le C_i + \rho_i^2, i = 1,2,\cdots,k,$$

where the constants $C_i, \rho_i, i = 1,2,\cdots,k$ are given in Assumptions HKS6 and HKS7. The theorem is then proved.

Proof of Theorem 9.10 Under the given conditions, the theorem follows from Theorem 4.11 of Chapter 4 immediately.

Proof of Theorem 9.11 Under the given conditions, by Theorem 9.8, as $n_{\min} \to \infty$, we have $\mathbf{z}(t) \xrightarrow{d} \mathrm{GP}_k(\mathbf{0}, \mathbf{\Gamma}_z)$. In addition, by (9.38), as $n_{\min} \to \infty$, we have $\mathbf{A}_n \to \mathbf{A} = (a_{ij})$. The theorem then follows from the continuous mapping theorem for random elements taking values in a Hilbert space (Billingsley, 1968, p. 34; Cuevas, Febrero, and Fraiman 2004) and Theorem 4.11 of Chapter 4 immediately.

Proof of Theorem 9.12 Notice that under the Gaussian assumption, we have $(n_i - 1)\hat{\gamma}_i(s,t) \sim \mathrm{WP}(n_i - 1, \gamma_i), i = 1,2,\cdots,k$, and they are independent of each other. It follows from Theorem 4.5 of Chapter 4 that

$$\mathrm{tr}((n_i-1)\hat{\gamma}_i) \stackrel{d}{=} \sum_{r=1}^{m_i} \lambda_{ir} A_{ir}, \quad A_{i1}, A_{i2}, \cdots A_{im_i} \stackrel{i.i.d.}{\sim} \chi^2_{n_i-1}; i = 1,2,\cdots,k.$$

To show the expression (9.54), by (9.53) and noticing that $\mathrm{Var}(A_{ir}) = 2(n_i - 1)$, we have

$$\mathrm{Var}(S_n) = \sum_{i=1}^k \left(\frac{a_{n,ii}}{n_i - 1}\right)^2 \sum_{r=1}^{m_i} 2(n_i-1)\lambda_{ir}^2 = 2\sum_{i=1}^k \frac{a_{n,ii}^2}{n_i-1}\mathrm{tr}(\gamma_i^{\otimes 2}),$$

as desired. The theorem is proved.

Proof of Theorem 9.13 Under the given conditions, the *ab* functional samples (9.66) are Gaussian and independent. The theorem follows immediately from Theorem 4.14 of Chapter 4.

Proof of Theorem 9.14 Notice that the components of $\mathbf{z}(t)$, $z_{ij}(t) = \sqrt{n_{ij}}[\hat{\eta}_{ij}(t) - \eta_{ij}(t)]$, $i = 1,2,\cdots,a; j = 1,2,\cdots,b$ are independent. Under the given conditions and by Theorem 4.15, as $n_{\min} \to \infty$, we have $z_{ij}(t) \xrightarrow{d}$

$\mathrm{GP}(0, \gamma_{ij}), i = 1, 2, \cdots, a; j = 1, 2, \cdots, b$. It follows that $\mathbf{z}(t) \xrightarrow{d} \mathrm{GP}_{ab}(\mathbf{0}, \Gamma_z)$, as desired. The theorem is then proved.

Proof of Theorem 9.15 Under the given conditions, the expression (9.79) follows from Theorem 4.16 of Chapter 4 immediately. It follows that as $n_{\min} \to \infty$, we have

$$\begin{aligned} \mathrm{E}\left[\hat{\gamma}_{ij}(s,t) - \gamma_{ij}(s,t)\right]^2 = \frac{\varpi_{ij}[(s,t),(s,t)]}{n_{ij}}[1 + o(1)], \\ i = 1, 2, \cdots, a; j = 1, 2, \cdots, b. \end{aligned}$$

The expression (9.80) follows immediately from the fact that

$$\begin{aligned} \varpi_{ij}\left[(s,t),(s,t)\right] \le \mathrm{E}[v_{ij1}^2(s)v_{ij1}^2(t)] + \gamma_{ij}^2(s,t) \le C_{ij} + \rho_{ij}^2, \\ i = 1, 2, \cdots, a; j = 1, 2, \cdots, b, \end{aligned}$$

where the constants C_{ij}, ρ_{ij}, $i = 1, 2, \cdots, a; j = 1, 2, \cdots, b$ are given in Assumptions HTW6 and HTW7. The theorem is then proved.

Proof of Theorem 9.16 Under the given conditions, the theorem follows from Theorem 4.11 of Chapter 4 immediately, keeping in mind that we now have double-digit subscripts.

Proof of Theorem 9.17 Under the given conditions, by Theorem 9.14, as $n_{\min} \to \infty$, we have $\mathbf{z}(t) \xrightarrow{d} \mathrm{GP}_{ab}(\mathbf{0}, \mathbf{\Gamma}_z)$. In addition, by (9.84), as $n_{\min} \to \infty$, we have $\mathbf{A}_n \to \mathbf{A} = (a_{ij,kl})$. The theorem then follows from the continuous mapping theorem for random elements taking values in a Hilbert space (Billingsley, 1968, p. 34; Cuevas, Febrero, and Fraiman 2004) and Theorem 4.11 of Chapter 4 immediately.

Proof of Theorem 9.18 Notice that under the Gaussian assumption, by Theorem 9.13, we have $(n_{ij} - 1)\hat{\gamma}_{ij}(s,t) \sim \mathrm{WP}(n_{ij} - 1, \gamma_{ij}), i = 1, 2, \cdots, a; j = 1, 2, \cdots, b$, and they are independent of each other. It follows from Theorem 4.5 of Chapter 4 that

$$\begin{aligned} \mathrm{tr}((n_{ij} - 1)\gamma_{ij}) \stackrel{d}{=} \textstyle\sum_{r=1}^{m_{ij}} \lambda_{ijr} A_{ijr}, \\ A_{ij1}, A_{ij2}, \cdots A_{ijm_{ij}} \stackrel{i.i.d.}{\sim} \chi^2_{n_{ij}-1}; i = 1, 2, \cdots, a; j = 1, 2, \cdots, b. \end{aligned}$$

To show the expression (9.99), by (9.98) and noticing that $\mathrm{Var}(A_{ijr}) = 2(n_{ij} - 1)$, we have

$$\begin{aligned} \mathrm{Var}(S_n) &= \textstyle\sum_{i=1}^a \sum_{j=1}^b \left(\frac{a_{n,ij,ij}}{n_{ij}-1}\right)^2 \sum_{r=1}^{m_{ij}} 2(n_{ij} - 1)\lambda_{ijr}^2 \\ &= 2\textstyle\sum_{i=1}^a \sum_{j=1}^b \frac{a^2_{n,ij,ij}}{n_{ij}-1} \mathrm{tr}(\gamma_{ij}^{\otimes 2}), \end{aligned}$$

as desired. The theorem is proved.

9.6 Concluding Remarks and Bibliographical Notes

In this chapter we reviewed some heteroscedastic ANOVA models for functional data. The L^2-norm-based, F-type, and bootstrap tests were described for the two-sample Behrens-Fisher problem, heteroscedastic one-way ANOVA, and heteroscedastic two-way ANOVA. These methods can be extended straightforwardly to heteroscedastic multi-way ANOVA for functional data. As before, the Welch-Satterthwaite χ^2-approximation was adopted for approximating the null distribution of the L^2-norm-based test, and the two-cumulant matched F-approximation was adopted for approximating the null distribution of the F-type test. The time consuming bootstrap test is recommended only when the functional data are not Gaussian and the sample sizes are small.

Two-sample Behrens-Fisher problems for functional data were first discussed by Zhang, Peng, and Zhang (2010) and Zhang, Liang, and Xiao (2010), among others, using various testing procedures including the L^2-norm-based and nonparametric bootstrap tests. For heteroscedastic one-way ANOVA for functional data, Cuevas, Febrero, and Fraiman (2004) proposed a parametric bootstrap approach that requires one to bootstrap several Gaussian processes repeatedly and hence is generally time consuming, especially when the sample sizes are large. For heteroscedastic two-way ANOVA for functional data, Cuesta-Albertos and Febrero-Bande (2010) proposed a random projection-based method that is based on the analysis of randomly chosen one-dimensional projections. The L^2-norm-based test and the F-type test described in this chapter are mainly due to Zhang and Xiao (2013a, b).

9.7 Exercises

1. Conduct an L^2-norm-based test to verify if the mean NOx emission level curves of working days and non-working days are equal using the bias-reduced method to approximate the null distribution of the L^2-norm-based test based on the information given in Examples 9.1 and 9.2.
   (a) Find the value of the test statistic T_n.
   (b) Compute the quantities $\hat{A}_1, \widehat{A_1^2}$, and $\hat{B}_1$.
   (c) Compute the bias-reduced estimators $\hat{\beta}_1$ and $\hat{d}_1$ so that you can specify the approximate null distribution of T_n.
   (d) Find the approximate P-value of the L^2-norm-based test and comment if the conclusion is similar to the one obtained in Example 9.2.

2. Conduct an F-type test to verify if the mean NOx emission level curves of working days and non-working days are equal using the bias-reduced method to approximate the null distribution of the F-type test based on the information given in Examples 9.1, 9.2, and 9.3.

(a) Find the value of F_n.
(b) Find the value of $\hat{d}_1$.
(c) Compute the bias-reduced estimator $\hat{d}_2$.
(d) Find the approximate P-value of the F-type test and comment if the conclusion is similar to the one obtained in Example 9.3.

3. Compute the following quantities which are defined in Section 9.3, based on the information given in Example 9.4 and Table 9.1.

(a) The naive and bias-reduced estimators of A_1, A_1^2, B_1, and B_2.
(b) The naive and bias-reduced estimators of β_1, d_1, β_2, and d_2.

4. Under the conditions of Theorem 9.9, show that as $n \to \infty$, we have

(a) $\text{tr}(\hat{\gamma}_i) \xrightarrow{p} \text{tr}(\gamma_i), i = 1, 2, \cdots, k$.
(b) $\text{tr}(\hat{\gamma}_i^{\otimes 2}) \xrightarrow{p} \text{tr}(\gamma_i^{\otimes 2}), i = 1, 2, \cdots, k$.

Chapter 10

Test of Equality of Covariance Functions

10.1 Introduction

In Chapter 5, various ANOVA models were studied under the equal-covariance function (ECF) assumption. When this ECF assumption is not satisfied, various heteroscedastic ANOVA models were studied in Chapter 9. In this chapter, we study how to test if two or several functional samples have the same covariance function. That is, we aim to test if the ECF assumption is valid for the functional samples involved in a functional ANOVA model. We start with the two-sample case (Zhang and Sun 2010) in Section 10.2. The methodologies are then extended for multi-sample cases (Zhang 2013b) in Section 10.3. Technical proofs of the main results are outlined in Section 10.4. Some concluding remarks and bibliographical notes are given in Section 10.5. Section 10.6 is devoted to some exercise problems related to this chapter.

10.2 Two-Sample Case

We use the following nitrogen oxides (NOx) emission level data example to motivate the two-sample ECF testing problem.

Example 10.1 *The NOx emission level data were introduced in Section 1.2.3 of Chapter 1. Figure 9.1 of Chapter 9 displays the NOx emission level curves for seventy-three working and thirty-six non-working days after six outlying curves were identified and removed using one of the methods described in Chapter 8. The NOx emission levels for each day were evenly recorded twenty-four times daily, once an hour, and the data were not smoothed. The sample covariance functions of the NOx emission level curves of working days and non-working days were displayed in Figure 9.2 of Chapter 9. It appears that the NOx emission level curves of working days are more variable than those of non-working days. Of interest is to test if this is the case. That is, we want to test if the two underlying covariance functions are different. This motivates a two-sample ECF testing problem for functional data.*

A general two-sample ECF testing problem for functional data can be formulated as follows. Suppose we have two functional samples

$$y_{11}(t), \cdots, y_{1n_1}(t) \overset{i.i.d.}{\sim} \text{SP}(\eta_1, \gamma_1), \quad y_{21}(t), \cdots, y_{2n_2}(t) \overset{i.i.d.}{\sim} \text{SP}(\eta_2, \gamma_2), \quad (10.1)$$

where $\eta_1(t)$ and $\eta_2(t)$ are two unknown mean functions, and $\gamma_1(s,t)$ and $\gamma_2(s,t)$ are two unknown covariance functions. We wish to test the following hypotheses:

$$\begin{array}{lll} & H_0 & : \gamma_1(s,t) \equiv \gamma_2(s,t), s,t \in \mathcal{T}, \\ \text{versus} & H_1 & : \gamma_1(s,t) \neq \gamma_2(s,t), \quad \text{for some } s,t \in \mathcal{T}, \end{array} \quad (10.2)$$

where $\mathcal{T}$ is a finite time interval of interest, often specified as $[a,b]$ with $-\infty < a < b < \infty$.

Recall that the sample mean and covariance functions of the two functional samples (10.1) are given by

$$\begin{array}{rcl} \hat{\eta}_i(t) & = & \bar{y}_i(t) = n_i^{-1} \sum_{j=1}^{n_i} y_{ij}(t), \\ \hat{\gamma}_i(s,t) & = & (n_i - 1)^{-1} \sum_{j=1}^{n_i} [y_{ij}(s) - \bar{y}_i(s)] [y_{ij}(t) - \bar{y}_i(t)], \\ & & i = 1, 2. \end{array} \quad (10.3)$$

10.2.1 Pivotal Test Function

To test the two-sample ECF testing problem (10.2) based on the two samples (10.1), a natural pivotal test function is

$$\Delta(s,t) = \sqrt{n_1 n_2 / n}[\hat{\gamma}_1(s,t) - \hat{\gamma}_2(s,t)], \quad (10.4)$$

where and throughout this section, $n = n_1 + n_2$ denotes the total sample size of the two samples. Notice that $\Delta(s,t)$ has a mean function

$$\text{E}\Delta(s,t) = \sqrt{n_1 n_2 / n}[\gamma_1(s,t) - \gamma_2(s,t)].$$

Under the null hypothesis in (10.2), we have $\text{E}\Delta(s,t) \equiv 0$. As before, let $\mathcal{L}^2(\mathcal{T})$ denote the set of all integrable functions on $\mathcal{T}$.

To study some procedures for testing the two-sample ECF testing problem (10.2), we first establish some properties of $\Delta(s,t)$ under various conditions. For this purpose, we impose the following assumptions:

Two-Sample ECF Testing Problem Assumptions (TS)

1. The two samples (10.1) are with $\eta_i(t) \in \mathcal{L}^2(\mathcal{T})$ and $\text{tr}(\gamma_i) < \infty, i = 1, 2$.
2. As $n_{\min} = \min(n_1, n_2) \to \infty$, we have $n_1/n \to \tau$ such that $\tau \in (0,1)$.
3. For each $i = 1, 2$, the subject-effect functions $v_{ij}(t) = y_{ij}(t) - \eta_i(t), j = 1, 2, \cdots, n_i$ are i.i.d..
4. The subject-effect functions $v_{i1}(t)$ satisfies $\text{E}\|v_{i1}\|^4 < \infty, i = 1, 2$.

5. The maximum variances $\rho_i = \max_{t\in\mathcal{T}} \gamma_i(t,t) < \infty, i = 1,2$.
6. For each $i = 1,2$, the expectation $\mathrm{E}\left[v_{i1}^2(s)v_{i1}^2(t)\right]$ is uniformly bounded. That is, for any $(s,t) \in \mathcal{T}^2$, we have $\mathrm{E}\left[v_{i1}^2(s)v_{i1}^2(t)\right] < C_i < \infty$, where C_1 and C_2 are two constants independent of any $(s,t) \in \mathcal{T}^2$.

Remark 10.1 *Under the null hypothesis in (10.2), the subject-effect functions $v_{ij}(t), j = 1,2,\cdots,n_i; i = 1,2$, are i.i.d. with mean function 0 and covariance function $\gamma(s,t), s,t \in \mathcal{T}$, where $\gamma(s,t)$ denotes the common covariance function of the two samples (10.1).*

Under proper assumptions, $\sqrt{n_1n_2/n}\,[\hat{\gamma}_1(s,t) - \hat{\gamma}_2(s,t)]$ is asymptotically a Gaussian process, as shown by the following theorem.

Theorem 10.1 *Under Assumptions TS1 through TS4 and the null hypothesis in (10.2), as $n_{\min} \to \infty$, we have*

$$\sqrt{n_1n_2/n}\,[\hat{\gamma}_1(s,t) - \hat{\gamma}_2(s,t)] \xrightarrow{d} GP(0,\varpi), \tag{10.5}$$

where $\varpi\,[(s_1,t_1),(s_2,t_2)] = E\{v_{11}(s_1)v_{11}(t_1)v_{11}(s_2)v_{11}(t_2)\} - \gamma(s_1,t_1)\gamma(s_2,t_2)$ with $\gamma(s,t)$ being the common covariance function of the two samples (10.1).

Under the null hypothesis in (10.2), the common covariance function $\gamma(s,t)$ of the two samples (10.1) may be estimated by the following pooled sample covariance function:

$$\hat{\gamma}(s,t) = [(n_1-1)\hat{\gamma}_1(s,t) + (n_2-1)\hat{\gamma}_2(s,t)]\,/(n-2), \tag{10.6}$$

where $\hat{\gamma}_1(s,t)$ and $\hat{\gamma}_2(s,t)$ are given in (10.3). By Theorem 10.1 and under the null hypothesis in (10.2), a natural estimator for $\varpi\,[(s_1,t_1),(s_2,t_2)]$ can then be constructed as

$$\begin{aligned}&\hat{\varpi}\,[(s_1,t_1),(s_2,t_2)] = \\ &n^{-1}\textstyle\sum_{i=1}^2\sum_{j=1}^{n_i}\hat{v}_{ij}(s_1)\hat{v}_{ij}(t_1)\hat{v}_{ij}(s_2)\hat{v}_{ij}(s_2) - \hat{\gamma}(s_1,t_1)\hat{\gamma}(s_2,t_2),\end{aligned} \tag{10.7}$$

where

$$\hat{v}_{ij}(t) = y_{ij}(t) - \bar{y}_i(t), j = 1,2,\cdots,n_i; i = 1,2, \tag{10.8}$$

are the estimated subject-effect functions of the two samples.

The following theorem shows that under some assumptions, the pooled sample covariance function is asymptotically a Gaussian process and it is uniformly consistent over $\mathcal{T}^2$.

Theorem 10.2 *Under Assumptions TS and the null hypothesis in (10.2), as $n_{\min} \to \infty$, we have*

$$\sqrt{n}\,\{\hat{\gamma}(s,t) - \gamma(s,t)\} \xrightarrow{d} GP(0,\varpi), \tag{10.9}$$

where $\varpi\,\{(s_1,t_1),(s_2,t_2)\} = E\{v_{11}(s_1)v_{11}(t_1)v_{11}(s_2)v_{11}(t_2)\} - \gamma(s_1,t_1)\gamma(s_2,t_2)$. In addition, we have $\hat{\gamma}(s,t) = \gamma(s,t) + O_{UP}(n^{-1/2})$ where, as before, O_{UP} means "bounded in probability uniformly."

The asymptotical expression (10.9) in Theorem 10.2 appeared in Theorem 5.3 of Chapter 5. We restate it here for easy reference in proving Theorem 10.3.

10.2.2 Testing Methods

When the sample sizes are large, based on Theorem 10.1, we propose an L^2-norm-based test for the ECF testing problem (10.2). Otherwise, the L^2-norm-based test may not work well. In that case, we propose a random permutation test.

L^2-Norm-Based Test The L^2-norm-based test for (10.2) uses the squared L^2-norm of the pivotal test function (10.4) as the test statistic:

$$T_n = \frac{n_1 n_2}{n} \int_{\mathcal{T}} \int_{\mathcal{T}} [\hat{\gamma}_1(s,t) - \hat{\gamma}_2(s,t)]^2 ds dt. \qquad (10.10)$$

When the sample sizes are large, its asymptotical null random expression is given in Theorem 10.3 below.

Theorem 10.3 *Under Assumptions TS and the null hypothesis in (10.2), as $n \to \infty$, we have*

$$T_n \xrightarrow{d} \sum_{r=1}^{m} \lambda_r A_r, \; A_r \overset{i.i.d.}{\sim} \chi_1^2,$$

where $\lambda_1, \lambda_2, \cdots, \lambda_m$ are all the positive eigenvalues of the covariance function $\varpi[(s_1,t_1),(s_2,t_2)]$ as defined in Theorem 10.1.

The above theorem indicates that T_n is asymptotically a χ^2-type mixture. Therefore, the null distribution of T_n can be approximated using the methods described in Section 4.3 of Chapter 4. By the Welch-Satterthwaite χ^2-approximation method described there, we have

$$T_n \sim \beta \chi_\kappa^2 \text{ approximately, } \quad \text{where } \beta = \frac{\text{tr}(\varpi^{\otimes 2})}{\text{tr}(\varpi)}, \; \kappa = \frac{\text{tr}^2(\varpi)}{\text{tr}(\varpi^{\otimes 2})}. \qquad (10.11)$$

In practice, the parameters β and κ must be estimated based on the two samples (10.1). Let $\hat{\beta}$ and $\hat{\kappa}$ be the estimators of β and κ. Then we have

$$T_n \sim \hat{\beta} \chi_{\hat{\kappa}}^2 \text{ approximately.} \qquad (10.12)$$

The L^2-norm-based test can then be conducted accordingly.

Remark 10.2 *When the two samples (10.1) are Gaussian, by Theorem 4.3 of Chapter 4 we have*

$$\varpi[(s_1,t_1),(s_2,t_2)] = \gamma(s_1,s_2)\gamma(t_1,t_2) + \gamma(s_1,t_2)\gamma(s_2,t_1). \qquad (10.13)$$

Then by some simple calculations, we have

$$tr(\varpi) = tr^2(\gamma) + tr(\gamma^{\otimes 2}), \;\; tr(\varpi^{\otimes 2}) = 2tr^2(\gamma^{\otimes 2}) + 2tr(\gamma^{\otimes 4}). \qquad (10.14)$$

Remark 10.3 *When the two samples (10.1) are Gaussian, by Remark 10.2, we can estimate* $\varpi\left[(s_1,t_1),(s_2,t_2)\right]$ *by*

$$\hat{\varpi}\left[(s_1,t_1),(s_2,t_2)\right] = \hat{\gamma}(s_1,s_2)\hat{\gamma}(t_1,t_2) + \hat{\gamma}(s_1,t_2)\hat{\gamma}(s_2,t_1), \tag{10.15}$$

where $\hat{\gamma}(s,t)$ *is the pooled sample covariance function as given by (10.6). Then the naive estimators of* $tr(\varpi)$ *and* $tr(\varpi^{\otimes 2})$ *are given by*

$$tr(\hat{\varpi}) = tr^2(\hat{\gamma}) + tr(\hat{\gamma}^{\otimes 2}), \quad tr(\hat{\varpi}^{\otimes 2}) = 2tr^2(\hat{\gamma}^{\otimes 2}) + 2tr(\hat{\gamma}^{\otimes 4}). \tag{10.16}$$

We can also define some bias-reduced estimators of $tr(\varpi), tr^2(\varpi)$, *and* $tr(\varpi^{\otimes 2})$ *in the following way:*

$$\begin{aligned} \widehat{tr(\varpi)} &= \widehat{tr^2(\gamma)} + \widehat{tr(\gamma^{\otimes 2})}, \quad \widehat{tr^2(\varpi)} = \left[\widehat{tr(\varpi)}\right]^2, \\ \widehat{tr(\varpi^{\otimes 2})} &= 2\left[\widehat{tr(\gamma^{\otimes 2})}\right]^2 + 2tr(\hat{\gamma}^{\otimes 4}), \end{aligned} \tag{10.17}$$

where

$$\begin{aligned} \widehat{tr^2(\gamma)} &= \tfrac{(n-2)(n-1)}{(n-3)n}\left[tr^2(\hat{\gamma}) - \tfrac{2tr(\hat{\gamma}^{\otimes 2})}{n-1}\right], \\ \widehat{tr(\gamma^{\otimes 2})} &= \tfrac{(n-2)^2}{(n-3)n}\left[tr(\hat{\gamma}^{\otimes 2}) - \tfrac{tr^2(\hat{\gamma})}{n-2}\right], \end{aligned} \tag{10.18}$$

are the unbiased estimators of $tr^2(\gamma)$ *and* $tr(\gamma^{\otimes 2})$, *respectively. Notice that in expression (10.17), the unbiased estimator of* $tr(\gamma^{\otimes 4})$ *is not incorporated as it is quite challenging to obtain a simple and useful unbiased estimator of* $tr(\gamma^{\otimes 4})$.

Remark 10.4 *When the two samples (10.1) are Gaussian, from Remark 10.3 we can propose a naive method and a bias-reduced method for obtaining* $\hat{\beta}$ *and* $\hat{\kappa}$. *The naive estimators of* β *and* κ *are obtained by replacing* $tr(\varpi), tr^2(\varpi)$, *and* $tr(\varpi^{\otimes 2})$ *in (10.11), respectively, with their naive estimators* $tr(\hat{\varpi}), tr^2(\hat{\varpi})$ *and* $tr(\hat{\varpi}^{\otimes 2})$:

$$\hat{\beta} = \frac{tr(\hat{\varpi}^{\otimes 2})}{tr(\hat{\varpi})}, \quad \hat{\kappa} = \frac{tr^2(\hat{\varpi})}{tr(\hat{\varpi}^{\otimes 2})}, \tag{10.19}$$

where $\hat{\varpi}\left[(s_1,t_1),(s_2,t_2)\right]$ *is given in (10.15). Under the Gaussian assumption TS2, the bias-reduced estimators of* β *and* κ *are obtained by replacing* $tr(\varpi), tr^2(\varpi)$, *and* $tr(\varpi^{\otimes 2})$ *in (10.11), respectively, with their bias-reduced estimators* $\widehat{tr(\varpi)}$, $\widehat{tr^2(\varpi)}$, *and* $\widehat{tr(\varpi^{\otimes 2})}$ *given in (10.17):*

$$\hat{\beta} = \frac{\widehat{tr(\varpi^{\otimes 2})}}{\widehat{tr(\varpi)}}, \quad \hat{\kappa} = \frac{\widehat{tr^2(\varpi)}}{\widehat{tr(\varpi^{\otimes 2})}}. \tag{10.20}$$

Random Permutation Test When the sample sizes of the two samples (10.1) are small, Theorem 10.2 is no longer valid so that the Welch-Satterthwaite χ^2-approximation for the null distribution of the L^2-norm-based test statistic is usually not accurate. In this case, we can approximate the critical values of the L^2-norm-based test statistic T_n by a random permutation method, resulting in the so-called random permutation test. The key ideas are described below.

Let $\hat{v}_l^*(t), l = 1, 2, \cdots, n$ be a random permutation sample generated from the pooled sample of the estimated subject-effect functions (10.8). That is, the random permutation sample of size $n = n_1 + n_2$ is obtained by randomly reordering the n estimated subject-effect functions (10.8). We then use the first n_1 functions to form the first permutation sample $\hat{v}_{1j}^*(t), j = 1, 2, \cdots, n_1$ and use the remaining n_2 functions to form the second permutation sample $\hat{v}_{2j}^*(t), j = 1, 2, \cdots, n_2$. Given the two original functional samples (10.1), both the conditional expectations of $\hat{\gamma}_1^*(s,t)$ and $\hat{\gamma}_2^*(s,t)$ are equal to the pooled sample covariance function (10.6) of the two original functional samples (10.1). That is, the two permutation samples conditionally satisfy the null hypothesis (10.2). Therefore, we can use the distribution of the following permutation test statistic

$$T_n^* = \frac{n_1 n_2}{n} \int_{\mathcal{T}} \int_{\mathcal{T}} [\hat{\gamma}_1^*(s,t) - \hat{\gamma}_2^*(s,t)]^2 \, dsdt$$

to mimic the null distribution of the test statistic T_n (10.10), where

$$\hat{\gamma}_i^*(s,t) = (n_i - 1)^{-1} \sum_{j=1}^{n_i} \hat{v}_{ij}^*(s)\hat{v}_{ij}^*(t), i = 1, 2.$$

Repeat the above process a large number of times to can get a sample of T_n^* and use its sample percentiles to approximate the percentiles of T_n. The associated random permutation test is then conducted accordingly.

Example 10.2 *For the NOx emission level curve data with six unusual observation excluded, we first obtained* $T_n = 1.5981e9$. *Under the Gaussian assumption, using the relationship (10.15) and by (10.19) and (10.20), we obtained* $\hat{\beta} = 1.3589e8, \hat{\kappa} = 8.1833$ *by the naive method and* $\hat{\beta} = 1.2959e8, \hat{\kappa} = 8.4718$ *by the bias-reduced method. Using (10.12), the associated naive and bias-reduced P-values were obtained as* 0.1733 *and* 0.1631, *respectively. The approximated P-values obtained by the random permutation test with* 100,000 *replicates is* 0.1631. *It appears that the difference between the covariance functions of working-days and non-working days are not significant.*

10.3 Multi-Sample Case

We now consider the ECF testing problem for several samples. This problem arises naturally in the homogeneous one-way and two-way ANOVA for functional data as discussed in Chapter 5 where we first need to determine if the covariance functions at all the levels of the one or two categorical variables are homogeneous; otherwise, it is preferred to use the heteroscedastic one-way or two-way functional ANOVA discussed in Chapter 9. We use the following two examples to motivate the ECF testing problem for several functional samples.

Example 10.3 *The Canadian temperature data (Canadian Climate Program 1982) were introduced in Section 1.2.4 of Chapter 1. They are the daily temperature records of thirty-five Canadian weather stations over a year (365 days), among which, fifteen are in Eastern, another fifteen in Western and the remaining five in Northern Canada. The reconstructed Canadian temperature curves were displayed in Figure 5.5 of Chapter 5. In Section 5.3 of Chapter 5, the equality of the mean temperature curves of the Eastern, Western, and Northern weather stations was tested under the assumption that the temperature curves of the Eastern, Western, and Northern weather stations have a common covariance function. However, Figure 9.3 of Chapter 9 indicates that the three groups of the temperature curves may have quite different covariance functions. That is why in Section 9.3 of Chapter 9, the equality of the mean temperature curves of the Eastern, Western, and Northern weather stations was re-tested without imposing the ECF assumption. To statistically verify if these Eastern, Western, and Northern weather stations have a common covariance function, we need to study how to test the equality of the covariance functions of several functional samples.*

Example 10.4 *The orthosis data were introduced in Section 1.2.7 of Chapter 1. They are the resultant moments of seven young male volunteers who wore a spring-loaded orthosis of adjustable stiffness under four experimental conditions: a control condition (without orthosis), an orthosis condition (with the orthosis only), and two spring conditions (spring 1, spring 2), each ten times, over a period of time (about ten seconds), which were equally divided into* 256 *time points. Figure 1.12 in Chapter 1 displays the raw curves of the orthosis data; each panel showing ten raw curves observed in the cell. In Chapter 9, this functional data set motivated and illustrated a heteroscedastic functional ANOVA model with two factors (Subjects: seven young male volunteers and Treatments: four conditions). This is because Figures 9.4 and 9.5 of Chapter 9 seem to indicate that the cell covariance functions may be not the same. This problem may be formally tested using the methods developed in this section for testing the equality of the covariance functions of several functional samples.*

Formally, we can define the ECF testing problem for several functional samples as follows. Suppose we have k independent functional samples:

$$y_{i1}(t), \cdots, y_{in_i}(t), \quad i = 1, \cdots, k, \tag{10.21}$$

satisfying

$$\begin{aligned} y_{ij}(t) &= \eta_i(t) + v_{ij}(t), \; j = 1, 2, \cdots, n_i, \\ v_{i1}(t), v_{i2}(t), \cdots, v_{in_i}(t) &\overset{i.i.d.}{\sim} \text{SP}(0, \gamma_i); \; i = 1, 2, \cdots, k, \end{aligned} \tag{10.22}$$

where $\eta_1(t), \eta_2(t), \cdots, \eta_k(t)$ are the unknown group mean functions of the k samples, $v_{ij}(t), j = 1, \cdots, n_i; i = 1, 2, \cdots, k$ are the subject-effect functions, and $\gamma_i(s,t), i = 1,2,\cdots,k$ are the covariance functions. We wish to test the following null hypothesis:

$$H_0: \gamma_1(s,t) \equiv \gamma_2(s,t) \equiv \cdots \equiv \gamma_k(s,t), \quad \text{for } s,t \in \mathcal{T}, \tag{10.23}$$

where again $\mathcal{T}$ is some finite time interval of interest, often specified as $[a,b]$ with $-\infty < a < b < \infty$. The above problem is known as the k-sample ECF testing problem for functional data, extending the two-sample ECF testing problem for functional data discussed in the previous section. We notice that the testing methodologies proposed in this section are natural generalizations of those methodologies proposed in the previous section for the two-sample ECF testing problem.

10.3.1 Estimation of Group Mean and Covariance Functions

Based on the k samples (10.21), the group mean functions $\eta_i(t), i = 1,2,\cdots,k$ and the covariance functions $\gamma_i(s,t), i = 1,2,\cdots,k$ can be unbiasedly estimated as

$$\begin{array}{rcl} \hat{\eta}_i(t) & = & \bar{y}_{i\cdot}(t) = n_i^{-1}\sum_{j=1}^{n_i} y_{ij}(t), \\ \hat{\gamma}_i(s,t) & = & (n_i-1)^{-1}\sum_{j=1}^{n_i}[y_{ij}(s)-\bar{y}_{i\cdot}(s)][y_{ij}(t)-\bar{y}_{i\cdot}(t)], \\ & & i = 1,2,\cdots,k. \end{array} \tag{10.24}$$

They are the sample mean and covariance functions of the k samples (10.21) respectively. Note that $\hat{\gamma}_i(s,t), i = 1,2,\cdots,k$ are independent and $\mathrm{E}\hat{\gamma}_i(s,t) = \gamma_i(s,t), i = 1,2,\cdots,k$. The estimated subject-effect functions are then given by

$$\hat{v}_{ij}(t) = y_{ij}(t) - \bar{y}_{i\cdot}(t), j = 1,2,\cdots,n_i; i = 1,2,\cdots,k. \tag{10.25}$$

In order to describe some procedures for testing the multi-sample ECF testing problem (10.23), we first establish the joint distribution of $\hat{\gamma}_i(s,t), i = 1,2,\cdots,k$ under various conditions. For this purpose, we list the following assumptions:

K-Sample ECF Testing Problem Assumptions (KS)

1. The k samples (10.21) are with $\eta_i(t) \in \mathcal{L}^2(\mathcal{T})$ and $\mathrm{tr}(\gamma_i) < \infty, i = 1,2,\cdots,k$.
2. As $n_{\min} = \min_{i=1}^k n_i \to \infty$, the k sample sizes satisfy $n_i/n \to \tau_i$, $i = 1,2,\cdots,k$ such that $\tau_1,\tau_2,\cdots,\tau_k \in (0,1)$, where and throughout this section $n = \sum_{i=1}^k n_i$ denotes the total sample size.
3. For each $i = 1,2,\cdots,k$, the subject-effect functions $v_{ij}(t) = y_{ij}(t) - \eta_i(t), j = 1,2,\cdots,n_i$ are i.i.d..
4. For each $i = 1,2,\cdots,k$, the subject-effect function $v_{i1}(t)$ satisfies $\mathrm{E}\|v_{i1}\|^4 < \infty$.
5. The maximum variances $\rho_i = \max_{t\in\mathcal{T}} \gamma_i(t,t) < \infty$, $i = 1,2,\cdots,k$.

6. The expectations $\mathrm{E}[v_{i1}^2(s)v_{i1}^2(t)], i = 1,2,\cdots,k$ are uniformly bounded. That is, for any $(s,t) \in \mathcal{T}^2$, we have $\mathrm{E}\left[v_{i1}^2(s)v_{i1}^2(t)\right] < C_i < \infty$, where $C_i, i = 1,2,\cdots,k$ are constants independent of any $(s,t) \in \mathcal{T}^2$.

The above assumptions are regularly imposed for studying the properties of the estimators $\hat{\gamma}_i(s,t), i = 1,2,\cdots,k$. First of all, we have the following theorem stating that the asymptotic joint distribution of $\hat{\gamma}_i(s,t), i = 1,2,\cdots,k$ is a k-dimensional Gaussian process.

Theorem 10.4 *Under Assumptions KS1 through KS4, as $n_{\min} \to \infty$, we have*

$$\begin{pmatrix} \sqrt{n_1-1}\left[\hat{\gamma}_1(s,t)-\gamma_1(s,t)\right] \\ \sqrt{n_2-1}\left[\hat{\gamma}_2(s,t)-\gamma_2(s,t)\right] \\ \vdots \\ \sqrt{n_k-1}\left[\hat{\gamma}_k(s,t)-\gamma_k(s,t)\right] \end{pmatrix} \xrightarrow{d} GP_k\left[0, diag(\varpi_1,\varpi_2,\cdots,\varpi_k)\right], \quad (10.26)$$

where $\varpi_i\left[(s_1,t_1),(s_2,t_2)\right] = E\{v_{i1}(s_1)v_{i1}(t_1)v_{i1}(s_2)v_{i1}(t_2)\} - \gamma_i(s_1,t_1)\gamma_i(s_2,t_2)$, $i = 1,2,\cdots,k$.

Under the null hypothesis (10.23), the common covariance function $\gamma(s,t)$ of the k samples can be estimated by the following pooled sample covariance function

$$\hat{\gamma}(s,t) = \sum_{i=1}^{k}(n_i-1)\hat{\gamma}_i(s,t)/(n-k), \quad (10.27)$$

where $\hat{\gamma}_i(s,t), i = 1,2,\cdots,k$ are given in (10.24). The following theorem gives the asymptotic joint distribution of $\hat{\gamma}_i(s,t), i = 1,2,\cdots,k$ when the null hypothesis is valid.

Theorem 10.5 *Under Assumptions KS1 through KS4 and the null hypothesis in (10.23), as $n_{\min} \to \infty$, we have*

$$\begin{pmatrix} \sqrt{n_1-1}\left[\hat{\gamma}_1(s,t)-\gamma(s,t)\right] \\ \sqrt{n_2-1}\left[\hat{\gamma}_2(s,t)-\gamma(s,t)\right] \\ \vdots \\ \sqrt{n_k-1}\left[\hat{\gamma}_k(s,t)-\gamma(s,t)\right] \end{pmatrix} \xrightarrow{d} GP_k\left[0, \varpi\mathbf{I}_k\right], \quad (10.28)$$

where $\varpi\left[(s_1,t_1),(s_2,t_2)\right] = E\{v_{11}(s_1)v_{11}(t_1)v_{11}(s_2)v_{11}(t_2)\} - \gamma(s_1,t_1)\gamma(s_2,t_2)$.

By Theorem 10.5 and under the null hypothesis (10.23), a natural estimator for $\varpi\left[(s_1,t_1),(s_2,t_2)\right]$ can then be constructed as

$$\begin{aligned} &\hat{\varpi}\left[(s_1,t_1),(s_2,t_2)\right] = \\ &n^{-1}\textstyle\sum_{i=1}^{k}\sum_{j=1}^{n_i}\hat{v}_{ij}(s_1)\hat{v}_{ij}(t_1)\hat{v}_{ij}(s_2)\hat{v}_{ij}(t_2) - \hat{\gamma}(s_1,t_1)\hat{\gamma}(s_2,t_2), \end{aligned} \quad (10.29)$$

a natural extension of (10.7), keeping in mind that n is now equal to $\sum_{i=1}^k n_i$. In the expression (10.29), the pooled sample covariance function $\hat{\gamma}(s,t)$ is given in (10.27). In the following theorem, we show that the pooled sample covariance function $\hat{\gamma}(s,t)$ is asymptotically a Gaussian process and is consistent uniformly over $\mathcal{T}^2$ under some mild assumptions.

Theorem 10.6 *Under Assumptions KS and the null hypothesis (10.23), as* $n_{\min} \to \infty$, *we have*

$$\sqrt{n-k}\,\{\hat{\gamma}(s,t)-\gamma(s,t)\} \xrightarrow{d} GP(0,\varpi), \tag{10.30}$$

where $\varpi\{(s_1,t_1),(s_2,t_2)\} = E\{v_{11}(s_1)v_{11}(t_1)v_{11}(s_2)v_{11}(t_2)\} - \gamma(s_1,t_1)\gamma(s_2,t_2)$. *In addition, we have*

$$\hat{\gamma}(s,t) = \gamma(s,t) + O_{UP}\left[(n-k)^{-1/2}\right]. \tag{10.31}$$

The asymptotical expression (10.30) in Theorem 10.6 is equivalent to the asymptotical expression (5.34) in Theorem 5.7 of Chapter 5. We restate it here in a slightly different form for easy reference in proving Theorem 10.7 given below.

Remark 10.5 *When the k samples (10.21) are Gaussian, the expressions (10.13) and (10.14) continue to be valid. In addition, the expressions (10.15), (10.16), and (10.17) continue to be valid except the expression (10.18) should now be replaced with*

$$\begin{aligned} \widehat{tr^2(\gamma)} &= \tfrac{(n-k)(n-k+1)}{(n-k-1)(n-k+2)}\left[tr^2(\hat{\gamma}) - \tfrac{2tr(\hat{\gamma}^{\otimes 2})}{n-k+1}\right], \\ \widehat{tr(\gamma^{\otimes 2})} &= \tfrac{(n-k)^2}{(n-k-1)(n-k+2)}\left[tr(\hat{\gamma}^{\otimes 2}) - \tfrac{tr^2(\hat{\gamma})}{n-k}\right], \end{aligned} \tag{10.32}$$

where $\hat{\gamma}(s,t)$ *is the pooled sample covariance function defined in (10.27).*

10.3.2 Testing Methods

We are now ready to describe an L^2-norm-based test and a random permutation test for the k-sample ECF testing problem (10.23). They are extensions of the L^2-norm-based test and the random permutation test for the two-sample ECF testing problem described in the previous section.

L^2-Norm-Based Test The test statistic of the L^2-norm-based test for the k-sample ECF testing problem (10.23) is defined as

$$T_n = \sum_{i=1}^k (n_i-1)\int_{\mathcal{T}}\int_{\mathcal{T}} [\hat{\gamma}_i(s,t)-\hat{\gamma}(s,t)]^2\,ds\,dt, \tag{10.33}$$

where $\hat{\gamma}_i(s,t), i = 1, 2, \cdots, k$ are the group sample covariance functions as defined in (10.24) and $\hat{\gamma}(s,t)$ is the pooled sample covariance function as defined in (10.27). On the one hand, when the null hypothesis (10.23) is valid, each of $\hat{\gamma}_i(s,t), i = 1, 2, \cdots, k$ should be close to their pooled sample covariance function $\hat{\gamma}(s,t)$ so that the test statistic T_n will be small. On the other hand, when the null hypothesis (10.23) is not valid, at least one of $\hat{\gamma}_i(s,t), i = 1, 2, \cdots, k$ should be very different from their pooled sample covariance function $\hat{\gamma}(s,t)$ so that the test statistic T_n will be large. Therefore, it is reasonable to use T_n to test the null hypothesis (10.23).

Remark 10.6 *According to the definition of $\hat{\gamma}(s,t)$ given in (10.27), in the expression of T_n, we use $n_i - 1$ instead of the usual n_i, aiming to simplify the derivation of the asymptotic distribution of T_n. This, however, will not have any impact on the asymptotic distribution of T_n.*

We first show that under some proper assumptions, the test statistic T_n is asymptotically a χ^2-type mixture as stated in the following theorem.

Theorem 10.7 *Under Assumptions KS and the null hypothesis (10.23), as $n_{\min} \to \infty$, we have*

$$T_n \xrightarrow{d} \sum_{r=1}^{m} \lambda_r A_r, \ A_r \overset{i.i.d.}{\sim} \chi^2_{k-1},$$

where $\lambda_r, r = 1, 2, \cdots, m$ are all the positive eigenvalues of $\varpi(s,t)$ as defined in Theorem 10.5.

By Theorem 10.7, the null distribution of T_n can then be approximated using the methods described in Section 4.3 of Chapter 4. By the Welch-Satterthwaite χ^2-approximation method described there, we have

$$\begin{array}{c} T_n \sim \beta \chi^2_{(k-1)\kappa} \text{ approximately}, \\ \text{where } \beta = \frac{\text{tr}(\varpi^{\otimes 2})}{\text{tr}(\varpi)}, \ \kappa = \frac{\text{tr}^2(\varpi)}{\text{tr}(\varpi^{\otimes 2})}. \end{array} \tag{10.34}$$

As before, in practice, the parameters β and κ must be estimated based on the k samples (10.21). Let $\hat{\beta}$ and $\hat{\kappa}$ be the estimators of β and κ. Then we have

$$T_n \sim \hat{\beta} \chi^2_{(k-1)\hat{\kappa}} \text{ approximately}. \tag{10.35}$$

The L^2-norm-based test can then be conducted accordingly.

Remark 10.7 *When $k = 2$, the expressions (10.34) and (10.35) will reduce to the expressions (10.11) and (10.12).*

Remark 10.8 *As in Remark 10.4, based on Remark 10.5, we can have a naive method and a bias-reduced method for obtaining $\hat{\beta}$ and $\hat{\kappa}$. The naive* *estimators of β and κ are obtained by replacing $tr(\varpi), tr^2(\varpi)$, and $tr(\varpi^{\otimes 2})$ in (10.34), respectively, with their naive estimators $tr(\hat{\varpi}), tr^2(\hat{\varpi})$, and $tr(\hat{\varpi}^{\otimes 2})$:*

$$\hat{\beta} = \frac{tr(\hat{\varpi}^{\otimes 2})}{tr(\hat{\varpi})}, \quad \hat{\kappa} = \frac{tr^2(\hat{\varpi})}{tr(\hat{\varpi}^{\otimes 2})}, \tag{10.36}$$

where $\hat{\varpi}[(s_1, t_1), (s_2, t_2)]$ is given in (10.29). When the k samples (10.21) are Gaussian, the bias-reduced estimators of β and κ are obtained by replacing $tr(\varpi), tr^2(\varpi)$, and $tr(\varpi^{\otimes 2})$ in (10.34), respectively, with their bias-reduced estimators $\widehat{tr(\varpi)}$, $\widehat{tr^2(\varpi)}$, and $\widehat{tr(\varpi^{\otimes 2})}$ given in (10.17):

$$\hat{\beta} = \frac{\widehat{tr(\varpi^{\otimes 2})}}{\widehat{tr(\varpi)}}, \quad \hat{\kappa} = \frac{\widehat{tr^2(\varpi)}}{\widehat{tr(\varpi^{\otimes 2})}}. \tag{10.37}$$

Random Permutation Test When the sample sizes $n_i, i = 1, 2, \cdots, k$ of the k samples (10.21) are not large, Theorem 10.6 is no longer valid so that the Welch-Satterthwaite χ^2-approximation for the null distribution of the L^2-norm-based test statistic T_n is usually not accurate. In this case, we can approximate the critical values of T_n by a random permutation method, resulting in the so-called random permutation test. The key ideas are along the same lines as those of the random permutation test for the two-sample ECF testing problem described in the previous section.

Let $\hat{v}_l^*(t), l = 1, 2, \cdots, n$ be a random permutation sample obtained by randomly reordering the pooled estimated subject-effect functions (10.25), where $n = n_1 + n_2 + \cdots + n_k$. We then form k permutation samples with sizes $n_1, n_2, \cdots, n_k$ from this permutation sample and record them as $\hat{v}_{ij}^*(t), j = 1, 2, \cdots, n_i; i = 1, 2, \cdots, k$. To mimic the L^2-norm-based test statistic T_n, the random permutation test statistic is then computed as

$$T_n^* = \sum_{i=1}^{k} (n_i - 1) \int_{\mathcal{T}} \int_{\mathcal{T}} [\hat{\gamma}_i^*(s, t) - \hat{\gamma}^*(s, t)]^2 \, dsdt,$$

where

$$\hat{\gamma}_i^*(s, t) = (n_i - 1)^{-1} \textstyle\sum_{j=1}^{n_i} \hat{v}_{ij}^*(s) \hat{v}_{ij}^*(t), i = 1, 2, \cdots, k,$$
$$\hat{\gamma}^*(s, t) = \textstyle\sum_{i=1}^{k} (n_i - 1) \hat{\gamma}_i^*(s, t)/(n - k).$$

Repeat the above process a large number of times so that we can get a sample of T_n^* and use its sample percentiles to approximate the percentiles of T_n. The associated random permutation test is then conducted accordingly.

Note that given the k original functional samples (10.21), the conditional expectations of $\hat{\gamma}_i^*(s, t), i = 1, 2, \cdots, k$ are equal to the pooled sample covariance function (10.27) of the k original functional samples (10.21). That is, the k random permutation samples conditionally satisfy the null hypothesis (10.23). Therefore, we can use the distribution of T_n^* to mimic the null distribution of T_n.

Table 10.1 *Tests of the equality of the covariance functions for the Canadian temperature data with resolution* $M = 1,000$.

		L^2-norm, naive			L^2-norm, bias-reduced			Rand. permu.
Time period	T_n	$\hat{\beta}$	$2\hat{\kappa}$	P-value	$\hat{\beta}$	$2\hat{\kappa}$	P-value	P-value
Spring	7.75e7	9.10e6	2.47	0.023	9.03e6	2.35	0.020	0.063
Summer	1.86e7	1.32e6	2.15	0.001	1.32e6	2.03	0.001	0.002
Fall	1.04e7	2.76e6	2.41	0.203	2.74e6	2.29	0.184	0.189
Winter	2.94e8	4.00e7	2.06	0.027	4.01e7	1.94	0.024	0.029
Whole year	7.82e8	9.05e7	3.18	0.040	8.86e7	3.07	0.034	0.051

Note: The estimated parameters $\hat{\beta}$ *and* $\hat{\kappa}$ *were obtained using (10.36) and (10.37) by the naive and bias-reduced methods, respectively. The P-values by the random permutation test were obtained by* 100,000 *replicates.*

Example 10.5 *We now apply the* L^2*-norm-based test and the random permutation test to check if the covariance functions of the Canadian temperature curves of the Eastern, Western, and Northern weather stations are the same over some time periods as specified in Table 9.1. Table 10.1 shows the test results of the* L^2*-norm-based tests with the null distribution approximated by the naive and bias-reduced methods, and the random permutation test. It is seen that the estimated values of the parameters* β *and* $d = 2\kappa$ *by the naive and bias-reduced methods are comparable and their associated P-values are also comparable although the P-values by the bias-reduced method are always smaller than those by the naive method. The P-values obtained by the random permutation test are slightly larger than those by the* L^2*-norm-based test. These P-values indicate that the covariance functions of the Canadian temperature curves of the Eastern, Western, and Northern weather stations are close to each other during the Fall season but there is some evidence showing that they are different during other seasons and the whole year.*

Remark 10.9 *As the sample sizes* $(n_1 = 15, n_2 = 15, n_3 = 5)$ *of the Canadian temperature curves of the Eastern, Western, and Northern weather stations are rather small, the conclusions based on the P-values of the random permutation test should be more liable.*

Example 10.6 *Figure 9.4 displays the sample cell covariance functions of the orthosis data introduced in Example 10.4. We now turn to check if the underlying cell covariance functions are the same. We obtain* $T_n = 1.5661\text{e}10$ *using (10.33). By the naive method, we obtain* $tr(\hat{\varpi}) = 2.9198\text{e}8$ *and* $tr(\hat{\varpi}^{\otimes 2}) = 5.1118\text{e}15$*. Based on these and using (10.36), we have* $\hat{\beta} = 1.7507\text{e}7, \hat{d} = 450.29$*. The associated P-value is* 0*. By the bias-reduced method, we obtain* $\widehat{tr(\varpi)} = 2.9051\text{e}8$ *and* $\widehat{tr(\varpi^{\otimes 2})} = 4.9242\text{e}15$*. Based on these and using (10.37), we have* $\hat{\beta} = 1.6950\text{e}7, \hat{d} = 462.75$*. The associated P-value is also* 0*. With* 100,000 *replicates, the P-value of* T_n *obtained by the random permutation test is also* 0*. These P-values indicate that the underlying cell covariance functions of the orthosis data are unlikely to be the same.*

10.4 Technical Proofs

In this section we outline the proofs of the main results described in this chapter.

Proof of Theorem 10.1 Under the null hypothesis in (10.2), let $\gamma_1(s,t) \equiv \gamma_2(s,t) \equiv \gamma(s,t), s,t \in \mathcal{T}$. Then

$$\sqrt{n_1 n_2/n}\,[\hat{\gamma}_1(s,t) - \hat{\gamma}_2(s,t)] = \\ a_n\sqrt{n_1}\,[\hat{\gamma}_1(s,t) - \gamma(s,t)] + b_n\sqrt{n_2}\,[\hat{\gamma}_2(s,t) - \gamma(s,t)],$$

where $a_n = \sqrt{n_2/n}$ and $b_n = \sqrt{n_1/n}$. Under the given conditions, by Theorem 4.16, we have

$$\sqrt{n_1}[\hat{\gamma}_1(s,t) - \gamma(s,t)] \xrightarrow{d} \text{GP}(0,\varpi), \quad \sqrt{n_2}[\hat{\gamma}_2(s,t) - \gamma(s,t)] \xrightarrow{d} \text{GP}(0,\varpi),$$

where $\varpi\{(s_1,t_1),(s_2,t_2)\} = \text{E}v_{11}(s_1)v_{11}(t_1)v_{11}(s_2)v_{11}(t_2) - \gamma(s_1,t_1)\gamma(s_2,t_2)$. The expression (10.5) follows immediately from the fact that $\hat{\gamma}_1(s,t)$ and $\hat{\gamma}_2(s,t)$ are independent and $a_n^2 + b_n^2 = 1$. The theorem is proved.

Proof of Theorem 10.2 The theorem is a special case of Theorem 10.6 and its proof is also a special case of the proof of Theorem 10.6.

Proof of Theorem 10.3 Under the given conditions, the theorem follows immediately from Theorems 10.1 and 10.2, the continuous mapping theorem for random elements taking values in a Hilbert space (Billingsley, 1968, p. 34; Cuevas, Febrero, and Fraiman 2004), and Theorem 4.2 of Chapter 4.

Proof of Theorem 10.4 Under the given conditions, by Theorem 4.16 of Chapter 4, we have

$$\sqrt{n_i}[\hat{\gamma}_i(s,t) - \gamma_i(s,t)] \xrightarrow{d} \text{GP}(0,\varpi_i), \quad i = 1,2,\cdots,k,$$

where $\varpi_i\{(s_1,t_1),(s_2,t_2)\} = \text{E}v_{i1}(s_1)v_{i1}(t_1)v_{i1}(s_2)v_{i1}(t_2) - \gamma_i(s_1,t_1)\gamma_i(s_2,t_2)$, $i = 1,2,\cdots,k$. The expression (10.26) then follows immediately from the fact that the k functional samples (10.21) are independent and $(n_i - 1)/n_i \to 1, i = 1,2,\cdots,k$ as $n_{\min} \to \infty$. The theorem is proved.

Proof of Theorem 10.5 Under the null hypothesis (10.23), we have $\gamma_1(s,t) \equiv \cdots \equiv \gamma_k(s,t) \equiv \gamma(s,t)$, for $s,t \in \mathcal{T}$, where $\gamma(s,t)$ is the common covariance function of the k samples (10.21). Then under the given conditions, by Theorem 10.4, we have

$$\sqrt{n_i}[\hat{\gamma}_i(s,t) - \gamma(s,t)] \xrightarrow{d} \text{GP}(0,\varpi), \quad i = 1,2,\cdots,k,$$

where $\varpi\{(s_1,t_1),(s_2,t_2)\} = \text{E}v_{11}(s_1)v_{11}(t_1)v_{11}(s_2)v_{11}(t_2) - \gamma(s_1,t_1)\gamma(s_2,t_2)$.

The expression (10.28) then follows immediately from the fact that the k functional samples (10.21) are independent and $(n_i - 1)/n_i \to 1, i = 1, 2, \cdots, k$ as $n_{\min} \to \infty$. The theorem is proved.

Proof of Theorem 10.6 Notice that by (10.27), we have

$$\sqrt{n-k}\,\{\hat{\gamma}(s,t) - \gamma(s,t)\} = \sum_{i=1}^{k} a_i \sqrt{n_i - 1}\,[\hat{\gamma}_i(s,t) - \gamma(s,t)],$$

where $a_i = \sqrt{(n_i - 1)/(n-k)}, i = 1, 2, \cdots, k$. Under the given conditions, the expression (10.30) follows from Theorem 10.5 and the fact that the k samples (10.21) are independent and $\sum_{i=1}^k a_i^2 = 1$ immediately. It follows that as $n_{\min} \to \infty$, we have

$$\mathrm{E}\,[\hat{\gamma}(s,t) - \gamma(s,t)]^2 = \frac{\varpi\,[(s,t),(s,t)]}{n-k}[1 + o(1)].$$

Under the null hypothesis (10.23), we have $\gamma(s,t) \equiv \gamma_i(s,t)$ and $\varpi\,[(s,t),(s,t)] = \varpi_i\,[(s,t),(s,t)]$. The expression (10.31) follows immediately from the fact that

$$\varpi_i\,[(s,t),(s,t)] \le \mathrm{E}[v_{i1}^2(s)v_{i1}^2(t)] + \gamma_i^2(s,t) \le C_i + \rho_i^2,\ i = 1, 2, \cdots, k,$$

where the constants C_i and ρ_i are given in Assumptions KS6 and KS7. It follows that

$$\varpi\,[(s,t),(s,t)] \le \max_{i=1}^{k}\left(C_i + \rho_i^2\right) < \infty.$$

The theorem is then proved.

Proof of Theorem 10.7 Under the null hypothesis in (10.23), we have $\gamma_1(s,t) \equiv \cdots \equiv \gamma_k(s,t) \equiv \gamma(s,t)$. Set $\mathbf{z}(s,t) = [z_1(s,t), z_2(s,t), \cdots, z_k(s,t)]^T$ with $z_i(s,t) = \sqrt{n_i - 1}[\hat{\gamma}_i(s,t) - \gamma(s,t)], i = 1, 2, \cdots, k$. Then we have

$$\begin{array}{rcl} T_n & = & \sum_{i=1}^k (n_i - 1) \int_{\mathcal{T}} \int_{\mathcal{T}} [\hat{\gamma}_i(s,t) - \hat{\gamma}(s,t)]^2 dsdt \\ & = & \int_{\mathcal{T}} \int_{\mathcal{T}} \mathbf{z}(s,t)^T \left[\mathbf{I}_k - \mathbf{b}_n \mathbf{b}_n^T/(n-k)\right] \mathbf{z}(s,t) dsdt, \end{array}$$

where $\mathbf{I}_k - \mathbf{b}_n\mathbf{b}_n^T/(n-k)$ is an idempotent matrix of rank $(k-1)$ with $\mathbf{b}_n = [\sqrt{n_1 - 1}, \sqrt{n_2 - 1}, \cdots, \sqrt{n_k - 1}]^T$. Under the given conditions, by Theorem 10.5, as $n_{\min} \to \infty$, we have $\mathbf{z}(s,t) \xrightarrow{d} \mathrm{GP}_k(\mathbf{0}, \varpi \mathbf{I}_k)$. The theorem then follows immediately from the continuous mapping theorem for random elements taking values in a Hilbert space (Billingsley, 1968, p. 34; Cuevas, Febrero, and Fraiman 2004) and Theorem 4.10 of Chapter 4. The theorem is proved.

10.5 Concluding Remarks and Bibliographical Notes

In this chapter, we studied how to test the two-sample and multi-sample ECF testing problems. For these problems, we described an L^2-norm-based test and a random permutation test. Real data examples were used to motivate and illustrate the methodologies.

To our knowledge, there is not much work in the literature for ECF testing. It is partly due to the fact that such kinds of problems are usually very challenging to deal with, even in the context of multivariate data analysis where the data dimension is finite and fixed. This chapter was written mainly based on Zhang and Sun (2010), who studied two-sample ECF testing problems by proposing an L^2-norm-based test and a nonparametric bootstrap test, and Zhang (2013b) who studied multi-sample ECF testing problems by proposing an L^2-norm-based test and a random permutation test. Further studies in this area are interesting and warranted.

10.6 Exercises

1. The progesterone data were introduced in Section 1.2.1 of Chapter 1. There are sixty-nine nonconceptive and twenty-two conceptive progesterone curves. These curves were reconstructed using the methods described in Chapter 3. To check if the two groups of progesterone curves have a common covariance function, we applied the L^2-norm-based test described in Section 10.2 under the Gaussian assumption. We obtained $T_n = 1.1107e5$ using (10.10). By the naive method, we found $\text{tr}(\hat{\varpi}) = 9.0708e5$ and $\text{tr}(\hat{\varpi}^{\otimes 2}) = 5.2632e11$.
   (a) Find the approximate null distribution of T_n using the Welch-Satterthwaite χ^2-approximation method.
   (b) Find the approximate P-value of the two-sample ECF test.
   (c) We also applied the random permutation test and found that the associated P-value was 0.99. Compare this permutation P-value with the approximate P-value you obtained in (b).
2. The Berkeley growth curve data were introduced in Section 1.2.2 of Chapter 1. They are the growth curves of thirty-nine boys and fifty-four girls. To check if the two groups of growth curves have a common covariance function, we applied the L^2-norm-based test described in Section 10.2 under the Gaussian assumption. We obtained $T_n = 1.3282e6$ using (10.10). By the bias-reduced method, we found $\widehat{\text{tr}(\varpi)} = 1.9919e6$ and $\widehat{\text{tr}(\varpi^{\otimes 2})} = 3.0664e12$.
   (a) Find the approximate null distribution of T_n using the Welch-Satterthwaite χ^2-approximation method.
   (b) Find the P-value of the two-sample ECF test.
   (c) We also applied the random permutation test and found that the associated P-value was 0.47. Compare this permutation P-value with the P-value you obtained in (b).

3. For the orthosis data, Figure 9.5 displays the sample cell covariance functions for Subject 5 at the four treatment conditions. It appears that these four sample cell covariance functions are not the same. To verify if this is the case, we applied the L^2-norm-based test for this four-sample ECF testing problem under the Gaussian assumption. We obtained $T_n = 6.1678e9$ using (10.33). By the naive method, we found $\text{tr}(\hat{\varpi}) = 1.0747e9$ and $\text{tr}(\hat{\varpi}^{\otimes 2}) = 9.2787e16$, and by the bias-reduced method, we found $\widehat{\text{tr}(\varpi)} = 1.0374e9$ and $\widehat{\text{tr}(\varpi^{\otimes 2})} = 7.3474e16$. The approximate null distributions of T_n were obtained by the Welch-Satterthwaite χ^2-approximation method.
    (a) Find the approximate null distribution of T_n with the parameters β and κ estimated by the naive method. Find the associated P-value.
    (b) Find the approximate null distribution of T_n with the parameters β and κ estimated by the bias-reduced method. Find the associated P-value.
    (c) Compare the P-values obtained in (a) and (b).

Bibliography

[1] Abramovich, F. and Angelini, C. Testing in mixed-effects FANOVA models. *Journal of Statistical Planning and Inference*, 136(12):4326–4348, 2006.

[2] Abramovich, F., Antoniadis, A., Sapatinas, T., and Vidakovic, B. Optimal testing in a fixed-effects functional analysis of variance model. *International Journal of Wavelets, Multiresolution and Information Processing*, 2(4):323–349, 2004.

[3] Aguilera, A., Ocana, F., and Valderrama, M. Estimation of functional regression models for functional responses by wavelet approximation. In *Functional and Operatorial Statistics*, Contributions to Statistics, pages 15–21. Physica-Verlag HD, 2008.

[4] Aguilera-Morillo, M., Aguilera, A., Escabias, M., and Valderrama, M. Penalized spline approaches for functional logit regression. *Test*, pages 1–27, 2012.

[5] Anderson, T. W. *An Introduction to Multivariate Statistical Analysis*. John Wiley & Sons, New York, 2003.

[6] Antoch, J., Prchal, L., Rosa, M., and Sarda, P. Functional linear regression with functional response: Application to prediction of electricity consumption. In *Functional and Operatorial Statistics*, Contributions to Statistics, pages 23–29. Physica-Verlag HD, 2008.

[7] Antoniadis, A. and Sapatinas, T. Estimation and inference in functional mixed-effects models. *Computational Statistics & Data Analysis*, 51(10):4793–4813, 2007.

[8] Baíllo, A. and Grané, A. Local linear regression for functional predictor and scalar response. *Journal of Multivariate Analysis*, 100(1):102–111, 2009.

[9] Behrens, B. Ein beitrag zur fehlerberechnung bei wenige beobachtungen. *Landwirtschaftliches Jahresbuch*, 68:807–837, 1929.

[10] Benhenni, K., Ferraty, F., Rachdi, M., and Vieu, P. Local smoothing regression with functional data. *Computational Statistics*, 22(3):353–369, 2007.

[11] Berkes, I., Gabrys, R., Horvath, L., and Kokoszka, P. Detecting changes in the mean of functional observations. *Journal of the Royal Statistical Society Series B (Statistical Methodology)*, 71(5):927–946, 2009.

[12] Besse, P. and Ramsay, J. O. Principal components analysis of sampled functions. *Psychometrika*, 51(2):285–311, 1986.

[13] Billingsley, P. *Convergence of Probability Measures.* Wiley, New York, 1968.

[14] Box, G. E. P. Some theorems on quadratic forms applied in the study of analysis of variance problems. II. Effects of inequality of variance and of correlation between errors in the two-way classification. *Annals of Mathematical Statistics*, 25(3):484–498, 1954a.

[15] Box, G. E. P. Some theorems on quadratic forms applied in the study of analysis of variance problems. I. Effect of inequality of variance in the one-way classification. *Annals of Mathematical Statistics*, 25(2):290–302, 1954b.

[16] Brumback, B. A. and Rice, J. A. Smoothing spline models for the analysis of nested and crossed samples of curves. *Journal of the American Statistical Association*, 93(443):961–976, 1998.

[17] Buckley, M. J. and Eagleson, G. K. An approximation to the distribution of quadratic forms in normal random variables. *Australian Journal of Statistics*, 30A:150–159, 1988.

[18] Burba, F., Ferraty, F., and Vieu, P. k-nearest neighbour method in functional nonparametric regression. *Journal of Nonparametric Statistics*, 21(4):453–469, 2009.

[19] Cai, T. T. and Hall, P. Prediction in functional linear regression. *The Annals of Statistics*, 34(5):2159–2179, 2006.

[20] Canadian Climate Programm. Canadian climate normals 1981–1980. Environment Canada, Ottawa,. 1982.

[21] Cardot, H., Crambes, C., and Sarda, P. Quantile regression when the covariates are functions. *Journal of Nonparametric Statistics*, 17(7):841–856, 2005.

[22] Cardot, H., Ferraty, F., and Sarda, P. Functional linear model. *Statistics & Probability Letters*, 45(1):11–22, 1999.

[23] Cardot, H., Ferraty, F., and Sarda, P. Spline estimators for the functional linear model. *Statistica Sinica*, 13(3):571–591, 2003.

[24] Cardot, H., Ferraty, F., Mas, A., and Sarda, P. Testing hypotheses in the functional linear model. *Scandinavian Journal of Statistics*, 30(1):241–255, 2003.

[25] Cardot, H., Goia, A., and Sarda, P. Testing for no effect in functional linear regression models, some computational approaches. *Communications in Statistics-Simulation and Computation*, 33(1):179–199, 2004.

[26] Cardot, H., Mas, A., and Sarda, P. CLT in functional linear regression models. *Probability Theory and Related Fields*, 138(3-4):325–361, 2007.

[27] Chen, K. and Müller, H.-G. Conditional quantile analysis when covari-

ates are functions, with application to growth data. *Journal of the Royal Statistical Society: Series B (Statistical Methodology)*, 74(1):67–89, 2012.

[28] Cheng, M.-Y., Fan, J., and Marron, J. S. On automatic boundary corrections. *The Annals of Statistics*, 25(4):1691–1708, 1997.

[29] Chiou, J.-M. and Li, P.-L. Functional clustering and identifying substructures of longitudinal data. *Journal of the Royal Statistical Society Series B (Statistical Methodology)*, 69(4):679–699, 2007.

[30] Chiou, J.-M. and Li, P.-L. Functional clustering of longitudinal data. In *Functional and Operatorial Statistics*, Contributions to Statistics, pages 103–107. Physica-Verlag HD, 2008.

[31] Chiou, J.-M. and Müller, H.-G. Diagnostics for functional regression via residual processes. *Computational Statistics & Data Analysis*, 51(10):4849–4863, 2007.

[32] Chiou, J.-M., Müller, H.-G., and Wang, J.-L. Functional response models. *Statistica Sinica*, 14(3):675–693, 2004.

[33] Cleveland, W. S. Robust locally weighted regression and smoothing scatterplots. *Journal of the American Statistical Association*, 74(368):829–836, 1979.

[34] Cleveland, W. S. and Devlin, S. J. Locally weighted regression: An approach to regression analysis by local fitting. *Journal of the American Statistical Association*, 83(403):596–610, 1988.

[35] Cox, D. D. and Lee, J. S. Pointwise testing with functional data using the Westfall-Young randomization method. *Biometrika*, 95(3):621–634, 2008.

[36] Crainiceanu, C. M., Staicu, A.-M., Ray, S., and Punjabi, N. Bootstrap-based inference on the difference in the means of two correlated functional processes. *Statistics in Medicine*, 31(26):3223–3240, 2012.

[37] Crambes, C., Kneip, A., and Sarda, P. Smoothing splines estimators for functional linear regression. *The Annals of Statistics*, 37(1):35–72, 2009.

[38] Craven, P. and Wahba, G. Smoothing noisy data with spline functions–Estimating the correct degree of smoothing by the method of generalized cross-validation. *Numerische Mathematik*, 31(4):377–403, 1979.

[39] Cuesta-Albertos, J. A. and Febrero-Bande, M. A simple multiway ANOVA for functional data. *Test*, 19(3):537–557, 2010.

[40] Cuesta-Albertos, J. A. and Nieto-Reyes, A. The random tukey depth. *Computational Statistics & Data Analysis*, 52(11):4979–4988, 2008.

[41] Cuevas, A., Febrero, M., and Fraiman, R. Linear functional regression: The case of fixed design and functional response. *Canadian Journal of Statistics-Revue Canadienne De Statistique*, 30(2):285–300, 2002.

[42] Cuevas, A., Febrero, M., and Fraiman, R. An ANOVA test for functional data. *Computational Statistics & Data Analysis*, 47(1):111–122, 2004.

[43] Cuevas, A., Febrero, M., and Fraiman, R. On the use of the bootstrap for estimating functions with functional data. *Computational Statistics & Data Analysis*, 51(2):1063–1074, 2006.

[44] Cuevas, A., Febrero, M., and Fraiman, R. Robust estimation and classification for functional data via projection-based depth notions. *Computational Statistics*, 22(3):481–496, 2007.

[45] de Boor, C. *A Practical Guide to Splines.* Springer-Verlag, New York, 1978.

[46] Donoho, D. L. and Johnstone, I. M. Ideal spatial adaptation by wavelet shrinkage. *Biometrika*, 81(3):425–455, 1994.

[47] Eilers, P. H. C. and Marx, B. D. Flexible smoothing with B-splines and penalties. *Statistics Science*, 11:89–102, 1996.

[48] Eubank, R. L. *Nonparametric Regression and Spline Smoothing.* Marcel Dekker, New York, 2nd edition, 1999.

[49] Fan, J. Design-adaptive nonparametric regression. *Journal of the American Statistical Association*, 87(420):998–1004, 1992.

[50] Fan, J. Local linear regression smoothers and their minimax efficiencies. *The Annals of Statistics*, 21(1):196–216, 1993.

[51] Fan, J. Test of significance based on wavelet thresholding and Neyman's truncation. *Journal of the American Statistical Association*, 91(434):674–688, 1996.

[52] Fan, J. and Gijbels, I. Variable bandwidth and local linear regression smoothers. *The Annals of Statistics*, 20(4):2008–2036, 1992.

[53] Fan, J. and Gijbels, I. *Local Polynomial Modelling and its Applications.* Chapman and Hall/CRC, Boca Raton, FL., 1996.

[54] Fan, J. and Lin, S.-K. Test of significance when data are curves. *Journal of the American Statistical Association*, 93(443):1007–1021, 1998.

[55] Fan, J. and Marron, J. S. Fast implementations of nonparametric curve estimators. *Journal of Computational and Graphical Statistics*, 3:35–56, 1994.

[56] Fan, J. and Yao, Q. Efficient estimation of conditional variance functions in stochastic regression. *Biometrika*, 85(3):645–660, 1998.

[57] Fan, J. and Zhang, J.-T. Two-step estimation of functional linear models with applications to longitudinal data. *Journal of the Royal Statistical Society, Series B (Statistical Methodology)*, 62:303–322, 2000.

[58] Faraway, J. J. Regression analysis for a functional response. *Technometrics*, 39(3):254–261, 1997.

[59] Febrero, M., Galeano, P., and González-Manteiga, W. Influence in the functional linear model with scalar response. In *Functional and Operatorial Statistics*, pages 165–171. Physica-Verlag HD, 2008.

[60] Febrero, M., Galeano, P., and González-Manteiga, W. Outlier detec-

tion in functional data by depth measures, with application to identify abnormal NO_x levels. *Environmetrics*, 19(4):331–345, 2008.

[61] Ferraty, F., Keilegom, I. V., and Vieu, P. Regression when both response and predictor are functions. *Journal of Multivariate Analysis*, 109(0):10–28, 2012.

[62] Ferraty, F., Goia, A., and Vieu, P. Nonparametric functional methods: New tools for chemometric analysis. In *Statistical Methods for Biostatistics and Related Fields*, pages 245–264. Springer, Berlin, Heidelberg, 2007.

[63] Ferraty, F., González-Manteiga, W., Martínez-Calvo, A., and Vieu, P. Presmoothing in functional linear regression. *Statistica Sinica*, 22(1):69–94, 2012.

[64] Ferraty, F. and Romain, Y. *The Oxford Handbook of Functional Data Analysis.* Oxford University Press, New York, 2011.

[65] Ferraty, F. and Vieu, P. The functional nonparametric model and application to spectrometric data. *Computational Statistics*, 17:545–564, 2002.

[66] Ferraty, F. and Vieu, P. *Nonparametric Functional Data Analysis: Theory and Practice.* Springer-Verlag, New York, 2006.

[67] Fisher, R. A. The fiducial argument in statistical inference. *Annals of Eugenics*, 6:391–398, 1935.

[68] Fraiman, R. and Muniz, G. Trimmed means for functional data. *Test*, 10(2):419–440, 2001.

[69] Fujikoshi, Y. Two-way ANOVA models with unbalanced data. *Discrete Mathematics*, 116:315–334, 1993.

[70] Gasser, T., Müller, H.-G., and Mammitzsch, V. Kernels for nonparametric curve estimation. *Journal of the Royal Statistical Society. Series B (Statistical Methodology)*, 47(2):238–252, 1985.

[71] Gasser, T. and Müller, H.-G. Kernel estimation of regression functions. In Gasser, T. and Rosenblatt, M., editors, *Smoothing Techniques for Curve Estimation*, volume 757 of *Lecture Notes in Mathematics*, pages 23–68. Springer, Berlin, Heidelberg, 1979.

[72] Gervini, D. Outlier detection and trimmed estimation for general functional data. *Statistica Sinica*, 22:1639–1660, 2012.

[73] Goldsmith, J., Bobb, J., Crainiceanu, C. M., Caffo, B., and Reich, D. Penalized functional regression. *Journal of Computational and Graphical Statistics*, 20(4):830–851, 2011.

[74] Goldsmith, J., Crainiceanu, C. M., Caffo, B. S., and Reich, D. S. Penalized functional regression analysis of white-matter tract profiles in multiple sclerosis. *NeuroImage*, 57(2):431–439, 2011.

[75] Green, P. J. and Silverman, B. W. *Nonparametric Regression and Gen-*

eralized Linear Models: A Roughness Penalty Approach. Chapman and Hall/CRC, Boca Raton, FL., 1994.

[76] Gromenko, O. and Kokoszka, P. Testing the equality of mean functions of ionospheric critical frequency curves. *Journal of the Royal Statistical Society: Series C (Applied Statistics)*, 61(5):715–731, 2012.

[77] Gu, C. *Smoothing Spline ANOVA Models.* Spring-Verlag, New York, 2002.

[78] Hall, P. and Hosseini-Nasab, M. On properties of functional principal components analysis. *Journal of the Royal Statistical Society. Series B (Statistical Methodology)*, 68(1):109–126, 2006.

[79] Hall, P., Müller, H.-G., and Wang, J.-L. Properties of principal component methods for functional and longitudinal data analysis. *The Annals of Statistics*, 34(3):1493–1517, 2006.

[80] Hall, P. and van Keilegom, I. Two-sample tests in functional data analysis starting from discrete data. *Statistica Sinica*, 17(4):1511–1531, 2007.

[81] Härdle, W. *Applied Nonparametric Regression.* Cambridge University Press, New York, 1990.

[82] Hastie, T. and Loader, C. Local regression: Automatic kernel carpentry. *Statistical Science*, 8(2):120–129, 1993.

[83] Herrmann, E., Gasser, T., and Kneip, A. Choice of bandwidth for kernel regression when residuals are correlated. *Biometrika*, 79(4):783–795, 1992.

[84] Hitchcock, D. B., Booth, J. G., and Casella, G. The effect of presmoothing functional data on cluster analysis. *Journal of Statistical Computation and Simulation*, 77(12):1043–1055, 2007.

[85] Hitchcock, D. B., Casella, G., and Booth, J. G. Improved estimation of dissimilarities by presmoothing functional data. *Journal of the American Statistical Association*, 101(473):211–222, 2006.

[86] Horvath, L., Kokoszka, P., and Reimherr, M. Two sample inference in functional linear models. *Canadian Journal of Statistics*, 37(4):571–591, 2009.

[87] Huang, J. Z., Shen, H., and Buja, A. Functional principal components analysis via penalized rank one approximation. *Electronic Journal of Statistics*, 2:678–695, 2008.

[88] James, G. S. Tests of linear hypotheses in univariate and multivariate analysis when the ratios of the population variances are unknown. *Biometrika*, 41(1-2):19–43, 1954.

[89] James, G. M. Generalized linear models with functional predictors. *Journal of the Royal Statistical Society, Series B (Statistical Methodology)*, 64:411–432, 2002.

[90] James, G. M., Hastie, T. J., and Sugar, C. A. Principal component

models for sparse functional data. *Biometrika*, 87(3):587–602, 2000.

[91] James, G. M. and Sood, A. Performing hypothesis tests on the shape of functional data. *Computational Statistics & Data Analysis*, 50(7):1774–1792, 2006.

[92] James, G. M., Wang, J., and Zhu, J. Functional linear regression that's interpretable. *The Annals of Statistics*, 37(5A):2083–2108, 2009.

[93] Johansen, S. The Welch-James approximation to the distribution of the residual sum of squares in a weighted linear regression. *Biometrika*, 67(1):85–92, 1980.

[94] Kayano, M. and Konishi, S. Sparse functional principal component analysis via regularized basis expansions and its application. *Communications in Statistics - Simulation and Computation*, 39(7):1318–1333, 2010.

[95] Kim, S.-J. K. A practical solution to the multivariate Behrens-Fisher problem. *Biometrika*, 79(1):171–176, 1992.

[96] Kneip, A. Nonparametric estimation of common regressors for similar curve data. *The Annals of Statistics*, 22(3):1386–1427, 1994.

[97] Kokoszka, P., Maslova, I., Sojka, J., and Zhu, L. Testing for lack of dependence in the functions linear model. *Canadian Journal of Statistics*, 36(2):207–222, 2008.

[98] Krafty, R. T. A note on pre-smoothing functional data in the smoothing spline estimation of a mean function. Unpublished manuscript, 2010.

[99] Krishnamoorthy, K. and Yu, J. Modified Nel and Van der Merwe test for the multivariate Behrens-Fisher problem. *Statistics & Probability Letters*, 66(2):161–169, 2004.

[100] Kshirsagar, A. M. *Multivariate Analysis*. Marcel Decker, New York, 1972.

[101] Laha, R. G. and Rohatgi, V. K. *Probability Theory*. Wiley, New York, 1979.

[102] Li, Y. and Hsing, T. On rates of convergence in functional linear regression. *Journal of Multivariate Analysis*, 98(9):1782–1804, 2007.

[103] Liu, R. Y. On a notion of data depth based on random simplices. *The Annals of Statistics*, 18(1):405–414, 1990.

[104] Malfait, N. and Ramsay, J. O. The historical functional linear model. *Canadian Journal of Statistics*, 31(2):115–128, 2003.

[105] Marron, J. S. and Nolan, D. Canonical kernels for density estimation. *Statistics & Probability Letters*, 7(3):195–199, 1988.

[106] Martínez-Calvo, A. Presmoothing in functional linear regression. In *Functional and Operatorial Statistics*, Contributions to Statistics, pages 223–229. Physica-Verlag HD, 2008.

[107] Mas, A. Testing for the mean of random curves: a penalization approach.

Statistical Inference for Stochastic Processes, 10:147–163, 2007.

[108] Mas, A. and Pumo, B. Functional linear regression with derivatives. *Journal of Nonparametric Statistics*, 21(1):19–40, 2009.

[109] McCullagh, P. and Nelder, J. A. *Generalized Linear Models.* Chapman and Hall, New York, 1989.

[110] Montgomery, D. C. *Design and Analysis of Experiments.* John Wiley & Sons, Inc., 6th edition, 2005.

[111] Muirhead, R. J. *Aspects of Multivariate Statistical Theory.* John Wiley & Sons, New York, 1982.

[112] Müller, H.-G. Smooth optimum kernel estimators near endpoints. *Biometrika*, 78(3):521–530, 1991.

[113] Müller, H.-G. On the boundary kernel method for non-parametric curve estimation near endpoints. *Scandinavian Journal of Statistics*, 20(4):313–328, 1993.

[114] Müller, H.-G. Functional modelling and classification of longitudinal data. *Scandinavian Journal of Statistics*, 32(2):223–240, 2005.

[115] Müller, H.-G., Sen, R., and Stadtmueller, U. Functional data analysis for volatility. *Journal of Econometrics*, 165(2):233–245, 2011.

[116] Müller, H.-G. and Stadtmuller, U. Generalized functional linear models. *The Annals of Statistics*, 33(2):774–805, 2005.

[117] Müller, H.-G. and Yang, W. Dynamic relations for sparsely sampled Gaussian processes. *Test*, 19:1–29, 2010.

[118] Munk, A., Paige, R., Pang, J., Patrangenaru, V., and Ruymgaart, F. The one- and multi-sample problem for functional data with application to projective shape analysis. *Journal of Multivariate Analysis*, 99(5):815–833, 2008.

[119] Munro, A. W., Ritchie, G. Y., Lamb, A. J., Douglas, R. M., and Booth, I. R. The cloning and DNA sequence of the gene for the glutathione-regulated potassium-efflux system *kefc of escherichia* coli. *Molecular Microbiology*, 5(3):607–616, 1991.

[120] Nadaraya, E. A. On estimating regression. *Theory of Probability and its Applications*, 9:141–42, 1964.

[121] Nair, D., Large, W., Steinberg, F., and Kelso, J. Expressive timing and perception of emotion in music: An fMRI study. *Proceedings of the 7th International Conference on Music Perception and Cognition*, pages 627–630, 2002.

[122] Nelder, J. A. and Wedderburn, R. W. M. Generalized linear models. *Journal of the Royal Statistical Society Series A-general*, 135(3):370–384, 1972.

[123] Ramsay, J. O. When the data are functions. *Psychometrika*, 47(4):379–396, 1982.

[124] Ramsay, J. O. and Dalzell, C. J. Some tools for functional data analysis. *Journal of the Royal Statistical Society, Series B (Statistical Methodology)*, 53(3):539–572, 1991.

[125] Ramsay, J., Hooker, G., and Graves, S. *Functional Data Analysis with R and MATLAB®*. Springer, London, 2009.

[126] Ramsay, J. O. and Li, X. Curve registration. *Journal of the Royal Statistical Society, Series B (Statistical Methodology)*, 60:351–363, 1998.

[127] Ramsay, J. O. and Silverman, B. W. *Functional Data Analysis.* Springer-Verlag, New York, 2005.

[128] Ramsay, J. O. and Silverman, B. W. *Applied Functional Data Analysis.* Springer-Verlag, New York, 2002.

[129] Rice, J. A. Functional and longitudinal data analysis: Perspectives on smoothing. *Statistica Sinica*, 14(3):631–647, 2004.

[130] Rice, J. A. and Silverman, B. W. Estimating the mean and covariance structure nonparametrically when the data are curves. *Journal of the Royal Statistical Society, Series B (Statistical Methodology)*, 53(1):233–243, 1991.

[131] Ruppert, D., Sheather, S. J., and Wand, M. P. An effective bandwidth selector for local least squares regression. *Journal of the American Statistical Association*, 90(432):1257–1270, 1995.

[132] Ruppert, D., Wand, M. P., and Carroll, R. J. *Semiparametric Regression.* Cambridge University Press, 2003.

[133] Satterthwaite, F. E. An approximate distribution of estimates of variance components. *Biometrics Bulletin*, 2(6):110–114, 1946.

[134] Satterthwaite, F. E. Synthesis of variance. *Psychometrika*, 6(5):309–316, 1941.

[135] Sentürk, D. and Müller, H.-G. Functional varying coefficient models for longitudinal data. *Journal of the American Statistical Association*, 105(491):1256–1264, 2010.

[136] Shen, Q. and Faraway, J. An F-test for linear models with functional responses. *Statistica Sinica*, 14(4):1239–1257, 2004.

[137] Shen, Q. and Xu, H. Diagnostics for linear models with functional responses. *Technometrics*, 49(1):26–33, 2007.

[138] Silverman, B. W. Incorporating parametric effects into functional principal components analysis. *Journal of the Royal Statistical Society, Series B (Statistical Methodology)*, 57(4):673–689, 1995.

[139] Silverman, B. W. Smoothed functional principal components analysis by choice of norm. *The Annals of Statistics*, 24(1):1–24, 1996.

[140] Simonoff, J. S. *Smoothing Methods in Statistics.* Springer-Verlag, New York., 1996.

[141] Solomon, H. and Stephens, M. A. Distribution of a sum of weighted

chi-square variables. *Journal of the American Statistical Association*, 72(360):881–885, 1977.

[142] Stone, C. J. An asymptotically optimal window selection rule for kernel density estimates. *The Annals of Statistics*, 12(4):1285–1297, 1984.

[143] Swihart, B. J., Goldsmith, J., , and Crainiceanu, C. M. Testing for functional effects. Technical report, Dept. of Biostatistics, Johns Hopkins University, Baltimore, M.D., 2012.

[144] Tuddenham, R. D. and Snyder, M. M. Physical growth of California boys and girls from birth to eighteen years. *University of California Publications in Child Development*, 1:183–364, 1954.

[145] Tukey, J. W. Mathematics and the picturing of data. *Proceedings of the International Congress of Mathematicians*, 2:523–531, 1974.

[146] van der Vaart, A. W. and Wellner, J. A. *Weak Convergence and Empirical Processes.* Springer, New York., 1996.

[147] Wahba, G. *Spline Models for Observational Data.* CBMS-NSF Regional Conference Series in Applied Mathematics. SIAM, Philadelphia, 1990.

[148] Wand, M. P. and Jones, M. C. *Kernel Smoothing.* Chapman and Hall, London, 1995.

[149] Watson, G. S. Smooth regression analysis. *Sankhya: The Indian Journal of Statistics, Series A*, 26:359–372, 1964.

[150] Welch, B. L. The generalization of 'student's' problem when several different population variances are involved. *Biometrika*, 34:28–35, 1947.

[151] Wu, H. and Zhang, J.-T. Local polynomial mixed-effects models for longitudinal data. *Journal of the American Statistical Association*, 97(459):883–897, 2002.

[152] Wu, H. and Zhang, J.-T. *Nonparametric Regression Methods for Longitudinal Data Analysis: Mixed-Effect Modelling Approaches.* Wiley-Interscience, 2006.

[153] Wu, Y., Fan, J., and Müller, H.-G. Varying-coefficient functional linear regression. *Bernoulli*, 16(3):730–758, 2010.

[154] Xu, H., Shen, Q., Yang, X., and Shoptaw, S. A quasi F-test for functional linear models with functional covariates and its application to longitudinal data. *Statistics in Medicine*, 30(23):2842–2853, 2011.

[155] Yang, X. and Nie, K. Hypothesis testing in functional linear regression models with Neyman's truncation and wavelet thresholding for longitudinal data. *Statistics in Medicine*, 27(6):845–863, 2008.

[156] Yang, X., Shen, Q., Xu, H., and Shoptaw, S. Functional regression analysis using an F-test for longitudinal data with large numbers of repeated measures. *Statistics in Medicine*, 26(7):1552–1566, 2007.

[157] Yao, F. and Müller, H.-G. Functional quadratic regression. *Biometrika*, 97(1):49–64, 2010.

[158] Yao, F., Müller, H.-G., and Wang, J.-L. Functional data analysis for sparse longitudinal data. *Journal of the American Statistical Association*, 100:577–590, 2005a.

[159] Yao, F., Müller, H.-G., and Wang, J.-L. Functional linear regression analysis for longitudinal data. *The Annals of Statistics*, 33(6):2873–2903, 2005b.

[160] Yen, S. S. C. and Jaffe, R. B. *Reproductive Endocrinology: Physiology, Pathophysiology, and Clinical Management.* W. B. Saunders, Philadephia., 1991.

[161] Yu, G., Zou, C., and Wang, Z. Outlier detection in functional observations with applications to profile monitoring. *Technometrics*, 54(3):308–318, 2012.

[162] Zhang, C., Peng, H., and Zhang, J.-T. Two sample tests for functional data. *Communications in Statistics–Theory and Methods*, 39(4):559–578, 2010.

[163] Zhang, J.-T. *Smoothed Functional Data Analysis.* PhD thesis, The University of North Carolina at Chapel Hill, 1999.

[164] Zhang, J.-T. Approximate and asymptotic distributions of chi-squared-type mixtures with applications. *Journal of the American Statistical Association*, 100:273–285, 2005.

[165] Zhang, J.-T. Statistical inferences for linear models with functional responses. *Statistica Sinica*, 21:1431–1451, 2011a.

[166] Zhang, J.-T. Two-way MANOVA with unequal cell sizes and unequal cell covariance matrices. *Technometrics*, 53(4):426–439, 2011b.

[167] Zhang, J.-T. Functional linear models with time-dependent covariates. Unpublished manuscript, 2013a.

[168] Zhang, J.-T. Multi-sample equal covariance function testing for functional data. Unpublished manuscript, 2013b.

[169] Zhang, J.-T. and Chen, J. Statistical inferences for functional data. *The Annals of Statistics*, 35(3):1052–1079, 2007.

[170] Zhang, J.-T. and Fan, J. Minimax kernels for nonparametric curve estimation. *Journal of Nonparametric Statistics*, 12(3):417–445, 2000.

[171] Zhang, J.-T. and Liang, X. One-way ANOVA for functional data via globalizing the pointwise F-test. Unpublished manuscript, 2013.

[172] Zhang, J.-T., Liang, X., and Xiao, S. On the two-sample Behrens-Fisher problem for functional data. *Journal of Statistical Theory and Practice*, 4(4):571–587, 2010.

[173] Zhang, J.-T. and Sun, Y. Two-sample test for equal covariance function for functional data. *Oriental Journal of Mathematics*, 4:1–22, 2010.

[174] Zhang, J.-T. and Xiao, S. Heteroscedastic one-way ANOVA for functional data. Unpublished manuscript, 2013a.

[175] Zhang, J.-T. and Xiao, S. Two-way ANOVA under heteroscedasticity for functional data. Unpublished manuscript, 2013b.

[176] Zhang, J.-T. and Xu, J. On the k-sample Behrens-Fisher problem for high-dimensional data. *Science in China Series A: Mathematics*, 52(6):1285–1304, 2009.

[177] Zhao, X., Marron, J. S., and Wells, M. T. The functional data analysis view of longitudinal data. *Statistica Sinica*, 14(3):789–808, 2004.

[178] Zhao, Y., Ogden, R. T., and Reiss, P. T. Wavelet-based LASSO in functional linear regression. *Journal of Computational and Graphical Statistics*, 21(3):600–617, 2012.

[179] Zhu, H. and Cox, D. D. A functional generalized linear model with curve selection in cervical pre-cancer diagnosis using fluorescence spectroscopy. *IMS Lecture Notes-Monograph Series*, 57:173–189, 2009.

Index